The Correspondence of John Tyndall

VOLUME II

THE CORRESPONDENCE OF JOHN TYNDALL

VOLUME II

The Correspondence, January 1869–February 1871

Edited by
Adrian Kirwan and Elizabeth Neswald

General Editors
Roland Jackson, Bernard Lightman, and Michael S. Reidy

Associate Editors
Michael D. Barton, Erin Grosjean, and Michael Laurentius

UNIVERSITY OF PITTSBURGH PRESS

Published by the University of Pittsburgh Press, Pittsburgh, Pa., 15260

Manufactured in the United States of America
Printed on acid-free paper
10 9 8 7 6 5 4 3 2 1

Cataloging-in-Publication data is available from the Library of Congress

ISBN 13: 978-0-8229-4682-3
ISBN 10: 0-8229-4682-3

CONTENTS

ACKNOWLEDGMENTS

The editors of this volume wish to thank the many people who have offered their advice and expertise as we worked on this volume. Roland Jackson was of particular help, not only in offering sage advice from his deep familiarity with the Tyndall archive but also through his biography of Tyndall, *Ascent of John Tyndall*, which was published just as we were beginning our work. We owe Roland, Michael Reidy, and Michael Barton additional thanks for their editorial support and their willingness to share their knowledge of all things Tyndall.

The project owes its success to the efforts of an expansive team of named and unnamed graduate students and scholars who have assisted with the project since its inception in 2007. The collection and collation of the letters for this and other volumes was the result of a considerable amount of work by the general editors, associate editors, and the team at the York office, while the early stages of transcription are the result of hours of labor by dozens of graduate students from the numerous universities associated with this project across the globe. The editors would like to thank Erin Grosjean, manuscripts coordinator, whose aid and organizational skills have been invaluable. We would like to thank, too, Caley McCauley, Emily Herring, and Victoria Mucilli, who transcribed, translated, and edited Tyndall's French correspondence, and Nicolas Sanchez-Guerrero and Sven Pinczewski, who along with Elizabeth Neswald and Tatyana Shestakov transcribed, translated, and edited letters from Tyndall's German correspondents. We owe a debt of gratitude to Jane Harrison, who obtained and scanned particular letters from the Royal Institution archive and answered our obscure archival questions with great patience; and to Jennifer Allan, who obtained scanned images for us from the British Library and other London archives.

Special thanks go to Michael Barton for his willingness to consult with us on questions of dating, aid in identifying correspondents, and help clarifying formatting issues, and to Alana Duggan for her assistance with the research for this volume and, especially, the biographical register.

LIST OF ABBREVIATIONS

ADB	*Allgemeine Deutsche Biographie*
Alum. Cantab.	J. A. Venn, *Alumni Cantabrigienses, 1752–1900*, 6 vols. (Cambridge: Cambridge University Press, 1940–54)
Amer. Assoc. Proc.	*Proceedings of the American Association for the Advancement of Science*
Amer. Phil. Soc. Proc.	*Proceedings of the American Philosophical Society*
Annal. Chim. et Phys.	*Annales de Chimie et de Physique*
Annal. Phys. und Chem.	*Annalen der Physik und Chemie/Poggendorff's Annalen*
ANB	*American National Biography*
Archives Nationales	Archives Nationales (France), Base Léonore (Légion d'honneur), http://www2.culture.gouv.fr/
Archives Sci. Phys. Nat.	*Archives des sciences physique et naturelle*
Ascent of John Tyndall	Roland Jackson, *The Ascent of John Tyndall: Victorian Scientist, Mountaineer, and Public Intellectual* (Oxford: University of Oxford Press, 2018)
BAAS	British Association for the Advancement of Science
BBAW	Berlin-Brandenburg Akademie der Wissenschaften
Berlin, Monatsber. Akad.	*Monatsberichte der Akademie der Wissenschaften zu Berlin*
BL	British Library
BMJ	*British Medical Journal*
BNA	British Newspaper Archive Online
Brit. Assoc. Rep.	*Reports of the British Association for the Advancement of Science*
Camb. Phil. Soc. Proc.	*Proceedings of the Cambridge Philosophical Society*
Catholic Encyclopedia	*The Catholic Encyclopedia*, ed. C. G. Herbermann, E. A. Pace, C. B. Pallen, T. J. Shanan, and J. J. Wynne, 16 vols. (New York: Robert Appleton, 1909)
CDSB	*Complete Dictionary of Scientific Biography*
CUL SC—add 7656	Cambridge University Library, Stokes Correspondence
Diamagnetism	John Tyndall, *Researches on Diamagnetism and Magne-crystallic Action* (London: Longmans, Green, 1870)
DIB	*Dictionary of Irish Biography*
Edin. Roy. Soc. Proc.	*Proceedings of the Royal Society of Edinburgh*

Edin. Roy. Soc. Trans.	*Transactions of the Royal Society of Edinburgh*
Faraday Correspondence	*The Correspondence of Michael Faraday*, ed. Frank A. J. L. James, 6 vols. (London: Institution of Electrical Engineers/Institution of Engineering and Technology, 1991–2012)
Faraday as a Discoverer	John Tyndall, *Faraday as a Discoverer* (London: Longmans, Green, 1868)
ForMemRS	Foreign Member of the Royal Society
Fragments of Science	John Tyndall, *Fragments of Science for Unscientific People: A Series of Detached Essays, Lectures, and Reviews*, 2nd ed. (London: Longmans, Green, 1871)
FRS	Fellow of the Royal Society
Glaciers of the Alps	John Tyndall, *The Glaciers of the Alps* (London: John Murray, 1860)
Hirst's Journal	Journal of Thomas Archer Hirst (RI MS JT/2/32d/)
HLS	*Historisches Lexikon der Schweiz*
Hours of Exercise in the Alps	John Tyndall, *Hours of Exercise in the Alps*, 1st ed. (London: Longmans, Green, and Co., 1871)
IC HP	Imperial College of Science, Technology and Medicine, College Archives, Huxley Papers
Irish Genealogy	Department of Culture, Heritage and the Gaeltacht, Irish Genealogy, https://www.irishgenealogy.ie/
Lay Sermons	Thomas Henry Huxley, *Lay Sermons, Addresses, and Reviews* (London: MacMillan, 1870)
Life of Faraday	Henry Bence Jones, *The Life and Letters of Michael Faraday*, 2 vols. (London: Longmans, Green, 1870)
NDB	*Neue Deutsche Biographie*
ODNB	*Oxford Dictionary of National Biography*
OED	*Oxford English Dictionary*
Paris, Comptes Rendus	*Comptes rendus hebdomadaires des séances de l'Académie des Sciences*
Phil. Mag.	*Philosophical Magazine*
Phil. Trans.	*Philosophical Transactions of the Royal Society of London*
RGO	Cambridge University Library, Royal Greenwich Observatory Archives
RI	Royal Institution of Great Britain
RN	Royal Navy
Roy. Inst. Proc.	*Notes and Proceedings of the Royal Institution of Great Britain*
Roy. Soc. Proc.	*Proceedings of the Royal Society of London*
RS	Royal Society of London
RS HS	Royal Society Letters, Herschel Papers
TRC Letters	Tennyson Research Centre

Sound (1867)	John Tyndall, *Sound: A Course of Eight Lectures Delivered at the Royal Institution of Great Britain* (London: Longmans, Green, and Co., 1867)
Tyndall Correspondence: Volume 8	Piers J. Hale, Elizabeth Neswald, Nathan N. Kapoor, and Michael D. Barton, eds., *The Correspondence of John Tyndall*, vol. 8, *The Correspondence, June 1862–January 1865* (Pittsburgh: University of Pittsburgh Press, 2020)
Tyndall Correspondence: Volume 10	Michael D. Barton, Ken Corbett, and Roland Jackson, eds., *The Correspondence of John Tyndall*, vol. 10, *The Correspondence, January 1867–December 1868* (Pittsburgh: University of Pittsburgh Press, 2021)
Wellesley's Index	*Wellesley's Index of Victorian Periodicals Database*
X Club	Ruth Barton, *The X Club: Power and Authority in Victorian Science* (Chicago: University of Chicago Press, 2018)

This portrait of Tyndall was taken circa 1872 by the portrait photographers Elliot & Fry of Portman Square, London. The firm was founded in 1863 by Joseph John Elliott (1835–1903) and Clarence Edmund Fry (1840–1897). The partners used the Fine Arts Copyright Law to their advantage: under this law, if the person photographed was of public interest, the photographers could retain rights to the image. The firm often photographed celebrities without charge, selling their images as visiting cards. Elliot & Fry specialized in photographing and selling these celebrity visiting cards (J. Hannevy, ed., *Encyclopedia of Nineteenth-Century Photography* [London: Routledge, 2008], 1:479–80). While it is not known whether Tyndall took advantage of this offer to be photographed without cost, as a stamp on the original photograph shows, Elliot & Fry retained their copyright to this image. Credit: Illustration used with permission of the National Portrait Gallery, London (NPG x13245; copyright, National Portrait Gallery, London).

INTRODUCTION TO VOLUME 11

The eleventh volume of *The Correspondence of John Tyndall* covers the period from January 1869 to the end of February 1871 and contains 430 letters with more than 130 individual correspondents, as well as letters to several newspapers. These years find Tyndall an internationally established scientist with broad influence and feeling increasingly confident in that role. They were highly productive research years, and Tyndall had a wide scope of interests, publishing in scientific journals, popular magazines, and newspapers on a variety of topics, including germ theory, comets, and atmospheric phenomena, as well as a collection of his papers on diamagnetism. The results of this research were presented in numerous papers and lectures in various venues and complemented by his regular public lecture series and annual Christmas lectures at the Royal Institution of Great Britain (RI).

Michael Faraday's death in 1867 continues to echo through this volume. Tyndall advised his colleague Henry Bence Jones on his Faraday biography and later reviewed the two-volume work, *Life and Letters of Michael Faraday* (1870). He also oversaw the translation of his own commemorative work, *Faraday as a Discoverer*, into German. This translation, undertaken by Anna Helmholtz, wife of physicist Hermann Helmholtz, only confirms the underappreciated and often invisible role of women, especially wives, in nineteenth-century science. Anna Helmholtz was among the most active English-to-German translators of physics in the nineteenth century and was a regular translator of Tyndall's publications.

Once again, Tyndall's address at a meeting of the British Association for the Advancement of Science (BAAS) reverberated widely. This time the lecture was "On the Scientific Use of the Imagination," and he delivered it on 16 September 1870 at the meeting in Liverpool. This discourse, which was printed in the newspapers and published as a pamphlet shortly after, was written during one of Tyndall's visits to the Alps, where, as he wrote, he had only his own mind, Johann Wolfgang von Goethe's *Farbenlehre* (Theory of colors), and Alexander Bain's recent publication on logic out of which to weave his lecture.[1]

Tyndall's advisory role with the Board of Trade on British and Irish Lighthouses continued, and he wielded his increasingly prominent position in attempts to influence government science policy and public debates. This volume also reveals Tyndall as a political animal, at least in scientific affairs. Elected to the Council of the Royal Society of London (RS) in 1869, Tyndall was deeply enmeshed in the internal affairs of the society and the letters also offer glimpses into the political mechanisms at work within this body and Tyndall's own political maneuvering.

As in previous years, Tyndall traveled extensively, but his travels in 1869 and 1870 were particularly adventurous. During a longer European tour in the summer of 1869, on which he visited numerous German friends and colleagues, he slipped in the Alps while bathing in a stream, injuring his leg. The ensuing wound and infection kept him bedridden for weeks and had a serious impact on his health and productivity in the following months. The summer of 1870 found him again in the Alps, this time caught up in the early phase of the Franco-Prussian War (19 July 1870 to 28 January 1871). The letters show not only Tyndall's strong sympathy for the German side but also his compassion for the friends, colleagues, citizens, and soldiers caught up in the hostilities on both sides. Letters from his French and German acquaintances provide poignant insights into their wartime experiences.

Finally, family affairs were never far from his mind. Although Tyndall's mother had passed away in April 1867, the distribution of her estate was not completely settled, and his impoverished Irish relations were the cause of much consternation in that regard. More positively, the health of his sister, Emily (Emma), which had suffered greatly after the death of their mother, had begun to improve.

Scientific Research

Tyndall engaged in several scientific research projects from 1869 to 1871. Publications in this period demonstrate his broad interests, with topics ranging from diamagnetism to the blue color of water and, at first sight unexpectedly, the germ theory of disease. He began experimenting on the chemical reactions produced by light in 1868, and this research continued to occupy him through 1869. He had already submitted his first paper on the topic, "On a New Series of Chemical Reactions Produced by Light," to the RS in the autumn of 1868.[2] A more substantial paper, "On the Blue Colour of the Sky, and the Polarization of Light," was read at the RS on 14 January 1869.[3] In this paper he argued that light scattered by fine matter in the air could produce the blue of the sky. For these experiments Tyndall used an apparatus that enabled him to manufacture artificial clouds, which he called "actinic clouds," in a tube. These vapor experiments continued throughout 1869, and on 27 January 1870 Tyndall once again read a paper to the RS, this time titled "On the Action of Rays of High Refrangibility upon Gaseous Matter."[4] Several letters in this volume cover his correspondence with George Gabriel Stokes, who critiqued the paper for the *Phil. Trans.*[5] The correspondence offers insight into the editorial process that developed within the RS in the nineteenth century.[6] Much of Stokes's editorial intervention centers on the use and definition of terminology for scientific phenomena.

This work took Tyndall in two additional directions. During these experiments Tyndall realized how much the "actinic clouds" reminded him of comet tails.[7] On 8 March 1869 he delivered a lecture on cometary theory to the Cambridge Philosophical Society. His central argument was that comets were clouds of vaporized matter and that their enormous tails were produced when the material was decomposed by solar light. The combination of vaporization and heating gave the comet's tail its specific shape. Demonstrating with his experimental "actinic clouds," which were produced with little vapor, he argued that a similar process produced the tails of comets. The paper was published in the *Phil. Mag.*, and Tyndall sent a copy to John Herschel, who recognized the importance of Tyndall's theories in this field.[8] Herschel followed up with a wide-ranging letter on the finer points of the theory in the hope of further refining it.[9] Tyndall, while agreeing with the need to further refine his theory, welcomed Herschel's and George Biddell Airy's support.[10]

This research took Tyndall in a second direction, which at first glance seems completely unrelated to blue skies and comet tails—the germ theory of disease. The germ theory generated considerable debate in Victorian Britain from the 1860s onward, slowly rising to a crescendo around 1875. Louis Pasteur's work in the early 1860s seemed to prove that fermentation was not due to spontaneous generation but was the result of microorganisms introduced from the external environment. The rinderpest cattle plague of 1865 had brought to the fore the importance of a scientific understanding of disease transmission not only for animal welfare but also for the economic welfare of Britain.[11] The debate surrounding the nature of infection was contested between those supporting the concept of spontaneous generation and those promoting the germ theory of disease, with Tyndall and his tight circle of professional friends, known as the X Club, at the forefront of the germ theory's supporters.

One of Tyndall's first forays into this debate was his lecture "On Dust and Disease," delivered at the RI on 21 January 1870. This work can be seen as a continuation of his earlier work on the "floating matter of the air," albeit developed in a specific direction. Dust was, of course, floating matter, and Tyndall had previously viewed it as a means to make sunbeams visible. The topic "dust and disease" was suggested by Henry Bence Jones, who had been a member of the Royal Commission on the Cattle Plague of 1865. Indeed, Tyndall relied on Bence Jones and Henry Holland in developing the lecture, in which he spoke highly of Pasteur and his work and in doing so placed himself firmly in the germ theory camp against the proponents of spontaneous generation. Tyndall was clear that he did not seek to claim a role in the genesis of germ theory and that his lecture was primarily concerned with its dissemination. Although the lecture brought praise, with this came criticism from Pasteur's opponents, and it provoked numerous publications on the matter as Tyndall sought to defend the germ theory.[12]

This engagement drew Tyndall into a very public dispute with Henry Charlton Bastian, British physician and physiologist, who was a firm supporter of spontaneous generation. Indeed, this was not just a debate about the cause of infections but a disagreement among supporters of Darwinian evolution. While those supporting spontaneous generation saw it as essential that life could begin from inorganic matter (for how else could the original organism that others evolved from appear?), the X Club members, and particularly Tyndall and Huxley (who had up to this point been somewhat of a mentor to Bastian) argued that spontaneous generation was not the origin of contemporary germs. They felt that physicians needed to accept the scientific methods and findings of Pasteur.[13]

Drawing on Joseph Lister's research in surgery, Tyndall argued that the lungs acted as a perfect filter to remove "seeds or eggs floating in the atmosphere," and in doing so, he argued, they removed the "associated phenomena of animalcule life." Tyndall supported Lister's approach of using cotton filtration. However, Bastian objected to Lister's findings, provoking further arguments from Tyndall. The dispute between the two was carried out in full sight of the British public in the pages of the *Times*[14] and was not the first—nor the last—time Tyndall engaged in acrimonious debates in this publication.

Tyndall's efforts to provide additional evidence for the germ theory were widely publicized, and his European network was an important source of support. He mobilized this asset both in the research on germs and in his efforts to promote germ theory. The great Pasteur sent Tyndall his personal thanks and encouragement.[15] Pasteur's French colleague and opponent Felix Pouchet shared his own research on the topic.[16] Pouchet also drew Tyndall's attention to the work of British scientific instrument maker and micro-photographer John Dancer.[17] At Tyndall's request, German physicist Rudolf Clausius contacted his colleague Max Schultze to ask for his view on the effects of carbolic acid on the germs that caused wound putrefaction.[18] The correspondence surrounding Tyndall's study of airborne particles and their significance for the germ theory of disease show the scientific community in action, warts and all: acrimonious public disputes and attacks on scientific credibility on one side; collaboration, support, and friendly communication and information transfer on the other.

Rounding out Tyndall's research publications in this period is his book *Researches on Diamagnetism and Magne-crystallic Action* (1870), the culmination of a long period of work beginning in the 1850s, when he built upon Faraday's discovery of diamagnetism and his early research. Roland Jackson argues that diamagnetism was the first of four major areas of research in Tyndall's career, the others being glaciology, molecular behavior in atmospheric conditions, and bacteriology. Indeed, Jackson states that diamagnetism was key to Tyndall establishing himself as a central scientific and social figure in Victorian society and that with the publication of his collected papers, Tyndall was, in his way, staking a claim to the subject against the emerging field theory.[19] Tyndall sent copies of this work to many friends, acquaintances, and colleagues, including Airy (who was in the process of writing his own book on magnetism).[20] These publication gifts between colleagues were reciprocal and were doubtless an important avenue for the circulation of scientific ideas and the solidification of scientific friendships.

Popular Works and Public Outreach

One central motivation for many of Tyndall's activities was his desire to influence and awaken the "inert mass" that was the English public to the benefits of science.[21] Demonstrating the importance of science and communicating it "to the public in a language they can understand" was an essential part of these efforts. He believed that only with public support and enthusiasm could scientists hope "to obtain fair play for it in all our educational institutions."[22] His lecture courses at the RI, the Christmas lectures, and presentations at other London societies and venues were complemented by contributions to newspapers and cultural magazines for lay audiences and published volumes of his RI public lecture series. In this volume, two contributions stand out, in particular, in regard to the amount of correspondence they generated—-the memorialization of Michael Faraday and Tyndall's discourse at the BAAS meeting in Liverpool in 1870.

While scientific research was at the forefront of Tyndall's publishing efforts in 1869 and 1870, a significant amount of correspondence concerns the biography and memorialization of Faraday. The number and nature of these letters demonstrate Tyndall's great interest in maintaining Faraday's legacy. His own book, *Faraday as a Discoverer*, had been published in 1868. Much of the correspondence in the following year concerned Henry Bence Jones's biography, *The Life and Letters of Faraday*, which was published in two volumes in 1870. The German translation of Tyndall's Faraday commemoration, which was done by Anna Helmholtz, was also completed in 1870, and letters between her husband, Hermann Helmholtz, Tyndall, and Bence Jones discuss integrating notes from Bence Jones's more comprehensive biography into this translation.[23] Indeed, Tyndall had already given Bence Jones much of the correspondence with Faraday's friends that he had collected during his own researches.[24]

He also invested much effort into his review of Bence Jones's Faraday biography. The review was first published as "Life and Letters of Faraday, by Dr. Bence Jones" in the *Academy* on 14 May 1870 and subsequently appeared as chapter 12, "Life and Letters of Faraday," in Tyndall's 1871 essay collection *Fragments of Science.*[25] Several letters between Tyndall, Faraday's widow, Sarah, and Faraday's niece, Jane Barnard, who cared for her elderly aunt, show the pains Tyndall took to ensure that the review met with their approval, a sign of his enduring respect for Faraday and his family.[26] This respect was obviously reciprocated, as Sarah willingly gave him access to personal correspondence between herself and her husband.[27]

Tyndall's efforts to memorialize Faraday went even further than these publications. A letter to Arthur Stanley, dean of Westminster Abbey, shows Tyndall inquiring as to whether there would be a place in the abbey for a statue of Faraday. A Sandemanian, Faraday had been interred in the Sandemanian plot in Highgate Cemetery. These thoughts must have been in their infancy, as Tyndall also inquired about potential cost.[28] Due to a disagreement about the funding and scope of the memorial, the plans did not proceed.[29] It took until 1931 for a memorial floorstone to be placed in the abbey in Faraday's honor.

Tyndall's public discourse for the BAAS meeting in Liverpool, held on 16 September 1870, was another frequently discussed subject in the letters.[30] Although he expressed difficulty in finding a topic in July 1870, by August he had settled on the title "On the Scientific Use of the Imagination."[31] His choice of topic might have been dictated by the fact that Tyndall, attempting to write it in the Alps, had brought only a volume by Johann Wolfgang von Goethe and one by Alexander Bain with him. This discourse was important to him, however, and he solicited advice and criticism from a number of associates, including Herbert Spencer, Heinrich Debus, Thomas Archer Hirst, and Charles Darwin.[32] The lecture was deemed a tremendous success and generated much interest, both at home and abroad. Tyndall received numerous congratulatory letters from British and foreign friends and colleagues, who had read reports of it in the newspapers or the later, published version.[33]

Advisory Activities

In this period Tyndall participated in numerous advisory bodies, committees, and councils, some with more enthusiasm (and correspondence) than others. His position on the council of the International School of London, for example, is only peripherally alluded to, and his intervention on the behalf of Joseph Dalton Hooker in a dispute over Hooker's position at the Royal Botanic Gardens, Kew, is the subject of only a few letters.[34] His role as government adviser for the Board of Trade on British and Irish Lighthouses and his attempts to influence RS affairs, in contrast, resulted in extensive correspondence.

Tyndall took over the post as scientific adviser to the Corporation of Trinity House, which was responsible for maritime safety, from an aging Faraday in 1866, though it was not official until after Faraday died in August 1867. The Elder Brethren of Trinity House had been established in 1514 to provide for safe navigation along the coasts of England and Wales. In 1836 the corporation took over all English and Welsh lighthouses and began an ambitious construction program, building new lighthouses and replacing old ones. For a maritime nation, improving naval safety along the UK's coast was becoming increasingly important. In his capacity as an adviser, Tyndall provided recommendations on lighthouse improvements and carried out research and inspections in Britain and Ireland.

In 1869–70, Tyndall advised Trinity House and the Commissioners of Irish Lights on converting from an oil to a gas lighting system at the Howth Baily Lighthouse in Dublin, as well as on the best lenses and lighting to fulfill their needs.[35] With several systems to choose from, Tyndall needed to weigh the cost, environmental conditions, and lighting needs of the site to determine the most suitable solution for the specific Irish lighthouses under review. He traveled regularly to Ireland to make observations and also made it a point to note other lighthouses on his country excursions. The reports of these inspections and experiments, as well as Tyndall's encompassing situational considerations, were contained in a series of reports sent to the board of commissioners.[36] Unfortunately, it was not always possible for Tyndall to combine these business trips to Ireland with a visit to family.

Tyndall's role as professor of natural philosophy and superintendent at the RI, the latter a role he assumed in 1867 upon the death of Faraday, meant that RI business kept him busy. He held his regular lecture series and the Christmas lectures in 1869, and organized—with Bence Jones, who was secretary of the RI from 1860 to 1872—lectures by other speakers, some affiliated with the RI and others from outside the institution. He and Bence Jones also consulted regularly on the mundane, everyday management of the institution and its premises and personnel.

While RI business is a continuous topic throughout the letters, Tyndall's attempts to influence RS affairs also generated considerable correspondence. Tyndall was elected to the council of the RS in 1869, and he used this position to good effect to try to shape the direction of the society and to promote the interests of his fellow X Club members. Ruth Barton has argued that election to the council opened the way for the X Club members to move into more RS positions.[37] Indeed, within a few years the X Club members were to hold significant power within the RS, occupying several posts on its council, a position they maintained into the 1880s.

Tyndall was particularly active in lobbying for positions within and awards from the RS for men whom he saw as worthy of such accolades and recognition. Even before his term on the RS council began in November 1869, he encouraged Airy to assume the presidency of the society. Airy, however, declined, citing his other commitments and concerns about the distance he would need to travel to RS meetings, as well as his age and hearing impairment, among other reasons. Despite Tyndall's persistence, Airy remained firm.[38] He eventually relented, and with the support of various members of the X Club was elected president of the RS in the fall of 1871, an event that falls outside the years of this volume.

When the role of RS treasurer opened up, Tyndall was eager to see a fellow X Club member nominated. William Allen Miller, the previous treasurer, had passed away unexpectedly, so possible successors had not yet been discussed within the society. Letters reveal discussions between Tyndall, Huxley, and Hirst about their favored candidates and the process of consultation.[39] That another potential candidate with internal RS support was marginalized in the process was the source of some bad blood, but the X Club members succeeded in getting their chosen candidate, William Spottiswoode, approved.[40]

Tyndall was also central to the RS awarding James Joule its Copley Medal in 1870. In the early 1860s Tyndall had been instrumental in championing the claim of German physician Robert Mayer against Joule for priority in the formulation of the mechanical theory of heat. This dispute, fought between Tyndall and Joule's proxies, William Thomson and Peter Guthrie Tait, had split the international physics community into national camps and led to an extended, rancorous, and increasingly public dispute in 1862–63.[41] Letters in 1870, in contrast, show how Tyndall lobbied prominent members of the RS, such as Edward Sabine and Thomson, in support of Joule for the Copley Medal.[42] Interestingly, Tyndall campaigned for Joule to be awarded the medal for his research into the dynamic theory of heat, the same research for which he had already been awarded the society's Royal Medal in 1852. David Cahan argues that this was strategic and intended to clear the way for the nomination of Robert Mayer for his work the following year.[43]

International Networks

Throughout this period, as in earlier years, Tyndall carried out considerable correspondence with a range of scientific figures in Europe and the United States. For Tyndall these were an important part of his efforts to promote science and to ensure a steady stream of scientific research exchange, particularly across the English Channel. He supported foreign colleagues like French physicist Joseph Plateau, Swedish physicist Anders Jonas Ångström, and German physician Robert Mayer for membership in the RS, sometimes with great persistence.[44] His keen interest in translating and disseminating works of continental scientists in English science journals continued, as his efforts with Emil du Bois-Reymond and Auguste de la Rive, among others, demonstrate.[45] This international network was additionally useful when Tyndall sought to promote his own work on the Continent.

Much of Tyndall's international correspondence concerned translations of his works into other languages, primarily French and German or, vice versa, his facilitation of translations of the work of others. Anna Helmholtz was a regular translator of Tyndall's works into German, with the assistance of her husband, Hermann, who brought his expertise as a physicist to the scientific detail. In 1869 she translated Tyndall's *Sound*, and in 1870 his commemoration of Faraday, *Faraday as a Discoverer*, which was published in German as *Faraday und seine Entdeckungen*.[46] Upon receiving copies of the translated work on Faraday, Tyndall wrote to Hermann immediately to offer his thanks and added that the translation surpassed the original.[47] While Anna's contribution can be seen as indicative of the hidden labor of many women affiliated with scientific men in the nineteenth century, it must be noted that she was compensated for her work and, as she wrote back to Tyndall, this was money she fully intended to spend on herself.[48]

French translation of Tyndall's works was the realm of Abbé François-Napoléon-Marie Moigno, the editor of several French popular science journals. Moigno had translated a number of Tyndall's books and papers in the past, and his translation of *Sound* was published in 1869. *Notes on the Lessons about Electricity*, which first appeared in the *Chemical News*, followed in 1870, translated and published in Moigno's journal *Les mondes*.[49] In turn, Tyndall facilitated the dissemination of the work of his international colleagues, occasionally translating short papers himself for publication in the newspaper or recommending them to journal editors.

Tyndall's international contacts were also important for providing access to continental instrument makers, and especially, the expert makers of physical and optical instruments in Germany. Through his European colleagues, he learned of new instruments and designs, and through contacts like du Bois-Reymond, he was able to order and secure apparatus such as a magnetic compass for a telescope from the renowned Berlin-based instrument maker Ferdinand Sauerwald.[50] The correspondence surrounding the order and completion of this instrument offers an indication of how useful Tyndall's continental network could be. The stand to hold the apparatus was of du Bois-Reymond's design, the telescope was constructed by the German physicist and astronomer Carl August von Steinheil, and the rest of the apparatus and its construction was in the hands of Sauerwald. Sauerwald's apparatus were in such demand that it took him six months to complete Tyndall's order. Thus, Tyndall was able to draw on the best German instrument makers in his quest for precision apparatus.[51]

Tyndall's Travels

Tyndall traveled regularly both for professional reasons and to recuperate from his heavy workload in London and at the RI. His travels in 1869 and 1870, however, stand out for being particularly influential. In these years, in addition to frequent short hiking outings and trips to lighthouses in Ireland and Britain for the Board of Trade,[52] he took three more extensive trips, each significant for a different reason (see maps 1, 2, and 3).

On 4 July 1869 Tyndall and his longtime friend Thomas Archer Hirst departed for continental Europe. They planned to spend a month traveling through Germany, visiting old friends and mentors from their university days and solidifying new professional connections.[53] They traveled by ship to Antwerp, where they did some sightseeing, then went on to Belgium, where they visited the battlefield and memorials at Waterloo. Tyndall had been there in 1848 and, he noted in a letter to Bence Jones, much had changed since then. The site no longer looked fresh; the effect of time was noticeable.[54]

From Belgium, Tyndall and Hirst continued on to Germany.[55] They visited Tyndall's friend and colleague, physicist Rudolf Clausius, who had recently accepted a position in Bonn, before they continued on to Koblenz, Marburg, and Berlin, managing to fit in several hikes along the way. The tour continued on to Halle, where Tyndall visited the physical instruments collection of his old colleague Hermann Knoblauch, and then on to Weimar, where he communed with the ghosts of Goethe and Friedrich Schiller.[56] From there the trip took them to Eisenach and Heidelberg, where Tyndall visited his old mentor Robert Bunsen and dined with his friends Helmholtz and Gustav Kirchhoff.[57] Tyndall and Hirst then traveled to Karlsruhe and parted at Strasbourg, with Hirst continuing on to Paris while Tyndall traveled to Switzerland, where he intended to meet Fred Pollock at Grindelwald.[58]

Tyndall hoped that hiking in the mountain air of Switzerland would help him recover his good health, which often suffered under the strain of his intensive work and research schedule. Unfortunately, his usual destinations did not provide relief this time, and he continued on until he reached Belalp. Between the hikes and a daily dip in the cold water of a nearby river, a new habit he had developed under Hirst's encouragement, he felt himself returning to strength and vigor.

While bathing in the river one morning, Tyndall slipped and severely bruised and lacerated his leg on a sharp stone.[59] The wound seemed to heal at first, but he developed a serious infection and abscess in his leg that required medical attention. Unable to walk, he had to be carried down the mountain on a stretcher, so that he could travel to Geneva to be treated by a physician. Recovery required weeks of bedrest in Geneva and then further weeks of convalescence at Bence Jones's country home at Folkestone. The months he spent recovering from the injury and infection affected Tyndall's productivity for the later months of 1869 and were, for obvious reasons, the topic of much of his correspondence for the rest of the year.[60]

The following summer Tyndall was again in the Swiss mountains. This time his trip was overshadowed not by a personal event, but by a political one. On 19 July 1870 Napoleon III declared war on Prussia. Although in Switzerland Tyndall was safely far away from the hostilities of the Franco-Prussian War, they still had a great effect on his time in Europe, in part because many of his friends and colleagues were unable to leave their European residences or travel freely through either country. Few letters exist from these months, since postal service was disrupted.[61] Tyndall hiked in the surrounding countryside, as usual, but between the bad weather in Switzerland that summer, general uneasiness due to the war, and Tyndall's own worries about his upcoming discourse at the meeting of the BAAS in Liverpool in September, his mood was more subdued than usual. He returned to England at the beginning of September, making it through France well before the Prussians and avoiding getting caught up in the siege of Paris, which began on 19 September 1870.

Tyndall's final large journey of the year took place in December, when he traveled to Oran in Algeria as part of the British expedition to view the solar eclipse. In part, he hoped to test his theory that solar dust was responsible for the color of the sun's corona. The steamship ran into a ferocious storm with strong winds and waves a few days after departing but arrived safely and in calm weather. Unfortunately, the eclipse was clouded over, and he was unable to test his theory. On the way back, however, he carried out experiments on the color of the sea that were to form a part of the material for a later paper in *Nature*, "On the Colour of Lake Geneva and the Mediterranean Sea."[62] He arrived back in Portsmouth safely on 5 January 1871. Another British ship that had traveled to Sicily to view the eclipse and that included several of his colleagues struck a rock and sank, fortunately with no loss of life.[63]

These three trips, Tyndall's preparation for them, and their consequences make up a considerable number of letters and recurring themes in this volume. They intersect with some of Tyndall's research interests, as well as his concerns for his international colleagues and his frustration with his own health.

Friends and Family

By 1869 Tyndall was one of the most well-known and influential men of science in Britain and internationally. He dined with lords and ladies, visiting them in their country homes and manors; was friends with Thomas Carlyle and Alfred Tennyson, two of Victorian Britain's most renowned creative minds; and he attended cultural events and soirées with famous singers and musicians. His social calendar was packed, and when he was not moving among the British elite he was entertaining visiting scholars or meeting with friends at the Athenaeum Club. Old friends like the Pollock family rounded out his lively and demanding social life, and he met regularly with his close friend Hirst for meals and country walks, in addition to their longer travels.

Tyndall's loyalty to his friends was demonstrated in his efforts to secure the position of chair of chemistry at Guy's Hospital for Heinrich Debus. Debus, like Tyndall, had studied in Marburg, and Tyndall had supported him for a post at Queenwood College in 1868. When Alfred Taylor resigned from his position at Guy's, Tyndall was approached by George Owen Rees, the senior physician, seeking advice on a possible replacement. He was quick to suggest Debus for the position and to act as an intermediary, advising Debus on how to negotiate for better conditions and Rees on the advantages of making these concessions and enabling both to reach a satisfactory arrangement.[64]

Tyndall had no lack of female friends and admirers either, to whom he freely distributed tickets to RI lectures and his popular publications. While he cultivated these platonic friendships with married women, including the wives of friends, he still had no luck in love. His friends the Lubbocks tried to set him up with a Miss Symonds, unsuccessfully.[65] His experience with serious illness after his bathing injury convinced him to take a chance and to declare his love to Mary Adair.[66] Unfortunately for Tyndall, she did not share his feelings, but they remained friends.

Family matters continued to be troublesome. Tyndall's mother had passed away in 1867, and family friend Elizabeth Dawson Steuart had been named executor. The correspondence between the two show their mutual frustration with the children of his uncle, Caleb Tyndall, who regularly pressured them for money and fought with each other over the distribution of his mother's estate.[67] Tyndall's sister Emma was another ground for concern. Her mental state deteriorated after their mother's death, and she was taken in by yet another branch of the Tyndall family at Gorey, County Wexford, where she gradually recovered.[68]

Through his correspondence and travels, Tyndall kept his international network and, especially, his connections to German scientists active. His tour of Germany with Hirst in 1869 enabled him to visit many old friends, some of whom had, like Clausius, accepted new positions in new towns. The Franco--Prussian War weighed heavily on him, since he had friends in both countries who were affected by the hostilities. Clausius went to the front with an emergency medical corps and was wounded in service.[69] Bunsen, while not on the front, reported on the wounded who were brought by train to emergency hospitals in Heidelberg.[70] Although Tyndall was clear in his support of the German side, at least in his English correspondence, he contributed generously to humanitarian aid for both sides.[71]

It was also a year of death and loss in Tyndall's circle of friends. His old mentor and rival Gustav Magnus passed away after a swift decline in 1870.[72] His friend Hermann Knoblauch, whom Tyndall had visited in 1869, lost a child to scarlet fever in the autumn of 1870. Mary Egerton's husband died in the summer of 1869, only days after Tyndall had been hiking with them in the Alps. The loss haunted her, and her melancholy letters to Tyndall in the following year express the persistence of her grief.[73] Even Tyndall's close friends the Pollocks struggled with the death of family members.[74] The severe loss of life in the Franco-Prussian War and the suffering of his European friends only added to the melancholy timbre of the year.

Conclusion

The letters in this volume provide much insight into an important period in Tyndall's life. They cover the continuation of his work into atmospheric phenomena and the beginnings of his interest in germ theory, a topic that was to consume his time in the coming years. In addition, his thinking around the importance of uses of the scientific imagination were solidified at his address to the BAAS. While the loss of some of his continental friends in this period was a personal tragedy, he continued to cultivate a strong network of scientific acquaintances both at home and in Europe. These proved useful in his mission to ensure the steady transmission of scientific knowledge and in allowing access to expertise and instruments. This network was also essential for the translation and publication of his own work, particularly into German.

Tyndall's efforts within the RS to gain awards and positions for his nominees demonstrate the strong position he held within London's scientific elite. At the end of this period the members of the X Club were also well positioned for the impending rise to the apex of the RS with the election of Spottiswoode to the society's treasurership. Likewise, Tyndall's successful efforts to gain foreign membership for Plateau, Joule's Copley Medal, and a post at Guy's Hospital for Debus are all indicative of his social and scientific standing.

Finally, while his relative position of wealth and prestige brought with it difficulties with his extended family in Ireland, he was also able to enjoy its benefits. He was able to donate considerable sums to his European colleagues during the Franco-Prussian War, as well as smaller sums in support of various British causes. In particular, he made several journeys to Europe to visit friends and engage his passion for the Alps, as well as to the northern coast of Africa to explore the solar eclipse. In comparison to the squabbling among his extended family in Ireland over the remnants of his mother's estate, these trips demonstrate to us and, perhaps also to Tyndall, how far he had come.

Notes

1. John Tyndall, *The Scientific Use of the Imagination* (London: Longmans, Green, 1870), 7.
2. John Tyndall, "On a New Series of Chemical Reactions Produced by Light," *Roy. Soc. Proc.* 17 (1868): 92–102.
3. John Tyndall, "On the Blue Colour of the Sky, the Polarization of Skylight, and on the Polarization of Light by Cloudy Matter Generally," *Roy. Soc. Proc.* 17 (1869): 223–33.
4. John Tyndall, "On the Action of Rays of High Refrangibility upon Gaseous Matter," *Phil. Trans.* 160 (1870): 333–65.
5. See letters 3269, 3271, 3280, 3284, and 3286.
6. Aileen Fyfe, Flaminio Squazzoni, Didier Torny, and Pierpaolo Dondio, "Managing the Growth of Peer Review at the Royal Society Journals, 1865–1965," *Science, Technology, & Human Values* 45, no. 3 (2020): 405–429.
7. *Ascent of John Tyndall*, 248–50.
8. John Tyndall, "On Cometary Theory," *Phil. Mag.* 37 (1869): 241–45.
9. See letter 2999 and *Ascent of John Tyndall*, 245–48.
10. Further discussion can be found in letters 2965, 2980, 3003, and 3166.
11. Terrie M. Romano, "The Cattle Plague of 1865 and the Reception of the Germ Theory in Mid-Victorian Britain," *Journal of the History of Medicine and Allied Sciences* 52, no. 1 (1997): 51–80.
12. See letters 3126 and 3127; John Tyndall, "Dust and Disease," *Pall Mall Gazette* 1547 (1870), 6; letter 3122; *Times*, 7 April 1870 (letter 3170); "On Haze and Dust," *Nature* 1 (27 January 1870): 339–42; "On Dust and Disease," *Fraser's Magazine*, March 1870, 302–10.
13. James Strick, "Darwinism and the Origin of Life: The Role of H. C. Bastian in the British Spontaneous Generation Debates, 1868–1873," *Journal of the History of Biology* 32 (1999): 51–92; James Strick, *Sparks of Life: Darwinism and the Victorian Debates over Spontaneous Generation* (Cambridge, MA: Harvard University Press, 2000).
14. *Times*, 7 April 1870 (letter 3170); *Times*, 21 April 1870 (letter 3177).
15. See letter 3188.
16. John Farley and Gerald L. Geison, "Science, Politics and Spontaneous Generation in Nineteenth-Century France: The Pasteur-Pouchet Debate," *Bulletin of the History of Medicine* 48 (1974): 161–98.
17. See letter 3208.
18. See letters 3283 and 3305.
19. Roland Jackson, "John Tyndall and the Early History of Diamagnetism," *Annals of Science* 72, no. 4 (2015): 436, 443.
20. See letter 3206.
21. See letter 3151.
22. See letter 3332.
23. See letters 2951 and 2953.
24. See letters 2951 and 2953.
25. John Tyndall, *Fragments of Science for Unscientific People: A Series of Detached Essays, Lectures, and Reviews* (London: Longmans, Green, 1871), 347–72.
26. See letters 3162, 3164, and 3165.
27. See letter 3159.
28. See letter 3098.
29. *Ascent of John Tyndall*, 209.
30. Tyndall, *Scientific Use of the Imagination*.
31. See letters 3238 and 3241.
32. See letters 3255, 3257, and 3259; and Ruth Barton, "Tyndall, John (1820–1893)," *Dictionary of Nineteenth-Century British Philosophers*, ed. W. J. Mander and Alan P. F. Sell, (Bristol: Thoemmes, 2002), 2:1139.

33. See letters 3256 and 3303.
34. This dispute is covered in more detail in the forthcoming twelfth and thirteenth volumes of *The Correspondence of John Tyndall.*
35. Roy M. MacLeod, "Science and Government in Victorian England: Lighthouse Illumination and the Board of Trade, 1866–1886," *Isis* 60, no. 1 (1969): 5–38.
36. See letters 2937, 3067, 3124, 3263, 3353, and 3360.
37. Ruth Barton, "'An Influential Set of Chaps': The X Club and Royal Society Politics, 1864–85," *British Journal for the History of Science* 23, no. 1 (1990): 53–81; and Ruth Barton, *The X Club: Power and Authority in Victorian Britain* (Chicago: University of Chicago Press, 2018).
38. See letters 3088, 3089, 3090, 3094, and 3095.
39. See letter 3274.
40. See letters 3292, 3294, 3299, and 3306; Barton, "Influential Set of Chaps," 65–67.
41. See *Tyndall Correspondence*, vol. 8, xvii–xix; and Crosbie Smith, *The Science of Energy: A Cultural History of Energy Physics in Victorian Britain* (Chicago: University of Chicago Press, 1999), ch. 9.
42. See letters 3181 and 3221.
43. David Cahan, "The Awarding of the Copley Medal and the 'Discovery' of the Law of Conservation of Energy: Joule, Mayer and Helmholtz Revisited," *Notes and Records of the Royal Society* 66 (2012): 125–39.
44. See letters 2975, 3138, 3155, 3168, 3179, 3181, and 3192.
45. See letters 3104, 3128, 3140, 3199, 3225, 3240, and 3242.
46. See letters 2951, 2953, and 2968.
47. See letter 3282.
48. See letter 2968.
49. See letters 2949, 3226, and 3359.
50. See letters 3028, 3061, 3104, and 3148.
51. See letters 3104 and 3148.
52. See letters 3262 and 3266.
53. For Tyndall's German university period, see Melinda Baldwin and Janet Browne, eds., *The Correspondence of John Tyndall*, vol. 2, *The Correspondence, September 1843–December 1849* (Pittsburgh: University of Pittsburgh Press, 2016), "Introduction," xxiv–xxvi; and *Ascent of John Tyndall*, ch. 4, pp. 43–58.
54. See letter 3020.
55. For a full description of Tyndall's itinerary, see Hirst's journal (RI MS JT/2/32d/1848–51); and *Ascent of John Tyndall*, 250–54.
56. See letter 3023.
57. See letters 3023 and 3024.
58. See letters 3023 and 3024.
59. See letter 3030.
60. Tyndall reports back to his friends the details of the accident, his injury, and his recovery in numerous letters through September and October 1869. See, in particular, letters 3031, 3037, 3048, 3049, and 3060.
61. For some discussion of the war, see letters 3243, 3249, 3251, and 3252.
62. John Tyndall, "On the Colour of Lake Geneva and the Mediterranean Sea," *Nature* 2, no. 51 (1870): 489–90.
63. See letters 3313, 3326, 3328, 3329, 3335, and 3336.
64. See letters 3202, 3209, 3211, 3213, 3214, and 3220.
65. See Hirst's journal (RI MS JT/2/32d/1842).
66. See letter 3059.
67. See letters 3318, 3322, and 3336.

68. See letters 3083, 3139, and 3310.
69. See letter 3252. Although Clausius does not mention it in the letter, he was wounded in the field (*NDB*). He later received an Iron Cross for his service.
70. See letter 3248.
71. See letters 3247, 3248, and 3305.
72. See letters 3140 and 3174.
73. See letters 3040, 3076, and 3077.
74. See letter 3253.

MAPS

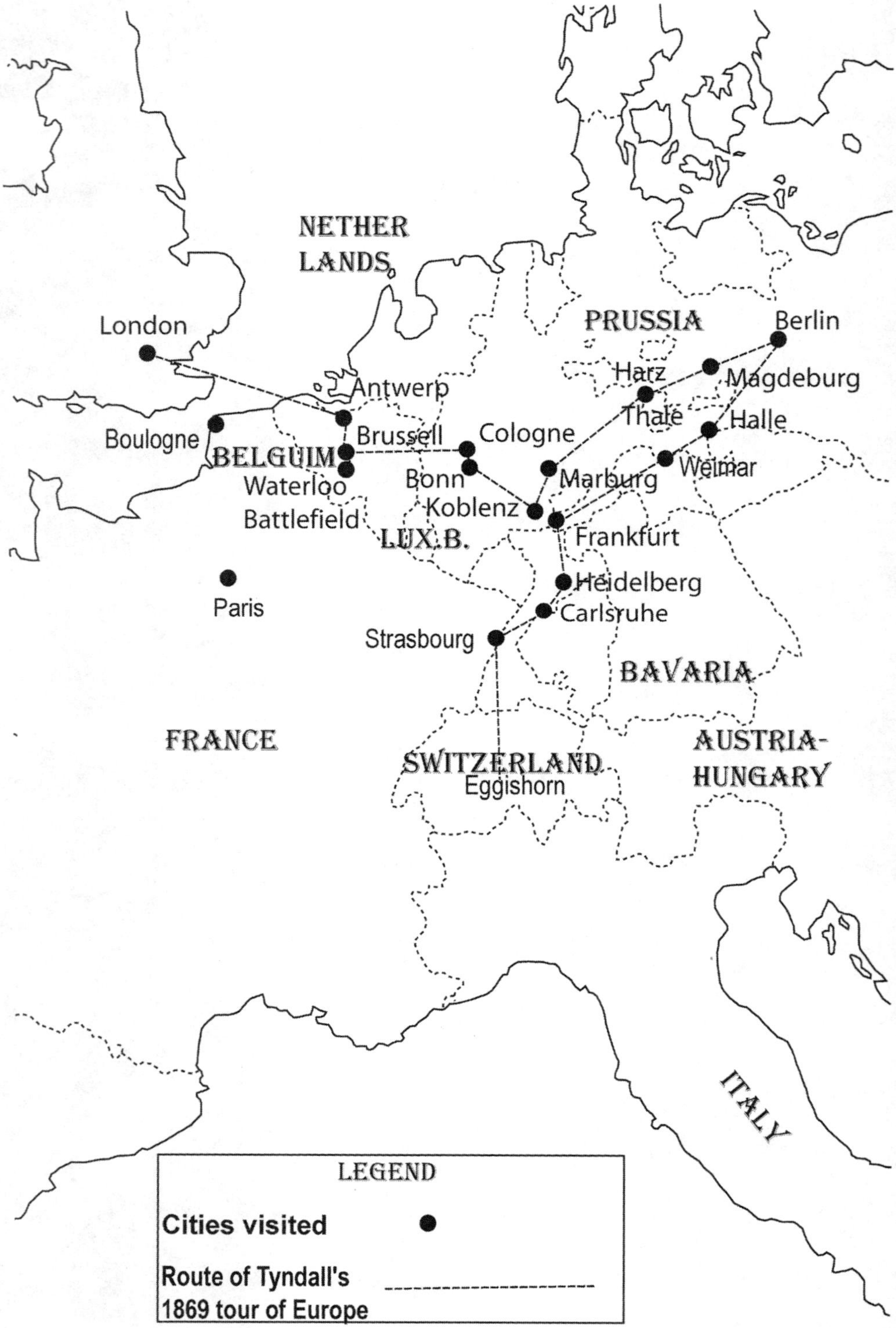

Map 1: Europe, showing the route of Tyndall and Hirst's July 1869 tour.

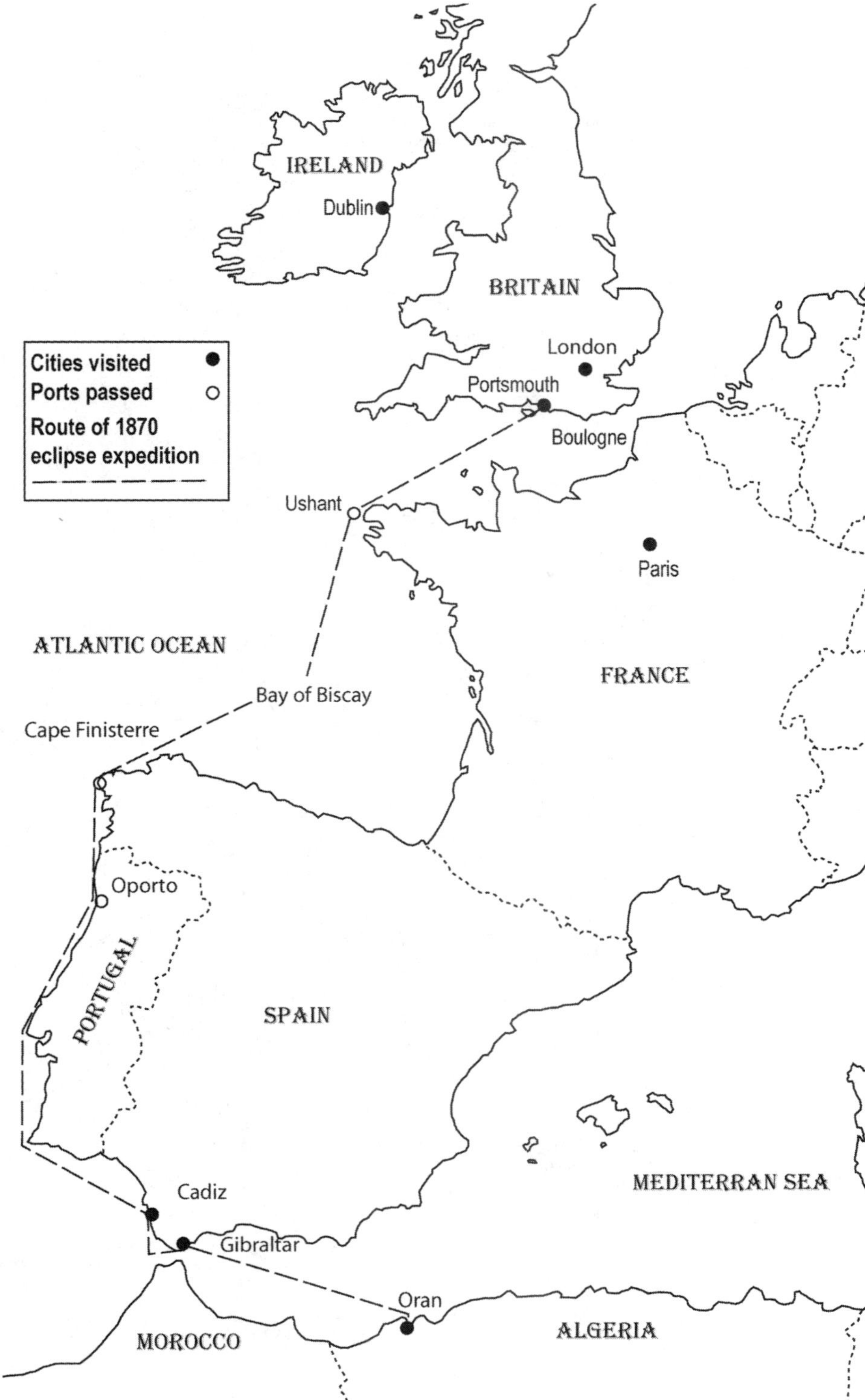

Map 2: Route of the solar eclipse expedition to Oran, December 1810–January 1871.

Map 3: Switzerland, showing locations mentioned in the letters published in this volume.

EDITORIAL PRINCIPLES

Our aim is to include all letters to and from John Tyndall that are currently extant. Some of the letters are from published sources, such as letters written by Tyndall to newspapers or journals, but the majority are either the originals or typescripts produced by Louisa Tyndall from originals. One central editorial principle has been to transcribe, wherever possible, from original letters rather than Louisa's typescripts. This is because of Louisa's conscious and unconscious editorial interventions, a point discussed in more detail below. We have also aimed to reproduce, as accurately as possible, the text of the letters as they were written. Spelling mistakes that have appeared in the handwritten letters have therefore been preserved; mistakes appearing where only the typescript letter has survived are silently corrected, as in our judgment they are almost always typographical errors. But, in general, nonstandard spellings (e.g., the use of dialect) has been retained even when in a typescript--only letter (though there is the possibility that the spelling is Louisa's, not her husband's). Where the transcription is based on a typescript transcription by Louisa (or one of her assistants), this will be indicated by "LT Transcript Only."

We have included illustrations that are part of the letter. In cases where it was possible to reprint the original handwritten image, we have done so. In other cases, volume editors have reconstructed drawings because it was not possible to reprint an image from the handwritten letter (due, for example, to deterioration of the original document).

The letters are presented in chronological order of writing. Where more than one letter has the same date, letters from Tyndall (in alphabetical order of addressee) come first, followed by letters to Tyndall (in alphabetical order of writer), with the exception that if the order of writing can be determined, that order takes precedence. Where letters cannot be dated precisely they are placed at the latest likely date of writing. (In some cases if it is possible, but unlikely, that they were written later, notes indicate such possibilities.) To indicate the beginning of a new year, the date of that year has been inserted in a large font before the first letter from that year. Readers should note that although volumes are arranged in chronological order, letters that are discovered from years already covered in published volumes will appear in the final volume of the correspondence.

For non-English-language letters, both the transcription of the original and the English translation are presented. The transcription of the non--English letter appears first, followed by the English translation in a smaller font. Editorial notes are inserted in the English translation. Some foreign letters are in an original hand, while others exist only as transcriptions by Louisa or others who often did not have a full command of the language. In the latter case these letters contain frequent spelling, grammatical, and interpretation errors that do not seem to have been committed by the original author. In these typescript-only cases, and only in these cases, spelling and grammatical errors have been corrected, as have poor interpretations where a better word choice was obvious from the context.

There is a standard format for each letter. The first line lists the author, if it is to Tyndall, or the recipient, if it is from Tyndall; the date; and the letter number. The second line lists any information supplied by the writer in their salutation, such as the place, date, or time of day when the letter was written. (In some cases, though, where the writer put the date at the foot of the letter we have left it there in order to reproduce the text accurately.) The opening salutation, the text of the letter, and closing then follow. Then come the source and editorial notes, the latter indicated in the text by arabic numerals.

Editorial notes are intended to provide the contextualization necessary to understand the contents of each letter. There are basically six types of editorial notes. Many of them are informative notes on persons mentioned in the letters. If a person referred to in a letter is not described in a note, they have been mentioned more than twice in the volume and they will have an entry in the biographical register at the back of the volume. For the rest—those mentioned once or twice—biographical information will appear in the editorial note appended to the letter in which he or she is first mentioned. When there is a second reference to that person a note will refer back to the letter in which he or she was first mentioned. Biographical notes on well-known figures such as Faraday or Huxley stress what that person was doing at the time when the letter was written rather than providing a full overview of the person's life.

A second type of editorial note identifies allusions and quotations. In the case of quotations, where possible, the first edition of the work is cited unless a different edition is mentioned in the letter. If a quote is not in English a translation is given. Glosses on obscure places, abstruse words, and words used in an unfamiliar sense constitute the third type of editorial note. The fourth type provides information about the context of the letter, drawing on such sources as contemporary publications, scholarly sources, Tyndall's personal journals and field notes, and other letters in the *Tyndall Correspondence*.

The fifth type of editorial note is cross-references to other letters in the correspondence, referring to the letter number. The last type is information on significant textual changes that Tyndall made to the original that is thought to be relevant to the meaning of the letter.

There are a series of conventions for indicating when editors have given additional information about the text of a letter:

When Louisa Tyndall has made an annotation or marginal insertion that the editors have deemed important enough to retain, we have included her annotation in an editorial footnote.

When editors other than she have made insertions, they are indicated by unitalicized square brackets; for example, [word].

When certain words are ambiguous and we have had to conjecture their meaning, they are indicated by italicized square brackets; for example, *[*words*]*; when we have considered such words illegible the number of illegible words is indicated in full italicization and square brackets; for example, *[3 words illeg]*.

If a word or series of words are not present or have been destroyed—by inkblots, holes, or mold, for example—they are indicated as missing, italicized and in angular brackets; for example, <*3 words missing*>.

In extremely rare cases where it is clear that Louisa herself has intentionally destroyed a few words, they are indicated as excised, italicized and in angular brackets; for example, <*3 words excised*>.

Finally, a brief word about Louisa. She was responsible for collecting together many of Tyndall's letters in order to write a biography of her late husband. This biography was not completed. It is important to recognize that Louisa was not always reliable as a transcriber. She often threw out the manuscript letter when she had made her transcription, but some original manuscripts remain, which indicate her editing processes. Like any transcriber, she was sometimes inaccurate, but corrections show that she proofread her transcriptions. However, she also edited Tyndall's writing to make his letters seem more consistent or grammatically correct. For example, she made punctuation more formal, corrected grammar, and, occasionally, corrected quotations. She consistently corrected Tyndall's usual (deliberate) lowercase for "god" to "God." Her typewriter was unable to insert the superscripts that Tyndall used in abbreviations, or the "&c" symbols that stood for an address at the foot of the letter. (Modern typefaces are unable to deal with all the variant scrawls that the "etc." symbol took and they are all formalized here to "&c.") Louisa also left out details that she considered too private or too salacious for public consumption. She equated postmark and date of writing and overlooked the time taken to write some letters. When she inferred dates from internal evidence, her reasoning was sometimes faulty. Similarly, her identifications of people are sometimes in error. Independent evidence has always been sought for her dates and identifications. Lastly, the patterns of punctuation differ between LT sources and manuscript sources. Tyndall himself used dashes more often than commas, semicolons, and stops, and he used dashes of many different lengths, which are all standardized here to em-dashes. It is often difficult to distinguish between very short dashes and commas or very short dashes and stops. Similarly, capitalization is often ambiguous; distinguishing *S/s*, for "science," for example, is a particular problem. Therefore, scholars for whom punctuation and capitalization are significant will need to consult the manuscripts. Users should also be aware that throughout this project the transcribers and editors have generally worked from scans and photographs rather than the original manuscripts.

NOTE ON MONEY

In the nineteenth century the currency used in England was the pound sterling, usually abbreviated to £ (or, occasionally, to *L*). A pound consisted of 20 shillings (abbreviated *s*) and each shilling contained 12 pennies (*d*). A guinea was 21 shillings; that is, £1 1s. A crown was 5 shillings (5s), and half a crown equaled two shillings and sixpence (2s 6d). The smallest coin was a farthing, one quarter of a penny (¼d).

Several forms of representation were in contemporary use. Sometimes the numbers for pounds, shillings, and pence were separated by dots, forward slashes, or spaces; thus, for example, 3 pounds, 4 shillings, and 5 pennies could be written as £3.4.5, £3/4/5, or £3 4 5. A zero could be represented by a dash; for example, 7 shillings was often written as 7/- or just 7/. A half crown was often written as 2/6.

Currencies mentioned in this volume are French francs, Belgian francs, United States dollars, and German Thaler. The franc was the national currency of France from 1795 until 1999, when France adopted the Euro. In 1864 the exchange rate between British and French currencies was approximately 25 francs per £.[1] A centime was a coin worth one-hundredth of a franc. Belgium adopted the franc as its national currency in 1832. The Belgian franc was equivalent to the French franc in value. In the mid-nineteenth century, one £ was worth approximately ten US dollars.[2] The Reichsthaler was a silver coin issued by Prussia from 1750 that became a common currency throughout much of Germany in the mid-nineteenth century before being replaced by the Vereinsthaler in 1857. In midcentury the exchange rate between the German and British currencies was approximately seven Thaler per £.[3]

Notes

1. Rodney Edvinsson, http://www.historicalstatistics.org/Currencyconverter.html.
2. Lawrence H. Officer, "Exchange Rates between the United States Dollar and Forty-One Currencies," MeasuringWorth, 2018, http://www.measuringworth.com/exchangeglobal.
3. Henry Vizetelly, *Berlin under the New Empire: Its Institutions, Industry, Monuments, Social Life, Manners* (London: Tinsley Brothers, 1879), 1:166.

TIMELINE OF JOHN TYNDALL'S LIFE

Year	Event
c. 1822	John Tyndall born at Leighlinbridge, County Carlow, Ireland
c. 1836	Attends John Conwill's National School in Ballinabranna
1839	Begins employment as civil assistant with Ordnance Survey of Ireland in Carlow
1840	Becomes civil assistant in the Ordnance Survey Office at Youghal, Cork County
1841	First appearance in print (poem in *Carlow Sentinel* under pseudonym "W[alter] S[nooks]")
1842	Joins English Ordnance Survey in Preston
1843	Leads written criticism of Ordnance Survey in *Liverpool Mercury*; dismissed November
1844	Home in Ireland, mostly unemployed, until obtaining a position with the firm of Nevins and Lawton, Surveyors, of Manchester
1845–47	Works for Richard Carter, Surveyor, of Halifax
1845	Meets Thomas Archer Hirst in Halifax
1847	Begins teaching mathematics and surveying at Queenwood College, Hampshire
1848–50	At University of Marburg, working for PhD with Robert Bunsen and Friedrich Stegmann
1850	Returns to England; meets Michael Faraday and William Francis and attends his first British Association meeting, before returning to Marburg
1851	Moves from Marburg to Berlin (April–June); then returns to Queenwood
1852	Elected fellow of the Royal Society
1852	Friendship developed with Thomas Henry Huxley
1853	First lecture at Royal Institution (RI)
1853	Meets Herbert Spencer
1853	Appointed professor of natural philosophy at RI
1853	Awarded Royal Medal for magnetic work; declines when controversy arises over award
1856	Meets Thomas Carlyle
1857	Climbs Mont Blanc for first time with Hirst
1859	Demonstrates existence of greenhouse gases and climatic implications
1859	Joins Government School of Mines as professor of natural philosophy
1860	*The Glaciers of the Alps* published
1860	Meets John Lubbock and his wife, Ellen Lubbock
1861	First ascent of the Weisshorn

Year	Event
1861	Delivers his first Christmas Lectures at RI
1862	*Mountaineering in 1861* published
1862	Begins to champion Mayer's priority in discovery of conservation of energy
1863	*Heat Considered as a Mode of Motion* published
1864	Awarded Royal Society's Rumford Medal
1864	First meeting of X Club
1865	*On Radiation* published
1865	Engages in public controversy over the efficacy of prayer
1866	Succeeds Michael Faraday as scientific adviser to Trinity House
1867	Faraday dies; Tyndall becomes superintendent of the RI
1867	*Sound: A Course of Eight Lectures* published
1868	Successfully climbs Matterhorn
1868	Discovers the cause of light scattering, to be known as "Tyndall Effect"
1868	*Faraday as a Discoverer* published
1868	"Scientific Materialism" lecture at the British Association
1868	Resigns from Royal School of Mines
1869	Joins Metaphysical Society
1870	"On the Scientific Use of the Imagination" lecture at British Association
1870	*Three Scientific Addresses* published
1870	*Researches on Diamagnetism and Magne-Crystallic Action* published
1871	Meets Louis Pasteur for first time while in Paris
1871	*Hours of Exercise in the Alps* published
1871	*Light and Electricity* published
1871	*Fragments of Science* published
1872	Debates over how to measure the efficacy of prayer, aka the "Prayer-Gauge Debate"
1872	*Contributions to Molecular Physics in the Domain of Radiant Heat* published
1872	*The Forms of Water in Clouds and Rivers, Ice and Glaciers* published
1872–73	USA Lecture Tour
1873	*Six Lectures on Light* published
1874	As president of the BAAS, delivers the "Belfast Address"
1876	Marries Louisa Charlotte Hamilton
1877	Develops "Tyndallization" (discontinuous heating and cooling process to destroy heat-resistant spores)
1877	Tyndall and Louisa build summer cottage, Alp Lusgen, at Belalp, northern side of Valais, above Brig
1877	*Fermentation and Its Bearings on Phenomena of Disease* published
1879	*Fragments of Science*, which had gradually expanded, first published in two volumes
1881	*Essays on the Floating Matter of the Air* published
1885	Tyndall and Louisa build Hindhead retreat, Surrey Downs
1887	Resigns from RI
1890	Fight with Gladstone in the *Times* over Irish Home Rule
1892	Hirst dies
1892	*New Fragments* published
1893	Dies at Haslemere, Surrey, accidentally poisoned

TIMELINE OF EVENTS IN JOHN TYNDALL'S LIFE SPECIFIC TO VOLUME 11

Month and year	Biographical details	Notable social, political, and cultural events
January 1869	Reads paper "On the Blue Colour of the Sky" at the RS (14th) Delivers Friday Evening Discourse at the RI, "On Chemical Rays, and the Light of the Sky" (15th) German translation of *Sound* (*Der Schall*) is published	Grigori Rasputin born (21st) Matthew Arnold publishes *Culture and Anarchy* Lewis Carroll publishes *Phantasmagoria and Other Poems*
February 1869	Publishes "On Chemical Rays, and the Light of the Sky" in *Fortnightly Review*	Largest gold nugget in the world weighing 64.75 kg discovered in Ballarat, Australia Robert Browning publishes final volume of *The Ring and the Book* Johannes Brahms's *German Requiem* premieres in Leipzig (18th)
March 1869	Delivers Friday Evening Discourse at the RI, "On Cometary Theory" (8th) Publishes first part of "Odds and Ends of Alpine Life" in *MacMillan's Magazine*	Ulysses S. Grant sworn in as eighteenth president of the United States (4th) Dmitri Mendeleev presents his periodic table to the Russian Chemical Society (6th) Neville Chamberlain born (18th)
April 1869	Publishes second part of "Odds and Ends of Alpine Life" in *MacMillan's Magazine*	Metaphysical Society (London) founded American Museum of Natural History in New York founded Hudson Bay Company cedes territory to Canada
May 1869	Discusses cometary theory with John Herschel	Folies Bergère opens in Paris (2nd) First transcontinental railroad in North America completed

Month and year	Biographical details	Notable social, political, and cultural events
May 1869		Susan B. Anthony and Elizabeth Cady Stanton form the National Woman Suffrage Association in New York (15th) John Stuart Mill publishes *The Subjection of Women*
June 1869	Contributes to the Eyre Defense Committee funds	2nd Ferndale Colliery explosion (10th) Boshin War in Japan ends (27th) Yasukuni shrine built in Tokyo
July 1869	Leaves for European continent with Thomas Archer Hirst	French chemist Hippolyte Mège--Mouriès patents margarine Daniel Dunglas Home publishes *Experiences in Spiritualism with D. D. Home*
August 1869	Travels to the Alps	Total solar eclipse (North America) (7th) Social Democratic Workers' Party founded in Germany Debtors' Act abolished Irish naturalist and astronomer Mary Ward dies (31st)
September 1869	Slips while bathing in river at Belalp Evacuates to Geneva for medical care	Richard Wagner's *Das Rheingold* premieres in Munich (22nd) US stock market crashes (Black Friday) (24th) Wallace Monument completed in Stirling, Scotland
October 1869	Travels to Folkestone to convalesce	Mohandas Karamchand Gandhi born in Gujarat (2nd) Henry M. Stanley receives an assignment from the *New York Herald* to find Dr. David Livingstone in Africa Alizarin dye, the first synthetic dye, patented
November 1869	Proposes to Mary Adair	American philanthropist George Peabody dies (4th) Suez Canal opened (17th) Norman Lockyer publishes first issue of *Nature* Gustave Flaubert publishes *Sentimental Education*
December 1869	Encourages George Biddell Airy to allow himself to be nominated for president of the RS; Airy declines	First Vatican Council opens in Rome (8th) Leo Tolstoy publishes *War and Peace*

Month and year	Biographical details	Notable social, political, and cultural events
December 1869		Haitian president Sylvain Salnave overthrown
January 1870	Reads paper "On the Action of Rays of High Refrangibility" at the RS (27th) Delivers Friday Evening Discourse at the RI, "On Dust and Disease" (21st)	Construction of Brooklyn Bridge begins Rockefeller establishes the Standard Oil Company Keble College, Oxford, designed by William Butterfield, opens
February 1870	Demonstrates experiments to the Chemical Society	British Elementary Education Act introduced John Ruskin delivers *Lectures on Art* at Oxford Esther Morris appointed first female justice of the peace (Wyoming, USA)
March 1870	Publishes fourth edition of *Heat as a Mode of Motio*n	African American men granted right to vote in the United States Paraguayan War ends PS *Normandy* sinks in the English Channel
April 1870	Writes letter to the *Times* defending Joseph Lister's germ theory against Henry Charlton Bastian Delivers a lecture course at the RI, "Electrical Phenomena and Theories"	Gustav Magnus dies (4th) Vladimir Lenin born (22nd) Antonio Guzmán Blanco made president of Venezuela
May 1870	Declines nomination for president of BAAS	Home Government Association established in Ireland Royal Academy of Arts Exhibition inaugural banquet (1st) Emergency conference on smallpox control in Paris
June 1870	Advises Heinrich Debus about position at Guy's Hospital Publishes *Diamagnetism*	Tianjin Massacre takes place Charles Dickens dies (9th) First telegraph message sent by submarine cable from Great Britain to India
July 1870	Travels to the Alps Arrives at the Belalp Experiences beginning of the Franco-Prussian War while in the Alps	Hudson's Bay and Northwest Territories become part of Canada; Manitoba becomes fifth province (15th) France declares war on Prussia (19th) New York City Orange Riot (12th)

Month and year	Biographical details	Notable social, political, and cultural events
August 1870	Donates funds to French and German colleagues to aid the war wounded	Sydney Intercolonial Exhibition opens UK Foreign Enlistment Act Married Women's Property Act Landlord and Tenant (Ireland) Act
September 1870	Delivers address "The Scientific Use of the Imagination" at BAAS meeting in Liverpool (16th) Travels to Dublin to inspect Irish lighthouses	Battle of Sedan Napoleon III deposed (4th) Siege of Paris
October 1870	Publishes "On the Colour of the Lake Geneva and the Mediterranean Sea" in *Nature* Supports William Spottiswoode for position of RS treasurer, alienating John Peter Gassiot	Siege of Metz First women admitted to University of Edinburgh Medical School begin taking classes Hurricane of San Marcos
November 1870	Begins preparations for participation in eclipse expedition	Battles of Amiens, Beaune la Rolande, Villiers and Havana Siege of Belfort Lydia Becker becomes first woman elected to a school board (Manchester, UK)
December 1870	Travels to Oran, Algeria, to view the solar eclipse	Alexandre Dumas dies (5th) Total solar eclipse (22nd) Wreck of HMS *Psyche* off the coast of Sicily
January 1871	Arrives back in London Delivers Friday Evening Discourse at the RI, "On the Colour of Water" (21st)	Commencement of the Irish Church Act Ruskin publishes first issue of *Fors Clavigera* William I of Prussia proclaimed German emperor
February 1871	Begins working on respirator for firemen	Charles Darwin publishes *Descent of Man* Great Gale of 1871 Treaty of Versailles (1871) signed

The Correspondence of John Tyndall

1869

To Robin Allen 4 January 1869 2937

Enclosure 1
Royal Institution, 4 January 1869.

Sir,

I HAVE been to Blackwall,[1] and have seen the performance of the magneto-electric machine of Mr. Holmes.[2] I have also, in company with Captain Nisbet,[3] Captain Weller,[4] and Mr. Douglass, observed the magneto-electric light[5] from the opposite side of the Thames.

I beg to recommend[6] that the following series of observations be made with the magneto-electric light:

That it be observed from a position near Charlton,[7] known to the gentlemen above named, for three or four nights in succession from the hour of 5 or 6, to the hour of 10 o'clock.

These observations are in the first instance to decide whether the third order dioptric apparatus, or the sixth order apparatus, is most suitable for the exhibition of the magneto-electric light.

The observations for each apparatus are to continue through intervals of 10 minutes; the character of the light during these 10 minutes to be recorded.

Intermissions of the light are sure to occur, and the times of their occurrence are to be noted, in order to determine how far they are due to the passage of vessels, or how far to the imperfections of the electric lamp.

Captain Nisbet, Captain Weller, and Mr. Douglass are aware of the device adopted to distinguish the lights of the two lanterns from each other; it consists in preceding the exhibition of the sixth order light by one minute's exhibition of red light.

Then in these first experiments, it would be a very great advantage if the lights from both the lanterns at Blackwall could be seen together.

This would necessitate the placing in position of Mr. Holmes' second magneto-electric machine. I recommend this to be done. The arrangements will enable us to make a far more satisfactory comparison of the lights than when they are seen in succession. It will also enable us to determine the gain of light which results from the convergence of the two machines upon a single lamp.

I have placed in the hands of Mr. Douglass, portions of carbon which I recommend for trial by Mr. Holmes in the presence of Mr. Douglass. It is desirable to diminish, if possible, the flickering of the light caused by the presence of foreign substances in the carbon.

Having decided which is the best form of dioptric apparatus for the exhibition of the magneto-electric light, it will be necessary to compare it with a dioptric light of the first order where the source of illumination is a flame of oil. To effect this comparison both lights must be seen at the same time from the position near Charlton to which I have referred.

The penetrative power of the magneto-electric light, as compared with that of the oil flame, may then be determined by means of Captain Nisbet's photometer or some similar arrangement.

It is also desirable to compare the magneto-electric light, as regards cost, convenience, and efficiency, with the voltaic light. But this comparison may be deferred until after the observations here recommended have been made.

I have &c. | (signed) *J. Tyndall.*

Robin Allen, Esq. | &c. &c. &c.

'Further Report from Corporation of Trinity House on Magneto-Electric Light,' HC 250, *Parliamentary Papers* (1868–69), pp. 2–3

---

1. *Blackwall*: area of East London and part of the Port of London. The machine Tyndall described was at the Lighthouse at Trinity Buoy Wharf at Blackwall. This was built by Trinity House (see n. 6) in 1846 as an experimental lighthouse to test equipment and train lighthouse keepers (J. Tyndall, 'The Electric Light', *Popular Science Monthly*, 14 (March 1879), pp. 553–72, on p. 561).
2. *Mr. Holmes*: Frederick Hale Holmes (1812–75?), Professor of Chemistry at the Royal Panopticon of Science and Art. Holmes patented a magneto-electric generator to power lights for lighthouses. One was installed in 1858 in the lighthouse at South Foreland. It was frequently used by Trinity House for experimental trials. Faraday lectured on Holmes' lighthouses and use of magneto-electricity at the RI in 1860 (*Faraday Correspondence*, Vol. 5, xxxi).
3. *Captain Nisbet*: Edward Parry Nisbet (1810?–99?), Member of the Committee of Elder Brethren at Trinity House (see letter 2969, n. 9) (F. Burkhardt, et al. (eds), *The Correspondence of Charles Darwin*, 27 vols (Cambridge: Cambridge University Press, 1993), vol. 8, p. 437).
4. *Captain Weller*: Captain Charles Weller (d. 1866?), served in the merchant navy until his retirement in December 1863. He was elected an Elder Brother of Trinity House on 22 December 1863.

5. *magneto-electric light*: see letters 2969, 2984, and 2989.
6. *I beg to recommend*: while Tyndall had assisted Michael Faraday since May 1866 in lighthouse matters, in July 1867 he officially took over for Faraday as Scientific Adviser to the Board of Trade and Trinity House. The Board of Trade was a British government department concerned with making policy and instituting regulations regarding economic activity in Britain and its empire. Trinity House was and remains the lighthouse authority for England and Wales, established by Royal Charter in 1514, by operating a fleet of vessels in waters around England, Wales, and the Channel Islands to ensure adequate maintenance of aids to navigation, wreck location and markings, towing, surveying, etc.; it also regulates pilotage on the River Thames and provides for aged mariners. For a history of Trinity House, see A. Adams and R. Woodman, *Light Upon the Waters: The History of Trinity House 1514–2014* (London: The Corporation of Trinity House, 2013). In this role as Scientific Adviser, Tyndall investigated and gave recommendations to Trinity House on matters of lighthouse illumination and, later, sound signaling. He was adviser until 1883. On Tyndall and his lighthouse work, see *Tyndall Correspondence*, vol. 10; and R. M. MacLeod, 'Science and Government in Victorian England: Lighthouse Illumination and the Board of Trade, 1866–1886', *Isis*, 60 (1969), pp. 4–38. For more correspondence related to his lighthouse work, see letters 2969, 2984, 2989, 3019, 3067, 3124, 3262, 3353, and 3360.
7. *from a position near Charlton*: Charlton, now a district of south east London, is across the River Thames from Blackwall.

From Rudolf Clausius 6 January 1869 2938

Würzburg, 6/1/69

Lieber Tyndall,

Ich danke Dir sehr herzlich für die freundliche Zusendung des Buches von Tait, sowie für die Mittheilung Deiner interessanten Untersuchungen über die Wirkungen des Lichtes und das Verhalten des reflectirten Lichtes.

Was zunächst das Buch von Tait anbetrifft, so ist es in seiner jetztigen Gestalt sehr verschieden von dem zuerst gedruckten Entwurf, wie Du besonders daraus sehen kannst dass der Entwurf nur 68 Seiten hatte, während das jetztige Buch ohne die Noten 117 Seiten hat. Die erste Partie bis Art. 41 ist nur wenig geändert, und es ist daher auch das, was er über Mayer gesagt hat, der Hauptsache nach beibehalten, indem nur die Form etwas gemildert ist. Ich habe vor längerer Zeit eine kurze Recension des Buches in welchem Mayer seine Abhandlungen gesammelt hat, veröffentlicht. Vielleicht interessirt es Dich sie zu lesen, und ich schicke sie Dir daher unter Kreuzband zu.

Von Art. 41 an wo ich zuerst erwähnt bin, ist das Buch vollkommen umgestaltet. Es ist mir klar, dass Tait meine Arbeiten früher gar nicht gelesen hatte, sondern sie nur aus Unterhaltungen mit Thomson kannte. Erst seit der englischen Übersetzung meines Buches erschienen ist, hat er gesehen, dass er über meine Arbeiten nicht so kurz hinweggehen konnte. Es ist auch jetzt noch deutlich zu sehen, dass der Zweck seines Buches ist, möglichst viel von dem Verdienste der neuen Theorie für Joule, Rankine und besonders für Thomson mit dem er speciell befreundet ist, in Anspruch zu nehmen. Er hat das Buch nicht als ein gerechter Richter, sondern als ein Advocat geschrieben, der es sich nur Aufgabe gesetzt hat, möglichst viel zu Gunsten seiner Clienten zu sagen. Aber er hat meine Arbeiten jetzt doch nicht mehr so sehr vernachlässigen oder entstellen können, wie früher.

Die Mittheilungen über Deine neuen Untersuchungen haben mich ausserordentlich interessirt. Du hast da wieder ganz neue Gebiete eröffnet, und hast in kurzer Zeit so viele neue und wichtige Thatsachen festgestellt, dass Deine Untersuchungen eine wahre Fundgrube für diejenigen welche sich weiter mit diesen Sachen beschäftigen wollen, sein werden.

Es würde wohl zu früh sein, jetzt schon bestimme theoretische Gründe für die verschiedenen von Dir beobachteten Erscheinungen auszusprechen, aber einen Punct möchte ich doch kurz erwähnen. Der Unterschied in dem Polarisationszustande des Lichtes welches von sehr feinem Dampfe oder Rauche reflectirt ist, von dem welches von dichterem Dampfe oder Rauche reflectirt ist, scheint mir, vorzugsweise durch folgenden Umstand begründet zu sein. Bei feinem Dampfe oder Rauche erhalten wir solches reflectirtes Licht, welches zum grössten Theile nur einmal reflectirt ist. Von dichterem Dampfe oder Rauche dagegen erhalten wir mit dem einmal reflectirten Lichte gemischt, auch viel solches Licht, was mehrere Reflexionen erlitten hat, indem es von einem Körperchen zu einem zweiten, von diesem vielleicht zu einem dritten u.s.f., reflectirt ist, und dann endlich nach der letzten Reflexion zu uns kommt. Da diese verschiedenen Reflexionen in verschiedenen Ebenen stattfinden, so muss dadurch der Polarisationszustand wesentlich modificirt werden. Ob dieser Grund zur Erklärung jener Erscheinungen ausreicht, kann natürlich nur aus der Vergleichung der einzelnen Resultate mit den speciellen Umständen, unter welchen diese Resultate gewonnen wurden, geschlossen werden.

Uns geht es allen ganz gut. Meine Frau hat sich vor Kurzem photographiren lassen mit Johnny neben sich und der kleinen Hedwig auf dem Schoosse, und ich schicke Dir einliegend ein solches Bildchen mit.

Hoffentlich hast auch Du das neue Jahr gut angetreten und wir wünschen Dir von Herzen dass das neue Jahr Dir nur Gutes bringe, und besonders dass Du immer recht gesund und munter bleibest.

Bitte sage auch Hirst unsere besten Grüsse und Glückwünsche zum neuen Jahre.

Dein | Clausius.

Würzburg, 6/1/69

Dear Tyndall,

My heartfelt thanks for kindly sending me Tait's book[1] and for communicating your interesting investigations into the properties of light and the behaviour of reflected light.

Regarding Tait's book first of all, in its present form it is very different from the first printed draft,[2] which you can see especially from the fact that the draft was only 68 pages long, while the present book is 117 pages not including notes. The first part to par[agraph] 41 is only slightly altered, and what he said about Mayer is therefore essentially the same and has only been somewhat softened in form. Some time ago, I published a short review[3] of the volume in which Mayer had collected his papers.[4] You may find it interesting to read, and I am therefore sending it to you separately as printed matter.

From par[agraph] 41, where I am mentioned for the first time, onward the book has been completely rewritten. It is clear to me that Tait had not previously ever read my works, and only knew about them from conversations with Thomson. Only since the publication of the English translation of my book,[5] has he realized that he could not so easily overlook my work. Even now it is clear that the purpose of his book is to claim as much credit for the new theory as possible for Joule, Rankine[6] and especially Thomson, with whom he is especially close. He wrote the book not as a fair judge, but as an advocate who has made it his task to say as much as possible in favour of his clients. But this time, he could not ignore or misrepresent my works as much as previously.

I found the communication about your new investigations[7] extraordinarily interesting. You have once again opened entirely new fields and in a short time established so many new and important facts that your investigations will be a real treasure trove to those who want to work on these topics in the future.

It would be too early to discuss definite theoretical reasons for the various phenomena you observed, but I would like to briefly mention one particular point. It seems to me that the difference between the polarisation state of light that is being reflected by very fine steam or smoke and that which is reflected by denser steam or smoke, is chiefly based on the following circumstance. With the fine steam or smoke, we obtain reflected light that for the most part has been reflected only <u>once</u>. With denser steam or smoke, on the other hand, in addition to light that has only been reflected once, we also obtain a great deal of light that has undergone <u>multiple reflections</u>, in that it is reflected from one particle to a second, from this perhaps to a third, and so on, before finally coming to us. Since these different reflections occur on different planes, the polarisation state must therefore be fundamentally modified. Whether or not this reason is sufficient to explain this phenomenon can naturally only be determined by a comparison of the individual results with the particular circumstances under which they were obtained.

We are all well. My wife has recently had photographs made with Johnny[8] at her side and little Hedwig[9] on her lap, and I send you enclosed one such little picture.[10]

I hope you have had a good start to the new year, and we hope that the new year will bring you only good things, and especially that you will stay healthy and cheerful.

Please also give our best regards and New Year's wishes to Hirst.

Your | Clausius.

RI MS JT/1/TYP/7/2311–13
LT Typescript Only

---

1. *Tait's book*: P. G. Tait, *Sketch of Thermodynamics* (Edinburgh: Edmonston and Douglas, 1868).
2. *first printed draft*: P. G. Tait, *Historical Sketch of the Dynamical Theory of Heat* (Edinburgh: Constable, 1867). This volume was a compilation of two essays on 'The Dynamical Theory of Heat' and 'Energy' that had been published in the *North British Review* in 1864, during the Mayer-Joule priority dispute. See 'Introduction' to *Tyndall Correspondence*, vol. 8, pp. xvii–xix.
3. *a short review*: not identified.
4. *Mayer had collected his papers*: J. R. Mayer, *Mechanik der Wärme*, 2 vols, ed. J. J. Weyrauch (Stuttgart: J. G. Cotta, 1867).
5. *English translation of my book*: R. Clausius, *The Mechanical Theory of Heat—with its Applications to the Steam Engine and to Physical Properties of Bodies*, ed. T. A. Hirst (London: John van Voorst, 1867).
6. *Rankine*: William John Macquorn Rankine (1820–72), a Scottish civil engineer and physicist. He attended classes in natural philosophy, natural history, and botany at the University of Edinburgh, while also studying chemistry outside the university. In 1838 under the tutelage of John Benjamin MacNeill, Rankine set out on a career as an engineer. He did extensive work on steam engines and was a founding contributor, with Rudolf Clausius and William Thomson, to the science of thermodynamics, particularly focusing on the first law of thermodynamics, also known as the law of energy conservation. He was elected FRS in 1853 and was a prominent figure in the BAAS (*CDSB*; *ODNB*).
7. *your new investigations*: Tyndall had written to Clausius in December 1868 about his work on the blue colour of the sky and the polarization of light. See letter 2931, *Tyndall Correspondence*, vol. 10.
8. *Johnny*: Rudolf John Clausius.
9. *Hedwig*: Hedwig Clausius (1866–1946), Rudolf Clausius' daughter. In total, Clausius had six children with his wife Adelheid Clausius: Mathilde (1861–1907), Helene (1863–1919), Rudolf John ('Johnny', 1864–1915?), Hedwig (1866–1946), Alfred (1868–1950), and Else (1872–1954).
10. *enclosed one such little picture*: enclosure missing.

To Juliet Pollock [8 January 1869][1] 2939

Friday

My dear Mrs Pollock

In the early morning hours between 5 & 8 I am casting off an article—a farewell one—entitled as "Odds & Ends of Alpine Life."[2]

I believe the only note that I made of the ascent of the Eiger[3] in 1867 was a letter I sent to you[4] Could you lay your hands on that letter & lend it to me? If so I should be thankful.

Yours ever | John Tyndall

RI MS JT/1/T/1205
RI MS JT/1/TYP/6/1/2093

---

1. *[8 January 1869]*: the date is given since the Friday preceding 15 January, when the proofs for this article were in the hands of the publisher, MacMillan, was 8 January. The articles appeared in the March and April issues of *MacMillan's Magazine* (see letter 2943).
2. *"Odds & Ends of Alpine Life"*: see letter 2943, n. 3.
3. *Eiger*: a mountain in the Bernese Alps near Grindelwald.
4. *a letter I sent to you*: see letter 2622, *Tyndall Correspondence*, vol. 10.

To Mary Somerville[1] 8 January 1869 2940

8th January 1869

My dear Mrs. Somerville

Accept my warmest thanks for your very beautiful New Years Gift.[2] I shall always preserve and cherish it. Those portions of it which I have been as yet able to read raise my astonishment at the extent of the knowledge which they indicate.

I have been very busy since it reached me with a subject to which your friend Sir John Herschel devoted much attention—the colour of the sky and the polarization of skylight: I can make a sky as beautiful as that which overspreads you at Naples, and out of this artificial sky all the phenomena can be extracted which have hitherto been observed with the natural air. The subject is to be brought before the Royal Society next Thursday[3] and as soon as my paper[4] is printed it will give me pleasure to send you a copy.

Please give my kind remembrance to your daughters,[5] and accept again and again my thanks for your gift.

Believe me | ever most truly yours | John Tyndall | A HAPPY HAPPY NEW YEAR.

Oxford Bodleian Library Dep c.372 (MST-1)

1. *Mary Somerville*: Mary Somerville (1780–1872), a talented mathematician who conducted original research on physical astronomy, but is best known as an influential author of popular and educational science books. *On the Connexion of the Physical Sciences* (London: John Murray, 1834) became one of the best-selling science books of the nineteenth century. In 1834 Somerville was elected Member of the Royal Irish Academy (MRIA) and the following year was made an honorary member of the Royal Astronomical Society (*ODNB*).
2. *New Years Gift*: probably a copy of her latest book, *On Molecular and Microscopic Science* (London: John Murray, 1869).
3. *brought before the Royal Society next Thursday*: on Thursday, 14 January 1869, Tyndall presented his paper 'On the Blue Colour of the Sky, the Polarization of Skylight, and on the Polarization of Light by Cloudy Matter Generally' (*Roy. Soc. Proc.*, 17 (1869), pp. 223–33).
4. *my paper*: see n. 3.
5. *your daughters*: Martha and Mary. Mary Somerville also had two sons, Woronzow and William George, from her first marriage to Samuel Grieg. Two of her children with her second husband, William Somerville, died very young. Martha compiled her mother's reminiscences. See M. Somerville, *Personal Recollections from Early Life to Old Age of Mary Somerville* (London: John Murray, 1873).

To Thomas Henry Huxley — 12 January 1869 — 2941

Royal Institution. | 12th. Jan. 1869.

My dear Huxley,

You will see from the enclosed[1] that I have not been altogether idle.

I think it very desirable that this programme[2] should be drawn out so that it may be printed and circulated; not only as a guide to the Teacher but as a statement of what is really taught in this college.[3]

It will therefore require thoughtful and careful drawing up. I have given you a sketch[4] of what I should propose to say in Physics and Chemistry. But I think it would be better to defer the final drawing up till next week when I could, without blasphemy, give it a little more attention.

I do not think that I should be prepared to sign any recommendation regarding the Theory of Law and one or two other points introduced by you. They might come in the form of supplementary suggestions, but ought not to be put on the same footing as science

Yours ever | John Tyndall.

You see I have given a portion of each year to Physics and a portion to Chemistry. I would divide the time in some such ratio as 3 to 2. It is better I think that they should be both worked at in the same year as they react beneficially upon each other.

If I had the time I would join the Committee[5] to day but I really have it not.

RI MS JT/1/TYP/9/2924–5
LT Typescript Only
IC HP 8.68
Typed Transcript Only

---

1. *the enclosed*: enclosure missing. See n. 2.
2. *this programme*: the programme for science and technology courses that Tyndall and Huxley developed is found in the Imperial College Huxley Collection, 42.48; 42.46 and 42.50. The latter was composed by Tyndall and is probably the programme referred to here. See also C. Bibby, 'A Victorian Experiment in International Education: The College at Spring Grove', *British Journal of Educational Studies*, 5:1 (1956), pp. 25–36.
3. *this college*: Tyndall and especially Huxley were involved in the development of a science curriculum for the London College of the International Education Society at Spring Grove, Middlesex. See n. 2.
4. *a sketch*: see n. 2.
5. *join the Committee*: after the Great Exhibition in Paris (1862), there was widespread enthusiasm for educating children of the different European nations together. The London Committee was charged with instituting the British campus for the 'international school'. Tyndall and Huxley joined others on the Committee, including chairman Abraham W. Paulton (1812–76), a politician and journalist who lectured for the Anti-Corn-Law League, William Ewart (1798–1869), M.P., and W. B. Hodgson (1815–80), who had served as Principal of the Liverpool Mechanics' Institute, among others.

From Jane Barnard 12 January 1869 2942

12 Barnsbury Villas | N. | 12 Jan[uary]. 1869

Dear Dr Tyndall

We thank you very much for the kind thought which prompted you to send us tickets for next Friday[1] & which are very welcome & we shall use.

My Aunt[2] joins me in kind *[regards]* & I am | yours very sincerely | Jane Barnard

RI MS JT/1/B/42

---

1. *tickets for next Friday*: on Friday, 15 January, 1869, Tyndall delivered a lecture on 'Chemical Rays and the Light of the Sky' at the RI (*Roy. Inst. Proc.*, 5 (1869), pp. 429–50). This was also printed as 'On Chemical Rays, and the Light of the Sky', *The Fortnightly Review*, 5:26 (1869), pp. 226–48. See also letter 2946.
2. *my Aunt*: Sarah Faraday.

From Alexander MacMillan[1] 15 January 1869 2943

Jan[uary] 15. 1869.

Dear Prof[essor]. Tyndall,

We have heard nothing from Youmans yet. But even if he does not begin his magazine,[2] may we not calculate on putting the first part in our March number.[3] He has pledged to pay for this whether he publishes or no—is he not? I propose giving you £50 for these papers.[4] I am a poor climber of mountains, and know little of science but a glance of the proofs charmed me.

Of course the pay is only for use in a magazine & the copyright remains wholly yours. I understand that what you have to give us will occupy from 25 to 30 pages of the mag[azine]. What a blessing for a man to have head to *[*endure*]* eyes to see, legs to carry you & tongue in face to utter all. Do you *[*find*]* that *[*wonderful*]* *[1 or 2 words illeg]* you *[*must*]* have that the normal man will be able to *[*call*]* them in the future years. *[2 words illeg]*.

Yours ever faithfully | A. MacMillan

BL Add MS 55842, f. 128

1. *Alexander MacMillan*: Alexander MacMillan (1818–96), a publisher. In 1843, Alexander and Daniel MacMillan founded what became a very successful publishing house in London. Among their authors were Charles Kingsley, Alfred Tennyson, and Thomas Henry Huxley. In 1868, the publishing house opened a branch in New York (*The MacMillan story. Bringing authors and readers together since 1843* (MacMillan Publishers, 2018), at https://read.macmillan.com/lp/the-macmillan-story [accessed 18 January 2020]).
2. *his magazine*: it is unclear whether the magazine referred to here is *Popular Science Monthly*, which Edward Youmans eventually founded in 1872.
3. *our March number*: the first part of Tyndall's 'Odds and Ends of Alpine Life' was published in *MacMillan's Magazine*, 19:113 (March 1869), pp. 369–85. The second part was published in the April issue, on pp. 465–79.
4. *giving you £50 for these papers*: see n. 3.

From William Frederick Pollock 16 January 1869 2944

The Athenaeum[1] | Sat[urday]. | 16th. January 1869

Dear Tyndall

You vanished so quickly last night from the scene of your success that I could not then tell you how much we were all delighted by your lecture[2]—I knew just enough of the matter, thanks to your kindness, to be able to form some notion of the difficulty of bringing it before the audience at the R.I—& over the difficulty you triumphed even more grandly than I expected you would—& my expectations about you are not pitched very low—at any time.

When you said that only 20 minutes of the hour remained I did not suppose that 20 minutes were gone—& you gave us all a notion of what the sporting world calls "improved form" very gratifying to your friends—there was a sense of reserved power more evident than on previous occasions—& as if you had in hand a high mettled horse whose performances were only beginning—& who had not yet been ridden to anything like the top of his pace—or the height or extent of his leap—

In short we were all perfectly delighted & grateful to you for it—

Yours ever truly | W. F. Pollock

RI MS JT/1/P/238
RI MS JT/1/TYP/6/2103

---

1. *The Athenaeum*: a private members' club located at 107 Pall Mall in London, particularly for those with intellectual interests, and accomplishments in art, science, or literature. The club was founded in 1824, at the instigation of John Wilson Croker. Humphry Davy became the first chairman and Michael Faraday the first secretary. Faraday became a member after a few months when he was able to resign the secretaryship. Tyndall became a member in 1860, Thomas Huxley in 1858, Joseph Hooker in 1851, and Thomas Hirst in 1866. On the club, see M. Wheeler, *The Athenaeum: More Than Just Another London Club* (New Haven, CT: Yale University Press, 2020).
2. *your lecture*: see letters 2942, n. 1, and 2946.

From Edward Sabine 16 January [1869][1] 2945

13 Ashley Place | Jan[uar]^{y} 16

Dear Tyndall

I am greatly concerned that you should have suffered so much from your kind exertion at the meeting of the Royal Society on Thursday[2] last: and I regret it the more, because the previous Thursday might quite as well have been arranged for your communication. The order in which the papers are to be read to the Society is understood to be subject to the Regulation of the President; but is a matter in which I very seldom interfere; and when I do so, I always endeavour to do so with the approval of the Secretaries; but there would be no difficulty in such a case as yours; and I can now only express my regret, that your desire to have the day changed from Thursday the 14th to Thursday the 7th was quite unknown to me until after your lecture on Friday last;[3] as the day of communication of your paper might just as well have been arranged for the 7th as for the 14th; and would have been so arranged had I known your wish:

Sincerely yours | Edward Sabine

RI MS JT/1/S/29
RI MS JT/1/TYP/4/1324

---

1. *[1869]*: the year is given by reference to the RS meeting. See n. 2.
2. *meeting of the Royal Society*: on 14 January 1869, Tyndall read a paper on 'On the Blue Colour of the Sky'. See letters 2940, n. 3, 2977, and 2978.
3. *your lecture on Friday last*: on 'Chemical Rays and the Light of the Sky' at the RI. See letters 2942, n. 1, 2946, 2955, 2977, and 3006.

To Juliet Pollock 17 January [1869][1] 2946

17th. January.

Many many thanks for your most cheerful & pleasant note.[2] How curiously different are often the opinions of the speaker & those of his audience. I had rather set my heart on making an exceedingly difficult subject plain, and hoped to keep my brain sufficiently clear & strong to realize this hope. But on Thursday night, I had to discourse before the Royal Society[3] & for hours after I went to bed I bathed my head in cold water to keep down the heat of the brain within it. I rose consequently on Friday morning cloudy & incohesive, without any real strength of intellectual grasp & in this enfeebled condition I had to attack a subject[4] needing all my powers. I thought I failed utterly, and in a kind of desperate indifference walked upstairs and smoked myself into stupidity. Had I known you could see me I would have waited & tried to persuade you to come upstairs, for you and your friend[5] would have done me much more good than the cigar.

But I have thrown the lecture into an article of which I am not ashamed. It will be published in the next number of the Fortnightly Review;[6] but I should much like your husband[7] to cast his eye over it, and say whether he thinks it can be made brighter & clearer. I expect proofs tomorrow (Monday) or Tuesday.

Curiously enough my strict and very competent critics[8] take your view of the lecture & not mine. I still think it a failure, but I am not so sure on the point as I was on Friday night.

Yours always dear friend | John Tyndall

RI MS JT/1/T/1162
RI MS JT/1/TYP/6/1950

1. *[1869]*: the year is given by reference to a discourse Tyndall gave at the RS on 14 January 1869 and a lecture at the RI on 15 January 1869. See letters 2942 and 2945.
2. *your most cheerful & pleasant note*: letter missing.
3. *discourse before the Royal Society*: see letter 2940, n. 3.
4. *attack a subject*: see letter 2942, n. 1.
5. *your friend*: not identified.
6. *article . . . published in the . . . Fortnightly Review*: see letter 2942, n. 1.
7. *your husband*: William Frederick Pollock.
8. *my strict and very competent critics*: not identified.

To William Frederick Pollock [17 January 1869][1] 2947

Sunday evening

My dear Pollock,

I have just this moment picked up your note:[2]—thanks, warm thanks my dear friend for this additional proof of your kindness. I know how much of it belongs to friendship. Had I had my fair average brain at command I should have approached what you here say I reached; but I was crippled. Ask Fred[3] what he would think of climbing the Eiger[4] on a broken leg!

Yours ever | John Tyndall.

RI MS JT/1/TYP/6/2106
LT Typescript Only

1. *[17 January 1869]*: this letter is undated, but the year has been added to the typescript by hand, presumably by LT. It is likely a response to letter 2944 (dated 16 January), in which William F. Pollock praises Tyndall for his Friday Evening Discourse of 15 January.
2. *your note*: letter 2944.
3. *Fred*: Frederick Pollock.
4. *Eiger*: see letter 2939, n. 3.

To Thomas Carlyle [22 January 1869][1] 2948

[Royal Institution] | Friday

General!

I thank you for the book![2]

There is no other book on Earth that I would accept in exchange for it.

Yours always | John Tyndall

Amid the stupidities, perversities, intellectual fopperies—and I must add the Reverences, of the world I have often occasion to speak of you & your work. We were doing so at Pollock's a night or two ago; and it was then proposed by M^rs^. Pollock that she, her husband & myself sh[oul]d run down to see you for half an hour on Wednesday evening.

Would this bother you?—Miss Aitken[3] will kindly write "yes" or "no"

Some fellow[4] said of you a few days ago "He does not advance." And I replied.

"Die unbegreiflich hohen Werke | Sind herrlich wie am ersten Tag."[5]

RI MS JT/1/T/166
RI MS JT/1/TYP/1

---

1. *[22 January 1869]*: the date was assigned by LT on the typescript. 22 January 1869 was a Friday.
2. *the book*: the book is not identified, however publication of an edition of Carlyle's collected works began in 1869.
3. *Miss. Aitken*: Mary Carlyle Aitken (1848–95), Carlyle's niece. She lived with him from 1866 until his death (F. Kaplin, *Thomas Carlyle. A Biography* [Berkeley: University of California Press, 1993]).
4. *Some fellow*: not identified.
5. *"Die unbegreiflich . . . ersten Tag"*: 'The lofty works, whose way none knowth, are grand, as on the primal day', from J. W. von Goethe, *Faust. Eine Tragödie* (1808), Prologue.

From François Napoléon Marie Moigno 25 January 1869 2949

Les Mondes | Revue Hebdomadaire | Des Sciences et de leurs applications aux arts et à l'industrie | par M. l'Abbé Moigno | Rédaction | 2, Rue d'Erfurth, 2.
Paris, le 25 janvier 1869

Mon cher Mr Tyndall

J'ai déjà traduit ce que l'Engineer a publié de votre belle science de Royal Institution combien je serais heureux si vous pouviez obtenir pour moi en le payant le cliché (block) qui représente l'amphithéâtre éclairé par votre belle lumière bleue, page 73 de l'Engineer.

Votre Son est trouvé admirable par tous nos professeurs et nos amateurs de science, vous pouvez compter sur un grand succès, sur un succès comparable à celui de l'édition anglaise.

Pressez vivement Mr Gauthier-Villars de commencer l'impression de la Chaleur que l'on demande de tous les côtés. Que je serais heureux si le volume était en vente dans trois mois.

Votre humble et dévoué serviteur et ami | l'abbé F. Moigno

Les Mondes[1] | Weekly Review | Of the sciences and their applications to the arts and the industry | by Father Moigno | Editorial offices: | 2 Erfurth street, 2. | Paris, 25 January 1869

My dear Mr Tyndall,

I have already translated what The Engineer[2] published[3] about your wonderful science of the Royal Institution. How happy I would be if you could obtain for me the picture (block) that represents the amphitheatre illuminated by your beautiful blue light, on page 73 of the Engineer.

All of our professors and science enthusiasts found your Sound[4] remarkable. You can count on a great success, on a success comparable to that of the English edition.[5]

Urge Mr Gauthier-Villars[6] to begin the printing of Heat,[7] which is being requested by all sides. How happy I would be if the volume were available for purchase in three months.

Your humble and devoted servant and friend | Father F. Moigno

RI MS JT/1/M/117

---

1. *Les Mondes*: after leaving the journal *Cosmos*, which he had edited since 1852, Moigno founded the popular science journal *Les Mondes: revue hebdomadaire des sciences* in 1863.
2. *The Engineer*: founded in 1856 by Edward Charles Healy, *The Engineer* covered topics mainly in engineering, industry, railways, and technological innovation.
3. *what The Engineer published*: in January 1869, *The Engineer* published a series of articles on the History of the Royal Institution, focusing on individual scientists. Tyndall was portrayed in the first issue of 8 January 1869.
4. *your Sound:* Tyndall's *Sound* (1867) was translated into French by Moigno and published in 1869 as J. Tyndall, *Le son* (Paris: Gauthier-Villars, 1869).
5. *the English edition*: J. Tyndall, *Sound: A Course of Eight Lectures Delivered at the Royal Institution of Great Britain* (London: Longmans, 1867).
6. *Mr. Gauthier-Villars*: Jean-Albert Gauthier-Villars (1828–98), a French engineer and publisher. Born to a family of printers in Lons-le-Saunier, he studied at the École Polytechnique (1848–50), became a telegraph engineer, and served in the military. In 1864 he purchased the Mallet-Bachelier printing house which had published the *Comptes rendus hebdomadaires des séances de l'Académie des sciences.* Under his direction the firm, renamed Gauthier-Villars, specialized in mathematical and scientific books and magazines.
7. *printing of Heat*: the first edition of J. Tyndall, *Le chaleur. Mode de movement* was published several years earlier, in 1864 by Gauthier-Villars. The second French edition, which was based on the fourth English edition, was published several years later, in 1874. This may refer to a reprint of the 1864 translation or possibly to a planned translation of the third English edition.

From Felice Giordano[1] 4 February 1869 2950

Corpo Reale | delle | Miniere | Florence le 4 février—69

Cher Mr le Prof. Tyndall

J'ai reçu votre gracieuse lettre du 30 janvier.—je suis très content que les sections géologiques du Mt Cervin que je vous avais envoyées, vous aient été agréables et que vous trouviez correct le résumé de mes peu d'observations sur les glaciers d'accord avec les beaux principes que vous avez posé. Quant aux sections du Mt Cervin vous êtes le maitre de vous en servir pour tout ce que vous voudriez et je me retiendrai fort honoré de cette remarque d'estime de votre part.—

Je suis à votre pleine disposition pour tout ce en quoi je pourrais vous servir—Agréez, cher M^r le prof., mes sentiments de profonde estime.

F. Giordano | 2 Via Venezia.Corpo Reale | delle | Miniere

Florence, February 4th—69.

Dear Prof. Tyndall,

I received your kind letter of January 30th[2]—I am very glad that the geological samples of Mt Cervin[3] that I sent[4] were useful to you and that you found that the summary of my few observations on the glaciers was correct, in accordance with the beautiful principles that you developed. As for Mt Cervin's samples, you can use them however you would like, and I will remain greatly honoured by your testament of respect.—

I am at your full disposal for everything that I can provide—please accept, dear Professor, my feelings of deep appreciation.

F. Giordano | 2 Via Venezia.

RI MS JT/1/G/7

---

1. *Felice Giordano*: Felice Giordano (1825–92), an Italian engineer and geologist. He was a founding member of the Italian Geological Society and the Italian Alpine Club. In 1868 he traversed the Matterhorn following the route previously taken by Tyndall.
2. *your kind letter of January 30th*: letter missing.
3. *Mt Cervin*: Cervin is the French name given to the Matterhorn, a mountain in the Alps bordering Switzerland and Italy.
4. *the geological samples of Mt Cervin that I sent*: see letter 2886, *Tyndall Correspondence*, vol. 10.

From Hermann Helmholtz 9 February 1869 2951

Heidelberg, 9 Febr 1869.

Verehrter Freund

die Übersetzung von "On Sound" werden Sie schon erhalten haben; meine Frau hat inzwischen auch das Manuscript der Übersetzung von "Faraday als Discoverer" vollendet. Inzwischen ist der Abriss seines Lebens von D^{r}. B. Jones in den Reports of the Royal Society erschienen in welchem mancherlei interessante Briefe und Thatsachen enthalten sind, die namentlich für ein fremdes Publikum den Charakter des Mannes anschaulicher und lebhafter hervortreten lassen. Ich glaube deshalb, dass es nützlich sein würde eine Reihe von Notizen daraus in Form von Anmerkungen oder vielleicht auch stellenweis als Einschaltungen (aber als solche bezeichnet durch Klammern [ ]) herüberzunehmen, vorausgesetzt, dass Sie und D^{r}. B. Jones nichts dagegen einzuwenden haben. Wenn Sie nur Ihre Einwilligung dazu geben, werde ich auch an D^{r}. B. Jones schreiben. Sollten Sie in Bezug auf die von letzterem gesammelten Notizen irgendwo eine abweichende Meinung haben, so bitte ich mir dies zu bezeichnen. Solche Dinge würde ich dann weglassen. Ich denke es würde niemand an diesem Verfahren Anstoss nehmen können. D^{r}. B. Jones hat ein Jahr länger Zeit gehabt, seine Notizen zu sammeln, als Sie hatten, und es ist daher natürlich, dass er mancherlei aufgefunden hat, was Ihnen noch fehlte.

Am Ende des vorigen Jahres wäre es beinahe dahin gekommen, dass ich die Professur der Physik in Bonn übernommen hätte. Schliesslich habe ich mich aber entschieden, hier in meiner bisherigen Stellung zu bleiben.

Bunsen geht es gut; seine sehr unangenehmen Kolikanfälle sind im letzten Jahre nicht wiedergekehrt. Kirchhoff geht noch immer an Krücken und jetzt ist nun gar auch seine Frau in Folge ihres letzten Wochenbetts an einem Brustleiden recht bedenklich erkrankt. In meiner Familie geht es gut, das kleinste Söhnchen macht uns viel Spass.

Für die Zusendung Ihrer Abhandlungen über die Bläue des Himmels und die merkwürdigen chemischen Lichtreactionen danke ich bestens.

Beste Grüsse von meiner Frau. | Ihr | H. Helmholtz.

Heidelberg, 9 Febr[uary] 1869.

Dear Friend!

You will have already received the translation of "On Sound";[1] in the meantime, my wife[2] has also completed the manuscript for the translation of "Faraday as Discoverer".[3] In the meantime, the sketch of his life[4] by D^{r}. Bence Jones has been published in the Reports of the Royal Society, which contains some interesting letters and facts which namely will allow the character of the man to appear in a more lively and characteristic manner for a foreign public. I believe therefore that it would be useful to include a series of notes from them in the form of annotations[5] or perhaps in places brought in as interpolations (but indicated as such by brackets [ ]), provided that you and D^r^. B. Jones have no objections to this. If you would give your consent, I shall also write to D^r^. B. Jones. Should you have a different opinion in regard to any of the notes gathered by the latter, I ask that you let me know. I would then leave out such things. I think that no-one could take exception to this procedure. D^r^. B. Jones has had a year longer to gather his notes than you had, and therefore it is natural that he has found various things that you were still missing.

At the end of the previous year, I almost reached the point of accepting the chair in physics[6] in Bonn. In the end, however, I decided to remain here in my present position.

Bunsen is well; his very unpleasant colic attacks did not recur last year. Kirchhoff is still walking on crutches and now his wife too has fallen quite seriously ill with a chest ailment as a result of her last confinement.[7] All is well in my family, the youngest wee son brings us much amusement.

I thank you sincerely for sending your papers[8] about the blue colour of the sky[9] and the curious chemical reactions of light.

Kind greetings from my wife. | Your | H. Helmholtz.[10]

RI MS JT/1/H/47
RI MS JT/1/H/66
Transcript Only

---

1. *"On Sound"*: Tyndall's *Sound* appeared in its first German edition in 1869, published as J. Tyndall, *Der Schall* (Braunschweig: F. Vieweg und Sohn, 1869). The translation was produced by Anna Helmholtz and supervised by Hermann Helmholtz and Gustav Wiedemann.
2. *My wife*: Anna Helmholtz.
3. *Faraday as Discoverer*: J. Tyndall, *Faraday as a Discoverer* (London: Longmans, Green, and Co., 1868), translated as J. Tyndall, *Faraday und seine Entdeckungen* (Braunschweig: F. Vieweg und Sohn, 1870).
4. *Sketch of his life*: H. B. Jones, 'Obituary notices of fellows deceased: Michael Faraday', *Roy. Soc. Proc.*, 17 (1868), i–lxviii. For correspondance between Tyndall and Bence Jones on the latter's *Life of Faraday* (1870), see letters 3021, 3029, 3038, 3043, 3053, 3057, 3236, 3245, and 3268. See also letter 2953.
5. *Heidelberg... annotations*: this portion of the letter is transcribed from RI MS JT/H/47.

6. *chair in physics*: after long negotiations, Helmholtz turned down the chair. It was then immediately offered to Rudolf Clausius, the second place contender, who accepted. See letter 2974. For the politics surrounding the negotiations, see D. Cahan, *Helmholtz: A life in Science* (Chicago: University of Chicago Press, 2018), pp. 348–53.
7. *her last confinement*: her last pregnancy. Hermann and Anna Helmholtz had three children: Robert von Helmholtz (1862–89), Ellen von Helmholtz (1864–1941), who in 1884 married Arnold von Siemens (1853–1918), and Friedrich Julius von Helmholtz (1868–1901) (D. Cahan, *Helmholtz: A Life in Science* (Chicago: University of Chicago Press, 2018).
8. *or perhaps . . . your papers*: this portion of the letter is transcribed from RI MS JT/1/H/66.
9. *your papers about the blue colour of the sky*: 'On the Blue Colour of the Sky' (see letter 2940, n. 3); 'Chemical Rays and the Light of the Sky' (see letter 2942, n. 1).
10. *about the blue . . . Your H. Helmholtz*: this portion of the letter is transcribed from RI MS JT/1/H/46.

To George Biddell Airy 11 February 1869 2952

Royal Institution of Great Britain | 11th Feb[ruary]. 1869

My dear Sir,

I send you by this day's post a report[1] of a lecture[2] the value of which you are far more competent to pronounce upon than I am. I hope it will interest you—

A number of new points not introduced in this report have been dealt with in the Proceedings of the Royal Society which I believe will be published this week.[3]

Yours very faithfully | John Tyndall

RGO MS 6/479

---

1. *report*: enclosure missing.
2. *lecture*: Tyndall read a paper at the RS on 14 January and gave a lecture at the RI on 15 January. Although it is unknown what report Tyndall sent to Airy, his RI lecture was reported in at least two newspapers ('An Evening at the Royal Institution', *Birmingham Daily Post*, 21 January 1869, p. 6; and 'Professor Tyndall on Chemical Rays', *Manchester Courier and Lancashire General Advertiser*, 26 January 1869, p. 2).
3. *published this week*: 'On the Blue Colour of the Sky, and the Polarization of Light' (see letter 2940, n. 3).

To Hermann Helmholtz 12 February 1869 2953

12th. Feb[ruary]. 1869.

My dear Helmholtz,

I have spoken to Bence Jones regarding his Memoir of Faraday:[1] and both he and I wish you to do whatever you like with both of our Memoirs.[2] I had shoals of letters offered to me by Faraday's friends but, they are all now in the hands of Bence Jones. In fact my intention was to give some account of his works, and I had not half time enough to do even that.

Remember me kindly to Bunsen and Kirchhoff. I am very sorry to hear that Kirchhoff is ill.[3]

In a few days you will receive an abstract of some experiments that I have been making on the colour and polarization of the sky:[4]—I hope they will interest you.

Would you do me the favour to hand the enclosed note[5] to Mrs. Helmholtz.[6]

Yours ever truly | John Tyndall.

BBAW NL Helmholtz, Nr. 477, BI_12
RI MS JT/1/T/486

1. *Memoir of Faraday*: *Life of Faraday*.
2. *both of our Memoirs*: Tyndall's *Faraday as a Discoverer* and Bence Jones's lengthy obituary notice in *Roy. Soc. Proc.* (see letter 2951, n. 4). Note that Bence Jones was working on his *Life of Faraday* (1870); see letters 3021, 2029, 3038, 3043, 3053, 3057, 3236, 3245, and 3268.
3. *Kirchhoff is ill*: see letter 2951.
4. *colour and polarization of the sky*: see letter 2940, n. 3.
5. *enclosed note*: enclosure missing.
6. *Mrs. Helmholtz*: Anna Helmholtz.

From Ernst Brücke 14 February 1869 2954

Währinger Gasse № 1 | 14 Febr. 1869

Geehrter Herr!

Sie haben mir, als es erschien, Ihr schönes Buch „on Sound" übersendet. Ich habe keine Hoffnung, und jetzt am wenigsten, Ihnen ein ähnliches Geschenk machen zu können und doch möchte ich mich irgend wie revengiren. Nun bin ich in Besitz von ein paar hübschen Stücken Steinsalz von Wieliczka gelangt: sie sind nicht fehlerfrei aber gehören unter dem, was man jetzt bekommt, doch zu den besseren. Ich kann natürlich nicht wissen, ob Sie nicht schon bessere besitzen und ob sie überhaupt noch für Sie von Interesse sind. Sollte das letztere der Fall sein, so bitte ich mich davon zu benachrichtigen und in Rücksicht auf Adresse und Verpackung dasjenige hinzuzufügen, was Sie für nöthig halten.

Ihr | E. Brücke.

Währinger Gasse, N° 1 | 14 Febr[uary]. 1869

Dear Sir!

You sent me your fine book "on Sound"[1] upon its publication. I have no hope, and even less so now, of being able to offer you a similar present but would like to reciprocate in some way. I am now in possession of a few beautiful pieces of mineral salt from Wieliczka:[2] they are by no means flawless, but they are among the better ones that one can obtain these days. Of course, I cannot know if you already have better ones and if they even are still of interest to you. Should the latter be the case, please inform me and add what you deem necessary with regard to address and packaging.

Your | E. Brücke.

RI MS JT/1/B/135

1. *"on Sound"*: presumably the German translation (see letter 2951, n. 1).
2. *Wieliczka*: the Wieliczka Salt Mine, in Wieliczka close to Kraków in southern Poland, dates to the thirteenth century. Tyndall used rock-salt plates for his experiments on the absorption of radiant heat by gases and vapours.

From George Biddell Airy 20 February 1869 2955

Royal Observatory, Greenwich, S.E. | 1869 February 20

My dear Sir

I am greatly obliged by your remembrance of me in sending the pamphlet "On Chemical Rays, and the Light of the Sky".[1] I deferred acknowledgment of it until I should have had opportunities of reading it more than once, and such opportunities to me do not abound.

In venturing to trouble you with some remarks on it, I will first say that I like your way of entering conjecturally into the wave-motion of molecules and their modifications from external and internal causes. And next I will say that that I do not at present feel myself able to express either belief or disbelief in the theories, but that I do not, on that account, think lightly of them, indeed for that very reason I desire to have them scrutinized more carefully.

Next I will critic[2] two points of mathematical character.

In page 16, you allude to the explanation of polarization of light 'by reflection made in air upon air' as if there was difficulty in it. There is no difficulty whatever.

To show this, I will call your attention to a cognate investigation, that of atmospheric refraction.[3] There we have 'refraction made in air upon air'. And the method of treatment, and which must be adopted in all similar cases, is this. We begin with assuming the air to be divided into a limited number of strata, each homogenous in itself, but with density varying by steps from one stratum to another. At each of these steps we can investigate the refraction. We can increase the number of steps more and more, and still we can investigate the refraction at each, the successive individual amounts of infraction will be smaller, but the number of such individual refractions will be greater, and the aggregate will not be materially altered. We increase the number of steps indefinitely and then we arrive at a hypothetical state of air which cannot in any way be distinguished from the real state of air, and, though we have all along computed 'refraction made in air upon air', we have arrived at a satisfactory and indubitable result.

Now to apply a similar investigation to the reflexion. Divide the air into a limited number of strata, at each dividing plane there will be reflection and polarization, and, as the index of refraction is very nearly 1, the angle of incidence for complete polarization is very nearly 45°. Increase the number of strata, at every dividing plane there will be reflection and polarization, but at each phase the index of refraction is more nearly 1, and the polarizing incidence is more nearly 45°: at each, there is less light reflected, but there is a greater number of reflecting *[power]*. And so with an unlimited number of dividing planes; we have, by the supposition of unlimited number of planes, reached a state of things which cannot be distinguished from that of the real atmosphere, and the polarizing incidence approaches without limit to 45°, and the gross amount of light reflected is as great as before.

Page 22. In speaking of the change of the plane of polarization, as indicated by change from white centered rings to black centered rings or vice versa, you have, as appears to me, laid too great stress upon a small phenomenon ('surprising'). Now the nature of the change is this. Resolve the total vibration into parts parallel to and perpendicular to the edge of the paper.

If, as in the first figure, the horizontal part is in the shorter, the light will present all the phenomena of polarization in the horizontal plane.

But if, as in the second figure, the horizontal part is the longer, the light will present all the phenomena of polarization in the vertical plane. So that all that is required is a very small change in the length of the vibrations in one plane as compared with the other, and, as I show by experiment, a very small change in the circumstances of the production of light will make this change in the resulting phenomena.

I am, my dear Sir, | Yours very truly | G B Airy

Professor Tyndall

RGO MS 6/479
RI MS JT/1/A/46
RI MS JT/1/TYP/1/36–7

1. *"On Chemical Rays . . . Sky"*: see letter 2942, n. 1.
2. *Next I will critic*: Tyndall replied to this letter on 22 February 1869 (letter 2956).
3. *atmospheric refraction*: for more correspondence related to this, see letters 2956, 2959, 2965, and 3283.

To George Biddell Airy — 22 February 1869 — 2956

Royal Institution of Great Britain | 22nd. Feb[ruary] | 1869

My dear Sir

I thank you very much for your letter,[1] which I have read with great interest.

I was not unaware of most of the considerations there set forth, nor of other conditions of the atmosphere than those to which you refer, which may & must produce to some extent atmospheric reflection & polarization[2]

These things were present to my mind in these enquiries. 1° the quantity of light reflected from the sky. 2° the colour of that light, and 3° Its polarization.

I should like exceedingly to have your opinion on these points:—

1°. Do you consider that reflection at the limiting surfaces of your atmospheric layers would give you the quantity of light which the sky sends to us.

2°. Do you consider that reflection at these surfaces w[oul]d give you the colour of the sky

3°. Do you consider that reflection at these surfaces would be competent to impress upon the light the amount of polarization which observation reveals?

I am my dear Sir | Yours very sinc[ere]ly | John Tyndall

RGO MS 6/479

1. *your letter*: letter 2955.
2. *reflection and polarization*: Tyndall had been working on the blue colour of the sky and polarization of light (see letter 2940).

From Jane Barnard 22 February 1869 2957

12, Barnsbury Villas, | Islington, | N.[1]
Monday. 22 Feb[ruary] 1869

Dear Dr Tyndall

We are very sorry that an engagement will prevent our having the pleasure of seeing you this evening, & we feel it unfortunate that either on your side or ours all other evenings this week are full, unless it is Saturday, which if suitable to you would be so to us. Will you kindly write us a line[2] & say whether that evening or one early next week would suit you for all are at present at liberty after that.

You see we are not willing to let you loose again, tho' it does seem a long time off.

Yours very sincerely | Jane Barnard
Dr. Tyndall | *[&c] [&c] [&c]*

RI MS JT/1/B/43

1. *Islington*: area in north London.
2. *write us a line*: letter 2958.

To Jane Barnard [22 February 1869][1] 2958

[Royal Institution] | Monday

My dear Miss Barnard

Let us say Saturday[2]—I am more unwilling to be "let off" than you are willing to hold me.

Yours ever | John Tyndall

RI MS JT/1/T/127
RI MS JT/1/TYP/3/4199

1. *[22 February 1869]*: this letter is a response to letter 2957, written on Monday, 22 February 1869, and written on the same Monday.
2. *Let us say Saturday*: see letter 2957.

From George Biddell Airy 27 February 1869 2959

Royal Observatory, Greenwich, London, S. E. |
1869 Feb[ruary] 27

My dear Sir

It appears to me that the questions in your letter of 22d.[1] cannot be answered in any very simple way, on account of the great differences in the state of aqueous vapour[2] in different localities. The black purple sky which I have seen in Sicily is very different indeed from the pale blue which I see here. But, taking our sky in its best state,

1. I do not imagine that reflection from layers of dry air would give the light of sky which we get here. But I doubt not that it would give the light of the Sicilian sky. On these points, photometry is wanting.

2. The colour of the sky really is very feeble even here, as is evident from this circumstance, that in spectrum experiments or depolarisation experiments you do not remark any preponderance of blue, such as it is, I think that it requires some other explanation.

3. The reflexion at the layer surfaces would certainly be competent to impress on the light the amount of polarization which we see. Theoretically, it would at 45^0 incidence or 90^0 change of course produce complete polarization: practically I know that the polarization is very great, but I never examined whether it is total or not.

I am, my dear Sir, | Yours very truly | G B Airy
Professor Tyndall

RGO MS 6/479
RI MS JT/1/A/47
RI MS JT/1/TYP/1/38

1. *your letter of 22*d.: letter 2956.
2. *state of aqueous vapour*: Airy and Tyndall were discussing his recent work on the colour of the sky and the polarization of light. See letters 2955 and 2956. See also letter 2940.

To Heinrich Rudolf Vieweg[1] 28 February [1869][2] 2960

28th Feb[ruary]

Gentlemen

The Messrs Longman[3] inform me that the portrait of Chladni[4] was not included in the estimate of 30 pounds.

Chladni was not sufficiently known in England, and therefore I thought it a duty to make him known. In Germany the case is different, and the portrait is to be found in an edition of his Akustik.[5] Hence I think the portrait might be omitted in the German translation.[6]

With regard to the figures I think in Lecture I. figs. 2 & 12 might be omitted.[7] It is also a question whether figs. 5 & 6 might not be omitted.[8] You appeal however to a field that likes figures and *[must]* therefore exercise your own judgment.

In lecture II fig. 14 might be omitted.[9]

In lecture III figs. 30 & 31 might I think be omitted.[10] The description would be sufficiently clear without them.

Fig. 81 might perhaps be omitted; perhaps also fig. 94.[11]

Fig 146 might be omitted or improved[12]

In fact I should be quite willing to agree to any changes of the figures suggested by Professor Helmholtz or Wiedemann.[13]

Yours very truly | John Tyndall

I quite agree with you that the illustrations executed in Germany are better than those executed in England; hence for my part

<page(s) missing>

TU Braunschweig/Universitätsbibliothek/UABS V 1 T: 63

---

1. *Heinrich Rudolf Vieweg*: while the letter is addressed to 'Gentlemen' the primary recipient is most likely Heinrich Rudolf Vieweg, who took over the family publishing firm, Friedrich Vieweg und Sohns, from his father Eduard Vieweg (1797–1869) in 1866.
2. *[1869]*: this letter refers to the German translation of *Sound,* which was published in 1869 (see letter 2951).
3. *Messrs Longman*: William (1813–77) and Thomas Longman (1804–79) were two brothers who ran the publishing firm Longmans after the sudden death of their father Thomas Norton Longman (1771–1849). Longmans was one of the largest publishing firms in nineteenth-century Britain, publishing books on history, theology, travel, sociology, novels, and science. Longmans published Tyndall's *Heat* (1863), *Sound* (1867), *Faraday as a Discoverer* (1868), and *Hours of Exercise in the Alps* (1871).

4. *portrait of Chladni*: Ernst Florens Friederich Chladni (1756–1827), a German physicist, musician, and pioneer of experimental acoustics based in Wittenberg. He is particularly known for his research on the vibration of rigid plate surfaces. He demonstrated that these vibrations were regular by showing that sand on a resonating plate distributed itself in regular forms (Chladni figures). His work served as the basis for the scientific understanding of sound (*CDSB*).
5. *his Akustik*: Chladni's comprehensive work on acoustics, *Die Akustik* (Leipzig: Breitkopf und Hartel, 1802).
6. *in the German translation*: the German translation of *Sound* (see letters 2949, n. 4, and 2951). The portrait of Chladni was omitted from the German translation.
7. *figs. 2 & 12 might be omitted*: figures 2 and 12 were included in the German version.
8. *whether figs. 5 & 6 might not be omitted*: figures 5 and 6 were included in the German version.
9. *fig. 14 might be omitted*: figure 14 was included in the German version.
10. *figs. 30 & 31 might I think be omitted*: figures 30 and 31 were included in the German version.
11. *81 might perhaps be omitted; perhaps also fig. 94*: figures 81 and 94 were included in the German version.
12. *Fig 146 might be omitted or improved*: figure 146 was included in the German version without alterations.
13. *Wiedemann*: Gustav Heinrich Wiedemann (1826–99), a German physicist especially known for his work on electromagnetism. He became editor of *Annal. Phys. und Chem.* after Poggendorff's death in 1877. Wiedemann's wife, Clara, translated scientific papers. Hermann Helmholtz and Wiedemann were co-editors of the German translation of *Sound* (J. Tyndall, *Der Schall: acht Vorlesungen gehalten in der Royal Institution von Grossbritannien* (Braunschweig: Vieweg, 1869)), with the translation by Anna Helmholtz.

To George Gabriel Stokes 2 March 1869 2961

2nd March 1869

My dear Stokes.

Has your pump <u>two</u> cylinders?[1] Also what is their size?

A perfect vacuum is not a necessity: but I should like to be able to exhaust with tolerable rapidity so as to save time.

The safest plan will be for me to send my assistant[2] down on Friday evening: he could then see the ‘pump on Saturday morning, and if necessary could inform me by telegram of its condition. He is clever and handy and will be able to execute small repairs.

He shall take with him every thing that we can think of, but needs &

requirements always arise in a case of the kind, and to meet them I shall be grateful for your assistance.

I suppose my assistant can get into the room on Saturday to unpack and to partially arrange the apparatus.

It is so probable that your pump will suit us that it would be worth while having it in the room by Saturday morning.

Yours very truly | John Tyndall

It will not be necessary to put a scale in the air pump—I will take a detached scale with me.

CUL SC—add 7656 T556

1. *your pump two cylinders?*: Tyndall gave a lecture to the Cambridge Philosophical Society on 8 March 1869 where he proposed actinic action as the cause of comet tails. It is likely that he used this pump to evacuate a 'long glass tube' used in the experiment. The rest of the letter is concerned with preparations for the lecture (J. Tyndall, 'On the Generation of Clouds by Actinic Action and the Reactions of Such Clouds upon Light', *Proceedings of the Cambridge Philosophical Society*, 2 (March 1869), pp. 136–40).
2. *my assistant*: John Cottrell.

To Elinor Colville 4 March [1869][1] 2962

Royal Institution of Great Britain | 4th March.

My dear Lady Colvile

You are very very good to me. It will give me very great pleasure to join you on the Sunday you mention. And let me say, much as I admire Joachim[2] the pleasure will not be altogether derived from hearing him. I am sure as a true gentleman he would not be offended with the avowal.

Yours very truly | John Tyndall

I do not by any means intend to lose the dinner.

[Envelope] Lady Colvile | 8 Rutland Gate | S.W.

BL Add MS 60634, ff. 123–4

1. *[1869]*: envelope is postmarked 1869.
2. *Joachim*: Joseph Joachim (1831–1907), a renowned European violinist. Born in Hungary, he moved to Germany, where he studied under Felix Mendelssohn. From the late 1850s, Joachim visited London annually to perform Popular Concerts with one of his string quartets (*NDB*).

From Edward Sabine 5 March [1869][1] 2963

Saturday March 5th

Dear Tyndall;

It is quite true that, not 40 but 50 years ago, I was interested and engaged in subjects which you recall to my recollection. The consequence of your note[2] has been that I have rummaged out of a closet where it must have been stowed for several years one of two Instruments which Newman[3] made for me in 1820, and which I employed in the Tropics in 1821–1822, and in the Arctic Circle in 1823.[4] I will bring this instrument with me to the Soirée tonight.[5] It is a differential Thermometer, boiling-freezing being divided into 1000 divisions—one of the balls is dark colored glass; the other is pellucid, protected by a casing of polished silver, not in absolute contact with it—the Thermometer itself was enclosed in a glass tube from which the air was excluded. The mode of using this contrivance was this; it was in a wooden case, and on removal of this and exposure to the Sun the colored liquid descended rapidly till it came to a stand, when the division on the stem, and the time of exposure were noted—varying in different Latitudes and different Altitudes. The other instrument was analogous, but measured absolute degrees of Fah[renhei]t, i.e. was not a differential instrument. The ball was pellucid, & enclosed in a polished silver casing, and the whole Therm[omet]er. was in glass case from which the air was removed. I employed with this a similar Therm[omet]er. with a blackened ball, & no silvered enclosure, both these were divided from 32° to 212°. They are both probably in a cupboard at Woolwich;[6] I have not seen them for years.

There is a notice of the differential Therm[omet]er. (the one which I shall take to Burlington House this evening) in my Quarto or "Pendulum & other Exp."[7] 1825. p. 506–9.; or more precisely in pages 508–9; where Exp[erimen]ts made at Jamaica are recorded. My Meteorological Journal[8] contained a great mass of observations at various stations, on many meteorological subjects—but by the time my book was finished, both my patience & my health were pretty well exhausted; and I have never taken up the subject, or rather the subjects for there were many subjects of enquiry, then or since. The book is I dare say, still in existence, though it must be many years since I saw it—but in respect to the main subject of your note to me, the employment of therm[omet]er with balls skreened by polished silver, the instrument itself which I shall hope to show you this evening, will be sufficient evidence.

Of course the instrument is somewhat out of repair; the silvered sphere has broken or separated, into two hemispheres as poor Murray,[9] were he alive, it is in a somewhat dilapidated state

Sincerely yours | Edward Sabine

RI MS JT/1/S/33
RI MS JT/1/TYP/4/1324

1. *[1869]*: the year is given by reference to the Soirée (see n. 5).
2. *your note*: letter missing.
3. *Newman*: John Newman, instrument maker at the RI.
4. *in the Tropics . . . in the Arctic Circle . . .* : in the 1820s, Sabine became interested in the science of geodesy, and especially with the degree to which the earth departs from perfect sphericity. In the Tropics and the Arctic Circle, he used pendulums to take measurements at different latitudes. The published results were the most accurate assessment of the earth.
5. *the Soirée tonight*: the RS hosted a Soirée (also called 'Conversazione') at Burlington House, London. As described in *The Athenaeum,* there was a 'collection of philosophical apparatus and instruments, and of mechanical contrivances, some showing a marked advance in principle, others improvements in form, or modifications in mode of application' ('The Royal Society', *The Athenaeum*, 2159 (13 March 1869), p. 375).
6. *Woolwich*: the Magnetic Department Office, under the Board of Ordnance, was located at the Royal Artillery Barracks, Woolwich, in south-east London.
7. *"Pendulum & other Exp."*: see E. Sabine, *An Account of Experiments to Determine the Figure of the Earth, by Means of the Pendulum Vibrating Seconds in Different Latitudes, as well as on Various other Subjects of Philosophical Inquiry* (London: John Murray, 1825).
8. *Meteorological Journal*: Sabine supervised the publication of and arranged the Appendix for Alexander Gordon Laing's *Travels in the Timannee, Kooranko, and Soolima Countries, in Western Africa* (London: John Murray, 1825), which included Sabine's recomputed results for the 'Meteorological Journal'.
9. *poor Murray*: presumably John Murray (1778–1843), publisher and proprietor of John Murray publishing house, which was founded by his father, John Murray (1737–93) and later passed on to his son, John Murray (1808–92). In 1824, Murray burned Lord Byron's autobiographical manuscript, considering it too salacious to publish.

To Thomas Archer Hirst 6 March 1869 2964

Royal Institution March 6 /69.

My dear Tom

I am too shaky to risk going to the festivities of Cambridge to day.[1] You are going I think you said to High Elms[2] tomorrow. Come down here and lunch at 12 ½ or 12 and I may run a little way into the country with you.

Affec[tionately] | John

RI MS JT/1/T/1003
RI MS JT/1/HTYP/549

1. *festivities of Cambridge to day*: presumably the meeting of the University Athletic Club, which commenced on Friday, 5 March, and finished on Monday, 8 March 1869 (*The Standard*, 8 March 1869, p. 2).
2. *High Elms*: High Elms is an extensive woodland estate that can be traced back to the Norman Conquest. In the nineteenth century it became the dwelling of the Lubbock family. Tyndall apparently did not go to High Elms that day, since Hirst noted in his journal that he walked to Acton with Tyndall on 7 March (*Hirst's Journal*, RI MS JT/2/32d/1843).

From George Biddell Airy 12 March 1869 2965

Royal Observatory, Greenwich, | London, S. E.[1]
1869 March 12 | evening

My dear Sir,

I have from time to time considered the question in your letter of Feb[ruary]. 28,[2] and have more particularly examined it this evening. And the result is, that I am unable to answer it to my satisfaction. I cannot trace the path of a beam, reflected as I conceive it must be at the dividing surfaces of an elastic atmosphere, so as to explain the visible phenomenon of skylight and its polarization. And I am greatly puzzled.

Remarking the great intensity of atmospheric light in the neighbourhood of the sun, I conceive that there must be molecular matter—say water—in the high air. But I do not profess to trace out all the consequences.

At great elevations there is much less of this light near the sun. When I was on the Görner Gratt,[3] I saw in the forenoon a 4-days moon below the sun. Such a thing was never seen here.

In my former letter,[4] I addressed my remarks simply to the assumed absurdity of reflection <u>in</u> air <u>upon</u> air, in which, as I endeavoured to explain, I see no absurdity. But I do not see how to bring it to bear upon the question of sky-polarization.

I am, my dear Sir, | Yours very truly | <u>G B Airy</u> | Professor Tyndall

RI MS JT/1/T/1003
RI MS JT/1/TYP/1/39

1. *Royal Observatory, Greenwich, | London, S.E.*: Royal Observatory headed notepaper.
2. *your letter of Feb[ruary]. 28*: letter missing.
3. *Görner Gratt*: the Gornergrat (3089 m) is a rocky ridge above Zermatt in Switzerland much visited by the tourists of the day. It offers views of many four-thousand-meter peaks, whose highest are Dufourspitze, Liskamm, Matterhorn, Dom and Weisshorn. Tyndall took yearly trips to the Alps from the summer of 1856 until the end of his life (*Ascent of John Tyndall*, p. 224).
4. *In my former letter*: letter 2959. See also letters 2940, 2955, and 2956.

To the Editors of the *Phil. Mag.* 15 March 1869 2966

XXXIII. *On Cometary Theory.* | By John Tyndall, *F.R.S. &c.* |
To the Editors of the Philosophical Magazine and Journal.

Gentlemen,

On the 8th of this month, in a lecture[1] delivered before the Cambridge Philosophical Society,[2] I ventured to enunciate a speculation regarding the origin and deportment of visible cometary matter. I had been led to reflect on the subject by those experiments, on the decomposition of vapours by light, which I have already described in abstract[3] in the Proceedings of the Royal Society. The speculation was introduced and communicated to the Philosophical Society in the following words:—

"In the course of my experiments on actinic action I have been often astonished at the body of light which a perfectly infinitesimal amount of matter, when diffused in the form of a cloud, can discharge from it by reflection. I have been repeatedly perplexed and led into error by the action of residues so minute as to be simply inconceivable. In order to get rid of these residues, my experimental tubes, after having been employed for any vapour, are flooded with alcohol, sponged out with soap and hot water, and finally flooded with pure water. Let me give you some idea of the quantities of matter that here come into play. The tube before you, which is 3 feet long and 3 inches wide, was so thoroughly cleansed that when filled with air, or with the vapour of aqueous hydrochloric acid, no amount of exposure to an intense light produced the least cloudiness. Having thus assured myself of the perfect purity of the tube, I took a small bit of bibulous paper, rolled it up into a pellet not the fourth part of the size of a small pea, and moistened it with a liquid possessing a higher boiling-point than that of water. I held the pellet in my fingers until it had become almost dry, then introduced it into a connecting-piece and allowed dry air to pass over it into this tube. The air charged with the modicum of vapour thus taken up was subjected to the action of light. A blue actinic cloud began to form immediately, and in five minutes the blue colour had extended quite through the experimental tube. For some minutes this cloud continued blue, and could be completely quenched by a Nicol's prism,[4] no trace of its light reaching the eye when the Nicol was in its proper position. But its particles augmented gradually in magnitude, and at the end of fifteen minutes a dense white cloud filled the tube. Considering the amount of the vapour carried in by the air, the appearance of a cloud so massive and luminous seemed like the creation of a world out of nothing.

"But this is not all; the pellet of bibulous paper was removed, and the experimental tube was cleared out by sweeping a current of dry air through it. *This current passed also through the connecting-piece in which the pellet of bibulous paper had rested.* The air was at length cut off and the experimental tube exhausted. Fifteen inches of hydrochloric acid were then sent into the tube through the same connecting-piece. Now it is here to be noted (1) that the total quantity of liquid absorbed by the pellet in the first instance was exceedingly small, (2) that nearly the whole of this small quantity had been allowed to evaporate between my fingers before the pellet was placed in the connecting-piece, (3) that the pellet had been ejected and the tube in which it rested rendered for some minutes the conduit of a strong current of pure air. It was part of such a residue as could linger in the connecting-piece after this process, that was carried into the experimental tube by the hydrochloric acid and subjected there to the action of light.

"One minute after the ignition of the electric lamp a faint cloud showed itself; in two minutes it had filled all the anterior portion of the tube and stretched a considerable way down it; it developed itself afterwards into a very beautiful cloud-figure; and at the end of fifteen minutes the body of light discharged by the cloud, considering the amount of matter involved in its production, was simply astounding. But though thus luminous, the cloud was far too fine to dim in any appreciable degree objects placed behind it. The flame of a candle seemed no more affected by it than it would be by a vacuum. Placing a page of print so that it might be illuminated by the cloud itself, it could be read *through* the cloud without any sensible enfeeblement. Nothing could more perfectly illustrate that 'spiritual texture' which Sir John Herschel ascribes to a comet than these actinic clouds. Indeed the experiments prove that matter of almost infinite tenuity is competent to shed forth light far more intense than that of the tails of comets. The weight of the matter which sent this body of light to the eye would probably have to be multiplied by millions to bring it up to the weight of the air in which it hung.

"And now will you bear with me for five minutes while I endeavor to apply these results to cometary theory? I am encouraged to do so by a remark of Bessel's,[5] who said that had any theory preceded his observations on Halley's comet, by fixing his attention either upon its verification or its confutation, it would have enabled him to return from his observations with a greater store of knowledge than he had actually derived from them. If time permitted, I should like to lead you by an easy gradient up to the view that I wish to submit to you; but time does not permit of this, and therefore the speculation must suffer from the baldness arising from the absence of such preparation.

"You are doubtless aware of the tremendous difficulties which beset cometary theory. The comet examined by Newton[6] in 1680 shot out a tail sixty millions of miles in length in two days. The comet of 1843, if I remember aright, shot out in a single day a tail which covered 100 degrees of the heavens.

This enormous reach of cloudy matter is supposed to be generated in the head of the comet and driven backwards by some mysterious force of repulsion exerted by the sun. Bessel[7] devised a kind of magnet polarity and repulsion to account for it. 'It is clear,' says Sir John Herschel, 'that *if we have to deal here with matter such as we conceive it, viz. possessing inertia, at all,* it must be under the dominion of forces incomparably more energetic than gravitation, and quite of a different nature.'[8] And in another place he states the difficulties of the subject in the following remarkable words:—

"'There is beyond question some profound secret and mystery of nature concerned in the phenomenon of their tails. Perhaps it is not too much to hope that future observation, borrowing every aid from rational speculation, grounded on the progress of physical science generally (especially those branches of it which relate to the aethereal or imponderable elements), may ere long enable us to penetrate this mystery, and to declare whether it is really *matter,* in the ordinary acceptation of the term, which is projected from their heads with such extravagant velocity, and if not impelled, at least *directed* in its course by a reference to the sun as its point of avoidance. In no respect is this question as to the materiality of the tail more forcibly pressed on us for consideration than in that of the enormous sweep which it makes round the sun *in perihelio,* in the manner of a straight and rigid rod, in defiance of the law of gravitation, nay, even of the received laws of motion, extending (as we have seen in the comets of 1680 and 1843) from near the sun's surface to the earth's orbit, yet whirled round unbroken—in the latter case through an angle of 180° in little more than two hours. It seems utterly incredible that in such a case it is one and the same material object which is thus brandished. [I would especially invite the reader's attention to these words in reference to the following theory.–J. T.] If there could be conceived such a thing as a *negative shadow,* a momentary impression made upon the luminiferous aether behind the comet, this would represent in some degree the conception such a phenomenon irresistibly calls up.'[9]

"I now ask for permission to lay before you a speculation which seems to do away with all these difficulties, and which, whether it represents a physical verity or not, ties together the phenomena exhibited by comets in a remarkably satisfactory way.

"1. The theory is, that a comet is composed of vapour decomposable by the solar light, the visible head and tail being an actinic cloud resulting from such decomposition; the texture of actinic clouds is demonstrably that of a comet.

"2. The tail, according to this theory, is not projected matter, but matter precipitated on the solar beams traversing the cometary atmosphere. It can be proved by experiment that this precipitation may occur either with comparative slowness along the beam, or that it may be practically momentary throughout the entire length of the beam. The amazing rapidity of the development of the tail would be thus accounted for without invoking the incredible motion of translation hitherto assumed.

"3. As the comet wheels round its perihelion, the tail is not composed throughout of the same matter, but of new matter precipitated on the solar beams, which cross the cometary atmosphere in new directions. The enormous whirling of the tail is thus accounted for without invoking a motion of translation.

"4. The tail is always turned from the sun for this reason:—Two antagonistic powers are brought to bear upon the cometary vapour,—the one an *actinic* power tending to produce precipitation, the other a *calorific* power tending to effect vaporization. Where the former prevails, we have the cometary cloud; where the latter prevails, we have the transparent cometary vapour. As a matter of fact, the sun emits the two agents here invoked. There is nothing whatever hypothetical in the assumption of their existence. That precipitation should occur behind the head of the comet, or in the space occupied by the head's shadow, it is only necessary to assume that the sun's calorific rays are absorbed more copiously by the head and nucleus than the actinic rays. This augments the relative superiority of the actinic rays behind the head and nucleus, and enables them to bring down the cloud which constitutes the comet's tail.

"5. The old tail, as it ceases to be screened by the nucleus, is dissipated by the solar heat; but its dissipation is not instantaneous. The tail leans towards that portion of space last quitted by the comet, a general fact of observation being thus accounted for.

"6. In the struggle for mastery of the two classes of rays a temporary advantage, owing to variations of density or some other cause, may be gained by the actinic rays even in parts of the cometary atmosphere which are unscreened by the nucleus. Occasional lateral streamers, and the apparent emission of feeble tails towards the sun would be thus accounted for.

"7. The shrinking of the head in the vicinity of the sun is caused by the beating against it of the calorific waves, which dissipate its attenuated fringe and cause its apparent contraction.

"Throughout this theory I have dealt exclusively with true causes, and no agency has been invoked which does not rest on the sure basis either of observation or experiment. It remains with you to say whether in venturing to enunciate it I have transgressed the limits of 'rational speculation.'

"If I have done so, surely I could not have come to a place more certain to ensure my speedy correction. If the theory be a mere figment of the mind, your Adams[10] and your Stokes (both happily here present), to whom I submit the speculation with the view of having it instantly annihilated by astronomy and physics, if it merit no better fate, will, I doubt not, effectually discharge that duty, and thus save both you and me from error before it has had time to lay any serious hold on our imagination."

The leniency with which this speculation was received at Cambridge induces me to submit it to the critical judgment of the readers of the Philosophical Magazine*.

I remain, Gentlemen, | Your obedient Servant, | JOHN TYNDALL. | Royal Institution. | March 15, 1869.

*There may be comets whose vapour is undecomposable by the sun, or which, if decomposed, is not precipitated. This view opens out the possibility of invisible comets wandering through space, perhaps sweeping over the earth and affecting its sanitary condition without our being otherwise conscious of their passage. As regards tenuity, I entertain a strong persuasion that out of a few ounces (the possible weight assigned by Sir John Herschel to certain comets) of iodide-of-allyl vapour, an actinic cloud of the magnitude and luminousness of Donati's comet[11] might be manufactured.

Phil. Mag., 37 (April 1869), pp. 241–5

---

1. *in a lecture*: see letter 2961, n. 1.
2. *Cambridge Philosophical Society*: the Cambridge Philosophical Society was founded in 1819 at the University of Cambridge, aiming to promote scientific inquiry and to facilitate the communication of new developments in the field. See S. Gibson, *The Spirit of Inquiry: How One Extraordinary Society Shaped Modern Science* (Oxford: Oxford University Press, 2019).
3. *described in abstract*: J. Tyndall, 'On the Blue Colour of the Sky, and the Polarization of Light', *Roy. Inst. Proc.*, 17 (1869), pp. 223–33.
4. *Nicol's prism*: a device normally made of calcite or Iceland spare which is used to produce plane polarized light, invented in 1828 by the Scottish physicist and geologist William Nicol (1770–1851) (*ODNB*).
5. *Bessel's*: Friederich Wilhelm Bessel (1784–1846), a German astronomer. Largely self-taught as an astronomer, Bessel recalculated the orbit of Halley's comet and later Encke's and Biela's comets. In 1810 he was named professor of astronomy at the newly founded German observatory at the University of Königsberg (*NDB*).
6. *comet examined by Newton*: Isaac Newton (1642–1727), an English natural philosopher and mathematician of outstanding ability who rose to the Lucasian Professorship of Mathematics at Cambridge (1669), Master of the Mint (1699), and President of the RS (1723) (*CDSB*; *ODNB*). Newton observed the comet discovered by German astronomer Gottfried Kirch in 1680 and demonstrated his law of universal gravitation and verified Kepler's laws by showing that it followed the path of a parabola.
7. *Bessel*: see n. 5.

8. "*It is clear . . .*": J. Herschel, *Outlines of Astronomy* (Philadelphia: Lea and Blanchard, 1849), p. 350.
9. "*There is beyond question . . .*": Herschel, *Outlines,* p. 357. See n. 8.
10. *your Adams*: John Couch Adams (1819–92), a British astronomer and mathematician. He made observations of Halley's comet in 1835 and, although he was only belatedly recognized, was co-discoverer of Neptune. Born in Cornwall, he was largely self-taught in mathematics until an inheritance enabled him to study at Cambridge. He later returned there as Professor of Mathematics in 1859 and became director of the Cambridge Observatory from 1861 (*Alum. Cantab.*; *ODNB*).
11. *Donati's comet*: Donati's comet, a celestial spectacle of the nineteenth century, was first observed by Italian astronomer Giovanni Battista Donati (1826–73) on 2 June 1858. The comet was visible to observers in both hemispheres between September 1858 and March 1859. See A. Gasperini, D. Galli, and L. Nenzi, 'The worldwide impact of Donati's comet on art and society in the mid-19th century', in D. Valls-Gabaud and A. Boksenberg (eds), *The Role of Astronomy in Society and Culture: Proceedings IAU Symposium No. 260* (Cambridge: Cambridge University Press, 2011), pp. 340–45. The painting of a coastal scene in southeastern England by William Dyce (1806–64)—*Pegwell Bay, Kent—a Recollection of October 5th 1858*—depicts the comet in daytime above people exploring the beach at low tide.

From Ernst Brücke 17 March 1869 2967

Währinger Gasse 1 | 17 März 1869

Besten Dank für Ihren freundlichen Brief. Das Steinsalz habe ich gestern franco durch die Post unter Ihrer Adresse—John Tyndall, Royal Institution, London—abgesendet. Mit meiner Gesundheit geht es wieder gut und morgen beginnen meine Osterferien die mir für 14 Tage etwas mehr Ruhe verschaffen als ich bisher gehabt habe. Damit habe ich zugleich die Hälfte meines Dekanats hinter mir und die Hoffnung in fünf Monaten zu meinen wissenschaftlichen Arbeiten zurückkehren zu können. Unsere medicinische Fakultät hat so grosse Dimensionen angenommen, dass das Dekanat dem, der es eben verwaltet, ein Jahr seines Lebens, wenigstens in Rücksicht auf seine wissenschaftliche Thätigkeit, vollständig raubt. Bence Jones bitte ich meine besten Grüsse zu sagen. Ich habe ihm vor einiger Zeit geschrieben. Mit der Bitte mich gelegentlich einmal wissen zu lassen ob die Sendung richtig und in gutem Zustande in Ihre Hände gekom[men] ist Ihr ergebener

E. Brücke

Herrn Prof. John Tyndall | Royal Institution London.

Währinger Gasse 1 | 17 March 1869

Many thanks for your kind letter.[1] Yesterday, I sent the mineral salt[2] postage paid to your address—John Tyndall, Royal Institution London. My health is improved and my Easter holidays begin tomorrow, which will give me for 14 days somewhat more rest than I have had until now. With this, I will also have brought half of my deanship behind me and hope that in five months I can return to my scientific work. Our faculty of medicine has grown so large that the deanship robs whoever administers it of a full year of life, at least as far as his scientific activity is concerned. I ask you to give my best regards to Bence Jones. I wrote to him some time ago. With the request that you let me know by and by if the delivery reached you in good condition, your humble

E. Brücke

Prof. John Tyndall | Royal Institution London.

RI MS JT/1/B/136

1. *your kind letter*: letter missing.
2. *sent the mineral salt*: see letter 2954.

From Anna Helmholtz 17 March 1869 2968

Heidelberg 17 März 1869.

Lieber Herr Professor

Ich sehe nicht ein, wozu ich mich mit englischen Briefen abquälen soll da sie solch schöne deutsche Episteln verfassen. Halten sie mich nur nicht für undankbar weil ich erst heute komme Ihnen für Ihren Verzicht zu unseren Gunsten zu danken. Ich finde immer noch dass unsere Arbeit in keinem Verhältniss zu diesem Honorar steht—und schäme mich einigermassen Ihr Opfer angenommen zu haben. Da es mir jedoch sehr unangenehm scheint wenn Jemand <u>meine</u> freundlichen Anerbietungen zurückweist, so dachte ich, es ist besser wenn wir uns Ihrem Willen fügen, und unsere geringe Mühe weit über Verdienst lohnen lassen—als dass wir uns auf das hohe Ross der strengen Tugend setzen.—Meine kleinen Buben kriegen aber nichts davon, sondern ich werde mir selbst irgend eine besondere Freude damit machen, welche dem ganzen Hause zu gut kommen soll.

Ich schliesse diesen wie jeden früheren Brief an Sie, lieber Herr Professor, mit der Bitte, dass Sie uns doch gewiss in diesem Jahre besuchen. Es ist nicht gut, dass wir uns so gar nicht kennen.—

Professor Clausius war gestern hier. Er ist jetzt im Begriffe nach Bonn zu ziehen. Er wünscht dass Kirchhoff sein Nachfolger in der Würzburger Stelle werde—allein Kirchhoff kann im Augenblicke keine Entschlüsse fassen, da seine Frau leider hoffnungslos an einem Lungenleiden erkrankt ist—dabei geht er an Krücken, kurz der Zustand im Hause Kirchhoff ist höchst jämmerlich. Mein Mann grüsst schönstens und ich danke Ihnen nochmals recht schön.

Ihre ergebene | Anna Helmholtz

Heidelberg 17 March 1869.

Dear Professor

I do not see why I should struggle to write letters in English, when you compose such beautiful epistles in German. Please do not think that I am ungrateful because I am only today thanking you for the sacrifice you made in our benefit. I still find that our work is not in any relation to this honorarium[1]—and I am somewhat embarrassed to have accepted your offering. But, since I find it unpleasant when someone rejects my friendly offers, I thought it better if we accepted your wish and allowed our modest efforts to be recompensed much more than they deserve, than if we sat on the high horse of severe virtuousness. My little boys will receive none of it, but I will grant myself some special pleasure, from which everyone in the house shall benefit.

I conclude this letter, like any other to you, dear Professor, with the request that you will surely visit us this year. It is not good that we do not know each other at all.—Professor Clausius came here yesterday. He is about to move to Bonn.[2] He desires that Kirchhoff become his successor in Würzburg[3]—but Kirchhoff cannot make any decisions at this point, as his wife is unfortunately suffering from an incurable lung disease; all the while, he is walking on crutches himself—in short, the situation at Kirchhoff's is quite wretched.[4] My husband sends his greetings and I thank you once again. Yours truly | Anna Helmholtz

RI MS JT/H/33

---

1. *this honorarium*: Anna Helmholtz translated Tyndall's *Sound* into German. The first German edition was overseen by H. Helmholtz and G. Wiedemann and published in 1869 by Vieweg. See letters 2951 and 2960.
2. *move to Bonn*: see letter 2951, n. 6, and letter 2974.
3. *Würzburg*: city in the German state of Bavaria. Clausius was professor of physics at the University of Würzburg from 1867 to 1869. Kirchhoff's wife, Clara, died in 1869. He remained at Heidelberg until 1875 when he accepted the chair of mathematical physics at Berlin.
4. *quite wretched*: see letter 2951.

To Robin Allen 19 March 1869 2969

Enclosure. 2
Royal Institution, 19 March 1869.

Sir,

THE proposed construction of a dioptric revolving apparatus for the Wolf Rock,[1] which should emit alternative flashes of red and white light, has brought forward the question of the proper proportions of the apparatus to be devoted to the respective lights.

The data here at our disposal are, I am sorry to say, exceedingly meagre. There is no way known to me of comparing with strict numerical accuracy, by means of a photometer, lights of difference colours.

M. Reynaud[2] has recorded some determinations of this character, at page 99 of his "Memoire sure l'Eclairage, &c.";[3] but his photometric results, at all events those which refer to the change in the relative intensities of red and white light as the distance from the photometer is augmented, seem so extraordinary as to lead me to think that some unrecognised source of disturbance has come into play in his experiments. I may feel it my duty at a future day to recommend a repetition of these experiments to the Elder Brethren.[4]

But even if such comparative experiments could be made close to the sources of illumination, they are not what we require. What is needed in the case now before us is that the white light and the red light should have, as nearly as possible, the same atmospheric range. Experiments through considerable distances of air are alone available here.

Learning from Mr. Douglass that the Rock Lighthouse[5] near Liverpool showed red and white light, which could be seen from the Point of Air[6] and from the Great Ormes Head,[7] and wishing to clear my mind as far as possible of uncertainty, I proposed a visit to these places. On the 11th March accordingly we visited the Rock Lighthouse, and found there ten reflectors employed for the red light against six reflectors employed for the white light. The light was reddened by chimneys of ruby glass.

Having instructed ourselves regarding the Rock apparatus we went to the Point of Air, and observed from that place the flashes of the Rock light. Some attention was requisite to distinguish the one flash from the other, though the atmosphere was singularly transparent at the time. I quitted the Point of Air with the conviction that the red of the Rock Lighthouse was not sufficiently deep in colour.

On questioning the keeper at the Point of Air I could not ascertain that he had ever, in any state of the atmosphere, seen one of the lights to the exclusion of the other. Whenever one light was seen, the other was seen also.

On the morning of the 12th we visited the Great Ormes Head, my principal object there being to test the statement that the red beam of the Rock Lighthouse was sometimes seen there when the white light was invisible. The head keeper, a highly intelligent man to all appearance, assured us that this was not the case. He was perfectly distinct in his statement that whenever the one light was seen the other was seen also.

Various other statements made by the keepers at the Rock Lighthouse regarding the superior penetrative power of their red light did not bear the test of strict inquiry. The result of our inspection was to render it certain that the white and red beams of the Rock Lighthouse were very nearly balanced in all states of the atmosphere.

Were the ruby glass chimneys of the proper colour the experience furnished by the Rock Lighthouse would be conclusive, and we should be justified in adopting the proportions, 10 to 6 for the red and the white light.

But the colour of the ruby glass is *not* such as I could recommend; and, with a view of more thoroughly examining this question, a camera has been constructed, and lenses and prisms have been furnished by Mr. Ladd.[8] This enables us to decompose the white beam of the magneto-electric light, and to display its constituent colours in a brilliant spectrum upon a white screen.

The red glass hitherto employed by the Trinity House[9] shows that the red rays of the spectrum had free transmission, while the remaining rays were practically extinguished. The yellow, whose admixture would be most prejudicial, was powerfully cut away by the Trinity glass. I consider it exceedingly well chosen.

The ruby glass chimneys employed at Liverpool produced very small amount of absorption. The experiments with the spectrum confirmed the results of direct observation, by showing, that this glass permits a large body of light to pass through it, which tends to destroy its distinctive character.

Having satisfied myself that the red glass of the Trinity House is at all events very nearly of the colour that I should recommend, I sought to compare the light transmitted by this glass with white light. On the 18th, I arranged a programme of observations with Mr. Jones,[10] in which the red and the white were to be compared in the following proportions:-

1st Experiment———-—- 1 red to 1 white.
2nd—-——-——-——-——- 2 " 1 "
3rd—-——-——-——-——- 3 " 1 "

Mr. Jones took charge of the lamps while I crossed the river to our temporary observatory at Charlton.[11]

The air was so charged with haze and smoke that none of the lights were visible to the naked eye. They were, however, visible in the telescope and opera glass. I found the opera glass best suited to the observations, because with it, when more than a single reflector was used, the lights blended together as if they came from a single reflector.

In the first experiment, viz. 1 to 1, begun at 6.30, the red light was not at all visible, while the white one was distinctly seen.

In the second experiment, viz. 2 to 1, the red was quite visible, but inferior to the white.

In the third experiment, viz. 3 to 1, the red appeared to have the upper hand.

I beg here to express my thanks to Mr. Jones for the extreme promptness and accuracy with which he carried out the programme agreed upon between us.

The results were afterwards checked by making the same experiments in the reverse order. In the repetition, the three red lamps did not show the same superiority as they did at first, but on looking through the telescope, two of them did not appear to be perfectly fitted. A very minute shifting would account for the difference.

The atmosphere on this occasion was favourable to the red light, for smoke is more transparent to the red than to any other colour of the spectrum. With a clear atmosphere, the single white light would have been nearly a match for the three red ones. If time permitted, I should like to make a more prolonged series of experiments on this subject. But taking everything into account, we can but commit an exceedingly slight error if we assume that the loss of light by transmission through the Trinity House glass is three-fifths of the total light. This proportion is somewhat higher than that taken by Mr. Chance[12] in his second specification. He there assumes the loss to be 12/21 parts of the whole, while I assume it to be 12/20 parts. This latter proportion is what I would recommend as a basis of construction.

I have, &c. | (signed) *John Tyndall.*

Robin Allen, Esq.

SUPPLEMENTARY REMARKS.

The purities and homogeneity of the glass in the apparatus of the Great Ormes Head struck me much. It was manufactured, I am told, by Mr. Chance, and, if so, it proves what Mr. Chance can accomplish. Nothing so excellent as this glass at Great Ormes Head has fallen to my lot to examine in Birmingham for some time past.

The next experiments to be made with the new magneto-electric machine must determine its capacity to maintain itself in action for 24 hours or more, without undue heating or other deterioration. The machine is not yet ready for this test.

I have been much impressed with the necessity of having the best possible electric lamp at Blackwall.[13] Much as I admire and respect the practical genius of Mr. Holmes,[14] I am not satisfied with his lamp, nor should I like to see it in the hands of a lightkeeper.

Of Serrin's[15] best lamps I have had no personal experience, but I am informed that they are very perfect instruments. They are dear; but it must be remembered that the entire success of the magneto-electric light depends upon this portion of the apparatus.

(signed) *John Tyndall.*

'Further Report from Corporation of Trinity House on Magneto-Electric Light', HC 250, Parliamentary Papers (1868–69), pp. 3–4

---

1. *Wolf Rock*: Wolf Rock Lighthouse, built by Trinity House in 1869, is on the Wolf Rock, a single rock located off the coast of Cornwall.
2. *M. Reynaud*: Léonce Reynaud (1803–80), a French architect and engineer. He also served as the Director of the Lighthouse and Beacons' Service (*directeur du service des phares et balises*) (*Times*, 17 February 1880, p. 8).
3. *"Memoire sure l'Eclairage, &c."*: L. Reynaud, *Mémoire sur l'éclairage et le balisage des côtes de France* (Paris: Imprimerie Impériale, 1864).
4. *The Elder Brethren*: the Committee of Elder Brethren at Trinity House (see letter 2937, n. 6) was an association of English mariners established in the early sixteenth century. They tended to be Master Mariners or Naval Officers. They oversaw the Lighthouse Board, along with various other maritime affairs.
5. *Rock Lighthouse*: the oldest lighthouse in the Liverpool area, built in 1827, was known as the Perch Rock Lighthouse.
6. *Point of Air*: the Point of Ayr Lighthouse, built in 1776 on the north coast of Wales, was replaced in 1844 with a metal pile lighthouse bearing a white light on the order of the Corporation of Trinity House.
7. *Great Ormes Head*: likely refers to the Great Orme Head Lighthouse, constructed in 1862 as the Llandudno lighthouse, on the northernmost point of the Orme, a limestone headland on the north coast of Wales.
8. *Mr. Ladd*: William Ladd (1815–85), an English optician and maker of scientific instruments with a London shop at 11–12 Beak Street, and 'Scientific Instrument Maker by Appointment to the Royal Institution of Great Britain' (see G. Clifton, *Directory of British Scientific Instrument Makers 1550–1851* (London: Zwemmer, 1995), pp. 125–6, 161).
9. *Trinity House*: see letter 2937, n. 6.
10. *Mr Jones*: Jones is described as the superintendent of the apparatus used to produce light during these experiments.
11. *Charlton*: see letter 2937, n. 7.
12. *Mr. Chance*: James Timmins, First Baronet of Chance (1814–1902), a glass manufacturer and lighthouse engineer. He specialized in lighthouse lenses. In 1860, he worked with Faraday on the Whitby Lighthouse. His lighthouse lamp designs rivalled the dominant French manufacture and replaced them. Chance refused to patent his inventions and instead made them public property. By 1869, he was the leading lighthouse lamp designer in Britain and long-term advisor to Trinity House (*ODNB*).
13. *Blackwall*: see letter 2937, n. 1.

14. *Mr. Holmes*: see letter 2937, n. 2.
15. *Serrin's*: Victor Serrin (1829–1905), a French engineer who developed an electric arc lamp that was used in lighthouses in the UK and France (*Nature*, 71 (2 February 1905), p. 325).

To Thomas Carlyle — 20 March [1869][1] — 2970

My dear General

I have three things to say & to do:—

1. Reasonable service—To give you heartiest thanks for the volumes which you have had sent me.[2]

2. To tell you an anecdote:—I went into the chambers of a mathematician[3] a day or two ago. "You will smile" he said "but I could not lay this book[4] down this morning—it fascinated me." "And what is it"? I asked. "I have been reading about the battle of Dunbar"[5] he said "and if you give me a protractor I will draw for you a plan of the battle field. I know & see every trench in it. And the writer knew all about erosion before you knew anything about it. One of the streams that played a part in the action digs its own ditch. No natural philosopher could exercise closer powers of observation or stick with more rigid fidelity, in the midst of his glorious composition, to the facts of the case." Only think of a tanned and curried mathematician—but he is really a fine fellow—being caught up in the air in that way. And for one thus caught up at present there will be 100 a century hence.

My faith is great in time.

3. There is an American gentleman;[6]—a gentleman I believe in the true sense, really gentle and really cultured who longs to see you. I met him I think at Charles Darwins, and in the house of Leslie Stephen.[7] He has a notion & a fear that you dislike Americans.[8] But I ventured to say that you liked men such as he all the world over. Will you let him come to see you sometime: If so, when?

My bad head & worse stomach are very shaky at present—I am going to run away for a few days[9] in the hope of getting a little strength for my lectures. I wish you were coming with me.

Yours ever | John Tyndall

RI MS JT/1/T/162
RI MS JT/1/TYP/1

1. *[1869]*: the year is given by relation to Carlyle's response (letter 2971).
2. *the volumes which you have sent me*: not identified.
3. *a mathematician*: not identified.

4. *this book*: possibly referring to Carlyle's *Oliver Cromwell's Letters and Speeches, with Elucidations*, 3 vols (London: Chapman and Hall, 1845–6).
5. *Battle of Dunbar*: the Battle of Dunbar (3 September 1650) was the conclusive battle in the English Civil Wars. English troops, led by Oliver Cromwell, defeated the Scottish army. As a result, Scotland was forcibly occupied by English troops.
6. *an American gentleman*: Tyndall was likely referring to the notable American author and social critic, Charles Eliot Norton (1827–1908) (*ANB*). This was a fortuitous meeting; Norton and Carlyle became friends, and Norton later edited some of Carlyle's correspondence. See letter 2996. Norton visited Charles Darwin in 1868 and 1869 (see Charles Darwin to Asa Gray, 1 June [1869], n. 12, in F. Burkhardt, et al. (eds), *The Correspondence of Charles Darwin* (Cambridge: Cambridge University Press, 2009), vol. 17, pp. 252–4, on 254; this is letter 6767 in the online edition of the Darwin correspondence).
7. *Leslie Stephen*: Leslie Stephen (1832–1904), a English writer, editor, and mountaineer. Stephen studied at Cambridge and was ordained a priest in 1859, but later became an agnostic. By 1869, he had left Cambridge and had become a well-known essayist, writing on literature, alpinism, and religious criticism. He was also a member of the Metaphysical Society (*ODNB*).
8. *you dislike Americans*: see letter 2971.
9. *run away for a few days*: Tyndall went to Exeter with Frederick Pollock on 25 March (*Hirst's Journal*, RI MS JT/2/32d/1844).

From Thomas Carlyle 20 March 1869 2971

Chelsea, 20 March, 1869

Dear Tyndall,

Bring y[ou][r] American[1] down on Monday Ev[enin][g] if you can (8½ p.m.);—tell him I am far from disliking Americans that are likable; on the contrary, I feel a nat[ura]l blood-relat[io][n]ship to them, and am sad when they are not of that pleasant category.

Two papers of y[ou][rs] I read;[2] clear as azure & sunlight, both of them, sequel to the 2[d] is one of the things I look forward to; and the first, with its clean spiritual radiancy, and pure sky-blue of subject and treat[men][t], was a kind of antidote to the Huxley Protoplasm,[3] one of vilest cleverest doggeries I have ever read; a true douche of stable droppings, from wh[ic][h] I am hardly yet quite washed. Pah!—

Tell me as soon as you can when y[ou][r] Lectures[4] do begin. I have a kind of reason; being bound for Melchet[5] soon

In superlative haste, and with such a hand!

Y[ou][rs] ever truly | T Carlyle

RI MS JT/1/C/14

---

1. *your American*: see letter 2970.
2. *Two papers of y[ou]rs I read*: probably 'On the Blue Colour of the Sky' and 'Chemical Rays' (see letters 2940, n. 3, 2942, n. 1, and 2946, n. 6).
3. *Huxley Protoplasm*: Huxley posited that the physical basis for life lay in a semi-fluid nitrogenous substance. By postulating an organic basis for life, he implicitly contradicted a divine origin. See T. H. Huxley, 'On the Physical Basis of Life', *The Fortnightly Review*, 5 (February 1869), pp. 129–45.
4. *y[ou]r lectures*: Tyndall delivered a lecture course at the RI, 'On Light', on Thursdays from 8 April to 3 June 1869 (*Roy. Inst. Proc.*, 5 (1866–69), p. 428; *Ascent of John Tyndall*, p. 250).
5. *Melchet*: Louisa Baring (Lady Ashburton) had a residence at Seaforth Lodge, Melchet Court in Romsey, Hampshire.

From Joseph Dalton Hooker [22 March 1869][1] 2972

Kew

Dear Tyndall,

Whether Mt Ararat[2] is 7000 ft or 17,000 I intend to go not measuring scientific interests by feet vertical. From what my wife told me I thought you were cool about it, and also from your never coming to the Lyell's[3] nor to Kew[4] (in any sane manner) nor to Phil[osophical] Club.[5]

I am writing to Trebizond[6] now for particulars.

Will you be at the Ethnological[7] tomorrow evening at Jermyn St. This damned Whig gov[ernmen]t will not pay my expenses to St Petersburg,[8] but nor is that to stop me. Nothing will I hope stop me but news from Trebizond that it is out of the question and then shall go to T. and see for myself. No-one has been up since Parrot[9] and it is doubtful whether he was on the right Mt. His account is most meagre. The former extension of glaciers in that region is of far more moment than reaching the top of Ararat. You are a mere acrobat. I want to be a scientific explorer! Go to, ye unwise weep and howl for the fix ye are in.[10]

Ever your aff[ectionately] | J. D. Hooker.

I dreamt last night that you and I and Huxley went up and that H. lost his life, and we were in the devil's own fix how to tell Mrs H.

RI MS JT/1/TYP/8/2599
LT Typescript Only

1. *[22 March 1869]*: this date is an estimate based on letter 2973 of 24 March 1869. Tyndall's response to this letter, which Hooker answered in letter 2973, is missing, so that a more precise dating is not possible.
2. *Mt. Ararat*: Mount Ararat, a double-peaked dormant volcano, is the highest and largest in Turkey.
3. *the Lyell's*: Tyndall was friendly with the extended family of Charles Lyell, but Hooker would most likely have been dining with Charles Lyell and his close family.
4. *Kew*: Kew Gardens was and remains an important home to diverse botanical specimens from all over the world. It was a centre for scientific research and was under the direction of Joseph Hooker's father, Sir William Jackson Hooker (1785–1865), before Joseph Hooker assumed the direction in 1865. Hooker lived on the premises, hence the sender address of his letter.
5. *Phil[osophical] Club*: Tyndall and Hooker were both members of the Philosophical Club of the RS, a dining club established in 1847, and to which all of the X Club members (George Busk, Edward Frankland, Thomas Huxley, William Spottiswoode, Thomas Hirst, and John Lubbock), with the exception of Herbert Spencer, belonged. On the X Club, see letter 3045, n. 5.
6. *Trebizond*: Trabzon (also called Trebizond) is a port city on the Black Sea in modern Turkey. It was founded by the Greeks, was under Roman, Byzantine, and Ottoman rule.
7. *Ethnological tomorrow*: meeting of the Ethnological Society of London. The Ethnological Society, founded in 1843, was devoted to anthropological research. Huxley served as President in 1869.
8. *will not pay my expenses to St Petersburg*: although originally pressured by the British government to be its representative at the International Botanical Exhibition, the Treasury decided against funding Hooker, and he travelled independently, instead of as a delegate (L. Huxley, *The Life and Letters of Joseph Dalton Hooker*, 2 vols (London: John Murray 1918), vol. 2, p. 85.)
9. *Parrot*: Johann Jacob Friederich Parrot (1791–1841), a German physician, professor of medicine, mountaineer, and adventurer. In 1829, he travelled with four students from the University of Dorpat, where he held a chair in physics, to Ararat. He reached the peak on his second try (F. Parrot, *Die Reisen zum Ararat* (Berlin: Haude und Spenerschen, 1834)).
10. *Go to ye unwise weep and howl for the fix ye are in*: a play on KJV 5:1 'Go to now, ye rich men, weep and howl for your miseries that shall come upon you'.

From Joseph Dalton Hooker 24 March 1869 2973

Kew | March 24th/69.

Dear old Tyndalides

Fiddle-de-dee. Who talks of not going to the top of Ararat.[1] What I say is that whether the top is 7000 or 17000, it is all the same to me, or, as you like it better, whether it be 17000 or 7000 it is all the same to me; up I go and to the top I go, and not being a mere acrobat like you, I am as pleased with the 7000 as the 17000 so long as it enables me to trace glacial action. I have written privately for leave to the office, "to visit the oak forests of Turkey".

The Treasury refuse to pay my expenses to St Petersburg![2] which is a sell, and I am selling some of my "Virtue" to raise the wind.

A precious complete man you are! Where are your wife and children? old Brain Sauce. Do come and spend Good Friday here and go up the pagoda!!! with us at 11.0 a.m. We shall have two of the Lyells,[3] Leonard[4] and Rosamond[5]—Now won't you? and I will coax you into a good humour.

Ever yours aff[ec]t[ionat]ely | J. D. Hooker.

RI MS JT/1/TYP/8/2600
LT Typescript Only

1. *Mount Ararat*: see letter 2972, n. 2. This letter is a response to a letter from Tyndall that is missing, and which itself was a response to letter 2972.
2. *The Treasury refuse to pay my expenses to St Petersburg!*: see letter 2972, n. 8.
3. *two of the Lyells*: both Hooker and Tyndall associated with the extended family of Charles Lyell.
4. *Leonard*: Leonard Lyell (1850–1926), Charles Lyell's nephew.
5. *Rosamond*: Rosamond Lyell (d. 1901), Charles Lyell's niece.

From Rudolf Clausius 25 March 1869 2974

Würzburg, 25/3/69.

Lieber Tyndall,

Bevor ich Würzburg verlasse, was in den nächsten Tagen geschehen wird, muss ich Dir doch noch einmal schreiben, um Dir von der neuen Wendung meines Schicksals Mittheilung zu machen. Man hat nämlich ganz unerwartet bei mir angefragt, ob ich die physikalische Professur der Universität Bonn annehmen, und Plückers Nachfolger werden wollte. Nach längerem Schwanken habe ich mich endlich überzeugt, dass ich in Bonn doch einen grösseren und schöneren Wirkungskreis haben werde als hier, und habe mich entschlossen, hinzugehen. Wir werden in der nächsten Woche von hier abreisen, erst auf kurze Zeit nach Berlin gehen, und in etwa 14 Tagen in Bonn eintreffen. Es ist uns nicht angenehm Würzburg nach so kurzem Aufenthalte wieder zu verlassen; aber der Ruf nach Bonn bot doch so viele Vorzüge, dass ich ihn nicht gut ablehnen konnte. Nach allem, was wir von Bonn gehört und gesehen haben, hoffen wir, dass es uns dort recht gut gefallen wird.

Ich habe lange von Dir keine Nachricht erhalten, und will hoffen, dass es

Dir gut geht. Wenn Du wieder einmal, was hoffentlich recht bald geschieht, an mich schreibst, so musst Du Deinen Brief schon nach Bonn, (Colmantstrasse), adressiren.

Mit herzlichen Grüssen von meiner Frau und mir an Dich und Hirst

Dein | Clausius.

Würzburg, 25/3/69.

Dear Tyndall,

Before I leave Würzburg,[1] which will happen in the next few days, I must write to you again to tell you about this new turn of my fate. I was unexpectedly asked if I wanted to accept the professorship of physics at the University of Bonn, and become Plücker's[2] successor.[3] After hesitating for a long while, in the end I convinced myself that I would have a broader and better sphere of activity in Bonn[4] than here, and decided to go there. We will depart Würzburg next week, go to Berlin for a few days, and in about 14 days arrive in Bonn. It is not pleasant for us to leave Würzburg again[5] after such a short stay, but the call to Bonn did promise so many advantages that I could not well turn it down. After everything that we have heard and seen of Bonn, we hope that we will enjoy our life there.

I have not received any news from you[6] for a very long time and hope that you are doing well. When you write me again, which I hope will happen quite soon, you will already need to address your letter to Bonn (Colmantstrasse).

With best wishes from my wife[7] and me to you and Hirst

Your | Clausius.

RI MS JT/1/TYP/7/2314
LT Typescript Only

---

1. *Würzburg*: see letter 2968, n. 3.
2. *Plücker*: Julius Plücker (1801–68), a German mathematician and experimental physicist and professor of physics at the University of Bonn (1847–68). After focusing on analytical geometry, he turned his attention to electricity and magnetism. He and Tyndall clashed in the 1850s, when Tyndall challenged Plücker's ideas on diamagnetism and the behavior of crystals in magnetic fields, but they later resolved their differences. See R. Jackson, 'John Tyndall and the Early History of Diamagnetism', *Annals of Science*, 72 (2015), pp. 435–89.
3. *Plücker's successor*: Clausius followed Julius Plücker as Professor of Physics at Bonn. Plücker, who died nearly a year earlier, on 22 May 1868, had built up an experimental laboratory and practices, and this made the position attractive to Clausius.
4. *Bonn*: in 1869 Clausius assumed the position of professor of physics at the University of Bonn, in western Germany, after Helmholtz turned it down. See letter 2951, n. 6.
5. *leave Würzburg again*: Clausius was professor in Würzburg from 1867 to 1869, after working at the Swiss Institute of Technology in Zurich for over a decade.
6. *I have not received any news from you*: Clausius wrote to Tyndall in January (letter 2938),

but it seems Tyndall did not return the New Year's greetings. He responded to this letter rather more quickly (letter 2980).

7. *my wife*: Adelheid Clausius.

From Joseph Plateau 25 March 1869 2975

Gand, 25 mars 1869.

Mon cher Monsieur Tyndall

J'ai l'honneur de vous adresser des exemplaires de ma 9me et de ma 10me série, pour vous et pour l'Institution Royale. Quant à la Société Royale, puisqu'elle continue à m'oublier, malgré l'espoir que vous aviez bien voulu me donner, je ne vois pas pourquoi je me souviendrais d'elle; je ne lui envoie donc pas d'exemplaires.

J'ai pris connaissance, avec un bien grand intérêt d'une partie de vos belles expériences relatives à l'action de la lumière sur les vapeurs, à la couleur bleue du ciel, et à la polarisation de l'atmosphère; je ne doute pas qu'elles ne vous conduisent à l'explication complète de ces phénomènes mystérieux.

Agréez, mon cher Monsieur Tyndall, l'assurance de mes sentiments d'affection. | J. Plateau

Ghent, 25 March 1869.

My dear Mr Tyndall,

I have the honour of sending you copies[1] of my 9th and 10th series,[2] for you and for the Royal Institution. As for the Royal Society, since it continues to forget about me,[3] despite the hope that you had kindly given me,[4] I do not see why I should remember it. Thus, I am not sending it any copies.

I have read with great interest some of your wonderful experiments[5] regarding the action of light on vapours, the blue colour of the sky, and the polarization of the atmosphere. I have no doubt that they will lead you to a complete explanation of these mysterious phenomena.

Please accept, my dear Mr Tyndall, the assurance of my deepest affection. | J. Plateau

RI MS JT/1/P/103

1. *sending you copies*: enclosure missing.
2. *copies of my 9th and 10th series*: 'Recherches expérimentales et théoriques sur les figures d'équilibre d'une masse liquid sans pesanteur, 9e série' and 'Recherches expérimentales et théoriques sur les figures d'équilibre d'une masse liquid sans pesanteur, 10e série', both in *Mémoires de l'Académie royale des sciences, des lettres et des beaux-arts de Belgique,* 37 (Brussels: Hayez, 1869), pp. 1–56 (9th), 1–52 (10th). Note that this volume (37) starts with Plateau's 8th series (pp. 1–102).

3. *it continues to forget about me*: Plateau had been chafing at the perceived lack of recognition from the RS for a while (to be elected ForMemRS).
4. *the hope that you had kindly given me*: Tyndall was campaigning to have Plateau elected ForMemRS. The effort was successful, as Plateau was elected on 24 November 1870 with Tyndall among those who signed the certificate (RS, Certificates of Election and Candidature, EC/1870/18). See letters 2995, 3138, 3155, 3168, 3179, 3181, 3184, and 3192.
5. *your wonderful experiments . . .* : probably 'On the Blue Colour of the Sky' (see letter 2940, n. 3).

From Alexander Buchan[1] 29 March 1869 2976

Scottish Meteorological Society | General Post Office Building.[2]
Edinburgh 29 March 1869.

Dear Sir

Accept my very hearty thanks for the two Tracts on "The Blue Colour of the Sky &tc"[3] which I received two months ago.

You have had the good fortune to light on a very remunerative vein which gives every promise of enriching most of the more abstruse departments of Meteorology. The blue colour of the sky and the polarization of light you have gone very far to give a complete exposition of. You will probably also be able to shew the origin and nature of the substratum on which the manifestations of the aurora are displayed. An explanation of the surprising observed facts of solar radiation and terrestrial radiation will follow in course. Here in all probability is the explanation of those brief intervals of the most intense solar heat which occur in cloudy and stormy weather when the sun shines out between two clouds in a patch of the darkest blue sky;—and of the great heat felt in the sun while east winds prevail in a clear sky, but hazy near the horizon,—and of the rapidity with which solar radiation diminishes in such a sky as the sun descends and his rays traverse a thicker screen of this haze.

I send you a short Paper[4] I recently read before our Royal Society,[5]—it *[debates]* on one of those topics through which Meteorology has received great injury at the hands of Scientific men. I have just finished the 13 Isobaric charts which when published will supply to travellers the zero point, approximately, of the sea level pressure in the interim of continents. I intend to expand it a little and offer it to the Roy[al]. Geog[raphical]. Society[6]

I gave a reading of your two Papers[7] to our friend,—a energetic opponent,—Russell[8] the Meteorological writer in the Scotsman.[9] He was charmed with them and begged me to say how much gratified he should be if you could spare him copies of them.

Believe me | Very Truly Yours | Alexander Buchan

Professor John Tyndall, L.L.D, &tc

P.S. I am, of course very ignorant of the subject of polarization of the sky, which ignorance I endeavoured to proclaim in the last clause of the note on page 327.[10] | AB

RI MS JT/1/B/137
RI MS JT/1/TYP/1/164

---

1. *Alexander Buchan*: Alexander Buchan (1829–1907), an eminent Scottish meteorologist. He began his meteorological career as Secretary of the Scottish Meteorological Society in 1860 and subsequently studied at the University of Edinburgh. He edited and contributed to the Society's journal and was named Fellow of the Royal Society of Edinburgh in 1869. Buchan is best known for his work on the deviations from normally expected temperatures occurring during certain seasons, now called 'Buchan spells' (*ODNB*).
2. *Scottish Meteorological Society | General Post Office Building*: Scottish Meteorological Society headed notepaper.
3. *two Tracts on "The Blue Colour of the Sky &tc"*: see letters 2942, n. 1, and 2940, n. 3.
4. *a short Paper*: A. Buchan, 'Notes on the Determination of Heights, Chiefly in the Interior of Continents, from Observations of Atmospheric Pressure', *Edin. Roy. Soc. Proc* 6 (1869), pp. 465–72.
5. *our Royal Society*: Royal Society of Edinburgh.
6. *Roy[al]. Geog[raphical]. Society*: A. Buchan, 'Note on the Determination of Heights, Chiefly in the Interior of Continents, from Observations of Atmospheric Pressure', *Proceedings of the Royal Geographical Society of London*, 13:5 (1868–9), pp. 360–65.
7. *your two Papers*: see note 3.
8. *Russell*: possibly Robert Russell (d. 1871) of Kilwhiss, Scotland. Russell was an agriculturalist and Fellow of the Royal Society of Edinburgh and published and spoke on North American agriculture and meteorology in the 1850s.
9. *The Scotsman*: a Scottish newspaper. Founded in 1817 as a weekly newspaper, it began daily production in 1855 when the newspaper stamp tax was abolished.
10. *the note on page 327*: A. Buchan, *Handy Book of Meteorology* (Edinburgh and London: Blackwood and Sons, 1868), p. 327.

From Raoul Jagnaux[1] 29 March 1869 2977

Paris 29 Mars 1869

Monsieur,

Je suis avec avidité vos savants travaux de Physique—Je les recherche avec empressement dès qu'ils paraissent à Paris et je *[suis]* gré à M^r^ L'abbé Moigno de me mettre à même de les connaître par ses bonnes traductions. J'ai lu avec beaucoup d'intérêt votre leçon sur les rayons chimiques et la lumière du ciel, faite à l'Institution Royale, et dont la Revue Scientifique a donné une traduction.

Je vous demande la permission de vous rappeler une note que j'ai trouvée dans le Répertoire d'Optique de l'abbé Moigno (tome 3—page 1293) et qui viendrait ajouter une nouvelle preuve à votre théorie sur la couleur bleue du ciel. Cette note est de Mr Dupasquier. En voici le contenu:

«Quand la lumière traverse une feuille d'or battu, on sait qu'elle prend une couleur bleue très *[prononcée]*, phénomène considéré jusqu'à présent comme caractéristique à l'égard de ce métal; car on le remarque également lorsque l'or, précipité chimiquement dans un liquide, y reste quelque temps à l'état de suspension.

«L'observation accidentelle de quelques faits isolés, m'a d'abord conduit à reconnaître que ce phénomène n'est pas spécial à l'or; plus tard, l'expérimentation m'a donné la certitude qu'il est général aux corps opaques; qu'il se produit avec plus ou moins d'intensité, toutes les fois que la lumière traverse un métal réduit à l'état de feuilles extrêmement minces, ou un corps quelconque considéré comme opaque. Lorsque, dans un grand état de division, le corps se trouve retenu quelque temps à l'état de suspension, soit dans un liquide, soit dans une vapeur, soit dans un gaz incolore. J'ai reconnu également par de nombreuses expériences, que ce phénomène est indépendant de la nature du fluide où s'opère la précipitation, et qu'il se manifeste d'ailleurs, quelque soit la couleur du corps solide, se précipitant à l'état de particules extrêmement ténues. La transparence ou la translucidité même imparfaite des corpuscules, suspendus en grand nombre dans un fluide, m'a paru seule être un obstacle à la production de ce phénomène.»

Mr Dupasquier entre ici dans le détail de ses expériences dont voici un résumé.

Pour reconnaître la coloration bleue, il faut se placer dans un lieu un peu obscur et dans lequel la lumière diffuse arrive par une ouverture plus élevée que la tête de l'observateur: on place la feuille métallique entre l'œil et la lumière. On aperçoit alors une coloration plus ou moins bleue selon la malléabilité du métal. Si le phénomène est plus tranché avec l'or c'est que ce métal est très malléable et que l'on peut obtenir avec lui des feuilles très minces. Mais ce phénomène est général.

On peut encore opérer en délayant dans un liquide le corps sur lequel on veut opérer. Il doit être rendu en poudre très fine. Il faut attendre que les particules les plus *[grosses]* se soient précipitées au fond du vase, avant d'observer la coloration bleue—Quand le corps est très lourd, on augmente la viscosité du liquide (pour l'eau on se sert de gomme arabique, pour l'alcool ou l'éther, d'une matière résineuse incolore, ou d'une substance grasse.

Des précipités blancs présentent une coloration bleuâtre, mais généralement faible, des substances organiques incolores peuvent même présenter ce phénomène (oxamide).

«Quant à la cause du développement de la couleur bleue dans ces expériences, tient-elle à ce que les particules des corps dits opaques, très divisés, ne se laissent traverser que par les rayons bleus de la lumière ou à ce que ces mêmes rayons, par l'effet d'une réfraction, glissent seuls entre les particules tenues en suspension? C'est là une question qu'il appartient aux physiciens de résoudre. En étudiant ces intéressants phénomènes, ils auront à rechercher aussi si les observations que je viens de présenter ne peuvent pas expliquer certaines colorations bleues que nous présente la nature, celle des glaciers, par exemple, qui pourrait être due à leur état de granulation, et celle des lacs, qui pourrait peut-être avoir pour causes des particules hétérogènes très subtiles, tenues en suspension dans une masse d'eau d'une grande épaisseur, etc.»

Veuillez agréer mes excuses, Monsieur, et pardonner à un débutant dans la science, d'avoir osé déranger un maître pendant quelques instants.

Raoul Jagnaux | 70 Boulevard du Prince-Eugène | Paris

Paris 29 March 1869

Sir,

I closely follow your scientific works in Physics—I look for them eagerly as soon as they appear in Paris, and I am grateful to Father Moigno that he allowed me to become acquainted with them through his wonderful translations.[2] I read with much interest your lesson on chemical rays and the light of the sky[3] given at the Royal Institution and for which la Revue Scientifique gave a translation.[4]

I ask that you permit me to remind you of a note that I found in Father Moigno's Répertoire d'Optique (volume 3—page 1293),[5] which would add further evidence to your theory on the blue colour of the sky. This note is written by Mr Dupasquier.[6] Here are the contents:

"When light passes through a sheet of beaten gold, we know it takes on a very dark blue colour, it is a phenomenon that was considered up until now to be characteristic of this specific metal; because we see the same thing when gold, chemically precipitated in a liquid, stays there for some time in a state of suspension.

"The accidental observation of some isolated facts first led me to recognize that this phenomenon is not particular to gold. Later on, experimentation assured me that it is common to all opaque bodies, and that it occurs with more or less intensity whenever light passes through a metal that is reduced to very thin sheets, or a body that is considered opaque. When in a state of great division, the body is kept for some time in a state of suspension in a liquid, vapour, or a colourless gas. I also realized through numerous experiments that this phenomenon is independent of the nature of the liquid where the precipitation takes place; and that in fact it manifests, regardless of the colour of the solid body, moving quickly towards a state of extremely thinly distributed particles. The transparency or even the imperfect translucency of the particles, suspended in large numbers within a liquid, seemed to be only an obstacle to the production of this phenomenon."

Here Mr Dupasquier gives the details of his experiments, which are summarized below.

In order to recognize the blue colour, we must be in a somewhat dark place where the diffused light comes from an opening that is above the head of the observer. We place the metallic sheet between the eye and the light. Then we see a more or less blue colour depending on the malleability of the metal. If the phenomenon is more distinct with gold, it is because that metal is very malleable and we can make very thin sheets out of it. However, this phenomenon is universal

We can still proceed by diluting the body that we want to examine in a liquid. It has to be ground into very fine powder. We have to wait until the largest particles precipitate at the bottom of the container before we can observe the blue colour.—When the body is very heavy we increase the viscosity of the liquid (for water we use Arabic gum, for alcohol or ether, a colourless resinous material, or a *[fatty]* substance.

The white precipitates have a bluish colour, though generally faint; organic colourless substances can also show this phenomenon (oxamide).

"As for the cause of the colour blue developing in these experiments, is it because the particles of said opaque bodies, very separated, do not allow for any rays of light to pass except for the blue ones; or is it because only these rays, due to refraction, slip between the particles kept in suspension? This is a matter for physicists to solve. By studying these interesting phenomena, they will also have to see if the observations that I just presented could perhaps explain certain blue colourations that nature gives us; that of glaciers, for example, which could be due to their state of granulation, and that of lakes, which could be due to very subtle heterogeneous particles that are kept in suspension in a mass of water of great thickness, etc."

Please accept my apologies, Sir, and forgive a beginner in science to have dared to disturb a master for a few moments.

Raoul Jagnaux | 70 Boulevard du Prince-Eugène | Paris

RI MS JT/1/J/3

---

1. *Raoul Jagnaux*: Raoul Jagnaux (1845–1908), a French engineer and Professor of Chemistry at the *la maison d'éducation de la légion d'honneur* at St. Denis, a famous French secondary school set up by Napoleon for girls whose male relations had been awarded the Legion of Honour. He authored several textbooks on chemistry, as well as a two-volume history of chemistry that sought to highlight French achievement ('Dossier de Légion d'honneur', *Archives nationales*, LH/1348/33, at https://cths.fr/an/savant.php?id=116827 [Accessed 27 Jan 2020]); B. Bensaude-Vincent, 'Between History and Memory. Centennial and Bicentennial Images of Lavoisier', *Isis*, 87 (1996), pp. 481–99).
2. *His wonderful translations*: Moigno translated Tyndall's works on heat, light, and sound into French.
3. *your lesson on chemical rays and the light of the sky*: see letter 2942, n. 1.
4. *a translation*: J. Tyndall, 'Les rayons chimiques et la lumière du ciel', *Revue des cours scientifiques de la France et de l'étranger*, 6:16 (1869), pp. 242–50.

5. *Répertoire d'Optique*: A. Dupasquier, 'Mémoire sur la lumière bleue, transmise par une feuille d'or. Généralité de ce phénomène observé avec tous les corps opaques', reproduced in M. L'Abbé Moigno, *Répertoire d'optique moderne ou analyse complète des travaux modernes relatifs aux phénomènes de la lumière, Troisième partie* (Paris: A. Franck, 1850), pp. 1293–6.
6. *Mr Dupasquier*: Gaspard Alphonse Dupasquier (1793–1848), a French pharmacist who taught pharmacy, chemistry, and medicine in Lyon (E. Clett, *Alphonse Dupasquier. Notice sur sa vie et ses traveaux* (Lyon: Savy, 1849)).

From Louis Soret 31 March 1869 2978

Genève 31 Mars 1869

Mon cher Tyndall

En vous remerciant de votre lettre et de l'opuscule que vous m'avez envoyé, je viens vous communiquer une observation qui pourra peut être vous intéresser.

J'ai été conduit à propos d'une autre question, à me demander si l'absorption de certaines espèces de rayons par l'eau et si la couleur bleue sont dues au liquide lui-même ou à des particules solides en suspension. C'est là un point qui a été souvent discuté et dont vous vous êtes occupé vous-même dans votre ouvrage "The Glaciers of the Alps"—Vos dernières publications sur la polarisation de la lumière du ciel, m'ont suggéré l'idée que si la couleur bleue de l'eau est due à des particules en suspension, il doit se produire des phénomènes de polarisation analogues à ceux que l'on observe sur la lumières du ciel.

Les eaux du lac de Genève, dont la belle couleur est bien connue, doivent être très favorables à des recherches de ce genre.

J'ai donc fait construire dans ce but un appareil très simple. Il se compose d'une sorte de lunette, dans laquelle l'objectif est formé d'une simple lame de verre à faces parallèles qui ferme hermétiquement le *[tuyau]* de la lunette de manière à ce que l'on puisse en plonger l'extrémité dans l'eau sans que celle-ci pénètre dans l'intérieur. L'oculaire est formé d'un prisme de Nicol.—Il est facile de comprendre qu'en plongeant dans l'eau l'extrémité objective de la lunette, l'œil placé à l'oculaire reçoit la lumière bleue émise par l'eau, et qu'en faisant tourner le Nicol on peut l'analyser.

En opérant ainsi j'ai trouvé en effet *[que]* que la lumière de l'eau de notre lac présente des phénomènes de polarisation tout à fait comparables à ceux de la lumière du firmament. Seulement l'observation *[en]* est moins facile, et je n'ai pas jusqu'ici les étudier d'une manière aussi complète que je l'aurais voulu.

Supposons que la surface NN de l'eau soit absolument plane, c'est à dire qu'il y ait un calme parfait un rayon solaire suivant la direction SI prendra après sa réfraction une direction IR. Si donc, *[se]* plaçant dans un bateau, on dirige la lunette ⊥ ⊥, dans le plan vertical passant par le Soleil,

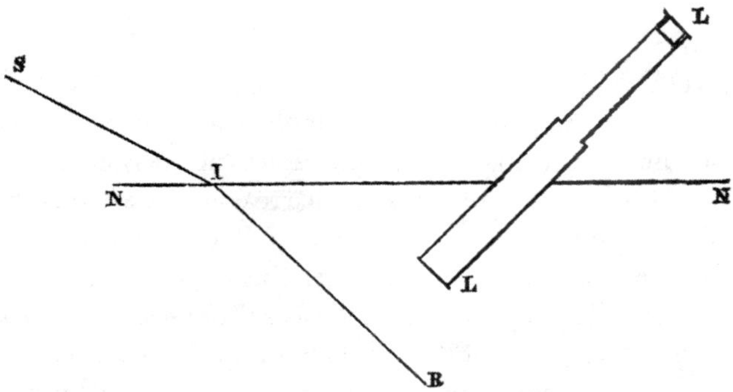

(soit dans le plan d'incidence) en l'inclinant de manière que son axe soit perpendiculaire à IR, la lumière que reçoit l'œil est émise perpendiculairement à la trace des rayons solaires dans l'eau. On est donc dans des conditions analogues à celles où l'on observe le maximum de polarisation de la lumière du ciel, c'est à dire quand on les regarde à angle droit du Soleil.

En observant de cette manière sur le lac de Genève et dans un endroit où la profondeur était suffisante pour que l'on ne pût pas voir la fond, j'ai reconnu une polarisation bien marquée; le plan de polarisation coïncidait avec le plan d'incidence.

Pour d'autres directions de la lunette, toujours dans le plan vertical passant par le Soleil on observe une polarisation d'autant moins prononcée que la direction de la lunette n'écarte plus de la perpendiculaire à IR. Ainsi *[en]* regardant verticalement de haut en bas je n'ai pas pu voir de traces de polarisation.

En plaçant la lunette dans un plan vertical à 90° du Soleil, la polarisation est d'autant plus *[nette]* que la direction de la lunette s'approche plus de l'horizontale, le plan de polarisation passe par la direction de la lunette et le Soleil.

Je n'ai pas déterminé jusqu'ici si le maximum de polarisation se manifeste exactement quand la lunette est dirigée à angle droit des rayons solaires.

Il ne vous échappera pas que les phénomènes sont ici plus complexes que pour la lumière du ciel.

D'abord il est évident que si la surface de l'eau est agitée, les rayons solaires après leur réfraction ne seront pas parallèles. Donc le phénomène sera d'autant moins prononcé que l'eau sois moins calme l'est en effet un peu *[près]* *[1 word illeg]:* la première fois que j'ai essayé mon appareil, le lac étant assez agité, n'ai pas vu la polarisation sensible. Ce qui *[devait]* peut être aussi à ce qui avait fait apercevoir la forte vente du Nord *[et] [pour]* l'eau n'était pas parfaite moins transparente. Dans deux autres jours, le lac étant bien peu agité, la polarisation était bien *[1 word illeg]*; enfin un jour de calme parfait *[1 word illeg]*

Geneva, March 31, 1869.

MY DEAR TYNDALL,

IN thanking you for your letter[1] and the pamphlet you sent me,[2] I take the opportunity of communicating to you an observation which may interest you.

Whilst dealing with a different subject, I was led to consider whether the blue colour and the absorption of certain rays of light by water are due to the liquid itself or to the therein suspended solid particles. This question has been often discussed by others, as well as by yourself in your work 'The Glaciers of the Alps.'[3] Your memoirs on the polarization of the blue light of the sky[4] have suggested to me the idea that if the blue colour of the water be due to suspended solid particles, phenomena of polarization will be produced analogous to those observed by you on the light of the sky.

The water of the Lake of Geneva, owing to the well-known beauty of its colour, is very favourable to investigations of this kind.

For this purpose I had a very simple apparatus constructed. It is a kind of telescope; a flat plate of glass with parallel surfaces, fitted hermetically at its one end, serves as *object-glass*. The instrument, therefore, can be immersed in water without the latter being able to enter it. The *eyepiece* consists of a Nicol's prism.[5] It is easy to understand that, by immersing the telescope in the water, the eye receives the blue rays of light emanating from the water, and that this light can be analyzed by turning the Nicol.

By proceeding in this manner, I found that the water of our lake really exhibits phenomena of polarization comparable to those observed on the light of the sky; only their observation is more difficult, and up to this time I have not been able to study them as well as I could have wished.

Supposing the surface of the water N N perfectly plane, which is the case in time of perfect

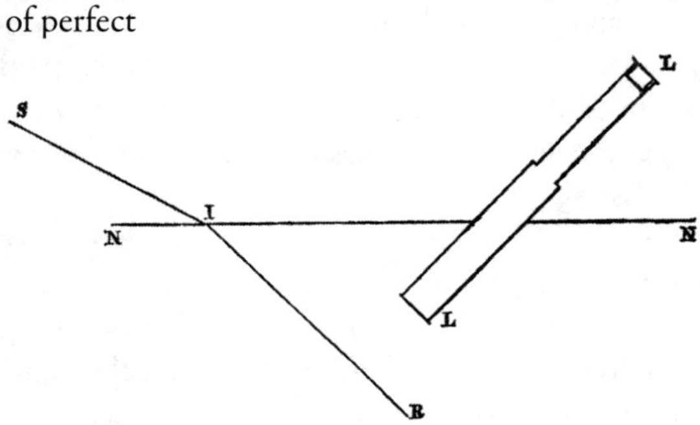

calm, a beam of solar light incident with the direction S I will be deviated to I R after the refraction. Now from a boat the telescope can always be placed in the vertical plane passing through the sun. If the telescope be inclined in such a manner that its axis becomes perpendicular to I R, then the light received by the eye is emitted perpendicularly to the direction of the solar rays in water. This arrangement is analogous to that where the maximum of polarization of the light of the sky appears—that is, when one is looking at a right angle from the sun.

In this manner I made a series of observations on the Lake of Geneva in a place where the water was sufficiently deep not to allow the ground to be seen; and I was able to perceive a marked polarization. The plane of polarization was coincident with the plane of incidence.

If the telescope, always in the same vertical plane passing through the sun, be inclined in any other direction, then the polarization is the less marked the more the direction of the telescope differs from the perpendicular to I R. If the direction of the telescope be vertical, so that one is looking downwards, no trace of polarization appears.

If the telescope be placed in a vertical plane perpendicular to the former one (that is, at 90° to the sun), the polarization is the more distinct the more the direction of the telescope becomes horizontal, and the plane of polarization passes through the direction of the telescope and the sun.

I have not been able as yet to prove that the maximum of polarization corresponds exactly to the position of the telescope where its axis is at a right angle to the direction of the solar rays.

It cannot escape your attention that the phenomena are more complicated in the case of water than in the case of the firmament.

In the first place, it is evident that, if the surface of the water is agitated, the solar rays cannot remain parallel after the refraction. The phenomenon, therefore, is the less distinct the more agitated the water is. This is exactly what I observed. When I tried my apparatus for the first time, the water was agitated, and I remarked no appreciable polarization—although this may be attributed as well to the fact that, for some days before, we had strong northerly winds and the water was not perfectly transparent. On two other days, when the water was only slightly agitated, an appreciable polarization could be observed. On a very fine and calm day at last,[6] the polarization was as marked as that of the sky, which, however, was not very blue at the time.

In the second place, the solar rays entering the water are already partially polarized by refraction; but when the telescope is placed in the vertical plane passing through the sun (that is, in the position most favourable for observation), it is easy to see that the rays already polarized must be extinguished (exactly as in the experiment where you produce a *blue cloud*[7] with a pencil of light already polarized).

Finally, the direct solar light is not the only light that enters the water. There are scattered rays of diffused light, which, impinging upon the water from all directions, produce after refraction a blue non-polarized light, or, properly speaking, an infinite number of rays polarized with different planes of polarization. I have satisfied myself that, when the sky is covered, no appreciable polarization is observed.

As I know of no prior description of this phenomenon, and as peculiar circumstances prevent for some time my pursuing this investigation, I communicate to you these results, although I regret that I was not able to finish the series of observations, and to repeat them with the use of artificial light.

Yours &c., | L. SORET.

'On the Colour of the Lake of Geneva', *Phil. Mag.*, 37 (1869), pp. 345–48[8]
St Bride Library, Authors' letters (French original)

---

1. *your letter*: letter missing.
2. *the pamphlet you sent me*: enclosure missing.
3. *"The Glaciers of the Alps"*: *Glaciers of the Alps*.
4. *Your memoirs on the polarization of the blue light of the sky*: see 'On the Blue Colour of the Sky' (letter 2940, n. 3); and 'Chemical Rays and the Light of the Sky' (letter 2942, n. 1).
5. *Nicol's prism*: see letter 2966, n. 4.
6. *at last*: the final page of the handwritten version is missing, and it ends here.
7. *the experiment where you produce a blue cloud*: see letter 2940, n. 3.
8. This transcription is based on the published version of Soret's letter, as the handwritten letter is incomplete.

From Emily Peel 2 April 1869 2979

Whitehall Gardens | April 2/69.

My dear Professor,

You have made me shudder & shiver by the perusal of your Alpine climbings[1]—& it is with horror that I think I may have been in any way responsible in urging you on to accomplish this marvellous feat. And now that you have done it, I hope you will be satisfied in future with tamer emotions & will take an occasional look at your friends in the Swiss Lowlands.[2]

I am very glad that we are to have the pleasure of seeing you at dinner on the 17th & I remain yours

very truly | Emily Peel

RI MS JT/1/P/49

1. *your Alpine climbings*: presumably 'Odds and Ends of Alpine Life' (see letter 2943, n. 3).
2. *your friends in the Swiss Lowlands*: Emily Peel lived with her daughters in Geneva, at Villa Lammermoor.

To Rudolf Clausius 11 April 1869 2980

11th April 1869

My dear Clausius

I heard with the greatest delight from Madame Helmholtz[1] that you had decided on coming to Bonn.[2] This brings you much nearer to me: and I doubt not that you and I will be able at times to make an expedition together among the Siebengeberge[3] or along

"Den Schönen Deutschen Rhein"[4]

It is a very great pleasure to me to have you at Bonn. Hirst[5] is also greatly gratified.

I have not I believe written to you since you sent me the photographs[6] which were excellent.

I am now at work at my lectures here, and am therefore obliged to give up my investigations for the present. After I little time I hope to be able to put what I have been doing in a connected form and to publish all in the Philosophical Transactions.[7]

I have been struck exceedingly by the deportment of some of these actinic clouds that I have been generating. It is impossible to examine them without being reminded of the texture of a Comet. I was asked to go down to Cambridge some time ago to make a communication to the Philosophical Society,[8] and as it is now some years since they elected me a member, and as I had never before then given them a communication I agreed to go. I send you an extract from the lecture then given.[9] Adams[10] the astronomer, and Stokes, Miller,[11] and others were present, but they did not overthrow the theory. In fact Miller told me afterwards that he found it impossible to pick a hole in it. I might show that it explains more than I have mentioned. Right or wrong it is very curious.

As a matter of fact I can dissipate actinic clouds by calorific rays, and augment their precipitation by the introduction of a layer of water which is transparent to the short waves & opaque to the long ones.

With kindest regards to your wife[12] dear Clausius I am
yours ever since[re]ly | John Tyndall

RI MS JT/1/T/203

1. *I heard . . . from Madame Helmholtz*: Anna Helmholtz. See letter 2968.
2. *coming to Bonn*: see letter 2974, n. 4.
3. *the Siebengeberge*: 'Seven Hills' is a hill range southeast of Bonn, overlooking the Rhine.
4. *Den Schönen Deutschen Rhein*: 'The beautiful German Rhine' from 'Der freie, deutsche Rhein' (1869) by German-American author Ernst Anton Joseph Zündt (1819–97). This poem was an adaptation of a song of the same name from Nikolaus Becker (1809–45), written in 1840.
5. *Hirst*: Thomas Hirst.
6. *you sent me the photographs*: see letter 2938.
7. *to publish all in the Philosophical Transactions*: the next work Tyndall published on this topic in the *Phil. Trans.* was 'On the Action of Rays of High Refrangibility upon Gaseous Matter', *Phil. Trans.*, 160 (1870), pp. 333–65.
8. *Philosophical Society*: Cambridge Philosophical Society. See letter 2966, n. 2.
9. *extract from the lecture then given*: see letter 2966.
10. *Adams the astronomer*: John Couch Adams (see letter 2966, n. 10).
11. *Miller*: William Allen Miller.
12. *your wife*: Adelheid Clausius.

From Mary Francis Egerton 12 April [1869][1] 2981

45 Eaton Place, | S. W. | April 12

My dear Mr. Tyndall,

You are good & true—& faithful! I scarcely thought you would remember the lecture[2] and was considering whether I should ever dare remind you of it after all the kindness you have done me. Very many thanks—<u>How</u> beautiful! "So beautiful that it <u>must</u> be true" may well be said of this! In fact it comes to one, more as a demonstrated fact than as a speculation; once stated one can hardly believe it can be otherwise. I hardly know how to express it but your comet's tail itself, seems to represent one of those "inspirations" as you truly call them, which, flashing forth, like the sun's rays, on pre-existing materials—draws them out into light & form, & beauty!

Who would antecedently have guessed the connection between your experiments on radiation through gases,[3] & the Cometary theory! and yet how the wonderful unity of nature leads, as you say, by "an easy gradient" from the one to the other.

But what a marvelous notion it gives one of the infinite <u>stillness</u> of space—that ever rippling ocean of ether—through which mass of mere vapour millions of miles in diameter can yet hold its course undissipated & undeflected, unless when the influences of a more solid body intervenes! But I must not go on—you do not want such thoughts as mine, only I considered your

<pages missing>

theory! I was surprised the other day, to see in his lectures how nearly he[4] had forestalled Faraday as to the influence of magnetism on polarization!

Y[ou]rs most truly | MF Egerton

I hope you are feeling fresher from your trip. I hope you did not think that P.S. was intended to be *[insensitive]*

RI MS JT/1/E/28

---

1. *[1869]*: the year is given by reference to a Tyndall lecture on comets. Tyndall delivered a lecture at the Cambridge Philosophical Society, 'On Cometary Theory', on 8 March 1869 (*Ascent of John Tyndall*, p. 249). It was published in the *Phil. Mag.* in April. See letter 2966.
2. *the lecture*: 'On Cometary Theory' (see n. 1).
3. *your experiments on radiation through gases*: published in J. Tyndall, 'On a New Series of Chemical Reactions Produced by Light', *Roy. Soc. Proc.*, 17 (1868), pp. 92–102; J. Tyndall, 'The Generation of Clouds' (see letter 2961); and later J. Tyndall, 'On the Action of Rays of High Refrangibility upon Gaseous Matter', *Phil. Trans.*, 160 (1870), pp. 333–65. See also letters 2940, 2942, and 2980.
4. *he*: not identified.

From Johann Conrad Bohn[1] 14 April 1869 2982

Hochverehrter Herr Tyndall!

Bald nachdem gestern mein Brief an sie abgegangen war, bemerkte ich, dass Sie allerdings für eine grössere Zahl von Flüssigkeiten und nicht nur für Wasser, die Absorption der Wärmestrahlen für verschiedene Dicken gegeben haben. Den ersten Theil Ihrer Abhandlung, in welchem diese Daten sich finden, habe ich schon vor längerer Zeit für den Jahresbericht bearbeitet, daher mir das Vorhandensein der in Rede stehenden Tabelle momentan aus dem Gedächtnis entschwunden war. Ich bitte Sie hierfür um Entschuldigung.

Einige der fraglichen Versuche habe ich soeben berechnet, allein zu meiner Ueberraschung gefunden, dass die Differenz der Logarithmen zweier durchgelassenen Wärmemengen, dividirt durch die Differenz der Dicken der entsprechenden Schichten keineswegs constante Werthe bei der Combination der verschiedenen Versuche gibt.

Ich kann nur diess nicht erklären; die Reflexion, welche ich zunächst nicht berücksichtigt hatte, hat keinen störenden Einfluss auf die Constanz jenes Quotienten; es muss also ein anderer, mir unbekannter Grund für die bemerkte Abweichung vorhanden sein.

Indem ich Sie nochmals bitte meine Leichtfertigkeit von gestern freundlichst entschuldigen zu wollen, verbleibe ich mit der ausgezeichnetesten Hochachtung

Ihr ergebenster | Dr CBohn

Giessen d. 14 April 1869.

Dear Mr Tyndall!

Soon after dispatching my letter[2] to you yesterday, I noticed that you had indeed noted the absorption of heat rays in a large number of fluids of different thickness, and not only for water. I had already worked on the first part of your memoir,[3] which contains these data, some time ago for the *Jahresbericht*,[4] which is why the existence of the mentioned table momentarily slipped my mind. I ask your forgiveness for this.

I have just calculated some of the experiments in question and found to my surprise that the difference of the logarithms of two amounts of conducted heat, when divided by the difference of the thicknesses of the corresponding layers, does not yield constant values when different experiments are combined.

I cannot explain this; reflection, which I initially did not consider, has no disturbing effect on the constant value of said quotient; therefore, there must be a different reason for the observed deviation that is unknown to me.

I ask you once again to kindly excuse my thoughtlessness from yesterday and remain with the utmost respect

Your most devoted | Dr CBohn

Giessen 14 April 1869.

RI MS JT/1/B/117

---

1. *Conrad Bohn*: Johann Conrad Bohn (1831–97), a German professor of physics. He studied mathematics in Marburg in 1850. He became professor at the University of Giessen in 1860, before moving to the Bavarian Central College for Forestry in Aschaffenberg in 1866. Bohn published on heat and light ray absorption in 1866 and on negative floresence and phosphorescence in 1868 (*NDB*).
2. *my letter*: letter missing.
3. *your article*: J. Tyndall, 'The Bakerian Lecture. Contributions to Molecular Physics. Being the Fifth Memoir of Researches on Heat', *Phil. Mag.*, 28 (1864), pp. 438–48.
4. *I had already . . . for the Jahresbericht*: Bohn wrote the physics section of *Jahresbericht über die Fortschritte der Chemie, Physik, Mineralogie und Geologie*. The annual report for 1864 (pub. 1865) includes a discussion of Tyndall's Bakerian Lecture cited from the *Phil. Mag.* (see n. 3). See *Jahresberichte über die Fortschritte der Chemie für 1864* (1865), pp. 24–32.

To Mr. Wright[1] [18 April 1869][2] 2983

Royal Institution of Great Britain

To Mr. Wright,

The methylated spirit may be used without fear of damage

John Tyndall

Experience must test whether it is equally good for cleaning purposes. It probably is.

RI MS DPM/1/F1

---

1. *Mr. Wright*: not identified, but possibly Richard Pears Wright (1829–92), an English mathematician and educator who taught engineering and surveying at Queenwood College from 1851. According to Hirst, who encouraged his development as a geometer, he had epilepsy. Wright left Queenwood in 1863. From 1870 until his death in 1892 he was a teacher of mathematics at University College School, London. Tyndall was friendly with Wright and his family. However, there are no other letters between Tyndall and Wright in the correspondence and the mode of greeting is awkwardly formal for two people who knew one another very well.
2. *[18 April 1869]*: the date had been added to the original letter by a second hand.

To Robin Allen 20 April 1869 2984

Enclosure 5.
Royal Institution, 20 April 1869.

Sir,

SUBSEQUENT to my report,[1] dated the 19th March, I wrote to the Deputy Master,[2] urging the desirability of repeating the observations on the comparative powers of red and white lights from the temporary observatory at Charlton.[3]

The lights and reflectors being duly fixed, under the superintendence of Mr. Douglass, in the experimental building of the Corporation at Blackwall, I accompanied a party of observation, which comprised the Deputy Master, to Charlton, and viewed the lights under favourable circumstances. From observations with the telescope, we assured ourselves of the proper adjustment of the reflectors; the atmosphere was also in good condition.

There was a slight diversity of judgment as to the comparative merit of the two lights; but we finally came to the conclusion that, employing red shades of the depth adopted by the Trinity House, the reddened beam of the Wolf

Rock Lighthouse[4] should, on starting, possess two and three-quarters times the intensity of the white one.

My last Report[5] made the proportion of red to white that of 10 to 4. This Report makes it 11 to 4, but by these proportions I trust we may safely and finally abide.

I have, &c. | (signed) *John Tyndall.*

Robin Allen, Esq.

'Further Report from Corporation of Trinity House on Magneto-Electric Light', HC 250, Parliamentary Papers (1869), p. 6

1. *my report*: see letter 2969. For Tyndall's work on lighthouses, see letter 2937, n. 6.
2. *Deputy Master*: the Deputy Master of Trinity House at this time was Frederick Arrow (1818–75), an English merchant navy commander (retired in 1853). He was elected an Elder Brother of Trinity House on 15 February 1859, and became Deputy Master in 1865 ('Obituary of Eminent Person', *Illustrated London News*, 24 July 1875, p. 95).
3. *Charlton*: see letter 2937, n. 7.
4. *Wolf Rock Lighthouse*: see letter 2969, n. 1.
5. *My last Report*: see letter 2969.

To Thomas Henry Huxley 22 April 1869 2985

Royal Institution. | 22nd. April, 1869.

My dear Huxley,

I have sent the card[1] to Mrs Huxley[2] herself.

I was wakeful on Wednesday evening and I knew the meeting[3] would make me more so. Hirst[4] took me to a theatre for an hour and I was able to do my work in a satisfactory manner yesterday.

The fact is I enjoy already as much liberty as I care to use, and I do not care one straw how the theologians use theirs. They are powerless and I am free, that is quite enough,

It is curious how one's feelings vary towards them: Some of them I could hew to pieces before the Lord in Gilgal.[5] but then others are so gentle, and noble; and their gentleness and nobleness are so intertwined with their theological wrappings. The problem of the future will be to detach the one from the other without killing the good.

Yours ever | John Tyndall.

RI MS JT/1/TYP/9/2925
IC HP 8.70
Typed Transcript Only

1. *the card*: letter missing.
2. *Mrs. Huxley*: Henrietta Huxley.
3. *the meeting*: the first meeting of the Metaphysical Society was held on Wednesday, 21 April 1869. The Society aimed to provide a space in which important questions of faith and science might be debated. Members included men of science such as Huxley and Tyndall, but also the Catholic Archbishop of Westminster, Henry Edward Manning, and other elites, including Sir James Fitzjames Stephen, Walter Bageshot, and Alfred Tennyson. On the Metaphysical Society, see B. Lightman, 'Science at the Metaphysical Society: Defining Knowledge in the 1870s', in B. Lightman and M. Reidy (eds), *The Age of Scientific Naturalism: Tyndall and His Contemporaries* (Pittsburgh: University of Pittsburgh Press, 2014), pp. 187–206; and C. Marshall, B. Lightman, and R. England (eds), *The Metaphysical Society (1869–1880): Intellectual Life in Mid-Victorian England* (Oxford: Oxford University Press, 2019).
4. *Hirst*: Thomas Hirst.
5. *Gilgal*: a place where the Israelites camped after they crossed the River Jordan (Joshua 4:19 NRSV).

From Friedrich Schotte[1] 23 April 1869 2986

Berlin 23 April 1869 | 32 Köpenicker Str.

Geehrter Herr.

Indem ich mir erlaube, Ihnen die ersten drei Hefte des von <u>mir herausgegebenen Repertoriums</u> zu übersenden, bitte ich Sie dieselben als schwaches Zeichen meiner Verehrung hochgeneigtest annehmen zu wollen. Im Falle Sie meine Bitte, wie ich hoffe, nicht ablehnen, werde ich nicht verfehlen, Ihnen die Fortsetzungen zur Zeit des Erscheinens pünktlich zuzusenden.

Genehmigen Sie, geehrter Herr, die Versicherung der vorzüglichsten Hochachtung mit der ich mich unterzeichne als

Ihr | ganz ergebenster | F. Schotte

Berlin 23 April 1869 | 32 Köpenicker Str.

Dear Sir.

I allow myself to send you the first three issues of the <u>repertorium that I edit</u>[2] and ask you hereby to accept them, with all respect, as a humble sign of my admiration. If, as I hope, you will not reject my request, I will not fail to punctually send you the forthcoming issues as soon as they are published.

Please accept, dear Sir, the expression of my highest respect, with which I sign

Your | most humble | F. Schotte

RI MS JT/1/S/56

1. *Friederich Schotte*: Friederich Schotte (1832–1912), a German engineer and editor. He served as librarian at the Königlichen Gewerbe-Akademie in Berlin (*NDB*). See the title page of his *Repertorium,* in n. 2.
2. *repertorium that I edit*: F. Schotte (ed), *Repertorium der technischen, mathematischen und naturwissenschaftlichen* (Leipzig: Quandt and Händel, 1869).

From George Hector Tyndale 1 May 1869 2987

Philadelphia May 1. 1869

My dear John

At his request I have taken the liberty of giving a letter to you introducing my friend Mr Julius E. Hilgard of Washington. Mr Hilgard is the Chief Superintendent in charge of the Office of the Coast Survey—United States. He is a gentleman of large scientific culture and stands high in the esteem of our best men—all over this country. He is of German birth 'though he has lived here since very early childhood. His acquaintance will I think be agreeable to you and his position here may make a valued correspondent and friend for you in him. I shall be much obliged to you for any attention or kindness you may show him for he is worthy of it. My eldest Brother Sharon,[1] married his Sister—she is now dead. We have known each other since boyhood. He sails for England next week I think.

Last Winter I received a short note from you[2] in reply to one I wrote you[3] just before—in which I spoke of an allusion you had made to a visit to this country and asked you to stay at my house so long as you might like should you come. As you do not refer to it in your last[4] I suppose you overlooked this so I renew it now. We, Mrs Tyndale[5] and myself, <u>will give you a hearty and glad welcome with plain comforts</u>. Still your stay will not be an unpleasant one to you and, should you come, it would not be unprofitable to you to have a central "point of departure" and friends who care for you <u>as friends</u>.

Still understand this as meaning to be entirely at <u>your</u> pleasure for while your coming will be of great cheer to us, we must not be thought to tie you up in any way or that there is any feeling of <u>obligation</u> in the matter at all. In other words come if you <u>wish</u> to and we shall be very glad of your coming.

All my folks here are well with no changes in our family. How is your Sister?[6] Does she still live with Dr: Tyndall?[7] Please make my kindest respects to her. Should I ever "cross the water" again I shall visit Ireland with some longer stay and will hunt up the Tyndale (-ll) settlement[8] therein.

How is Hirst[9]? What a nice thing t'would be if he would come over with you. Why not? Put the question to him and try it. It would be very jolly to have you both here—by George it warms me up to think of it.

No man would be more welcome in my house than Tom Hirst, and there is plenty of room for both of you i.e. a room a piece. Do come. How is Debus[10] and where?

By the way it was only yesterday in the rooms of the Am[erican] Phil[osophical] So[ciety] to which they did me the honor last Autumn of elect[in]'g me a Member, that I read your admirable suggestions as to Cometary matter.[11] I think them grand. Profs Frazer[12] & Lesley[13] spoke to me of your experiments upon gases with light—but I have not yet seen the publication. I came across your suggestions in Taylor & Francis' Journal,[14] for April I think.[15]

I have, also, read your "Odds and Ends of Alpine Life".[16] I sincerely trust that you have made an end of your Alpine trips—at least the Matterhorn[17] and such like peaks. Why I could get up a vast dizziness only from reading of the thing. Nevertheless it is pleasant reading. With regards to Tom Hirst and to Debus, and with regards of Mrs Tyndale[18] and of Mr. Mitchell[19]

I am dear John affectionately your Cousin | Hector Tyndale

Did I tell you in my last, that during the election canvass of last year, President Johnson,[20] though I opposed his administration appointed me, and sent commission as Brevet Major General of U.S. vols.

Quite a compliment and unsolicited or thought of by me.

A Major General here, I believe, ranks with a Lieutenant Gen[era]l in most if not all the Countries of Europe.

RI MS JT/1/T/61
RI MS JT/1/TYP/5/1678–9

---

1. *eldest Brother Sharon*: Sharon Tyndale (1816–71), a brother of Hector, a civil engineer and surveyor, and the Secretary of State of Illinois from 1865–9. He was robbed and murdered outside his home in Springfield, Illinois, in the early morning hours of 29 April 1871. His murder was never solved (N. Bateman and P. Selby, *Historical Encyclopedia of Illinois and History of Tazewell County*, 2 vols (Chicago: Munsell, 1905), 2 vols., p. 532).
2. *a short note from you*: letter missing.
3. *one I wrote you*: letter missing.
4. *it in your last*: letter missing.
5. *Mrs. Tyndale*: Julia Tyndale (née Nowlen, 1823–97) of Philadelphia. The two were married in August 1842.
6. *your sister*: Emily Tyndall.
7. *Does she still live with Dr. Tyndall*: after his mother's death in 1867, Tyndall lodged Emily, who had lived with her mother but suffered under mental illness, under the care of Dr. John Tyndall, a relative in Gorey, County Wexford, Ireland.
8. *hunt up the Tyndale (-ll) settlement:* Hector and Tyndall shared an interest in genealogy. See letter 1956, *Tyndall Correspondence*, vol. 8.
9. *Hirst*: Thomas Hirst.
10. *Debus*: Heinrich Debus.

11. *Cometary matter*: Tyndall's lecture, 'On Cometary Theory' (see letter 2966).
12. *Frazer*: John Fries Frazer (1812–72), an American chemist and educator at the University of Pennsylvania. He took part in the Geological Survey of Pennsylvania in the 1830s but he is remembered for his work in evaluating and publicizing the work of other scientists from the United States and Europe (*ANB*).
13. *Lesley*: Joseph Peter Lesley (1819–1903), an American geologist and topographer. Born in Philadelphia, Lesley graduated from the University of Pennsylvania in 1838. He joined the first state Geological Survey of Pennsylvania and was director of the second. Throughout his tenure in the American Philosophical Society, he held several executive positions including that of vice president (1887–98) (*ANB*; *CDSB*).
14. *Taylor & Francis' Journal*: the *Phil. Mag.*
15. *your suggestions*: see n. 11.
16. *'Odds and ends of Alpine life'*: see letter 2943, n. 3.
17. *Matterhorn*: the Matterhorn is one of the highest peaks in Switzerland, located on the Swiss-Italian border between the Aosta Valley and the canton of Valais. Tyndall made the seventh ascent and first traverse of the Matterhorn, from the Italian side to Switzerland, from 26–8 July 1868 (*Hours of Exercise in the Alps*,) pp. 273–96; letters 2856, 2869, and 2905, *Tyndall Correspondence*, vol. 10).
18. *Mrs. Tyndale*: see n. 5.
19. *Mr. Mitchell*: Edward P. Mitchell (1812–80), a business partner and brother-in-law of Hector Tyndale. Mitchell married Elizabeth Tyndale (*c.* 1812–66) and joined Hector as a partner in the china and glassware firm established by Elizabeth and Hector's father Robinson Tyndale (1775–1842).
20. *President Johnson*: Andrew Johnson (1808–75), the 17th president of the United States. Johnson assumed the presidency after the assassination of Abraham Lincoln. Johnson's Reconstruction politics were highly controversial. He was impeached in 1868 but acquitted, although some senators may have been bribed for their vote (*ANB*).

To Thomas Henry Huxley [3 May 1869][1] 2988

Royal Institution. | Monday.

My dear Huxley,

You will, I dare say, be glad to hear that the Managers elected Foster[2] to day,[3] firstly by a very large majority, and secondly by a unanimous vote.

Yours ever | John Tyndall.

RI MS JT/1/TYP/9/2925
Imperial College Huxley Collection 8.71
Typed Transcript Only

1. *[3 May 1869]*: the date is given since Michael Foster was elected Fullerian Professor of Physiology at the RI General Monthly Meeting held on 3 May 1869 (*Roy. Inst. Proc.*, 5 (1866–69), p. 549).
2. *Foster*: Michael Foster (1836–1907), a British physiologist and politician. Foster studied medicine at University College, London before working for several years as a ship's surgeon and in private practice. He began teaching medicine and physiology in 1867 and worked as a demonstrator for Huxley. In 1869 he became Huxley's successor as Fullerian Professor of Physiology at the RI. He was named Professor of Practical Physiology at Trinity College, Cambridge in 1870, where his evolution-based research program came to be known as the Cambridge School of Physiology. Foster was instrumental in establishing physiology as a profession in Britain (*ODNB*).
3. *the Managers elected Forster to-day*: as Fullerian Professor of Physiology at the RI.

To Robin Allen 4 May 1869 2989

4 May 1869.

Sir,

WITH reference to the experiments[1] made at Blackwall,[2] and witnessed from Charlton[3] by the Deputy Master,[4] Captains Close,[5] Nisbet,[6] Weller,[7] and Lambert,[8] Mr. Douglass, and myself, we are unanimously of the opinion that the electric light,[9] as exhibited by the third order apparatus, was decidedly superior to the same light exhibited by an apparatus of the sixth order.

I have, &c. | (signed) *John Tyndall.*

Robin Allen, Esq.

'Further Report from Corporation of Trinity House on Magneto-Electric Light', HC 250, Parliamentary Papers (1869), p. 7

1. *the experiments*: there was an experimental lighthouse on the Trinity Wharf at Blackwall that could be viewed from Charlton across the Thames. See letters 2937 and 2984. For Tyndall's work on lighthouses, see letter 2937, n. 6.
2. *Blackwall*: see letter 2937, n. 1.
3. *Charlton*: see letter 2937, n. 7.
4. *Deputy Master*: Frederick Arrow (see letter 2984, n. 2).
5. *Captain Close*: Captain M. C. Close served in the merchant navy until his retirement in May 1853. He was elected an Elder Brother of Trinity House on 1 January 1856. Close had served as one of the British Commissioners to the Paris Universal Exposition in 1867 producing a Report on Lighthouses and Coast Illuminators (see *Reports on the Paris Universal Exhibition, 1867* (London: G. E. Eyre and William Spottiswoode, 1868), vol. 4, pp. 665–98).

6. *Nisbet*: Edward Nisbet (see letter 2937, n. 3).
7. *Weller*: Charles Weller (see letter 2937, n. 4).
8. *Lambert*: Captain G. P. Lambert served in the merchant navy until his retirement in January 1862. He was elected an Elder Brother of Trinity House on 22 December 1863.
9. *electric light*: Tyndall was assessing the use of magneto-electric light in lighthouses. See letters 2937, 2969, and 2984. For Tyndall's work on lighthouses more generally, see letter 2937, n. 6.

To John Murray[1] 4 May [1869][2] 2990

4th May[3]

Dear Sir,

I called yesterday for a copy of the Glaciers of the Alps[4] & was confirmed that it is out of print. This seems so unlikely as it went off so slowly that I write to ask you whether it is really the case.[5]

Yours very truly | John Tyndall[6]

National Library of Scotland, John Murray Archive, Ms. 41214

1. *John Murray*: John Murray (1808–92), an English publisher and the eldest child of Anne Elliot and publisher John Murray. Murray began his career in the family firm at 50 Albemarle Street, London, in 1828 and became head of the firm after his father's death in 1843. Murray published the *Quarterly Review*, a popular series of *Handbooks for Travellers*, and numerous scientific publications. On 24 November 1859 Murray published both Darwin's *On the Origin of Species* and Samuel Smiles's *Self-Help*. He became a member of the Alpine Club in 1858 and in 1860 published Tyndall's *Glaciers of the Alps* (*ODNB*).
2. *[1869]*: the year of this letter has been added in a second hand to the verso side of the letter.
3. *4th May*: repeated on the verso side of the letter, probably for Murray's record-keeping: '1869. May 4. Tyndall. Prof.'.
4. *Glaciers of the Alps*: *Glaciers of the Alps*.
5. *really the case*: Tyndall was dissatisfied with the way Murray managed his publications and reimbursement. He published his other books with Longmans. See letters 2866, 2868, and 2872, *Tyndall Correspondence*, vol. 10.
6. *John Tyndall*: written in pencil in another hand at the bottom of the page: 'June 30. 1868 | £22–9–11 Sept 4 | & a cheque for £10–10–0'.

To Thomas Archer Hirst 10 May 1869 2991

Royal Institution | 10th. May. /69

Dear Tom

Let us go to Bushy Park[1] to see the chestnuts in bloom next Sunday.[2] We both need a little freshening—I especially for this worthless head of mine is very very bad.

I have been trying to write since half past 4 this morning; but I am good for nothing.

Yours affectionately | John

RI MS JT/1/T/674
RI MS JT/1/HTYP/549

1. *Bushy Park*: Bushy Park is the second largest of London's Royal Parks. Linked to Hampton Court Palace via the Longford River, it was and remains home to an annual celebration of Chestnut blossoms.
2. *next Sunday*: on Saturday, 15 May, Hirst and Tyndall took the train to Canterbury where they dined with the Egertons (*Hirst's Journal*, RI MS JT/2/32d/846). The following day they walked to Ramsgate (see letter 2992).

To Thomas Archer Hirst 11 May 1869 2992

Royal Institution | Tuesday May 11th. /69.

My dear Tom,
I am ready for *[1 word illeg]* if you prefer it.[1]

But Bushy Park[2] will be at its maximum splendour next Sunday.

I shall dine at the Athenaeum[3] to day—but do not hope to be there by 7. I am going down to witness some experiments at South Kensington[4]

Yours affect[tionately] | John.

RI MS JT/1/T/674
RI MS JT/1/HTYP/549

1. *if you prefer it*: see letter 2991.
2. *Bushy Park*: see letter 2991, n. 1.
3. *Athenaeum*: see letter 2944, n. 1.

4. *South Kensington*: the South Kensington Museum was founded in 1857. It included facilities for teaching science to science teachers. Tyndall taught there and was responsible for assessing examinations.

From George Biddell Airy 15 May 1869 2993

Royal Observatory, Greenwich, | London, S.E.[1]
1869 May 15

My dear Sir,

I ought to have acknowledged long since your little paper[2] in which you treat (among other things) of the tails of comets. I think your hit excellent, and I think that it tends much to support your general theory. I have never doubted that the tail of a comet is an effect which the nucleus produces on something pervading space (as the movement of the tail defies all mechanical laws), but I never could guess the nature of the effect. You have given us something worth considering.

En revanche,[3] I send you a copy of Syllabus of Magnetic Lectures[4] which I have lately given at Cambridge. I am perhaps the prime mover there in the direction of physical science: and, to support my words with deeds, first I wrote my Sound,[5] secondly I gave these Lectures. The preparation was laborious, as, in want of books, I had to write out enormous sheets of mathematics (called "cards" in the Syllabus), which my men copied with great avidity. I made the small disturbances of small magnets visible by a magic lantern constructed to throw images on the ceiling of the room. I am,

my dear Sir | Yours very truly | G B Airy | Professor Tyndall

RI MS JT/1/A/49
RI MS JT/1/TYP/1/38

1. *Royal Observatory, Greenwich, | London, S.E.*: Royal Observatory headed notepaper.
2. *Your little paper*: 'On Cometary Theory' (see letter 2966).
3. *En revanche*: on the other hand (French).
4. *Syllabus of Magnetic Lectures*: G. B. Airy, *Syllabus of Lectures on Magnetism* (London: Macmillan and Co., 1869).
5. *my Sound*: G. B. Airy, *On Sound and Atmospheric Vibrations, with the Mathematical Elements of Music* (London: Macmillan and Co., 1868).

To Leonhard Schmitz[1] 15 May [1869][2] 2994

15th May

Dear Dr Schmitz

A nephew of Dr Stegmann,[3] Professor of Mathematics in the University of Marburg has been very strongly recommended to me by his Uncle. He is much in want of employment: Could you find him any thing to do in the way of teaching?[4]

Yours very truly | John Tyndall

The Waller Manuscript Collection, Uppsala University, gb-01757

---

1. *Leonhard Schmitz*: Leonhard Schmitz (1807–90), a classics scholar. Born in Eupen, Prussia (now Belgium), Schmitz studied at the University of Bonn, before marrying an Englishwoman and moving to Yorkshire. He worked as a tutor, translator of German classics scholarship, and editor of *Classical Museum: a Journal of Philology and of Ancient History and Literature* in the 1840s. From 1845 to 1867 he taught classics at the Royal High School in Edinburgh, before becoming the first headmaster of the International School at Isleworth and an examiner at the University of London. Tyndall would have known him through both the International School, where he sat on the board, and the University of London examinations ('Dr. Schmitz', in E. Burke (ed), *The Annual Register: A Review of Public Events at Home and Abroad for the Year 1890* (London: Longmans, Green, and Co, 1891), pp. 158–9).
2. *[1869]*: although the year of this letter is not given, it must be either 1868 or 1869. These limits are based on the opening date of the International School in June 1867 and the fate of Stegmann's nephew in August 1869, as described in letters 3027, 3051, and 3054.
3. *A nephew of Dr Stegmann*: the name of the nephew is Scheffer (*Hirst's Journal*, RI MS JT/2/32d/1847). Little more is know about him.
4. *in the way of teaching*: Tyndall is presumably inquiring whether Schmitz might employ the nephew as an instructor at the International School.

From Joseph Plateau 18 May 1869 2995

Gand, 18 Mai 1869.

Mon cher Monsieur Tyndall

Mon ouvrage est enfin terminé, et j'ai l'honneur de vous adresser, pour vous et pour l'Institution Royale, des exemplaires de ma onzième et dernière série. Quant à la Société Royale, je viens de voir les nominations de membres étrangers qu'elle a faites, et je n'y suis pas compris; mais je ne veux pas *[bouder]*, et je joins mes exemplaires ci-dessus aux des 9me, 10me, et 11me séries, en vous priant de vouloir bien les offrir de ma part à cette Société.

Agréez, mon cher Monsieur Tyndall, mes compliments affectueux. | J. Plateau

J'ai pris connaissance de votre théorie cométaire, qui me paraît résoudre une énorme difficulté; seulement je n'ai pas vu que vous ayez cherché à découvrir la nature de cette singulière action de la lumière sur les vapeurs; mais vous y arriverez sans doute.

Ghent, 18 May 1869.

My dear Mr Tyndall,

My book[1] is finally completed, and I have the honour of sending you, for you and for the Royal Institution, copies of my eleventh and final series.[2] As for the Royal Society, I have just seen the nominations of foreign members, and I was not among them.[3] But I do not wish to sulk[4] and thus I am adding the above copies to the 9th, 10th, and 11th series, and ask that you to kindly offer them on my behalf to that Society.

Please accept, my dear Mr Tyndall, my warmest affections. | J. Plateau

I read about your theory of comets,[5] which seems to resolve an enormous problem; except that I did not see that you had attempted to discover the nature of this singular action of light on vapours, but you will doubtlessly come to it.

RI MS JT/1/P/104

---

1. *My book*: J. Plateau, *Recherches expérimentales et théoriques sur les figures d'équilibre d'une masse liquide sans pesanteur* (Brussels, 1849–69).
2. *eleventh and final series*: 'Recherches expérimentales et théoriques sur les figures d'équilibre d'une masse liquide, sans pesanteur, 11e série', *Mémoires de l'Académie royale des sciences, des lettres et des beaux-arts de Belgique*, 37 (1869), pp. 1–56. For the ninth and tenth series referred to in the same paragraph, see letter 2975, n. 2.
3. *I was not among them*: Alphonse Pyrame de Candolle, Charles-Eugène Delaunay, and Louis Pasteur were the three ForMemRS nominated in 1869. Tyndall was campaigning to have Plateau elected ForMemRS (see letter 2975, n. 4).
4. *I do not wish to sulk*: Plateau had been sulking for a while about the perceived neglect by the RS. See letter 2904, *Tyndall Correspondence*, vol 10; and letter 2975 in this volume.
5. *your theory of comets*: Tyndall's lecture, 'On Cometary Theory' (see letter 2966).

To George Hector Tyndale 19 May 1869 2996

Royal Institution | 19th May 1869

My dear Hector

Your friend Mr. Hilgard[1] shall be my friend when he comes to London.[2] Even apart from the fact that he comes recommended from you I should be the most ungrateful of men did I fail to show every attention to my American brethren, for assuredly they have shown civility and kindness to me.

With regard to going out there[3] it is a vague design and desire. I should like to do it, but decision will not converge itself—something always interposes whenever I think of carrying the intention out. The most formidable barrier is the weariness and apathy which beset me when I have worked some months in London.

I gave your letter[4] to Tom Hirst last night well knowing that it would give him pleasure. He is still Professor of Mathematics at University College,[5] respected by every body.

We have just returned from a three days trip to Canterbury, Ramsgate & Margate:[6] we were both cloudy & the little run did much to dispel the clouds.

Debus is at a large school in Clifton[7]—Clifton College near Bristol.[8]—They pay him well, but he has to slave terribly and has no time to pursue his own researches. This is the chief hardship for a man who has capacity for original work.

I am glad you like the comet theory.[9] I have worked round it and through it without discovering a hole in it; even the few objections that have been raised against it can be converted into arguments in its favour. I think the speculation will be remembered. Airy calls it "an excellent hit."[10] and says he always thought that the action of the nucleus had something to do with the production of the tail, but he could not grasp what. I will send you the other little papers[11] now that I learn you take an interest in Science. They will be all collected and expanded into a large memoir[12] by and by.

I gave as I have just said your letter to Tom[13] and I am trusting to my memory—a very bad one—in replying to it.

Last week I had the pleasure of meeting Professors Lyman[14] & Newton[15] of Yale College.[16] We dined together at the Athenaeum;[17] but I regret that my illness[18] prevented me from seeing them on Thursday last when they called upon me.

There is an exceedingly nice fellow named Norton[19]—a friend of Emersons & of Lowell's[20]—in London now. He is perfectly charming—I went down with him to Chelsea[21] some time ago to pay a visit to Carlyle. It was very pleasant.

Let me now at the end of my note, thank you and M[rs]. Tyndale[22] for your offer of hospitality[23]—I will seek your shelter should I go to America. Kindest regards to your wife and M[r]. Mitchell.[24]

My sister[25] is still with D[r]. Tyndall.[26] He sent me yesterday a number of testimonials of a son of his[27] who has just taken out his Medical degree in Dublin.

I am going to Ireland in a few days on the business of the Board of Trade[28] & shall take a run down to Gorey.[29]

RI MS JT/1/T/1449
RI MS JT/1/TYP/4/1680–1

1. *M[r]. Hilgard*: Julius Erasmus Hilgard.
2. *when he comes to London*: see letter 2987.
3. *going out there*: Tyndall was invited numerous times to visit the United States in a professional capacity. Hector invited him as well (see letter 2987). Tyndall eventually traveled to the United States in 1872 to deliver a series of lectures on light in Boston, Philadelphia, Baltimore, Washington, DC, New York, Brooklyn, and New Haven, CT. He departed from Liverpool on 28 September 1872, and arrived in New York on 9 October. He returned to England on 19 February 1873 after fourteen days at sea. His five months in America brought him in contact with numerous notable Americans, dazzled audiences, involvement in a debate on the efficacy of prayer, a side trip to Niagara Falls, the opportunity to set up a fund for pure science research in America using his profits from the lecture tour, and a lavish farewell banquet in New York before his departure from American soil. He published the lectures as J. Tyndall, *Six Lectures on Light delivered in America in 1872–1873* (London: Longmans, Green, and Co., 1873). Letters from during this trip will be published in the forthcoming thirteenth volume of *The Correspondence of John Tyndall*.
4. *your letter*: letter 2987.
5. *University College*: Hirst was appointed Chair of Mathematics at University College London in 1867 and would go on to become Assistant Registrar in 1870.
6. *Canterbury, Ramsgate & Margate*: three towns in South East England. Ramsgate and Margate are both seaside towns and thus common holiday destinations for Londoners. Tyndall and Hirst travelled to Canterbury on 15 May 1869. See letters 2991, n. 2, and 2992.
7. *Clifton*: located in South West England, in the eighteenth century, many of the rich left the industrial center of Bristol for Clifton. It became a one of the more affluent areas of the city.
8. *Clifton College near Bristol*: founded in 1862, Clifton College was a boys' school near Bristol with a special focus on science. It became co-educational in 1987.
9. *the comet theory*: Tyndall's lecture 'On Cometary Theory' (see letter 2966). For Tyndale's comments, see letter 2987.

10. *"an excellent hit"*: see letter 2993.
11. *the other little papers*: not identified.
12. *a large memoir*: there are three possibilities for this: *Diamagnetism* (published in 1870); *Fragments of Science* (1871); or *Contributions to Molecular Physics in the Domain of Radiant Heat* (London: Longmans, Green, and Co., 1872).
13. *Tom*: Thomas Hirst.
14. *Professor Lyman*: Chester S. Lyman (1840–90), an American clergyman and astronomer who became a Corresponding Member of the BAAS in 1869 ('Chester Smith Lyman', *The Sidereal Messenger*, 9 (November 1890), pp. 400–03; 'Prof. Chester O. Lyman', *The New York Times*, 30 January 1890, p. 5).
15. *Newton*: Hubert Anson Newton (1830–96), Professor of Mathematics and Astronomy at Yale College. Newton is best known for his work on the orbits of comets, meteors, and meteor showers (W. Gibbs, 'Biographical Memoir of Hubert Anson Newton, 1830–1896', in *Biographical Memories of the National Academy of Science for 1896*, 4 (1987), pp. 99–24).
16. *Yale College*: Yale College in New Haven, Connecticut, later Yale University.
17. *Athenaeum*: see letter 2944, n. 1.
18. *my illness*: Tyndall had been complaining to Hirst that his head was bothering him in letter 2991, but it is unclear whether this is what he is referring to.
19. *Norton*: Charles Eliot Norton (see letter 2970, n. 6).
20. *Lowell's*: John Amory Lowell (1798–1881), an American businessman, philanthropist, and lawyer. Lowell was interested in natural history and botany and donated substantial funds to the Harvard Botanical Garden. He was trustee of the Lowell Institute in Boston, Massachusetts, which had been founded by his cousin as an institute for popular education. (Anon, 'John Amory Lowell', *Proceedings of the American Academy of Arts and Sciences*, 17 (1881–82), pp. 408–11). Lowell made several attempts to recruit Tyndall for scientific lectures at the Institute (see letters 2160 and 2188, *Tyndall Correspondence*, vol. 8).
21. *Chelsea*: Chelsea is an area of South West London bordering the Thames where Thomas Carlyle lived with his wife.
22. *M^rs^. Tyndale*: Julia Tyndale (see letter 2987, n. 5).
23. *your offer of hospitality*: see letter 2987.
24. *Mr. Mitchell*: Edward P. Mitchell (see letter 2987, n. 19).
25. *My sister*: Emily Tyndall.
26. *still with D^r^. Tyndall*: Emily Tyndall was lodged under the care of Dr. John Tyndall, a relative from Ireland (see letter 2987, n. 7).
27. *son of his*: not identified, but one of three sons (see letter 3139, n. 4).
28. *Board of Trade*: see letter 2937, n. 6.
29. *Gorey*: Gorey, County Wexford in southeast Ireland. Dr. John Tyndall lived in Gorey. Tyndall was going to visit his sister. See n. 25.

From Mary Francis Egerton [19 May 1869][1] 2997

45 Eaton Place | Wednesday

Dear Mr Tyndall,

May wants to ask you a favour, but if there is any, the least difficulty about it, pray say so at once. Her friend Miss Kingsley[2] is coming to town today, and, we know, would so much enjoy hearing one even, of your lectures,[3] if she was able to do so, that Mary wishes me to ask whether you could give her an order for tomorrow, & the following one, in case she is still in town? Of course we have said nothing about this before asking you, so that we do not know whether she will be able to go with us, so do not hesitate to refuse; especially if the */*starch Committee*/* have anything to say in the matter!

I shall stay after the lecture tomorrow to have a little discourse reflecting the British Museum project which you did not quite throw cold water upon. I happened to see Mrs Maskelyne[4] the other day & she said her husband[5] w[oul]d be there, & delighted to offer the crystal cases for us, & recommended a private day. There is one glowing block, with rainbows at every corner, that I'll like you to renew acquaintance with, if you have not seen it lately!

Y[ou]rs very kindly, | Mrs Egerton

RI MS JT/1/E/72

---

1. *[19 May 1869]*: the dating of this letter is tentative. Egerton mentions the British Museum visit on 1 June in letter 3007. The lectures ended on 3 June, so it must have been written at least two weeks earlier, since May had enquired about two lecture tickets for her friend. The museum visit is not mentioned in Tyndall's letter of 29 May (letter 3000), but he did visit Egerton at her country home near Margate on 17 May (*Hirst's Journal*, RI MS JT/2/32d/1846), where they may have spoken about the possibility. 19 May 1869 was a Wednesday.
2. *Miss Kingsley*: a daughter of Charles Kingsley.
3. *your lectures*: see letter 2971, n. 4.
4. *Mrs Maskelyne*: Thereza Mary Dillwyn-Llewelyn (1834–1926), wife of Nevil Story Maskelyne (see n. 5), married in 1858.
5. *her husband*: Nevil Story Maskelyne (1823–1911), an English geologist, mineralogist, and politician. Maskelyne was professor of mineralogy at Oxford from 1856 to 1895, and Keeper of Minerals at the British Museum from 1857 to 1880. In 1869 he gave a series of well-received lectures on crystallography in London and Oxford (*ODNB*).

To Thomas Archer Hirst 20 May 1869 2998

Royal Institution | 20th May 1869.

My dear Tom

You will be at the Athenaeum[1] to day probably (full of happy recollections).[2] Pray arrange with Mr. Webster[3] for the invitation of Mr. Hilgard[4] who was recommended so strongly by Hector.[5]

Yours affect[ionately] | John

RI MS JT/1/T/676
RI MS JT/1/HTYP/550

---

1. *Athenaeum*: see letter 2944, n. 1.
2. *full of happy recollections*: Tyndall may have been referring to their retreat to Canterbury, Ramsgate, and Margate in South East England (letter 2996).
3. *Mr. Webster*: possibly James Claude Webster (1830–1908), Fellow of the Society of Antiquaries and Barrister-at-Law.
4. *Mr. Hilgard*: Julius Erasmus Hilgard, who was not a member of the Athenaeum and would have required a member's invitation to dine there. See letters 2987 and 2996.
5. *Hector*: Hector Tyndale. See letters 2987 and 2996.

From John Frederick William Herschel 28 May 1869 2999

Collingwood | May 28/69

My dear Sir,

Pray accept my best thanks for the exceedingly interesting & important paper[1] you have been so kind as to send me.[2] You have opened out quite a new vista in scientific enquiry and one which gives the promise of what now we may call unutterable things—but for which we shall one day find expression though at present misty & confused—"cloudy & indistinct like form in feverish dream".[3]

There is something wonderfully taking in your comet speculation[4]—there is much in it which will call for a good deal of further development as details of the process—but the general notion is perfectly siez-able and gives the best—because at present the only prospect of a solution of that great mystery if repulsive force be abandoned. What at present seems to me not very clear in your idea is this—I do not quite see where you suppose your real material comet to terminate. The actinic cloud forming the tail at millions of miles off from the comet itself must I suppose be produced by the action of the Suns rays, on all but infinitely attenuated vapourous matter occupying all the interplanetary space and in fact consisting of the whole of the unabsorbed nebula out of which the whole system was produced (of course in a state of circulation about the Sun). The all but infinite attenuation of such matter must of course be conceded and its appropriateness to form such an actinic cloud is conceivable. I can understand too, accepting your view of the antagonistic power of the Suns light and heat, how the whole interplanetary space is not always one vast actinic cloud—viz: because the Suns heat and light act together on it and ergo there is no cloud formed.* But then the tail comes to be considered as nothing but an actinic shadow of the head—and the question of course arises where does the head terminate and the tail begin. Comets no doubt are very large bodies—but still some limit must be assigned—infinitely short of filling the whole star system and yet this seems to be called for by the three first propositions (1) (2) (3) in page 4 which contains the heads of your theory. Then again—(granting that the already all but infinitely attenuated vapour of which the material comet & its coma consist (for as to the little solid? nucleus that may be neglected as of no consequence to the present speculation) is competent to arrest and absorb so large a portion of solar heat while all but perfectly transparent to the light)—the heat-shadow must conform in shape to the same geometrical rules as the light-shadow. It should therefore consist of a cone and penumbra—(or what corresponds to a cone taking account of the irregular form & density of the head & coma). There ought therefore to be formed within the tail produced by the penumbra, a much brighter interior tail of a conical form having the head of the comet for its base and coming to a point at a moderate distance behind—whereas in point of fact the space immediately behind the head is more commonly dark.

Pray understand that I by no means want to throw cold water on your theory which, as I said before, does really seem to hold out the promise of some kind of rational explanation of the facts—but only to call attention to conditions which will have to be considered in carrying out its application.—Your terminal note certainly does seem to imply that a portion of the actual material substance of a comet is in that part of space whence the light of the tail is radiated to our eyes. If so the question still arises—How did it get there? and thus the repulsive force crops up again.

I remain | my dear Sir | Yours very truly | J F W Herschel

*But why then should there be any blue sky?

P.S. Would not the calorific power of the sun's rays be (not "absorbed" but) rendered latent in the very act of effecting the vaporisation of the matter of the head and coma and their dilatation to the immense magnitude which they certainly do attain?

RI MS JT/1/H/113
RI MS JT/1/TYP/2/559–60
RS HS 406a
Transcript Only

1. *important paper*: 'On Cometary Theory' (see letter 2966).
2. *to send me*: letter missing.
3. *"Cloudy and indistinct, as feverish dream"*: W. Scott, *The Lay of the Last Minstrel: A Poem* (London, 1805), Canto VI, Section XVIII, line 4.
4. *your comet speculation*: see letter 2966.

To Mary Francis Egerton 29 May 1869 3000

Royal Institution | 29th. May.

My dear Lady Mary

I have just had a beautiful letter from Sir John Herschel.[1] He takes the cometary theory to pieces, but in such a manner as to enable me to strengthen it; and he says this:—"You have opened out a new vista in scientific enquiry, and one which gives promise of what now we may call unutterable things. There is something wonderfully taking in your comet speculation—There is much in it which will call for a good deal of further development in the details of the process—but the general notion is perfectly seiz-able and gives the best—because at present the only prospect of a solution of that great mystery." Now I call that very nice.

When you were at Eversley[2] I went down to Tring;[3] met there a great Archeologist;[4] walked with him through green lanes and over closeshaven Downs[5] to the "Becon".[6] There turned and walked through Ashbridge:[7]—the loveliest place it seems to me that I ever saw. And it was in all its glory that Sunday. Lord Brownlow[8] owned the land and trees—but I owned the beauty for the time being. It made me drunk with delight, and I needed the spur for I was very low.

Yours ever | John Tyndall
1869.

RI MS JT/1/TYP/1/389
LT Typescript Only

1. *a beautiful letter from Sir John Herschel*: see letter 2999.
2. *Eversley*: a village located in northeast Hampshire on the southern coast of England.
3. *Tring*: a small town in Hertfordshire in the south of England. It is around 60 miles north of Eversley.

4. *a great Archeologist*: not identified.
5. *closeshaven Downs*: 'closehaven' may be a misreading in the typescript ('close shaven'?). There does not seem to be a record of a 'Closehaven Downs', but there were a number of 'Downs' and paths in the woods around Tring, including Dunstable and the Chiltern Hills.
6. *the "Becon"*: Ivinghoe Beacon is on a hill a few miles north-east of Tring in the Chiltern Hills.
7. *Ashbridge*: a few miles from the Ivinghoe Beacon, Ashbridge had been the Egerton family estate from the 1600s.
8. *Lord Brownlow*: Charles Brownlow, 2nd Baron Lurgan (1831–82), an Anglo-Irish Liberal politician in the House of Lords.

To John Frederick William Herschel 29 May 1869 3001

Royal Institution | 29th May 1869.

My dear Sir John,

I am exceedingly gratified by your letter,[1] and I thank you most heartily for the suggestions and remarks with which you have favoured me.

I have done my best to look into the comet speculation with unclouded vision, but the more I think of it the more it commends itself to my understanding as at present informed. No doubt it will have to be worked out in detail. The outline given was published rather as a feeler than otherwise—I wished to know whether minds like your own could detect in it any blunders which had escaped me. Mr. Airy has also sent me a very agreeable letter[2] in which he calls the speculation an "excellent hit"; and says he never doubted that the comet's tail "(which defies all mechanical laws) was produced by an action of the nucleus on something pervading space".

I think I shall have no difficulty in accounting for the envelopes: In fact I have reproduced them. They are due in my case to the rotations set up within the cloud by slight differences in the energy of the chemical or calorific action at various points. These differential motions cause the clouds to assume the most astonishing forms. The likeness in some cases to complex animal organisms is simply terrific! Now such motions must exist on a large scale in the actual Comet and I am strongly inclined to think that by taking them duly into account many points raised in your letter will be satisfactorily accounted for. But I refrain from pushing the matter further at present.

With regard to the power of the nucleus to stop the calorific rays; only think what a column of ammonia—the lightest of all gases—three feet long can do. It is almost as black as pitch for all rays emitted by bodies under a state of incandescence. Think also of what transparent water or alcohol can do with the calorific rays.

Yours ever faithfully | John Tyndall.

RS HS 406b
RI MS JT/1/TYP/2/562

1. *your letter*: letter 2999. This letter presumably crossed with letter 3002 from Herschel.
2. *a very agreeable letter*: letter 2993.

From John Frederick William Herschel — 29 May 1869 — 3002

Collingwood May 29/ 1869

My dear Sir,

I forgot to mention in my letter of yesterday[1] that the case of the great comet of 1843 presents very considerable difficulties in the way of an explanation by a negative shadow or actinic cloud formed in the shadow of the chemical rays. The tail was of immense length & strongly visible in the twilight on the 24th of March, when the head corona & nucleus were only to be seen through a 20 inch achromatic telescope as not brighter altogether than a star 5 mag[nitude]. But the next night the tail remaining nearly or quite as bright & long as before the head was reduced to a faint nebula visible only in a 7 feet Newtonian telescope—and the next (the 26th) even this had disappeared the bright long tail remaining though diminishing in brightness without change of situation.

I remain | my dear Sir | Yours very t[rul]y | J F W Herschel

RI MS JT/1/H/114
RI MS JT/1/TYP/2/561

1. *letter of yesterday*: letter 2999.

From John Frederick William Herschel — 30 May 1869 — 3003

Collingwood | May 30/69

My dear Sir,

It may seem obtrusive & impertinent in any one to suggest experiments to you but one or two have occurred to me which it seems to me worth while to make.

1. In reference to polarization in your artificial blue sky.—Query whether it would not be worth while to examine the phenomena resulting when not an ordinary but a <u>polarized</u> beam is introduced as the efficient light—and to enquire <u>whether</u> & if so <u>how</u> this affects the resultant polarization of the dispersed rays.

2) To examine whether the cloud-producing power of a beam is diminished by being first passed through divers media—e.g. Sulphate of Quinine & others of that stamp—as also through non-trans-calescent media—&c. But this must have suggested itself to you already.

3) As a matter of kaleidoscopic effect—to try the effect of introducing into your vapours a broad beam divided into squares, triangles, circles &c &c &c by intermission through an opaque diaphragm cut out in regular patterns thus:

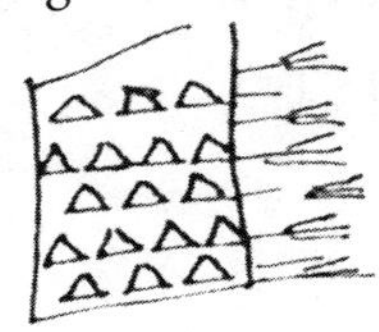

If the effect of a single beam be to produce regular & surprising cloud-convolutions what may not be expected from such combinations?—The interference patterns produced in the image of a star by limiting the aperture of a telescope by such a diaphragm are wonderfully beautiful & various.

Looking at my figures of Halley's comet in its recess from the sun I observe the formation & gradual increase of an interior bright tail which will do very well for the umbra

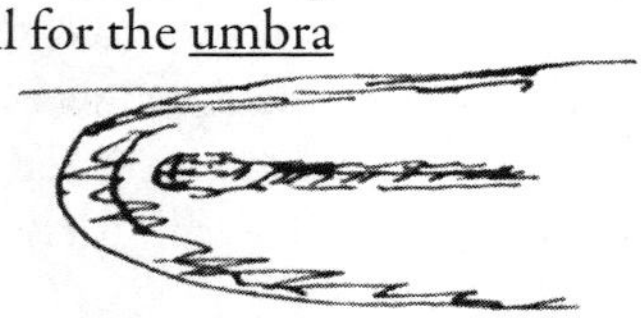

of a head obstructing the actinic power (in reference to my remark about the interior cone of light in my former letter)[1]

The comet of 1744 had six tails spread out like an immense fan extending to a distance of nearly 30° in length (25° = 10 millions of french leagues (Cassini)).[2] I cannot remember where I have seen a figure of it but it was something like this

Now such a form would be very difficult to account for on the shadow principle.

I really must apologize for writing such crudities but they will serve to show you how great a degree of interest this speculation of yours has excited in my mind.

Believe me | my dear Sir | yours very truly | J F W Herschel

P.S. Lalande[3] (Astr. V01.3.[4] art. 3209) says "La comète de 1744 s'est montrée de nos jours avec une queue d'un éclat extraordinaire; Elle était divisée en plusieures branches comme un immense eventail. Elle s'étendait le 19 fev[r] jusqu'à 30°. On en voit la figure dans le Traité de la Comète de 1744 de Cheseaux à Lausanne et à Geneva in 8° 1744 p.155."[5]—Perhaps this work may be in the R[oyal] S[ociety] library or that of the R[oyal] Inst[itutio][n]—or R[oyal] A[stronomical] Soc[iety] or in the B[ritish] Museum.—It certainly was not in that book[6] that I saw the picture—with six tails. Grant[7] says[8] it was seen in full sunshine & was more bright than Sirius.

Looking into Grant more narrowly I find that he has consulted Cheseaux's book[9]—& he says[10] (p.297) "Cheseaux states that the 6 tails of the comet of 1744 varied in length from 30° to 44°" and further on (p. 298)—"Each of the 6 tails of the comet of 1744 were 4° broad" If this be correct the whole spread of the fan must have been far wider than any penumbra could account for, unless seen very much foreshortened by perspective but then what an enormous length the tail would have been perpendicularly!

It is singular that Cassini[11] in his memoir on this comet (in the Mem[oire]. de l'Acad[emie].[12] 1744) says nothing about the extraordinary appearance of its tail.

RI MS JT/1/H/115
RI MS JT/1/TYP/2/563–4
RS HS 407a

1. *my former letter*: letter 2999.
2. *Cassini*: Jacques Cassini (1677–1756), a French astronomer and son of Giovanni Domenico Cassini. The younger Cassini succeeded his father as director of the Paris Observatory in 1712. Cassini measured and calculated the meridian arc between Dunkirk and Paris and published extensively on the figure of the earth, astronomy, and tables of Saturn's rings (*CDSB*).
3. *Lalande*: Joseph Jérôme Lefrançais de Lalande (1732–1807), a French astronomer whose primary research focused on the solar system and the solar parallax. Tremendously successful as an astronomer and administrator, Lalande wrote several widely used astronomical textbooks, integrated innovative mathematical techniques into astronomy, and was involved in numerous astronomical controversies, including on calculating the orbit of comets (*CDSB*).
4. *Astro. V01.3.*: J.-J. Lalande, *Abrégé d'astronomie* (Paris: Libraires Associés, 1775).
5. *La comète . . . p.155*: 'The comet of 1744 has appeared in our time with a tail of extraordinary brilliance; it was split into several branches like an enormous fan. It lasted from Feb. 19 to 30. You can see the figure in the Traité de la Comète from 1744, published by Cheseaux at Lausanne and Geneva in 8° 1744 p. 155.'
6. *that book*: see n. 4.

7. *Grant*: Robert Grant (1814–92), a Scottish astronomer, Professor of Astronomy at the University of Glasgow from 1860, and Fellow of the Royal Astronomical Society. Grant found his way into astronomy through compiling popular works, before learning observational techniques. He spent the 1860s–70s compiling star charts (*ODNB*).
8. *Grant says*: R. Grant, *History of Physical Astronomy from the Earliest Ages until the Middle of the Nineteenth Century* (London: Henry G. Bohn, 1852), p. 305.
9. *Cheseaux's book*: J.-P. Loys de Chéseaux, *Traité de la Comète* (Lausanne, Geneva: Marc-Michel Bousquet & com., 1744). Jean-Philippe Loys de Chéseaux (1718–51), a Swiss astronomer and theologian. He was a corresponding member of a number of European learned societies, including the RS. He established an observatory on his father's estate, in the Swiss village of Cheseaux, and his observations there from 1743 to 1747 provided the material for *Traité de la Comèt*. He also produced a second unpublished manuscript, 'Nouvelles méthodes'. *Traité de la Comèt* was based on his observations of comet C/1743 made from 1743 t01744 (V. Trimble and K. Bracher (eds), *The Biographical Encyclopedia of Astronomers* (New York, London: Springer, 2006), p. 714).
10. *he says*: see n. 7.
11. *Cassini*: see n. 2.
12. *Mem. de l'Acad*: J. Cassini, 'Sur la Comete de 1744', in *Histoire de L'Académie Royale des Sciences avec les Mémoires de Mathématique & de Physique pour la même Année* (1744), pp. 32–4.

From Pierre-Paul Deherain[1] May 1869 3004

Mai 1869

Monsieur,

J'ai été bien touché de l'envoi que vous avez bien voulu me faire et rien ne pouvait m'être plus agréable que de recevoir la seconde édition de votre beau livre. Vous avez, monsieur, rendu à tous les amis des sciences, et leur nombre s'accroit chaque jour, un véritable service en popularisant la nouvelle théorie de la chaleur et vous avez prouvé une fois de plus qu'il n'est pas de sujet si délicat qu'un esprit *[net]* et *[1 word illeg]* n'arrive à éclairer au point que chacun puisse s'y reconnaître.

Je pense que vous aurez été satisfait du succès que votre livre a obtenu en France, quant à moi je n'ai rencontré que des personnes enchantées de l'avoir lu et il n'a *[suscité]* que des éloges. Votre style rapide *[animé]* la *[forme]* vivante que vous avez donné à vos leçons en fait un de ces ouvrages qui resteront comme des *[1 word illeg]* qu'il faut s'efforcer d'imiter.

Je me permets, monsieur, de joindre à cette lettre quelques uns des ouvrages que j'ai publiés récemment. Je n'ai pas cru devoir y joindre l'*[1 word illeg]* scientifique car je pense que vous l'avez entre les mains et que c'est la lecture de l'article sur la Chaleur solaire et forces terrestres qui m'a vu l'honneur d'entrer en relation avec vous.

Agréez, monsieur, l'expression de la très haute considération et très vive sympathie de votre très humble serviteur

P. P. Deherain | 10 rue *[*Jay*]*. Paris

May 1869

Sir,

I was touched by the mail you sent me[2] and nothing could have been more pleasing than to receive the second edition of your beautiful book.[3] You have, Sir, done all the friends of sciences, and their number increases each day, a true favour by popularizing the new theory of heat and you have proven once again that there is no subject so delicate that a neat and steady mind cannot enlighten to the point of being accessible to everyone.

I think that you must have been satisfied of the success of your book in France,[4] as for me I have only met people delighted to have read it and it has only received compliments. Your animated, rapid style, the lively *[*form*]* that you have given to your lectures makes it one of the kind of book that will remain *[1 word illeg]*, that one must seek to imitate.

I permit myself, sir, to attach to this letter some of my recent publications.[5] I did not believe it was necessary to add the scientific *[1 word illeg]* since I think you must have it in your hands and that it is the article on solar heat and terrestrial forces that gave me the honour of being in touch with you.

Accept sir, the expression of the highest consideration and the profound regard from your very humble servant

P. P. Deherain | 10 Jay street, Paris

RI MS JT/1/D/86

---

1. *Pierre Paul Dehérain*: Pierre-Paul Dehérain (1830–1902), a French agronomist who specialized in plant physiology. He was appointed a professor at the Collége Chaptal at the age of 26. In 1860 he published *Recherches sur l'Emploi Agricole des Phosphates* and published research in a number of journals. He held the post of Professor of Vegetable Physiology at the Muséum d'Histoire Naturelle from 1880 until his death ('Science Gossip', *The Athenaeum*, 3920 (13 December 1902), p. 799).
2. *the mail you sent me*: letter missing.
3. *second edition of your beautiful book*: J. Tyndall, *Heat considered as a Mode of Motion*, 2nd edn (London: Longman, Green, Longman, Roberts, & Green, 1865).
4. *your book in France*: Tyndall's *Heat* had been translated into French in 1864.
5. *my recent publications*: enclosure missing. These most likely included P.-P. Deherain, 'Annuaire scientifique', *Les Progrès des sciences en 1867* (Paris: Charpentier, 1868). Since most of Deherain's publications covered narrower agricultural questions, it is unclear which publications he would have sent to Tyndall.

To Louisa Baring [Spring 1869][1] 3005

My dear Lady Ashburton.

I had the pleasure of meeting Miss Coutts Trotter[2] yesterday, and with the rapid energy of a powerful imagination she at once planned a system of telegraphy between my housekeeper and myself where by chance I was to obtain such trappings as would raise me out of the condition of pleasant savagery in which I now find myself; have them sent to Romsey[3] or Salisbury,[4] pick them up there on my way to London, and go at once to Melchet.[5]

But I am compelled to go to London first. I shall do what in me lies to get down on Friday–but certainly if you let me, I will go on Saturday. If you would say this to your friend "the Great man",[6] and tell him that I will nurse him all the way to Melchet, and charm all the demons away from him, he might be induced to accompany me.

I failed to reach the Edystone[7] though we tried to do so yesterday–there was no landing upon the stone the sea was so rough. This morning I proceed to the Start point by Dartmouth to visit the lighthouse[8] there. Some of these lighthouses are very grand things. I wish you could see the Sunlike blaze that we produce sometimes at Blackwall.[9] I have to give counsel in the matter of lighthouses to the Board of Trade and the Trinity Corporation: hence in part the interest that I feel in them.

Miss Coutts Trotter told me that Lord Houghton[10] is staying at Melchet. This induces me to confide to you a theory, or rather the rudiments of a theory—of the blood of St. Januarius.[11] Lord Houghton has witnessed its liquefaction and he will be able to say whether the theory has any basis in fact. It is all a case of Naples soap! I bought two bladders of the substance at Naples last year and have been over and over again astonished at its hygroscopic properties It becomes hard in dry weather; almost as soft as cream in humid weather. The simple aqueous vapour of the air causes the most wonderful variations in the consistency of this soap. Were I a priest or a scoundrel with a slight turn for natural philosophy the soap would give me a basis out of which to dissolve the blood with ease & certainty. This then is the theoretic rudiment that I wish you to criticize with the aid of Lord Houghton's experience.

Always yours | John Tyndall.

National Library of Scotland, Ashburton Papers Acc.11388 N0.101

---

1. *[Spring 1869]*: this letter can only be roughly dated. Tyndall visited Naples only once, in April 1868, and the letter is probably also fairly early in his work for the Board of Trade studying lighthouses (see letter 2937, n. 6), since he felt compelled to explain his lighthouse work to Baring. Tyndall worked on the lights in Blackwall that year, but since he

does not mention the significant events that occupied him in the later part of the year, this is presumably from the spring of 1869.

2. *Miss Coutts Trotter*: Margaret Coutts Trotter (d. 1882), a friend who shared a passion for collecting art, and in the 1860s shared romantic affection for each other that would waver in the early 1870s; she also sometimes helped Baring care for her daughter Mary (Virginia Surtees, *The Ludovisi Goddess: The Life of Louisa Lady Ashburton* (Michael Russell, 1984), pp. 129–30, 155–6; and M. Vicinus, *Intimate Friends: Women Who Loved Women, 1778–1928* (Chicago: University of Chicago Press, 2004), pp. 87–8).
3. *Romsey*: Baring's estate Melchet Court was located in Romsey, Hampshire.
4. *Salisbury*: a city to the northwest of Romsey.
5. *Melchet*: see letter 2971, n. 5.
6. *"the Great man"*: Thomas Carlyle
7. *Edystone*: Eddystone lighthouse on the Eddystone Rocks in the English Channel.
8. *Start Point . . . lighthouse*: the Start Point lighthouse is at the end of the peninsula on Start Bay, near Dartmouth, in South Devon. It was erected in 1836 and operated by Trinity House (see letter 2937, n. 6).
9. *the Sunlike blaze that we produce sometimes at Blackwall*: for Tyndall's experiments on electric lights at Blackwall, see letters 2969, 2984, and 2989.
10. *Lord Houghton*: Richard Monckton Milnes (1809–85), first Baron Houghton, a traveler, poet, author, and politician. He was a member of the famous Apostles Club, an intense group of undergraduates at Trinity College, Cambridge who attempted to make sense of the religious changes that were taking place in mid-Victorian Britain in a spirit of freedom and openness. Members—which then included Alfred Tennyson, Arthur Hallam, Richard Chenevix Trench, Joseph Williams Blakesley, and many others—discussed their concerns about natural science, natural theology, and Samuel Taylor Coleridge's distrust of natural theology (*ODNB*).
11. *the blood of St. Januarius*: Januarius supposedly lived in the third century and was executed by the Roman Emperor Diocletian. His relics eventually were housed in Naples, where, in the fourteenth century a vial of his dried blood miraculously liquefied. This liquification reportedly takes place two or three times a year and sometimes at special occasions. Januarius is the patron saint of Naples.

To John Frederick William Herschel — 1 June 1860 — 3006

Royal Institution | 1st June 1869.

My dear Sir John,

I do not like to offer an explanation in reference to the comet of 1843;[1] without full reflection.[2] But I think the effects to which you refer may be accounted for without introducing any violent supposition into the speculation. Were the <u>faintness</u> of the head and nucleus an indication of its <u>non-</u>

existence—did the reality correspond with the visible appearance then the objection would be unanswerable. But there is no reason to suppose that the eye sees all. The mass of the head and nucleus might preserve its relation to the mass in which the visible tail is scattered without being themselves visible. A head and nucleus which according to hypothesis absorb the calorific rays may undergo great changes of visible appearance without suffering corresponding changes of mass. However, as I have said I will ponder the matter as soon as I can take it up again.

At the lower part of page 18 of the memoir[3] on "Chemical Rays and the Light of the Sky" you will find some reference to experiments made with polarized light. They are very striking. What I have called the "residual blue" is the most gorgeous that can be imagined. Even when the cloud has passed the blue stage and ceased to polarize common light normally, the crossed Nicol cuts out a bar of intense blackness along the axis of the experimental tube. The other experiments you suggest I have not made, but I will make them as soon as I take the subject up again.

I can say without the least affectation that your letters so far from being "crudities" are of the greatest aid and encouragement to me.

Yours very faithfully, | John Tyndall

Sir John Herschel Bart. | &c. &c. &c.

RS HS 407b
RI MS JT/1/TYP/2/565

1. *comet of 1843*: the Great March Comet of 1843 (C/1843 D1) was noted for its particularly bright and large tail.
2. *full reflection*: Tyndall and Herschel discussed the subject of comets and Tyndall's recent paper on comets in letters 2999, 3001, 3002, and 3003.
3. *the memoir*: 'Chemical Rays and the Light of the Sky' (see letter 2942, n. 1).

From Mary Francis Egerton 1 June [1869][1] 3007

45, Eaton Place, | S.W. | June 1st

Dear Mr. Tyndall,

I congratulate you heartily on having two of the best Astronomers[2] on your side! Dear old Sir John Herschel's is real sympathy, and may well be pleasant to you. It is beautiful to see the eye of the imagination undermined with years gazing with such friendship of delight into the new "vistas" opening to science!

I was very near writing to his daughter[3] some time ago to ask about his opinion as to the Comets, but I was afraid of bothering them, they have so much Correspondence. It is very good of you to have taken the trouble to write out what he says for me.[4]

That beautiful Ashbridge[5] was nearly coming into the Egerton family some years ago, by a curious will of the late Lord Bridgewater,[6] which however, after various ups & downs, was finally upset by the House of Lords.

I suppose you have not had much time to give to the "points of no polarization"? We tried them a few times at Eversley[7] & found them at somewhere about 90° from the Sun's position in the morning, & about 140° at ½ to 5 in the afternoon; but it was only one day the Sky was blue enough to show much. You will have to find out someday what they depend on.
I read there the "Bekanntschaften *[reiner]* schönen Seelen"[8] you were speaking of. Much of it is certainly beautiful, but to me it is the beauty of a fancy picture, not of one taken in the living sunlight & the fine air of heaven. It suggested[9] many thoughts, some of which I feel much disposed to send you, but it is a foolish notion, & they shall stay in my own note book. I have also brought back some of Fichte's popular lectures & his life; they had not the *[1 word illeg] [*John*]* which I most wanted.

Do you think Tuesday next the 8th would do for the British Museum that is if it really would not bore you to go? but please say so if it would. If you do come, how much time can you spare? because if you have not been there of late, it w[oul]d. be worth while just to walk through the new rooms with the beautiful Greek & Carian sculptures & I would write to *[*dine*]* with Mr. Maskelyne[10] so as to allow time for that before his crystals.

If that wooden model is in your possession, could you bring it with you? We could leave it in the carriage, but I think it would make one understand the optic axis &c better than anything. What time should we call for you? ½ to 2? or later? I return the other valuable document![11] I wish you w[oul]d. send somebody to personate you & find out what this grand discovery is! One had supposed that, what between the stars & the compass, we already had a general notion of "North & South in space"!! I hope Ireland will have had as good an effect as Tring; you looked so much better on Thursday.

Forgive this huge note. I wish I would write concisely!

Y[ou]rs most truly | M F Egerton[12]

RI MS JT/1/E/36
RI MS JT/1/E/28

---

1. *[1869]*: this letter is a response to letter 3000, of 29 May 1869.
2. *two of the last Astronomers*: John Herschel and George Biddell Airy.
3. *his daughter*: Herschel had several daughters, but this was likely Isabella Herschel (1831–93), who assisted her father in various administrative tasks.
4. *write out what he says for me*: see letter 3000.

5. *Ashbridge*: see letter 3000, n. 7.
6. *Lord Bridgewater*: Francis Henry Egerton, eighth Earl of Bridgewater (1756–1829), a manuscripts collector and patron of scholarship. During his life, Egerton lived mainly in Paris, where he spent substantial amounts of money, fought with family members and fathered illegitimate children. He is best known for endowing funds for the publication of works unifying science and natural theology, known as the Bridgewater Treatises and for the bequest of his manuscripts to the British Library (*ODNB*).
7. *Eversley*: see letter 3000, n. 2.
8. *"Bekanntschaften [reiner] schönen Seelen"*: (incorrect) German: "acquaintance with *[*pure*]* beautiful souls". Allusion unclear.
9. *45 Eaton place . . . it suggested*: this portion of the letter has been transcribed from RI MS JT/1/E/36.
10. *Mr. Maskelyne*: Nevil Story Maskelyne (see letter 2997, n. 5).
11. *I return the other valuable document!*: enclosure missing.
12. *many thoughts . . . M F Egerton*: this portion of the letter has been transcribed from RI MS JT/1/E/28.

From Heinrich Debus 8 June 1869 3008

University of London, | Burlington House, W. | June 8th 1869.

My dear John,

I hoped to have the pleasure of taking luncheon with you to day, and of having some conversation with you, and of seeing your face once more, but all hopes it appears are dashed to the ground. For, the servant whom I sent to the Institution[1] says that you are not there, that you are gone to Ireland,[2] and that you are not likely to return for some time to come. So, nothing is left to me, but to return to Clifton[3] without having enjoyed the pleasure of seeing or speaking to you. I am now at the University[4] engaged with the examination of two Gentleman who intend to receive the DSc degree. To morrow evening I am obliged to return to Clifton. Since the end of April I have had no direct or indirect communication with you. Whether you are well or unwell happy or unhappy, cheerful or traurig,[5] all remains a mystery around here. From myself I have little to say. Weeks pass by like minutes and my body and my mind are getting more and more like a machine. There is nobody like you, or Tom,[6] or Williamson[7] at Bristol,[8] there is no scientific conversation, or interest which is independent of curiosity & fashion. Now a mans life is like a string on a piano, its motions have their own law, yet they are also excited and dependent on the vibrating sound. Now, whether constant life amongst boy's, and with men who are always amongst boy's, produces that ring (klang)[9] in one's soul which is the most desirable may be a question, I believe it is not, under present conditions. Our poor schoolmaster gets too much of it, one note overpowers all the rest.

What are you going to do at the end of the Session? Are you going abroad again? My holidays commence on 1 of August & last until the 18 of September. If I cannot see you before Aug 1, I expect at least to see you at Exeter.[10]

Always yours sincerely | H. Debus

Excuse my bad writing

RI MS JT/1/D/38

1. *the Institution*: RI.
2. *you are gone to Ireland*: Tyndall was in Ireland from 5 to 9 June 1869 studying the use of gas lighting in Irish lighthouses for the Board of Trade (see letter 2937, n. 6, and letter 3019).
3. *Clifton*: Debus was Science Master at Clifton College (see letter 2996, n. 8).
4. *the University*: Debus was at the University of London.
5. *traurig*: 'sad' (German).
6. *Tom*: Thomas Hirst.
7. *Williamson*: Alexander William Williamson (1824–1904), an influential English chemist. Williamson, who was disabled, studied with Justus Liebig before setting up his own chemistry laboratory in Paris. In 1849 he accepted the chair of practical chemistry at University College, London. Williamson developed a dynamic atomic theory of matter against Daltonian conceptions of atoms. His research on alcohols and ethers helped to clarify the molecular structure of organic compounds (*ODNB*).
8. *Bristol*: location of Clifton College.
9. *klang*: ring/tone (German).
10. *see you at Exeter*: the annual meeting of the BAAS was held in Exeter from 18–25 August 1869 (*Brit. Assoc. Rep. 1869*). Tyndall did not attend this meeting.

To Mr. Sedgwick[1] 10 June [1869][2] 3009

10th June

Dear Mr. Sedgwick

I am unhappily out of London at present: making observations nightly on the Lighthouse of the Hill of Howth—But when I return to London I will avail myself of your kind suggestion.[3]

Yours very truly | John Tyndall

RI MS JT/1/T/1282

1. *Mr. Sedgwick*: probably John Bell Sedgwick (d. 1895). Sedgwick had numerous ties to the

RI: attended meetings, appeared regularly in visitors lists, and donated books and funds to the institution from the 1860s until his final bequeathment of 1895 in aid of experimental research. He was Secretary of the Royal College of Physicians of London in the 1840s and on the board of the Leavesden Lunatic Asylum in the 1880s. Sedgwick's wife, Frances Lambert (1799–1880), was well-known for her publications on knitting and embroidery (*Lancet,* 52:1309 (1848), p. 362; *Roy. Inst. Proc.*, 7 (1874), pp. 299, 469; *Science*, 2:48 (1895), p. 732).

2. *[1869]*: Tyndall made observations at the Howth Baily Lighthouse for the Commissioners of Irish Lights in June 1869. See letter 3019. For Tyndall's work on lighthouses more generally, see letter 2937, n. 6.
3. *your kind suggestion*: letter missing.

From Wilhelm Haidinger[1] 13 June 1869 3010

Dornbach near Vienna, 13th June 1869
[On the polarization of light by air mixed with Aqueous Vapour | by Prof Haidinger | To Prof. Tyndall[2]]

My dear Sir

Your late experiments and reports on the polarization of light by cloudy matter[3] (Proceedings of the Roy. Soc. No 108, V.XIII. p.223 &c.) Jan. 14. 1869) have made a deep impression on my mind.

Permit me to advert to an ancient observation of mine, relating to a subject of the kind, but under circumstances widely different, which nevertheless I now very much should wish, you may think worthy of a glance in the development of your further inquiries.

I have observed the polarization of light by air mixed with watery vapour. I gave an account[4] of it in Poggendorff's Annals for 1846, Band LXVIII, page 73–(77). Abbé Moigno, likewise from Poggendorff gave a report of it in the 4th volume of his "Répertoire d'Optique Moderne" 1850.[5] Page 1338 & 1339. Both were accompanied with diagrams. In the vapour bath of course I had no optical apparatus with me, but having shortly before been struck with the appearance of the brushes of polarized light, or of polarization (Polarisations büshel), I was well prepared to test or to recognize polarized light under certain circumstances with the naked eye, by trying whether I could not see these brushes.

It is perhaps hardly discreet of me to demand you should be at the trouble of searching out old volumes, so I beg you will permit me just to translate that portion of one of my old papers,[6] which refers to the subject.

"Brushes of polarization observed in watery vapour.

White bows, or nebulous arches (Nebelbogen) have been observed in fogs or mist, having nearly the apparent diameter of rainbows. The light of the rainbow has been found to be polarized light by Biot[7] and Sir David Brewster,[8] conformably to the well known explanation by single reflexion of the light of the sun for the interior rainbow, and by double reflexion for the exterior rainbow.

I had an opportunity to observe the white vapour bows or arches in the vapour baths of the "Sophienbad",[9] a most meritorious establishment, conducted by M. Morawetz[10] in the suburb Landstrasse in Vienna. Since my observation a new building has been raised on the east side, so that it is no longer possible there to repeat the observation.

The sun shone bright, at 7 o'clock in the morning, under a small elevation through the window into the vapour. A beautiful circular arch presented itself to the eye the centre of which was the shadow of the head. I endeavoured to represent it in the diagram Fig. 1., AB:CD being the projection of the window upon the wall on the opposite side of the room.

Fig. 1.[11]

The colour of the arch f is a pale blueish white. It is slightly fringed on both sides with a pale orange or brownish yellow not over bright. The space e without and the space g within the arch is inferior in light and of a grey rather reddish colour. Opposite to the eye, the sun just grazing the eye, there appears a brighter circular spot a, fringed at b with the slight yellowish or reddish taint. To begin from b the light is distinctly polarized. The brushes of polarization are quite visible, if the eye from one place or direction is quickly directed to another. The brushes have a direction corresponding to the radius in the whitish arch, and a tangential direction in the spaces within and without of it. The light of the arch appears then to be polarized by reflexion from the surface of the particles of vapour or water. The spaces without and within the arch appear therefore to be polarized by transmission, perpendicularly to the polarization of the arch. The blueish-white and the reddish tints may be faint mixtures of the blueish or reddish fringes of diffraction combined with the direct reflexion from the watery particles floating in the air.

It is well known that a real rainbow may be produced on a small scale by taking some water in the mouth and then forcibly spouting or puffing it out reduced to the finest watery dust or powder. I availed myself of this method to ascertain at least approximately the diameter of the nebulous arch, being without any other apparatus in a vapour bath. The nebulous arch still continued visible, as in Fig. 2, but the first or interior rainbow now became visible and was situated pretty much in the central line of the nebulous arch, the exterior rainbow visible only in faint traces appeared beyond the nebulous arch.

Fig. 2.[12]

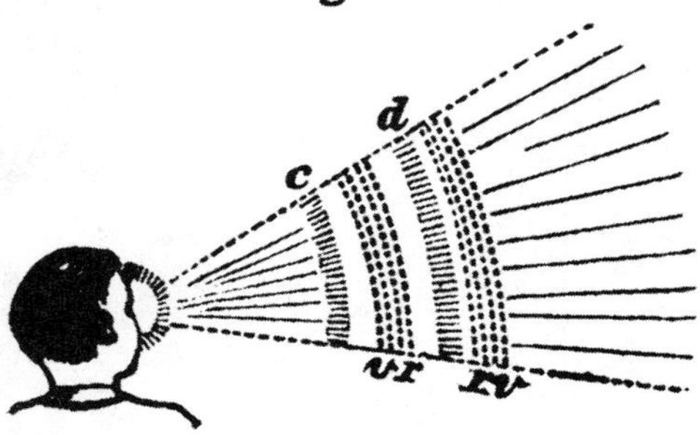

The angular values of the semidiameters being for the red of the interior rainbow 42°2', for its breadth 1°45', for the red of the outer rainbow 50°58', and its breadth 3°10', for the distance of the two rainbows 8°15', the breadth of the nebulous arch follows to be equal about to 12°, its central line being nearly at the angular distance of 41° from the centre. But I must claim for these angular values only the character of approximations, as I could only note the data from memory, and did not succeed in getting another sight of the phenomenon.

In the situation Fig. 3, looking at column of air, loaded with vapour and obliquely illuminated by the sun, entering through a small window, the transverse brushes of polarization produced by transmission, were distinctly visible at a, while from the wet boards of the floor at b as distinctly the polarization of reflexion was visible in the longitudinal brushes.

Fig. 3. [13]

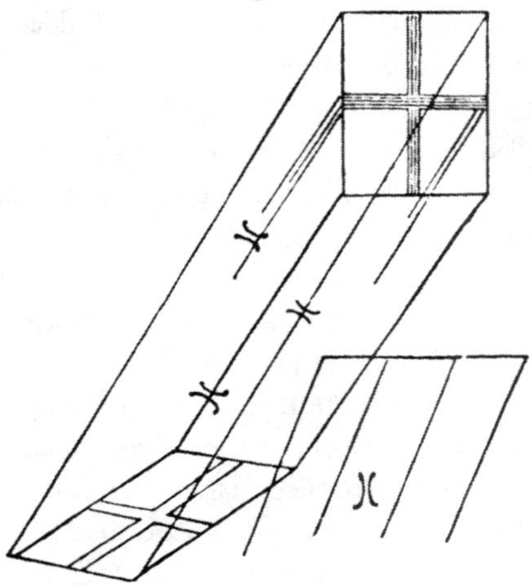

In a manner somewhat analogous to the preceding observations the tangential or transverse brushes of polarization may be observed near the sun in vapoury air, while the sun itself is screened from the eye of the observer by intervening objects."

You see, my dear Sir, I have reported only on the bare observation, and that only for the sake of following up the "brushes of polarization." But I have not found myself either sufficiently prepared, nor prompted by circumstances, to follow up the subject of studying the subject itself in the manner it well deserves. You are now in the course of the most interesting inquiries, and I should be happy to find, that you would give some glance at my own long ago brought forward and now nearly antiquated endeavours.

I still retain the most lively recollection of your friendly visit at my house in Vienna in 1856,[14] when I still was laid up in my bed from the cold I had caught the first day of the opening of our scientific association.[15] And greatly interested I was at so many of your investigations, several of which I had the good luck to quote as supports in my humble contributions. Perhaps even I should have written this letter in German, so completely you were master of my own language, but at last I thought, this would more agree with the impression of your daily general use and intercourse.

Believe me ever, my dear Sir | yours very truly | W[ilhelm] Haidinger

RI MS JT/1/H/9
Phil. Mag., 38 (July 1869), pp. 54–6.

---

1. *Wilhelm Haidinger*: Wilhelm Karl Ritter von Haidinger (1795–1871), an Austrian geologist and mineralogist. Haidinger began his mineralogical career with the curation of a

private collection. With the help of his well-situated brother-in-law, he founded the Geological Reichsanstalt (Imperial Geological Institute) of Vienna in 1849. Haidinger published prolifically on crystallography and minerology, as well as on topics in palaeontology and meteorology (*NDB*).

2. *On the polarization . . . Tyndall*: this letter was published in the *Phil. Mag.*, 38 (1869), pp. 54–6. Title and author name were added to the handwritten letter by a second hand, probably Tyndall's, in preparation for submission to the *Phil. Mag.*
3. *polarization of light by cloudy matter*: see letter 2940, n. 3.
4. *I gave an account*: W. Haidinger, 'Beobachtung der Lichtpolarisationsbüschel im gradlinig polarisirten Lichte', *Annal. Phys. und Chem.*, 144 (1846), pp. 73–87.
5. *4th volume of his "Répertoire d'Optique Moderne" 1850*: F.-N.-M. Moigno, 'Analyse des travaux de 1847, 1748 et 1849', in F.-N.-M. Moigno, *Répertoire d'optique modern, quatrième partie* (Paris/Leipzig: A. Franck, 1850), pp. 1297–1780, on pp. 1338–9.
6. *one of my old papers*: the following passage is a free translation from excerpts of Haidinger, 'Beobachtung der Lichtpolarisationsbüsche im geradlinig polarisirten Lichte' (see n. 4).
7. *Biot*: Jean-Baptiste Biot (1774–1862), a French physicist, astronomer, and mathematician. Biot was educated at the École Polytechnique, Paris, and later became professor of physics at the Collège de France. He conducted experiments on balloon flights and also made a close study of meteorites. He is primarily known for his research on the polarization of light, for which he was awarded the RS's Rumford Medal in 1840 (*CDSB*). Tyndall met Biot in Paris in July 1857 while attending Anna Hirst's funeral.
8. *Brewster*: David Brewster (1781–1868), a Scottish mathematician, astronomer, editor, and educator. Brewster was known for his work in optics and the polarization of light, for which he received the RS's Copley Medal and Rumford Medal, and for his role in founding the BAAS. He became a regular contributor to numerous periodicals, and served as an editor on several, by which he mostly earned his living. He was one of three founding editors of the *Edinburgh Encyclopaedia.* He was appointed Principal of the United College of St Salvator and St Leonard at the University of St Andrews in 1838, and Principal of the University of Edinburgh in 1859 (*ODNB*; *CDSB*).
9. *"Sophienbad"*: the published version of this letter mistakenly identifies this as the 'Sorbienbad'. The Sophienbad in the Landstrasse neigbourhood of Vienna was a bathhouse with concert hall and ballroom and one of mid-century Vienna's most well-known and stylish venues.
10. *Morawetz*: Franz Morawetz (1789–1868), a Jewish businessman and textile manufacturer. The textile business was not successful, and after Morawetz lost his eyesight, he converted the building into a Russian-style bathhouse, the Sophienbad, named after the Archduchess Sophie (M. J. Vogel, *Das Sophienbad des Franz Morawetz in Wien,* 3rd edn (Vienna: Peter Rohrmann, 1849)). See. n. 9. The published version of this letter mistakenly identifies him as 'M. Marawetz'.
11. *Fig. 1*: this figure was not included in the handwritten letter but is found both in the original German publication and in Tyndall's republication of the letter in the *Phil. Mag.*, 38 (1869), p. 55.

12. *Fig. 2*: this figure was not included in the handwritten letter but is found both in the original German publication and in Tyndall's republication of the letter in the *Phil. Mag.*, 38 (1869), p. 55.
13. *Fig. 3*: this figure was not included in the handwritten letter but is found both in the original German publication and in Tyndall's republication of the letter in the *Phil. Mag.*, 38 (1869), p. 56.
14. *Vienna in 1856*: Tyndall attended the annual meeting of the Society of German Natural Researchers and Physicians (Gesellschaft Deutscher Naturforscher und Ärtze) in Vienna in September 1856. See letters 1272, 1280, and 1282, in W. H. Brock and G. Cantor (eds), *The Correspondence of John Tyndall, Volume 5: The Correspondence, January 1855–October 1856* (Pittsburgh: University of Pittsburgh Press, 2018). There is no mention of the meeting with Haidinger in the letters.
15. *our scientific association*: Society of German Natural Researchers and Physicians.

From Henry Young[1] 15 June 1869 3011

To Professor Tyndall

Sir,

I regret I have to communicate to you that the Special Commissioners of the Public Schools Act[2] have not yet approved of the Statute framed by the Governors of Harrow School for the Constitution of the new Governing Body of that school under the Public Schools Act of 1868

The objection taken to it has been on legal grounds alone, the Commissioners having been advised by the Attorney-General at an advanced period of their proceedings that the power of existing Governors of the School was restricted to the establishment of a Constitution for forming the new Governing Body and not the nomination of the Members themselves of such Body.

The names inserted by the Governors of Harrow School in their Statute were I believe fully approved of by the Commissioners—but such nomination was at last considered to be beyond the power vested in the Governors.[3]

The matter therefore at present rests with the Special Commissioners.

I am, Sir, | Yours faithf[ull]y | Henry Young

12 Essex Street | 15th. June, 1869.

[Similar Letter to the above to the Rev[eren]d. Dr. Merivale,[4] the Rev[eren]d J. W. Blakesley[5] & Professor Bernard[6]][7]

Harrow School Archives *02/3—Governor's Letter Book, 1856–1871*
Used with the kind permission of Harrow School Governors
Transcript Only

1. *Henry Young*: Henry Young was the solicitor for Harrow School. Tyndall served as governor there from 1871 until 1892 (*Ascent of John Tyndall*, pp. 291–4).
2. *Special Commissioners of the Public Schools Act*: The Public School Commissioners were established as part of the changes to the school governing bodies in the wake of the Public Schools Act of 1868. The Act itself was the result of an investigation into the nine most prestigious large public schools in Britain, including Harrow. It made some changes to the governing bodies and modernized the curriculum, 'but it envisaged no public accountability or external scrutiny through statutory inspection. It effectively closed access to the elite public schools to all but those privileged by birth and wealth.' (*Ascent of John Tyndall*, p. 292).
3. *the Governors*: following the reforms introduced by the Public Schools Act of 1868, the old governing body attempted to elect Tyndall as a governor of Harrow, but as this letter illustrates this was rejected by the Public School Commissioners. He was elected in 1871 as the RS's nominee. Although a sporadic attendee at the board of governors' meetings, he did attempt on numerous occasions to influence and strengthen the teaching of science at the school. This was successful, but Tyndall's efforts were somewhat frustrated by the slow progress in enacting reforms (*Ascent of John Tyndall*, p. 291–2).
4. *Rev[eren]^{d}. D^{r}. Merivale*: Charles Merivale (1808–93), an English clergyman and historian. Merivale wrote a seven-volume *History of the Romans under the Empire* (1850–64), among other works on classical history and acted as chaplain for the House of Commons from 1863 to 1869, when he accepted the deanery of Ely (*ODNB*). Merivale was not listed as Governor in the *The Harrow Almanack, 1872–73* (Harrow: Crossman and Clarke, 1872).
5. *Rev[eren]^{d} J. W. Blakesley*: Joseph William Blakesley (1808–85), an English clergyman. Blakesely was ordained deacon in 1833 and priest in 1835. In 1845 he became Vicar of Ware and became known for the social and political issues he discussed in local newspapers. Although a critical theologian, he became Dean of Lincoln in 1872 (*Alum. Cantab.*). Blakely was not listed as Governor in the *The Harrow Almanack, 1872–73* (Harrow: Crossman and Clarke 1872).
6. *Professor Bernard*: probably Mountague Bernard (1820–82), an English jurist and international lawyer. He studied at Trinity College, Oxford, graduating in 1842 and was called to the bar at Lincoln's Inn on 30 April 1846. He was elected to the Chichele chair of international law at Oxford in 1859 and served on several state commissions, including the 1886 royal commission that investigated the cattle plague, and was elected a professorial fellow at All Souls College, Oxford. He was subsequently elected a governor at Harrow and served alongside Tyndall (*ODNB*). Bernard was listed as Governor in *The Harrow Almanack, 1872–73* (Harrow: Crossman and Clarke 1872), the same year as Tyndall.
7. *[Similar . . . Bernard]*: the letter was transcribed into a letterbook. This note followed the signature but was not part of the letter itself.

To Roderick Murchison 16 June 1869 3012

16th June 1869.

My dear Sir Roderick

I do not think that I shall be able to be present today at the meeting of the Eyre Defence Committee.[1] But were I present I should advocate earnestly that whatever funds remain in hands, should with the consent of the subscribers—which I should imagine to be all but universal—be handed over to Mr. Eyre.

Yours very since[re]ly | John Tyndall

BL Add MS 46128, f. 223

---

1. *Eyre Defence Committee*: Edward John Eyre (1815–1901) was appointed Governor of Jamaica in 1862. Some prominent members of British society (including John Stuart Mill, Charles Darwin, Thomas Huxley, and Herbert Spencer) wanted Eyre tried for his role in what they deemed was an excessive suppression of a rebellion in Morant Bay in 1865. Thomas Carlyle set up the Governor Eyre Defense and Aid Committee in support of Eyre. Tyndall, Charles Dickens, Charles Kingsley, John Ruskin, and Alfred Tennyson joined Carlyle's Committee. Tyndall spoke in Eyre's behalf. See *Tyndall Correspondence*, vol. 10, letters 2561, 2571, 2572, and 2576. Vol. 10, letter 2566, n. 4. summarizes some of the exchanges.

From Frédéric Zurcher[1] 20 June 1869 3013

Toulon 20 Juin 1869

Cher Monsieur,

Nous avions déjà lu, mon beau frère M. Margollé et moi, avec un grand plaisir, votre bel ouvrage sur la Chaleur dans la traduction de M. l'abbé Moigno. Mais nous sommes très heureux de le recevoir de son illustre auteur et de pouvoir le lire dans le texte original. Nous vous en témoignons notre cordiale reconnaissance. Nous espérons avoir bientôt l'occasion de parler de cette nouvelle édition et des importantes additions qu'elle renferme.

La manière élevée dont vous exposez vos admirables découvertes scientifiques nous est extrêmement sympathique. En révélant les lois de la nature vous atteignez à la plus haute poésie et à une philosophie vraiment religieuse.

Nous venons *[d'achever]* la traduction de la petite Géographie physique du commandant Maury. Elle contient d'excellentes citations de vos théories et contribuera beaucoup, nous l'espérons, à faire naître en France le désir de les étudier.

Voudriez-vous avoir la bonté, quand vos travaux vous laisseront un moment de loisir, de jeter les yeux sur une théorie relative à la variation séculaire des saisons mentionnée, à propos de la période glaciaire, dans *[mon]* petit article sur les hautes régions de l'atmosphère (Annuaire scientifique).

Cette théorie a été élaborée par un éminent philosophe de notre pays que nous avons eu le malheur de perdre récemment, M. Jean Reynaud, dans une note de son ouvrage Terre et Ciel (3ème édition). Votre grand astronome, Sir J. Herschell avait écrit à M. J. Reynaud pour approuver ses idées. Il m'a semblé qu'elles pourraient être importantes dans la discussion élevée sur l'étendue des anciens glaciers, et cela *[m'enhardit]*, malgré mon ignorance, à les signaler à un savant aussi compétant que vous dans cette question.

Veuillez agréer, cher Monsieur, l'assurance de nos sentiments dévoués. | F. Zurcher

Toulon 20 June 1869

Dear Sir,

We have already read, my brother-in-law Mr. Margollé[2] and I, with great pleasure, Father Moigno's translation[3] of your wonderful book on Heat.[4] But we are thrilled to *[receive]* it from its illustrious author and to be able to read it in its original format. We extend our sincere gratitude. We hope to soon have the opportunity to speak of this new edition and of the important additions that it contains.

We very much appreciate the elevated manner in which you present your admirable scientific discoveries. By revealing the laws of nature, you reach for the highest poetry and for a truly religious philosophy.

We have just finished the translation of the brief Physical Geography[5] by Commander Maury.[6] It contains excellent citations of your theories and will contribute significantly, we hope, to inspiring in France the desire to study them.

Would you be kind enough, when your work allows you a moment of spare time, to glance at a theory related to the secular variation of the seasons, regarding the glacial period, mentioned in my short article on the Upper Regions of the Atmosphere[7] (Scientific Directory).

That theory was elaborated by an eminent philosopher of our country, whom we had the misfortune of losing recently, Mr. Jean Reynaud,[8] in a note from his book Terre et Ciel[9] (3rd Edition). Your great astronomer, Sir J. Herschel wrote to Mr. J. Reynaud to approve his ideas. It seems to me that they could be important in the elevated discussion on the expanse of ancient glaciers, and this emboldens me, despite my ignorance, to report them to a scholar as knowledgeable as you in the matter.

Please accept, dear Sir, the assurance of our most devoted sentiments. | F. Zurcher

RI MS JT/1/Z/4

1. *Frédéric Zurcher*: Frédéric Zurcher (1816–90), a French popularizer of science who wrote, with Elie Margollé, numerous books on a variety of scientific subjects including *Les Glaciers* (Paris: L. Hachette, 1868) (Archives Nationales).
2. *Mr. Margollé*: Élie Philippe de Margollé (1816–84), a French popularizer of science who co-authored numerous books with Frédéric Zurcher (Archives Nationales).
3. *Moigno's translation*: J. Tyndall, *La chaleur: Mode de mouvement*, trans. M. Moigno (Paris: Gauthier-Villars, 1864). See also letter 2949, n. 7.
4. *your wonderful book on heat*: J. Tyndall, *Heat Considered as a Mode of Motion: Being a Course of Twelve Lectures Delivered at the Royal Institution of Great Britain in the Season of 1862* (London: Longman, Green, Longman, Roberts, & Green, 1863).
5. *Physical Geography*: M. F. Maury, *Géographie physique de la mer*, trans. F. Zurcher and E. P. Margollé (Paris: J. Hetzel, 1866).
6. *Commander Maury*: Matthew Fontaine Maury (1806–73), an American naval officer whose observations and involvement in diverse international organizations resulted in significant contributions to astronomy, oceanography, cartography, navigation, and geology. His *Physical Geography of the Sea,* published in 1855, is considered one of the foundational works of physical oceanography (*ANB*; *CDSB*).
7. *my short article on the Upper Regions of the Atmosphere*: F. Zurcher, 'Les hautes regions de l'atmosphère', *Annuaire scientifique* (Paris: Charpentier, 1865), pp. 104–10.
8. *Mr. Jean Reynaud*: Jean Ernest Reynaud (1806–63), a French politician, philosopher, and mine engineer who founded, with Pierre Leroux, the *Encyclopédie nouvelle* in 1833 (F. Hoefer (ed), *Nouvelle Biographie Générale*, 46 vols (Paris: Firmin Didot fréres, 1852–66), vol. 42, pp. 79–81). The encyclopedia reflected their socialist philosophy.
9. *Terre et Ciel*: J. Reynaud, *Philosophie religieuse: Terre et Ciel* (Paris: Furne Éditeur, 1854).

To John Frederick William Herschel [22 June 1869][1] 3014

Royal Institution Tuesday

My dear Sir John,

I hope you will find nothing objectionable in the enclosed reference[2] to one of your letters to me.[3] I ought not to have printed the extract[4] without your first seeing it: but a similar allusion[5] having already passed through your hands I thought it hardly worth while troubling you.

Yours very sincerely | John Tyndall.

RS HS 414
RI MS JT/1/TYP/2/566

1. *[22 June 1869]*: the date is given by relation to letter 3017, dated 28 June 1869, which is a

response to this letter. The preceding Tuesday was 22 June 1869.

2. *in the enclosed reference*: see letter 3014encl.
3. *one of your letters to me*: letter 2910, *Tyndall Correspondence*, vol. 10.
4. *printed the extract*: the passage in the enclosure is from 'On Chemical Rays' in the *Fortnightly Review,* p. 237 (see letter 2946, n. 6).
5. *a similar allusion*: the same passage is included verbatim in an earlier version of the paper in *Roy. Soc. Proc.* (see letter 2942, n. 1).

To John Frederick William Herschel [22 June 1869][1] 3014encl

And here[2] let me briefly say that, were it not for the stimulus imparted to me by the private correspondence of a celebrated man, I should not have entered upon the investigation of these subjects so soon. In reference to the effects of light which you have just witnessed, Sir John Herschel wrote to me[3] thus:—"It is a class of relations eminently calculated to set one thinking, and it seems to have had that effect upon you to excellent purpose. I am glad it has brought you into contact with the blue colour of the sky; still more so if it should lead you to any satisfactory explanation of the polarization of sky-light." The letter went on to treat "this mysterious and beautiful phenomenon" in a manner which excited in me the strong desire to throw, if possible, some certain light upon a question regarding which the most divergent opinions and speculations are afloat among out most eminent scientific men.

RI MS JT/1/H/116

1. *[22 June 1869]*: the date is given as this was enclosed with letter 3014, dated (22 Janaury 1869).
2. *And here . . . scientific men*: the enclosed passage is a clipping from Tyndall's essay 'On Chemical Rays' in the *Fortnightly Review* (see letter 2942, n. 1).
3. *Sir John Herschel wrote to me*: letter 2910, *Tyndall Correspondence*, vol. 10.

From Frederick Pollock[1] 25 June 1869 3015

Hatton | Hounslow | 25th. June 1869

My dear Professor

I really cannot find words to express how much I am gratified & obliged by your last communications[2]—what you say about the tails of comets[3] is most astonishing and yet is very satisfactory I shall not complain of age as long as I can take delight in such studies—I wish I was active enough to join

in the pursuit and investigation
ever your much obliged and sincere friend | Fred Pollock.

RI MS JT/1/P/161

1. *Frederick Pollock*: (Jonathan) Frederick Pollock (1783–1870), first baronet, judge and father of William Frederick Pollock. He graduated from Cambridge as senior wrangler and first Smith's prizeman in 1807 and MA in 1809 and was called to the bar in 1807. In 1831 he entered parliament as a Tory and and served twice as Attorney General (1843–45, 1841–44). He became Lord Chief Baron of the Exchequer from 1844 to 1866 and received a baronetcy in July 1866 following his retirement. He was elected FRS in 1816 and communicated three mathematical papers to the RS (*ODNB*). His death in 1870 hit the family hard (see letter 3253).
2. *your last communications*: letter missing.
3. *what you say about the tails of comets*: see letter 2966.

To Edward Frankland 28 June [1869][1] 3016

28th June

My dear Frankland
Mr. Dumas, who describes you as having reproduced the elements of the Sun[2] upon the earth, has /requested/ me to send you, with his compliments a medal struck at his suggestion[3] on the occasion of the discovery of the 100th little planet between Mars and Jupiter.
Yours very since[re]ly | John Tyndall
[Envelope] Prof. Frankland | Royal College of Chemistry | Oxford Street

John Rylands Library, University of Manchester, Frankland Letters, 2171

1. *[1869]*: Jean-Baptiste Dumas was in London in June 1869 to give a lecture to the Chemical Society, held at the RI on 17 June 1869, as first recipient of the Faraday Lectureship (*BMJ*, 1:440 (5 June 1869), p. 522). See also *Hirst's Journal*, RI MS JT/2/32d/1847.
2. *reproduced the elements of the Sun*: in October 1868 Frankland identified helium by using spectroscopy (C. A. Russell, *Edward Frankland: Chemistry, Controversy and Conspiracy in Victorian England* (Cambridge: Cambridge University Press, 1996), pp. 435–7). He shares the discovery of helium with his countryman, Norman Lockyer (see letter 3026, n. 7), with whom he collaborated, and the French astrophysicist, Pierre Jules César Jansson (1824–1907).
3. *a medal struck at his suggestion*: from 1868 to 1870 Dumas was Master of the Royal Mint in France ('Jean-Baptiste-Andre Dumas', *Science*, 3 (1884), pp. 750–2).

From John Frederick William Herschel 28 June 1869 3017

Collingwood | June 28/69

My dear Sir,

I cannot have the slightest objection to the Paragraph you send[1] (reinclosed)[2]—only there is an evident misprint of radiations for relations or perhaps reactions.

I suppose it is a draft report of your lecture[3] that you are preparing. How I regret on such occasions the necessity I am under of foregoing such an intellectual treat as it would have been to have been present at its delivery.

Believe me | my dear Sir | Ever yours truly | J F W Herschel

RI MS JT/1/H/116
RI MS JT/1/TYP/2/567

1. *the Paragraph you send*: see letters 3014 and 3014encl.
2. *(reinclosed)*: letter 3014encl was returned to Tyndall with this letter.
3. *your lecture*: 'On Chemical Rays' (see letter 2942, n. 1). The report and essay had been published several months earlier.

To Heinrich Debus 29 June 1869 3018

29th June 1869

My dear Heins.

I almost envy you that strength and purity of soul which enables you to go out into Nature & to be a sharer of its peace & beauty. I used to take large draughts of this blessedness, but they are not mine now so frequently as formerly; and the loss of them is not to be atoned for, or compensated, by any worldly success, however high.

I am aiming to get away with Tom[1] at the end of this week.[2] We shall both think of you when we enter Marburg.[3] I wish you could be there along with us—it would be such a pleasure.

My health is very low, but not at all so low as it was last year. Perhaps it may improve as I grow older, and enable me to fulfil to some small extent the promise of my youth.

Yours ever | John Tyndall

RI MS JT/1/T/264
RI MS JT/1/TYP/7/2370

1. *Tom*: Thomas Hirst.
2. *get away with Tom at the end of this week*: Tyndall and Hirst left England via London on Sunday, 4 July 1869, arriving in Antwerp for a tour of the Continent. See letters 3020, 3022, 3023, and 3024.
3. *Marburg*: Marburg was the main administrative centre of the Prussian province of Hessen-Nassau. Tyndall and Debus studied chemistry at the University of Marburg between 1845 and 1848, with Tyndall also studying physics and completing a Ph.D. thesis in mathematics. Hirst studied primarily mathematics there a few years later.

To Thomas Henry Farrer[1] 29 June 1869 3019

REPORT OF PROFESSOR TYNDALL, F.R.S. | H. 2651.
§. 1.
Royal Institution, June 29, 1869.

SIR,

I READ with attention the Reports and Appendices submitted by the Commissioners of the Irish Lights to the Board of Trade[2] regarding the employment of gas for lighthouse purposes as illustrated in the lighthouses of Howth Baily[3] and of Wicklow Head.[4]

With the view of ascertaining by personal inspection the merits of this new system of illumination, I have visited the establishments referred to,[5] and by a series of measurements and observations conducted during several successive days and nights, established the facts, and came to the conclusions embodied in the following report.

With a view of checking the observations I availed myself of the services of Mr. Valentin, the principal assistant in the Royal College of Chemistry,[6] a gentleman of great experience both in the chemistry of gas and in the practice of photometry. On Saturday the 5th of June Mr. Valentin, at my request, executed at Howth Baily a series of experiments in which the illuminating power of the gas flame was compared with that of the most powerful oil lamp employed by the Commissioners of Irish Lights.[7]

This lamp consists of four concentric wicks, which give rise to four concentric cylinders of flame, the interior flames radiating their light through the exterior ones. A greater body of light is thus sent forth than could possibly be obtained from a single wick. Such a four-wick lamp was placed in a dark room, and by means of a Bunsen's photometer[8] 36 feet long was compared with the gaslight now employed at Howth Baily.

The gas burner devised and constructed by Mr. Wigham consists of a series of concentric fish tail jets. The three central rings embrace a group of 28 jets, and this is the light employed under ordinary circumstances at Howth Baily.

To this central group can be added in succession four other circles of burners, each embracing 20 jets. Thus the lowest light employed is emitted by 28 jets, the next in power by 48, the next by 68, the next by 88, and the next by 108 jets of gas. It is possible, therefore, to employ lights of five different powers.

The following tabular statement exhibits the results obtained by Mr. Valentin. In all cases the four-wick oil lamp is taken as the unit, and the illuminating power of the gas-flame is expressed in terms of that unit.

Four-wick lamps.	Gas-flame.	Number of jets.
———	———	———
1	2½	28
1	4¼	48
1	7¼	68
1	9¾	88
1	13	108

The superiority of the gas over the oil-flame is rendered very conspicuous by these experiments. The 28-jet burner possesses 2½ times, the 48-jet burner 4¼ times, the 68-jet burner 7¼ times, the 88-jet burner 9¾ times, and the 108-jet burner 13 times the illuminating power of the four-wick flame.

On Saturday evening the 5th of June I reached Dublin, and proceeded immediately to the Howth Baily lighthouse. The experimenters who had been employed there during the day had departed, and my arrival was unexpected by the keeper. The 28-jet burner was in action, and the light it emitted appeared to be a very fine one. I caused the keeper to pass to the 48-jet and the 68-jet burners in succession. The ease with which he did so convinced me that complete practical mastery over the apparatus, if not already attained, is certainly attainable.

On Monday the 7th of June I accompanied Capt. Roberts[9] and Mr. Wigham to Howth Baily, and there executed with the same lights the series of measurements which had been carried out by Mr. Valentin on the previous Saturday. With one exception I found his results to be sensibly correct. The exception occurred in the case of the 68-jet burner, which Mr. Valentin made 7¼ times, and which I made a little over 6 times the power of the four-wick lamp. The difference was no doubt due to a temporary change in the light of the oil lamp.

It is difficult to make the intensity of a light an object of memory, but it sometimes appeared to me that the oil-flame at Howth Baily did not quite equal in intensity the flame of the four-wick lamp which I have sometimes operated on at the Trinity House.[10] But taking every thing into account I think it may be safely affirmed that the results embodied in the Report submitted by the Irish Light Commissioners to the Board of Trade are not exaggerated.

When the larger burners were employed the powerful rush of the flame into its mica-chimney, and the body of radiant heat emitted, at first startled me. Further investigation satisfied me that all the powers of the burners were perfectly under the control of the keeper.

I could discover no defect in the dioptric apparatus arising from this heat, nor do I anticipate that any damage is likely to arise from it.

§ 2.

I have thus far limited myself to the observations made at the lighthouse. But the atmosphere itself, through which the light has to pass to assist the mariner, is the best photometer. I accordingly arranged with Capt. Roberts and Mr. Wigham a programme of observations from a distant position. The place chosen, which had been previously employed for similar observations, was a point on the North Wall near Dublin, about 6 miles distant from the Howth Baily Lighthouse.

The observations began at 40 minutes past 9 P.M. on the 7th of June. The 28-jet burner, embracing the three central rings, was first compared with the oil-flame. The superiority of the gas was strikingly manifest. The fourth, fifth, sixth, and seventh rings of burners were successively added, an interval of seven minutes being allowed for the observation of each. The oil-flame became relatively more and more eclipsed, and, when the full energy of the gas was invoked, its light would, in common language, be pronounced incomparably more powerful than that of the oil.

And in fact this language would be scientifically correct as applied to one of the experiments, in which two layers of calico, the one oiled and the other not, together with a sheet of muffled glass, were placed in front of the two lights. This arrangement roughly represents a fog. The oil-light was completely extinguished, while the gas continued to be distinctly visible. The full power of the gas-burner was here employed, 108 jets being ignited.

No doubt could rest upon the mind that, at the point of observation chosen, the gas flame from first to last was superior to the oil flame, being with the higher powers of the gas burner *exceedingly superior.* I did not, however, think it safe to limit myself to this particular point of observation; and to meet my wishes Capt. Roberts was good enough to engage a steamer, which enabled me to proceed down the river, and to pass across it from side to side between the North Wall and the South Wall,[11] thus varying the point of observation. It soon became manifest that the oil-flame at Howth Baily varied in intensity with our position, and that the direction of *minimum* intensity corresponded almost exactly with that in which our observations had been made on the previous evening from the North Wall.

No safe conclusion therefore could be drawn from the observations made on the evening of the 7th, for it was manifest that some cause existed which prevented the oil-lamp from displaying its full power in the direction of the North Wall, thus giving the gas a relative superiority which in reality it did not possess.

It is to be borne in mind that the oil-lamp in these experiments was placed in the first order dioptric apparatus of the lighthouse, the apparatus being as usual surrounded by its glazed lantern. The gas flame, on the contrary, was placed in a temporary hut at some distance below the lantern. A refracting panel, similar in all respects to those of the dioptric apparatus, was placed at the proper distance in front of the gas-flame, and, to make the conditions alike, the upper and lower reflecting prisms of the apparatus were shut off. It was therefore the lights transmitted through the two lenticular belts that were compared together.

In company with Capt. Roberts I again visited Howth Baily on Wednesday the 9th. We were preceded by Capt. Hawes[12] and Mr. Wigham, who were instructed to examine with all care whether any obstruction was offered by the lantern to the passage of the light towards our station on the North Wall. *It was soon found that one of the vertical sashes of the lantern was directly interposed between the light and our point of observation, and that to this obstruction the enormous apparent superiority of the gas-flame over the oil, manifested on the evening of the 7th, was to be ascribed.* A very obvious inadvertence, I may add, led Capt. Roberts to the choice of the North Wall as a station.

Another line of observation had, therefore, to be chosen, and that between Howth Baily and Poolbeg Lighthouse[13] appeared free from objection. The refracting panel placed in front of the gas-flame was shifted to suit this new line, Capt. Roberts, Mr. Wigham, and myself returned to Dublin, and steamed from the North Wall to Poolbeg. Capt. Hawes remained at Howth Baily to see that the oil-flame was maintained at its maximum brilliancy. In short every precaution was taken to show the oil-lamp fair play.

The oil-lamp was lighted at sunset, and when observed from Poolbeg a little after 9 P.M. it showed a very brilliant light. At 9.40 the 28-jet gas burned was ignited, when the wisdom of our previous scrutiny became at once manifest. *From the new station it was impossible to say which of the lights was best.* Capt. Roberts, Mr. Wigham, and myself, were unanimous upon this point.

At 9.50 an oiled calico screen was placed in front of each light; the gas flame had slightly the advantage over the oil in penetrative power. The same feeble superiority was manifested when an additional screen of dry calico and a sheet of muffled glass were added to the first.

At 10 P.M. we passed from the 28-jet to the 68-jet burner. The decided superiority of the gas-flame was immediately apparent; it exhibited a very brilliant globe of light.

At 10–10 the 6th and 7th rings of the burner were added, 108 jets being now employed; the light was truly splendid; the oil-flame, which, when burning alone, shone brilliantly, appeared to shrink and pale before the flame of the 108-jet burner.*

§. 3.

There is a manifest discrepancy between the measurements with Bunsen's photometer and the observations from Poolbeg. The former assign to the 28-jet gas-flame 2½ times the illuminated power of the oil-flame, while, to an eye placed at Poolbeg, the lights appear equally intense.

The discrepancy may be in part due to the better draught, and consequent brisker combustion of the oil in the lantern of the dioptric apparatus; or it may be due to a portion of the gas-flame being out of focus. Indeed Mr. Wigham has since found that the focusing of the gas-flame was imperfect. The 28-jet burner consumes 50 cubic feet of gas per hour. Captain Roberts and Mr. Wigham have found by photometric measurement that a consumption of 33 feet per hour produces a light fully equal to that of the 4-wick lamp, and that this light, when properly placed with in the dioptric apparatus, maintains its equality from all points of observation. I have no reason to doubt but very strong reason to accept as correct, this result, which has been recently communicated to me by Mr. Wigham.

§ 4.

The coal employed for the manufacture of the Howth Baily gas is a Scotch cannel, called Stockwell cannel. It is delivered at Howth at a cost of 35*s.* a ton. It yields an exceedingly rich gas. A fish-tail flame of London gas is equivalent to 12 sperm candles, while the flame of the Howth Baily gas is equivalent to 38 such candles. The retorts containing the cannel are heated with Ayrshire coal, which is delivered at Howth at 22*s.* a ton, and with the coke of previous charges. Thus an 8 stone charge of cannel coal is distilled off at an expense of 10 stones of Ayrshire coal, together with the coke left by a previous 8 stone charge of cannel. This proportion between the cannel distilled and the fuel employed in distilling it is quite exceptional. With further practice a less amount of fuel will probably be found sufficient.

From data collected for me with all possible care by Mr. Valentin, I infer that 10,000 cubic feet of 38 candle gas are manufactured at a cost of raw material alone of 3*l.* 2*s.* 7½*d.* Taking the expenses of purification, labour, and renewal of retorts and furnaces at 50 per per, cent. additional, which, according to my data, is a close approximation to the truth, the total cost of the production of 10,000 cubic feet of Howth Baily gas is, in round numbers, 5*l.*, or 10*s.* per 1,000 cubic feet.

Assuming the number of hours of illumination in a year to be 3,780, the first power of Mr. Wigham's burner, which consumes 50 cubic feet per hour, would require, per annum, 189,000 cubic feet of gas, which, at a cost of 10*s.*

per 1,000 cubic feet, would amount to £94·5

The second power of the burner which consumes 105 cubic feet per hour, would consume annually 396,900 cubic feet of gas, at a cost of £198·45

The third power, which consumes 148 cubic feet per hour, would consume in a year 559,440 cubic feet, at a cost of £279·7

These sums do not include the interest on the original outlay. Taking that outlay at 800*l*., and allowing 4 per cent. interest, 321. will have to be added to each of the foregoing sums. This would make the annual cost of the respective burners:—

			£
Power I.	-	-	126·5
Power II.	-	-	230·0
Power III.	-	-	311·0

It is the first of these powers—the "standing order burner"—that ought to be compared with the oil lamp. Taking the annual consumption of oil at 880 gallons, and the price 3*s*. 10½*d* per gallon, being the average of all the prices paid at Howth Baily during the last five years, the cost of the four-wick lamp for oil alone would be 155*l*. Adding to this 13*l*. for wicks, cylinders, and burners,* we have the total annual cost of the four-wick lamp, 168*l*. A clear annual saving of 41*l*. 5*s*. therefore accrues from the employment of the gas.

Assuming the result communicated to me by Mr. Wigham to be correct, that a consumption of 33 feet per hour yields a light equal to that of the four-wick lamp, an additional annual savings of 32·13*l*. would be effected by the employment of this lower power; making a total saving of 73·63*l*. In fact, with this low power, which Mr. Wigham proposes to call "the equality burner," the annual expenditure would be 94·37*l*. as contrasted with 168*l*. for the four-wick lamp.

As already stated, the data on which this conclusion is founded are the best I could obtain. Relying on the figures as to expenditure furnished in Dublin, I think the result may be regarded as a close approximation to the truth.

But besides the advantage of greater cheapness, the great increase of illuminating power which the employment of gas places at the disposal of the keeper on foggy nights is a consideration of the very highest importance. This power is not drawn upon until it is needed, and it is then evoked in a moment of time. In connexion with this point, I would strongly recommend the continued comparison of the oil-flame with the higher powers of the composite burners in foggy weather. Such observations would yield results of far greater practical value and significance than those obtained with an ordinary photometer.

Whether gas can be profitably applied in any particular lighthouse or not, must depend upon the locality. In Rock lighthouses it would be out of the question. But there are doubtless various important points upon our coasts where its introduction would not be difficult. I do not think that either the manufacture of the gas, or the construction of the burner has, at present, reached its maximum perfection, and, with a view of attaining this perfection and of extending the basis of our induction, I would venture to recommend to the Board the further encouragement of this system of illumination *in Ireland.*

§ 5.

The observations and remarks regarding the Howth Baily light have been permitted to follow each other without interruption; I however snatched a day, the 8th of June, from them to visit the lighthouse at Wicklow Head. The light here is also produced from gas. It is an intermittent light, being rendered so by a machine which cuts off at intervals the main supply, leaving open a byeway which is barely sufficient to keep the tips of the flames alight. The apparatus when wound up is self-acting, and it appears to do its work admirably. The intervals of darkness are of 3 seconds, during which there is no waste of light, and the luminous page intervals of 10 seconds duration. The light is cut off with exceeding promptness, the dark intervals being sharply separated from the light ones.

Throughout the observations, both at Howth Baily and at Wicklow Head, I was much struck with the perfect mastery of the keepers over the apparatus entrusted to them.

Nor can I omit to mention the promptness and the skill with which I was seconded by both Capt. Roberts and Mr. Wigham. My suggestions were attended to, my requests complied with, and my instructions carried out with fidelity and cheerful despatch. To both these gentlemen I feel personally indebted, as also to the Commissioners of Irish Lights for their courteous interpretation of my wishes regarding the mode of observation.

I have, &c. | (Signed) JOHN TYNDALL.

T. H. Farrer, Esq., | Secretary to the Board of Trade.

* Since my return to London Captain Roberts, at my request, placed the gas flame in the dioptric apparatus and the oil flame in the hut below. From the North Wall the oil flame was brightest; thus affording additional evidence, if such were needed, as to the influence of the obstruction offered by the sash of the lantern.

* See page 6.

'Papers relative to proposal to substitute gas for oil as illuminating power in lighthouses', C 4210, Parliamentary Papers (1868–9), pp. 16–20

---

1. *Thomas Henry Farrer*: Thomas Henry Farrer (1819–99), and English civil servant and statistician. He worked briefly as a barrister at Lincolns Inn. In 1849 he provided legal assistance to the Board of Trade following the repeal of the navigation laws and the following

year he was appointed secretary of the Board's new marine department. His upward career trajectory continued with his promotion to assistant secretary of the Board (1853), full joint secretary (1865), and eventually first sole permanent secretary (1867). He was central to the Board of Trade's development in this period and was a proponent of a minimalist state and free trade (*ODNB*).

2. *Reports and Appendices submitted by the Commissioners of the Irish Lights to the Board of Trade*: presumably the 'Documents relative to the applicability of gas as an illuminating power for lighthouse purposes, forwarded by the Commissioner of Irish Lights to the Board of Trade' in 'Papers relative to proposal to substitute gas for oil as illuminating power in lighthouses', C 4210, Parliamentary Papers (1868–9), pp. 5–16. For Tyndall's work on lighthouses, see letter 2937, n. 6.
3. *Howth Baily*: in 1865 the Commissioners for Irish Lights received approval from the Board of Trade for a gas light to be installed on Howth Baily, in County Dublin (Ireland).
4. *Wicklow Head*: Wicklow Head, so named because it is a headland near the town of Wicklow, in County Wicklow, located in the east of Ireland. The light at Wicklow Head Lighthouse was converted to oil after the success at the Baily Lighthouse.
5. *visited the establishments referred to*: Tyndall was invited by the Board of Trade to visit the Howth Baily lighthouse in 1869 as part of his work for the Commissioners for Irish Lights.
6. *Royal College of Chemistry*: the Royal College of Chemistry was an institution devoted to the teaching of practical chemistry. It was merged into the Royal School of Mines in 1853.
7. *Commissioners of Irish Lights*: the Commissioners of Irish Lights held supervisory jurisdiction over public lighthouses in Ireland. In 1854 efforts were made to centralise the Commissioners, Trinity House, and the Scottish Board. However, the three boards continued to operate semi-autonomously. Trinity House, due to its seniority, was given the power to approve requests from the Irish and Scottish boards for improvements but appeals and the power to sanction expenditure was in the remit of the Board of Trade. This complex system meant that that action was often a slow process. On Trinity House and the Board of Trade, see letter 2937, n. 6.
8. *Bunsen's photometer*: an apparatus for comparing the intensity of two light sources. The sources are placed at either end of a bar *c.* 80–100 inches long. A small disk of partially oiled paper in a moveable frame is moved along the bar until it is equally illuminated on both sides. The reading is made from gradations on the bar (*Johnson's Universal Cyclopaedia: A Scientific and Popular Treasury of Useful Knowledge* (New York: Appleton and Co., 1896), vol. 6, p. 593).
9. *Capt. Roberts*: Capt. E. F. Roberts, R.N., was the Inspector of Lights for the Commissioners of the Irish Lights.
10. *Trinity House*: letter 2937, n. 6.
11. *North Wall and the South Wall*: the North and South Bull Walls were constructed on either side of the River Liffey and extend into Dublin bay. The South Bull Wall was built in the late eighteenth century to provide protection to shipping. Despite this, access to the port of Dublin was often restricted by the presence of a sand bar at the mouth of the port.

At the start of the nineteenth century the North Bull Wall was constructed. This in conjunction with the South Wall created a scouring action as the tide receded which halted the build up of sand. The North Wall is located approximately six miles away from Howth Bailey Lighthouse.

12. *Capt. Hawes*: Elliot H. Hawes, RN, was the Assistant Inspector of Lights for the Commissioners of the Irish Lights. He succeeded Roberts as Inspector upon his death.
13. *Poolbeg Lighthouse*: recognizable by its bright red brick facing, the Poolbeg Lighthouse was built on the Great South Wall in Dublin Bay at the Port of Dublin.

To [Henry Bence Jones][1] 5–8 July 1869 3020

Hotel del Europe Antwerp | 5th July 1869

My dear friend

I have been wandering among the pictures[2] till quite wearied out. I have come home and lain down for an hour, for last night was spent in the deck of the Baron Osy.[3] I have had a doze, and as the brain refuses to doze more I amuse myself by writing to you. I started from Catherine's Wharf[4] or thereabouts yesterday at noon; down the Thames, dodging the vessels in our way, gliding past Greenwich, Blackwall & Woolwich—onto Nore.[5] past the distant North Foreland[6] and into the sea. Just enough of tossing to make the squeamish stomachs uncomfortable; highly enjoyable for even those of average tenacity. Also we dined—a nameless fish and boiled mutton. Various other things, but these were very *[1 word illeg]*, and a moderate draught of 4c claret. I mean 4c at first. It did no harm. The breeze very stiff but clean and bracing: the sun sank in about half cloud half smoke. It is marvellous what a *[*trail*]* and *[2 words illeg]* came after it in the air, marvellous the altercation of the matter where you came to think of it; and yet what is it to other matter! I think I told you what I told them at University College[7] that I believe the blue light in the sky to be due to refraction from small particles; and that I also believed it to be demonstrable that if all the particles, necessary and sufficient to swathe the entire earth with a firmament of deepest azure were swept together with a broom, they might be held as a pinch of snuff between the finger and thumb—that the folding up of the visible heavens a scroll would amount to this and no more; The sky polarization was defective throughout the day; but improved towards evening. A coarser <u>Dunst</u>[8] diffused uniformly through the air damaged the polarization of the pure firmamental blue by the admixture of unpolarized light. The stars came out at the proper time, and I walked under them with Hirst[9] until 11.10 when we thought it time to turn in. We had each a berth in a small cabin which was filled to saturation with humanity. I took off my coat and boots & lay down; but the wisk of the paddles was augmented by the resonance of the cabin & the heat was too great, and altogether I thought a free life on deck preferable. So I rose and went up again. A first impassable spray which had been licked up by the sea during the day now *[sheeted]* the deck with moisture; but I made friends of the *[1 word illeg]* of *[1 word illeg]* *[*fellowship*]* and by tipping the waiter with half a crown :nanaged to get a mattress and some coverlets on deck. I would have had a couple of hours sleep. And the *[*keen*]* air which searched out the pores of the coverlet made everything clean and brisk and pleasant. I rose about 4 and had some breakfast at about 6, and reached Antwerp at about 8½; and at 10 after a wash, feeling slightly nauseous & not liking the thought of waiting 7 hours for the 50 *[1 letter illeg]* table d'hote,[10] I had a little cheese. Then we rambled through the town, and went to the Cathedral,[11] but too soon to see the pictures: so we went to another gallery, and thence to the nursery, and there back again to the Cathedral, where we saw those glorious works of Reuben.[12] What a worker the fellow must have been. I must mount a good Nicol's prism[13] for Boxall[14] when I return I found a small one very useful today in *[*quenching*]* the glimmer from the varnish. It is odd to see the picture which had been previously masked by the shimmer coming solidly out when the Nicol is turned into its proper position. I am high up, and the room is very hot, and my fingers are not very steady—I am not like Dean Stanley who I am told writes badly out of malice prepense. I would write well if I could. Today at 7 I go on to Brussels: sleep there tonight and take a scamper tomorrow near Waterloo.[15] There is a weight of sin upon my conscience; for I promised to write to Lady Tankerville[16] when I came from Ireland—She is so gracious and good to me that it would affect me greatly if she thought that I did not appreciate her goodness. Put in a good word for me if you have a chance.

Monday 5th July cont[inue]d.

Tom[17] lay sound asleep: I rather weary of working and decidedly simplified by the heat turned out in the air *&* read "In a gondola."[18] Then to the cathedral door, when the thought of going to the top of the spire arose. I paid my frank and bolted up the 600 steps. The air brisk at the top though dead calm was below; but the heat of motion more than compensated the meteorologic cold. The view was naturally very wide and the flat country round about: through it the Scheldt[19] wound in a serpentine bed. Waited for a time: then distant patter of feet was heard & after a time the keeper of the spire—a younger one than she to whom I paid my frank appeared. She brought a key which opened the door of a small central tower. Then a second door which opened a middle shaft, and in this the steps ended. It is close to the top of the cathedral. Halted for a time and swept the horizon; and looked on all that is embraced: field, woods and edifices. Heard the bells strike and sample the hour. It was half past four: my conductress took me everywhere & *[seemed]* desirous to shew me everything. As table d'hote was at 5; hastened down: I was moist with perspiration so had a *[*spring*]* of cold water. Dined and at 7 oC[lock]. departed for Brussels: the carriage in which we travelled horribly hot: it was the thick breezeless unmitigated heated or air oven; with walls warm around. I disliked it intensely. At Brussels first up at the Hotel de l'Europe a stroll in the park in the even[in]g : a small Champs Elysses;[20] weather lovely: stars aloft and the light of lamps below: the air itself serves a source of illumination; and the heavy glow thus looked upon was delicate & beautiful. At about 11 went to bed and had a very poor night.

Tuesday 6th July

Rose and had an extremely light breakfast: drove to Hougemont:[21] walked round the place—time advance[in]g and the marks of Waterloo fighting like every thing else are be[ginn]ing to be effaced. The change was great from 1848.[22] Then all was as fresh as if it had occurred a year ago; now the marks though plain are much less distinct. The filled */banks/* have crumbled & what was the unmistakable *[1 word illeg]* of grape & musquetry in 1848 might now be mistaken for general decay. The daughter of *[1 word illeg]* & the niece of Cotton[23] keeps the old museum of relics; being now an appendage to the Museum Hotel. This close to the Lions Mound;[24] went up the Mound; I rested there on the field, and lunched afterwards: then walked to La Haye Sainte[25] and La Belle Alliance.[26] The fact that Col. Gordon[27] was 29 years of age when he was mortally wounded remained in my brain for 21 years; and I mentioned it to Tom today before I saw the stone. The occurrences of the world become more phantom-like as I[28] grow older. I peopled the ground today with fighting columns, where are they? Where the glory of Wellington[29] as far as Wellington himself is concerned; where the agony of Napoleon as far as Napoleon is concerned? Brought Cotton's voice from Waterloo;[30] and Cotton himself lies under a stone in the garden of Hougemont.

"Bald ruhest du auch"[31]

Returned by 6 to Brussels: slight dinner and a walk before going to bed. Brussels is beautiful.

Wednesday

Travelled to Cologne and read Cotton. Mr. Kane[32] Lady Kane,[33] and two daughters.[34] They halted at Aix la Chapelle.[35] A short halt at Cologne & thence to Bonn. Put up at the Golden Star—induced to do so by an old gentleman who was loud in its praise. Had a little dinner & called on Clausius.[36] Saw the children all asleep. Clausius & his wife kind and affectionate as ever. Tried then Assmaushouse and found it good. Slept pretty well through many noises afterwards.

Thursday, 8th July.

Quitted the Golden Star for the Bellevue: a mistake. We ought to have gone to the Royal. The Golden Star only wants quiet to make it an excellent place. The Bellevue is a faded establishment. The people however were kind & civil. Went with Clausius to the Koenigswinter[37] & climbed the Drachenfels,[38]

had a little luncheon at the inn, and clambered afterwards to the top, all lava. The whole Siebengeberge[39] were once volcanos. The view exceedingly beautiful. Enjoyed it & descended; took a small boat & two men—rowed first up the Koenigswinter side of the Rhine,[40] struggling at intervals with the rapids; then crossed the river & were drawn up to Rolandseck.[41] Went up the Roderberg[42]—an extinct crater: the ruin evident all round there & then the lava was exposed. The crater itself leaves heavy crops & a farmhouse is planted in the middle of it. Exceedingly interesting. Walked all round the crater, and descended in the ruin of Rolandseck. The view of Siebengeberge from this point is strikingly beautiful. Reached the station in time for the 5.49 train & returned to Bonn. Supped with Clausius—saw the children; beautiful and exceedingly loveable little things. Met Max Schulze and liked him. Prof. Pflueger,[43] an eminent physiologist
<Letter ends here>

BL Add MS 63092, ff. 39–42

---

1. *[Henry Bence Jones]*: although the addressee is not named, it is presumably Henry Bence Jones, to whom Tyndall regularly sent long reports of his travels. See letter 3021.
2. *the pictures*: according to *Hirst's Journal* (RI MS JT/2/32d/1848), he and Tyndall visited the Picture Galleries, presumably the Royal Museum of Fine Arts, which had been founded in 1810.
3. *the Baron Osy*: an iron steam ship built for the Antwerp Steam Navigation Company, launched in 1851.
4. *Catherine's Wharf*: St Katharine docks in central London.
5. *Greenwich . . . Nore*: Tyndall's ship traveled from the docks of central London past Greenwich in South East London, home of the Royal Observatory and the Naval Academy. Following the Thames curve, he passed the town of Blackwall, a part of East London, and Woolwich, site of the Royal Arsenal. Nore is a sand dune that marks the point where the Thames River flows into the North Sea.
6. *North Foreland*: chalk headland on the Isle of Thanet on the southeastern coast of England. It offers a clear view of the North Sea. North Foreland was the site of a lighthouse managed by Trinity House (see letter 2937, n. 6).
7. *University College*: University College London.
8. *<u>Dunst</u>*: mist (German).
9. *Hirst*: Thomas Hirst.
10. *table d'hote*: restaurant or hotel meal served at a set time with a set menu and where diners shared a table (*OED*).
11. *The Cathedral*: the Cathedral of Our Lady located in Antwerp, constructed in the fourteenth century.
12. *Reuben*: Peter Paul Rubens (1577–1640), a Flemish painter of the Baroque period. Rubens lived in Antwerp. His four paintings, *The Assumption of the Virgin*, *The Raising of the*

Cross, The Descent from the Cross, and *The Resurrection of Christ* hang in the Cathedral of Antwerp.

13. *Nicol's prism*: see letter 2966, n. 4.
14. *Boxall*: Sir William Boxall (1800–79), a British painter. In 1866 he was appointed second director of the National Gallery. Boxall was instrumental in expanding the gallery's collection of seventeenth-century Dutch masterpieces (*ODNB*).
15. *Waterloo*: a town in Belgium, Waterloo was the site one of the most decisive battles of the Napoleonic wars, when armies of the multinational Seventh Coalition, led by the Duke of Wellington, defeated the French troops. Both sides suffered enormous casualties.
16. *Lady Tankerville*: Olivia Bennett, Countness of Tankerville (1830–1922) and wife of Charles Bennett, 6th Earl of Tankerville, British peer and Conservative politician.
17. *Tom*: Thomas Hirst.
18. *"in a gondola"*: 'In a Gondola', an 1842 poem by the English poet Robert Browning (1812–89).
19. *the Scheldt*: a navigable river flowing through France and Belgium.
20. *a small Champ-Elysses*: possibly the Parc de Bruxelles and Rue Royale.
21. *Hougemont*: Château d'Hougoumont is a farmhouse on an escarpment, where the coalition's forces and French troops faced off during the Battle of Waterloo.
22. *change was great from 1848*: Tyndall had visited the battlefield on 20 July 1848 (RI MS JT/2/14, pp. 364–8).
23. *Cotton*: Edward Cotton (1792–1849), Sargent-Major in the 7th Hussars. Cotton fought at the Battle of Waterloo and later compiled its history. After retiring from the Army, he returned to Waterloo, married, and opened an Inn there ('Plan of the Battle of Waterloo created by Sergeant Major Edward Cotton, January 1846', at https://collection.nam.ac.uk/detail.php?acc=1965-01-46-1 [accessed 9 Feb 2020]; C. Dalton, *The Waterloo Roll Call, with Biographical Notes and Anecdotes*, 2nd edn (London: Eyre and Spottiswoode, 1904), p. 269). See also n. 29.
24. *Lions Mound*: the Lion's Mound is a manmade hill which was constructed between 1823 and 1826 to commemorate the Allies' defeat of Napoleon. It is supposedly located at the place where William III, Prince of Orange, was hit in the shoulder by a musket ball.
25. *La Haye Sainte*: a farmhouse at the center of the Waterloo battlefield.
26. *La Belle Alliance*: an inn where the Arthur Wellesley, first Duke of Wellington and German General Gebehard Leberecht von Blücher met after the Battle of Waterloo to mark the end of fighting. The name signified the partnership of the countries making up the Seventh Coalition.
27. *Col. Gordon*: Lieutenant-Colonel Alexander Gordon (1786–1815), a Scottish Officer in the British Army. Gordon was chief aide-de-camp to Arthur Wellesley, first Duke of Wellington, who led the British and Allied troops against the French. He was killed during the Battle of Waterloo (*ODNB*).
28. *I*: 'I' is inserted above 'we'.
29. *Wellington*: Arthur Wellesley, first Duke of Wellington (1769–1852), a British military officer and later prime minister. Wellesley's military and civil service career had him

serving in India, as chief secretary for Ireland, and in the Peninsular War, before he was given charge of the Allied forces against Napoleon. Under his leadership, the Allied forces defeated the French troops, and Wellington became a national hero. He became involved in British politics and was briefly Prime Minister from 1828 to 1830 (*ODNB*).

30. *Cotton's voice from Waterloo*: E. Cotton, *A Voice from Waterloo: Being an abridged history of the battle, founded on authentic documents, official returns, and the combined testimony of eye witnesses; illustrated by the plan and key and embellished with portraits of the leading characters* (Brussels, 1846). Cotton's account was published in numerous editions.
31. "*Bald ruhest du auch*": German: 'Soon thou too shalt rest' (German), from 'Wanderers Nachtlied II', a 1780 poem by Johann Wolfgang von Goethe (1749–1832).
32. *Mr. Kane*: Robert Kane (1809–90), an Irish chemist, editor, and educator. In 1834 he was appointed lecturer in natural philosophy at the Royal Dublin Society and traveled to Giessen to study chemistry under Liebig in 1836. Kane received a medal from the Royal Society of London for his work on the composition of dyes from lichens in 1841 and became FRS in 1849. He was the Irish editor of the *Phil. Mag.*, to which he regularly contributed. Kane strongly promoted the expansion of science and industry for Ireland (*DIB*; *ODNB*).
33. *Lady Kane*: Katherine Sophia Kane (neé Baily, 1811–86), daughter of Henry Baily of Newbury, Berkshire, and a niece of the astronomer Francis Baily. Katherine Kane was interested in botany and is presumed to be the anonymous author of *The Irish Flora* (1833), the first comprehensive flora of Ireland. She was the first woman to be elected to the Royal Botanical Society of Edinburgh 1836. She married Robert Kane in 1838 (E. C. Nelson, 'Katherine Sophia Baily (Lady Kane) and The Irish Flora (1833)', *Archives of Natural History*, 46 (2019), pp. 44–57).
34. *two daughters*: not identified, but two of their seven surviving children.
35. *Aix la Chapelle*: Aachen, a town in the northwest of Germany.
36. *Clausius*: Rudolf Clausius.
37. *Koenigswinter*: a town on the Rhine River at the foot of the Siebengeberge in the contemporary German state of North Rhineland-Westphalia.
38. *Drachenfels*: a small mountain in the Siebengeberge near Koenigswinter. It was formed by lava pushing towards the surface of the earth but failing to break through to form a volcano.
39. *Siebengeberge*: see letter 2980, n. 3.
40. *Rhine*: the Rhine River, a major European river that has its source in Switzerland and runs through Germany and the Netherlands to the North Sea.
41. *Rolandseck*: a castle on a hill overlooking the Rhine. It offers a clear view of the Siebengeberge and the Drachenfels.
42. *Roderberg*: the crater of an extinct volcano near Rolandseck.
43. *Prof. Pflueger*: Eduard Friederich Wilhelm Pflüger (1829–1910), a German physiologist and professor of medicine. In 1858 Pflueger accepted the chair of physiology at the University of Bonn after it had been vacated by Helmholtz. In 1868 he founded the *Archiv für die gesamte Physiologie des Menschen und der Tiere* (*Archive for the General Physiology of Humans and Animals*), which was known as 'Pflügers Archiv' (*NDB*).

From Henry Bence Jones 17 July 1869 3021

My dear friend

Many thanks for your letter.[1] I wish you would give a lecture on the material helps to discovery in the laboratories abroad. Look at Ludwigs[2] laboratory[3] at Leipsic[4] as well as Hofmanns at Berlin[5] & see what they are in arrangements. It is too late for you to see Kühne at Amsterdam.[6] You might make a very interesting Friday evening; not your first however for in that I hope you will tell us what *[wit]* without material helps can do. Our institution laboratory ought to be the type for this country instead of being an adapted back kitchen. However <u>if it pleased you</u> the managers would I am sure do far more than they have as yet thought of doing. The South Kensington laboratory[7] is stopped for want of funds but when the money is voted I suppose it will be very perfect. The new larger theatre is I suspect a failure as far as being good for hearing.

Mr Berkeley[8] has settled with me as to the extent of our charges & by the time you come home I hope the work will be nearly done. Our total expenditure comes to about £85 & next year if the managers wish to make up for the lost settings we shall have a plan for a new gallery that will cost much but give back more than we have taken away.

Oh how hot you will be in Berlin my dear friend the baking from Antwerp to Brussels will be nothing to it I fear. Tell me all about your doings for they interest me much.

Pray give all kinds of kind words to duBois[9] & Hofmann if you see them from me. I wish I could be at your side during your visit. To hear & see all you will hear & see but alas I am fixed from within & have to learn to be quiet which is not easy for me

Since you went poor or rather I sh[oul][d] say lucky Barlow[10] has gone to rest. I saw him the day previously his mind quite clear & intending to have his motionless body moved to Brighton[11] but his phlegm would not be moved up his bronchial tubes & the machine stopped quite easily & a great release it must have been to him. I should think at least it would have been so to me if I had been in his state. I have nothing else new the Faraday Life[12] is gone to press, now I wish I would have victimised you by making you read the proofs but it is no use. I cannot change things after I have settled them. Miss Barnard[13] & Mrs Faraday are out of my course & they will not trouble me by corrections &c and will take what I say without approving or disapproving. I wish I had something better to talk to you about, but alas I have nothing. Ask in Berlin what is that last word on results of taking Chloral as a medicine by mouth or otherwise. Have any good results been obtained since duBois wrote to me. Here it seems as if chloral was not to be got easily I can as yet get none for experiments.

I got your line[14] about Lord Stanley[15] the money comes in very slowly. The total subscriptions at present do not reach 700. In fact without asking very few give and the purely voluntary system in church or science or any thing else is not a paying system for the receivers tho' no doubt very beneficial by reflection on the givers. Good for their spirits tho' bad for their pockets they give far surpassing their lots, not being indeed comparable if they could but see it so.

There have been one or two excellent articles[16] on Mr Lowes[17] law[18] in the Telegraph[19] & I hope it will not drop but I have not yet seen the Editor. I expect to do so

& now dear friend good bye.

I hope you will come back in Love with *[1 word illeg]* water, which is all one can take in this broiling weather.

I hope to be at Folkestone[20] in a fortnight.

There is another side to the *[*Institution*]* question of which you must have heard something in London & elsewhere. Be thankful that you escape the bad even tho' you have not the good. They altogether change a mans life & make him a money getter whether he will or no in England at least

Yours ever Dear J[ohn] | H Bence Jones | July 17. 1869

RI MS JT/1/J/57

1. *Your letter*: see letter 3020.
2. *Ludwig*: Carl Ludwig (1816–95), a German physician and physiologist. With Hermann Helmholtz, Ernst Brücke, and Emil du Bois-Reymond, Ludwig founded a new school of empirical and experimental physiology that treated physiology as a form of organic physics. Ludwig is considered to be one of the founders of modern physiology (*NDB*).
3. *Ludwigs laboratory*: Carl Ludwig's experimental laboratory at the University of Leipzig in the German state of Saxony focused on physiological research and was the most advanced of its kind in the world.
4. *Leipsic*: Leipzig, a city in eastern Germany.
5. *Hofmanns at Berlin*: August Wilhelm von Hofmann (1818–92), a German organic chemist. Hofmann studied with Justus Liebig at the University of Giessen. On the recommendation of Liebig and Prince Albert he was appointed director of the newly-established Royal College of Chemistry in Oxford Street in 1845. Hofmann became Professor of Chemistry at the University of Berlin 1865. The large laboratory was built under his direction and completed in 1869 (*ODNB*; *NDB*).
6. *Kühne at Amsterdam*: Wilhelm Friedrich Kühne (1837–1900), a German physiologist. After working as Rudolf Virchow's assistant in Berlin, he became Professor of Physiology at the University of Amsterdam in 1868. Kühne worked on muscle physiology and protein metabolism. In 1871 he became Helmholtz's successor at the University of Heidelberg (*NDB*).

7. *South Kensington laboratory*: the laboratory was completed in 1871. It was renamed the Normal School of Science in 1881 and subsequently the Royal College of Science. It is now the Henry Cole Wing of the Victoria and Albert Museum.
8. *Mr. Berkeley*: George Berkley.
9. *duBois*: Emil du Bois-Reymond.
10. *Barlow*: John Barlow (1799–1869), an Anglican clergyman and socialite who served as Secretary of the RI. Barlow studied at Trinity College, Cambridge, and was ordained in 1823. From 1842–60 he was the secretary of the RI, and together with Michael Faraday he undertook the day-to-day running of the institution. Barlow lived near the RI and entertained many of its wealthy and aristocratic supporters as well as fellow scientists, including Tyndall, who was a frequent guest (*Alum. Cantab.*).
11. *Brighton*: a town located on the southern coast of England.
12. *Faraday Life*: *Life of Faraday.*
13. *Miss Barnard*: Jane Barnard.
14. *your line*: letter missing.
15. *Lord Stanley*: Edward Henry Stanley (1826–93), the 15th Earl of Derby and a well-known British Conservative statesman, who expressed Britain's 'Splendid Isolation' policy in 1866. He was Foreign Secretary from 1866 to 1868 and again from 1874 to 1878 (*ODNB*).
16. *two excellent articles*: not identified.
17. *Mr. Lowe*: Robert Lowe, Viscount Sherbrooke (1811–92), a British politician. After studying law, Lowe began his political career in Australia. Returning to Britain, he was appointed to the Board of Control in 1852 and to the Privy Council in 1859. In 1868 he was appointed to the University of London. Lowe had a liberal position and worked on educational, electoral, and medical reform (*ODNB*).
18. *Mr Lowes law*: Robert Lowe was appointed Chancellor of the Exchequer in December 1868. In early April 1869, he proposed abolishing the one shilling registration duty on corn.
19. *The Telegraph*: the *Daily Telegraph* newspaper.
20. *Folkestone*: an important shipping and harbour port town on the English Channel in south-east England. Bence Jones had a residence there.

To the Editor of the *Times* 22 July 1869 3022

SWIMMING BATHS. | TO THE EDITOR OF THE TIMES.

Sir,—Yesterday, in Berlin, I had the great pleasure of reading your article[1] on the establishment of swimming baths in the Thames. Only a few minutes previously, my friend, Professor Hirst,[2] and myself, returning from a plunge in the "melancholy Spree,"[3] were speaking of the immense benefit derived from the frequent baths which we have taken since our departure from London a fortnight ago. At Bonn and Coblentz[4] we dived into the Rhine;[5] at Limburg[6] and Marburg[7] into the Lahn;[8] at Blankenburg[9] in the Hartz,[10] through lack of anything better, into a marl-pit; at Thale[11] into a mill-race; at Berlin into the Spree; and at Halle[12] into the Saale.[13] The "creatures of God," flesh and wine, which we have taken inwardly of late, would certainly have never been blessed to our use as they actually have been, were it not for this large external acquaintance with God's creature, water. London now possesses splendid facilities for the establishment of swimming baths, and it is to be earnestly hoped that the metropolitan press will not relax its efforts until every man of sound mind and flexible limbs among the millions upon its banks shall be able to indulge in a "header" into the Thames. The mere "tub" is a delusion in comparison with the free and joyful use of the muscles in the water.

JOHN TYNDALL. | Halle, July 22.

Times, 26 July 1826, p. 11
RI MS JT/6/5/1

---

1. *your article*: this article was published anonymously and without title in *Times*, 19 July 1869, p. 9.
2. *Professor Hirst*: Thomas Hirst.
3. *melancholy Spree*: the source of this quote has not been identified, but it seems to have been a regularly used phrase. The Spree flows through the Czech Republic, Saxony, Brandenburg, and Berlin.
4. *Coblentz*: see letter 3023, n. 2.
5. *Rhine*: the Rhine is a major European river that runs from Switzerland through Germany and the Netherlands to the North Sea.
6. *Limburg*: Limburg an der Lahn is a town is western Germany on the river Lahn. It was under Prussian control after the Austro-Prussian War of 1866.
7. *Marburg*: see letter 3018, n. 3.
8. *Lahn*: the Lahn is a tributary of the Rhine that extends for more than two hundred kilometers through western Germany.
9. *Blankenburg*: see letter 3023, n 7.
10. *Hartz*: the Harz is a mountain range with the highest elevations in northern Germany.
11. *Thale*: see letter 3023, n. 10.
12. *Halle*: see letter 3023, n. 12.
13. *Saale*: the Saale is a river in Germany. It is the second longest tributary of the Elbe after the Vltava.

To Juliet Pollock 23 July 1869 3023

Weimar 23rd July 1869

My dear Mrs. Pollock

Early in the morning of what will be a scorching day I am about to write to you. I raised my Ebenezer[1] yesterday in this famous little town. Thus far the lord had helped me. In reaching this place I scudded over Antwerp, Brussels, Waterloo, Bonn, Coblentz,[2] my old university nest Marburg[3]—so lovely it appeared and so lovely it was—Thence to Andreasberg in the Hartz,[4] then up the Brocken[5] where I had a glorious blaze of sunsets illuminating the dusky air as a fire at night reddens the atmosphere of London. Afterwards a howling storm and drifting fog over the crown of the hill. Thence to Wernigerode[6] through odorous pinewoods; then to Blankenburg[7] then to the Rosstrappe[8] and Hexentanzplatz,[9] through the most glorious rock and woodland scenery—This bit of the Hartz is of inexpressible beauty; finally we shot away from Thale[10] and the Hartz through Magdeburg to Berlin.

Here life changed wholly; a sandy flat with its heaps of edifices took the place of the mountain slopes and the mountain pines; while the philistines that swarmed over the Hartz gave way to a set of as fine fellows, intellectually considered, as it would not be easy to meet. We (that is Hirst[11] and myself) spent two pleasant and profitable days among our friends, and on Wednesday we quitted Berlin for Halle.[12] I should not have thought of halting there were it not that the man in whose cabinet,[13] and indeed in whose company, I made my first important investigation is now Professor there—I found him and received the memories of "Auld Lang Syne". And yesterday we came to Weimar.[14]

At 6 oC[lock]., when the sun had become feeble, we wandered out; first to the duplex statue of Goethe and Schiller which stands before the theatre.[15] There is calm grandeur in the face & pose of Goethe; while a bright, free, eager spirit looks forth from the feature & attitude of his companion. Then we went to Schiller's house. I wish humanity could be sifted and the philistines kept away from such places. We seemed fortunate, for nobody was there when we arrived. While I was sitting at Schiller's table a pair of chattering visitors came, and as the maiden who performed the part of showwoman was determined to kill all birds with one stone she sought to involve Hirst and myself in the exposition with which she favoured the chatterers. This would by no means go down with me. So I waited long until they disappeared and until the air behind them had subsided into peace, and then I had my own hour with Schiller.

Thence to the Church, where there is a noble painting of the crucifixion by Cranach;[16] the painting comprising portraits of Luther and of Cranach himself. There were other portraits also of Luther[17]—As monk in cape and then with fasting—as Ritter, in a Knight's disguise as he lived at the Wartburg where he translated the bible—As reformer where he stormed the world. The Church was full other interest: Here Luther preached; here Goethe had been often seen by my conductor. Here lay the bones and dust of celebrated men and women—Herder[18] among the former, for he was preacher in this church. "Licht, Liebe, Leben"[19] is the motto on his stone, and on a scroll held by the hand of his statue outside the words are to be seen. Here also lay the bones of the man who had shielded Luther, when courage was needed to shield him.

And then we went to the Fridenhof[20]—the place of peace—to see where Goethe and Schiller rest. We walked through an old burying ground—all quiet now and clothed in lasting peace. The sun was low. Gold was sprinkled over the leaves of the Accacias,—gold over the grass of the graves. We entered the temple raised over the remains of former grand Dukes and their ducal spouses & children. From the floor of the temple we looked down through a grating into a vault containing them all. By a series of stone steps we descended into this vault. A series of huge coffins were ranged like packed trunks around it. Near the door were two sarcophagi—I did not look at the inscription. I waited with expectancy to hear the names of those to whom they belonged; for by a kind of instinct I felt that they were those of Goethe & Schiller. Well the names were uttered and my instinct proved correct. I cannot tell you how the place moved me; it required a very hard twist of the screw to keep my feelings down. I turned from the dead men to the dead wall of the vault and asked myself why I was so affected—But I could render no account of it. Goethe & Schiller had been long in my brain—This my antecedent state, added to the present, produced the observed effect, that is the only explanation. I left them in their gray twilight, and ascended to the world alone. The gold was still showered upon the graves; and from a distance, like the Engelchor in Faust,[21] sweet voices came—such doubtless as Goethe & Schiller had often heard. We passed the singers: they were a group of beautiful girls amusing themselves in their own garden. By the way all Germany I have not seen so many fair faces as have shone upon me during this snatch of a visit to Weimar.

Then we wandered until nearly 9 oC[lock] when the earthy man requiring it we came home & had a little supper. The sun had sunk and left a belt of crimson round the west—while facing it was the full moon. We wandered to the park; for an hour & more we walked along the avenue to the Belvedere;[22] and at length finding it nearly 11 oC[lock]. we returned. The moon was to our right, and we, on the shadow side of the trees, walked in gloom. But against the boles and branches of the opposite row of trees the light was poured, and they shone like bits of the moon itself. The trees were Linden trees, and they filled all the air with odour. I give you bits and scraps of imagery out of which you may construct a picture of as perfect a heaven, as lovely an earth, and as glorious a memory as ever delighted the hearts of two solitary ramblers to whom rich things are dear.

We depart to day and will probably spend the next two days in the Thuringer Wald.[23] Afterwards we go to Heidelberg[24] and Carlrsruhe[25] to see our friends. Then to Basle[26] and the mountains. I hope to meet Fred[27] at Grindelwald[28] in the first week of August.

Yours ever as of yore | John Tyndall

My health which has been greatly shaken during the last couple of years is getting slowly & steadily better.

BL Add MS 63092, ff. 222–5
RI MS JT/1/TYP/6/2108–10

---

1. *I raised my Ebenezer*: from the book of Samuel (1 Sam. 7:12): Samuel sacrifices and prays to God that he and the Israelites may be granted divine protection against the Philistines. In the passage, after resisting the Philistines, Samuel sets up a stone between Mizpah—a city of the tribe of Benjamin—and Shen, and names the stone Eben-Ezer (Hebrew for 'the stone of help') and says: 'Thus far the Lord has helped us'). Raising one's Ebenezer means to acknowledge God's interventions in one's life.
2. *Coblentz*: Koblenz is a German city less than 60 miles southeast of Bonn also situated on the banks of the Rhine.
3. *Marburg*: see letter 3018, n. 3.
4. *Andreasberg in the Hartz*: Saint-Andreasberg is a village situated in the Harz mountain range in modern Lower Saxony.
5. *The Brocken*: the Brocken is the highest peak in the Harz mountain range.
6. *Wernigerode*: located on the north slopes of the Harz mountain range, Wernigerode sits at the foot of the Brocken.
7. *Blankenburg*: also on the northern edge of the Harz mountains, Blankenburg is a town east of Wernigerode.
8. *Rosstrappe*: the Rosstrappe is a crag in the Harz mountains with views into the Bode Gorge.
9. *Hexentanzplatz*: opposite the Rosstrappe in the Harz mountains, the Hexentanzplatz is a plateau lying above the Bode Gorge.
10. *Thale*: Thale is a town that borders on the northeastern limit of the Harz mountains.
11. *Hirst*: Thomas Hirst.
12. *Halle*: Halle is a city in the former Prussian province of Saxony (modern state: Saxony-Anhalt). Just outside of Leipzig, it sits along the Saale river.
13. *the man in whose cabinet*: referring to Hermann Knoblauch (1820–95) who was Tyndall's physics professor at the University of Marburg and who accepted a Professorship at the

University of Halle in 1853 (*CDSB*; *NDB*). In letter 3024, Tyndall praises Knoblauch's cabinet of instruments.

14. *Weimar*: a city in central Germany that was an important center for the German Enlightenment.
15. *duplex statue of Goethe and Schiller which stands before the theatre*: in front of the Court Theatre in Weimar stands the bronze statue of Johann Wolfgang von Goethe and Friedrich Schiller, two of Germany's most important literary figures of the late eighteenth and early nineteenth centuries. Schiller (1759–1805) was a German philosopher and poet, whose works focused on ethics, aesthetics, and notions of the sublime. Tyndall became acquainted with the works of Goethe during his Marburg studies and frequently quoted from them.
16. *noble painting of the crucifixion by Cranach*: Lucas Cranach, father (1472–1553) and son (1515–86) of the same name, painted the altarpiece for St. Peter and St. Paul's church featuring the crucified Jesus Christ.
17. *Portraits of Luther*: a triptych of Luther is attributed to a Weimar artist named Veit Thiem. Luther is depicted as a monk, a paster, and as Junker Jörg (when Luther's life was endangered because of his Theses, he grew a beard and disguised himself as an aristocratic) (J. L. Koerner, *The Reformation of the Image* (Chicago: University of Chicago Press, 2004), pp. 389–90).
18. *Herder*: Johann Gottfried von Herder (1744–1803), a German philosopher, poet, and theologian. Close friend and mentor to Goethe, Herder is also associated with the *Weimarer Klassik* movement (*NDB*).
19. *"Licht, Liebe, Leben"*: 'Light, Love, Life' (German). Herder's self-proclaimed motto is based on the Gospel of John.
20. *Fridenhof*: *Friedhof*, meaning cemetery (German).
21. *the Engelchor in Faust*: the choir of angels in Goethe's drama, *Faust* (1808). At the end of the play, the angels come down from heaven to lead Faust's soul to heaven.
22. *the Belvedere*: the *Schloss* (English: estate, palace, chateau) Belvedere is located just outside of Weimar.
23. *Thuringer Wald*: Thüringer Wald (Forest) is a forested mountain range in central Germany.
24. *Heidelberg*: located in south-west Germany, Heidelberg is famous for its university, Germany's oldest. There they would visit Bunsen, Kirchhoff, and the Helmholtzes.
25. *Carlsruhe*: Carlsruhe is located between Heidelberg and the Franco-German border. Tyndall's friend Gustav Wiedemann taught at the University of Carlsruhe.
26. *Basle*: a Swiss city located at the convergence of the French, German, and Swiss borders.
27. *Fred*: Frederick Pollock.
28. *Grindelwald*: a village in the Bernese Alps located in west-central Switzerland. Many foreigners, including Tyndall, came to the village to climb the peaks (i.e., Schreckhorn, Wetterhorn, Eiger, among others) surrounding it.

To Henry Bence Jones [1][1] August 1869 3024

Grindelwald | *[1]* August 1869

Dear friend

You almost cause me to return to my old habit of keeping a promise. From Berlin I went to Halle[2] to visit Knoblauch. If you wanted a specimen of methodical arrangement you would find it here. His apparatus is so beautiful and its organization so excel. It is quite a treat to look at Knoblauch's cabinet. This has always been a characteristic of his. From Halle we went to Heidelberg,[3] halting at Frankfurt where we had a plunge into the Main.[4] And at Heidelberg we found Bunsen and his house and laboratory. Nothing could be more strongly contrasted than Bunsen and Knoblauch. With the one order is every thing; the other seems to revel in confusion. my books and papers are symmetry itself compared with those of Bunsen. His laboratory is a fine one; full of strokes of genius, but of a genius to whom more systematic arrangement has but a limited value. In fact Bunsen is too full of thought to pay attention to the appearance of things save only so far as if they *[1 word illeg]* to the realization of his thoughts. With regard to my own laboratories your remarks would apply to the chemical one: for chemistry has its required processes and methods of investigation. You have here your contributions & the furnaces necessary for them; your methods of purification, distillation, crystallization, volumetric analysis, gas analysis—all methods well established and consolidated, and the applications of which consumes the chief part of the time devoted to an investigation. Here there is an opportunity for organization and arrangement. But with physics the case is totally different; each fresh investigation needs new modes of proceeding; new apparatus, and new combinations of apparatus. You have here no <u>firmament-substratum of processes</u> as in chemistry, and consequently you could not in the same sense have a model physical "research" laboratory. Were <u>teaching</u> our object a model laboratory or cabinet might be aimed at; for here the apparatus employed in the whole science of physics would be evoked, and methods of teaching the use of it might be readily devised. But the model physical <u>research</u> laboratory is that which furnishes the worker with all the means of carrying out his ideas, means which are incessantly changing with the professor's investigation and the direction of thought.

But even with regard to the chemical laboratory what can you do in rivalry with your competitors? What can you do in competition with[5] Frankland with all the appliances of South Kensington[6] at his back? I think discovery and discovery alone ought to be your aim in Albemarle Street;[7] that all your powers ought to be directed to the promotion of discovery, that whatever means you employ ought to have this for their end. It cannot however be doubted that the laboratories may be greatly improved even with reference to this object; but with your limits I hardly think you will ever be able to raise the chemical laboratory to the dignity of a model laboratory; and of a model physical research laboratory I have no distinct notion.

We spent two pleasant days at Heidelberg. Supped with Bunsen on the evening of our arrival, and dined with him next day, our party including Helmholtz and Kirchhoff. We had a butter brod[8] with Helmholtz in the evening. The following day we quitted Heidelberg for Carlsruhe where I rushed to see Wiedemann. But he was absent, and his very intelligent wife was also on the point of departing. Madame Helmholtz is handsome, & full of spirit and intelligence. From Carlsruhe I went on to Strasbourg where I parted[9]
<letter ends here>

BL Add MS 63092, ff. 43–4
BL Add MS 63092, f. 55r

---

1. *[1]*: the day of the letter is illegible, but this letter is referred to by Bence Jones in letter 3026 of 4 August.
2. *Halle*: see letter 3023, n. 12.
3. *Heidelberg*: located in south-west Germany, Heidelberg is famous for its university, Germany's oldest. Tyndall and Hirst visited Bunsen, Kirchhoff, and Hermann and Anna Helmholtz there. See *Hirst's journal* (RI MS JT/z/32d/1849–51).
4. *Main*: a large river running through Frankfurt.
5. *Grindelwald . . . competition with*: this portion of the letter is transcribed from BL Add MS 63092, ff. 43–4.
6. *South Kensington*: Frankland was professor at the Royal College of Chemistry. The College had not yet moved to South Kensington. See letter 3021, n. 7.
7. *Albemarle Street*: location of the RI.
8. *butter brod*: buttered bread (German), i.e. a light supper.
9. *Frankland . . . parted*: this portion of the letter is transcribed from BL Add MS 63092, f. 55r.

To Thomas Archer Hirst 2 August 1869 3025

2nd Aug[ust] 1869

My dear Tom

Miss Savage[1] (our housekeeper at the R.I.) writes to me saying there is a heavy letter[2] from the Board of Trade, and a second one[3] from Dublin[4] waiting for me in London. Like a good fellow open them, and if necessary reply to them—at all events let me know what they contain.

If you are in London I wish you would look at the other papers also & see whether any of them demands immediate attention. If you are not in London then be good enough to write to Miss Savage Royal Institution Albermarle Street and desire her to send you the two papers referred to.

I am not yet better than when we parted; but I am not without hope. My life needs a revolution to set it right. I rose this morning at 20 minutes before 1 oC[lock]. with a view of scrambling up the Matterhorn.[5] I wished to do it in a single day, and hence the necessity for an early start. But the heavens looked doubtful. Clouds were aloft, and sheet lightning played in the distance. And though there were prophets which prophesised fine weather it was too doubtful in my estimation to justify the attempt. So after scanning the firmament & the hill tops for an hour I went to bed again, and now the rain falls heavily & determinedly.

Monday I came here a body of guides had gone to seek a man who had slipped upon the Shreckhorn.[6] I believe he was a fine fellow, but very rash. He was between his two guides unroped—he refused to be roped—when standing on a step hewn in a steep ice-slope he thought he could reach by a spring some rocks which were at hand. He touched the rocks, but did not grasp them with sufficient firmness to hold himself. He glided past the guide behind him, who seized him by the arm but was unable to stop him. So he went down to the end of the slope where the precipices began and there disappeared. He was found clear of the mountains in the Lauteraar glacier[7] and must have fallen through a vertical height of at least 3000 feet. His name was Elliot:[8] he was a clergyman, and it was his traces that I found on the Matterhorn last year.

If the weather were not so abominable I would go up to the place where he slipped. He was buried the day before yesterday, the mountains round his grave firing shots of thunder at the time.

Saturday I walked up the Wengern Alp[9] & on my return had to wade knee deep in places which I had passed dry shod in the morning. A portion of the Upper Grindelwald glacier[10] had fallen in and had dammed the stream which flows from the glacier; the barrier although gave way & hence the swelling of the river. Near the end of the glacier was a plank bridge, and on this bridge stood a gentleman his wife and son, and their guide. A block of ice struck the bridge and carried it away, and with it the people upon it. The two men were saved, the woman & the boy were lost. The bodies of both have now been found.

What must the agony of this poor man be. Still the capacity to feel agony is a sign of health. There are things in this world worse than woe and grief, when the human life is emptied of risk and hope, and ceases to care about anything.

I shall stay here till the end of the week when I expect Fred Pollock. I shall then probably cross to the AEggishhorn,[11] and think you had better address your reply to this letter to me.

Hotel Jungfrau | AEggishhorn | Lei Viesch | Canton de Valais | Switzerland
Goodbye dear Tom | Yours affectionately | John

In looking over[12] an entirely new landscape we sometimes seem to illustrate Socrates' doctrine of reminiscence.[13] We discern in its features something that we seem to have seen before. And in observing the play of the human countenance a similar reminiscence is from time to time revived. I sat quite recently beside a lady[14] at dinner. I was silent for a time and she was occupied in conversation with her husband.[15] The talk & laughter subsided and then my own stillness became an object of selfconsciousness. I made some colourless remark and she immediately turned to speak with me. We continued to talk and I felt more and more interested in her speech and in her face. She was young and seemed rather delicate; dark hair; dark eyes which looked up at you at times from under a drooping forehead with an amazing depth of expression. And sometimes as if ignited by a sudden thought she would raise her head and with a flush of tropic sunshine on her cheeks would look you full in the face. Then she would turn aside her head musingly as if led away by the reflexions which the topic suggested, and then again eagerly return and throw upon the subject of a conversation inexplicable light and grace. I hardly dared to let her eyes play upon me. There was an abyss of feeling behind them which threatened to swallow me up. At times I permitted my vision to range across her face as over the sun where it is too bright to be looked at directly. And when we parted—and after we parted—and even when she was at the opposite side of the room the same deep splendour came from those eyes to me. The whole thing produces within me a cross fire of reflexions and reminiscences—it was so like the realization of something that I had seen and heard years ago in a dream. A match can fire a cannon, and the slightest spark of intellect or feeling suffices to fill those surcharged natures with a glow incomparably deeper and fuller than its exciting cause. And how beautiful is that glow so long as it remains unconscious; but how difficult where habit renders it the instrument of the will! There now I have filled half an hour of a dreary day with my last experience of womankind.

BL Add MS 60634, ff. 50r–51v
RI MS JT/1/HTYP/551–2

---

1. *Miss Savage*: Sarah Savage.
2. *a heavy letter*: letter missing. For Tyndall's work on lighthouses for Trinity House under the Board of Trade, see letter 2937, n. 6.
3. *a second one*: letter missing.

4. *Dublin*: probably a reference to the Commissioner for Irish Lights, which was located in Dublin at the time.
5. *Matterhorn*: see letter 2987, n. 17.
6. *Shreckhorn*: the Schreckhorn, a mountain in the Swiss Alps with an altitude of 4078 m.
7. *Lauteraar glacier*: a glacier is in the Swiss canton of Bern. It is a tributary of the Unteraar Glacier.
8. *Elliot*: Julius Marshall Elliott (1841–69), Reverend at St Mary's Chapel in Brighton. An experienced mountaineer, Elliott climbed Mont Blanc, the Dom, and the Weisshorn. In 1868 he successfully climbed the Matterhorn, the second ascent on the Swiss side. In July 1869 he slipped while climbing the Schreckhorn and died as a result of the deep fall (Anon., *The Churches of Brighton, Descriptive Sketches of their Past History and Present Characteristics* (London: H. J. Infield, 1882), pp. 403–6).
9. *Wengern Alp*: an alpine meadow located between Wengen (a mountain village) and Kleine Scheidegg (a mountain pass) in the Swiss Alps.
10. *Upper Grindelwald Glacier*: the Upper Grindelwald Glacier and its counterpart the Lower Grindelwald Glacier are valley glaciers near Grindelwald in the canton of Berne. The Upper Grindelwald Glacier is located between the Schreckhorn and the Wetterhorn.
11. *AEggishhorn*: the Eggishorn is a mountain in the Swiss Alps located in the canton of Valais.
12. *In looking over . . . womankind*: this passage is on separate sheets of paper from the main letter.
13. *Socrates' doctrine of reminiscence*: from Plato's Socratic dialogue *Meno* (81e) in which Socrates argues that human beings have knowledge of certain things that they could not have experienced. He claims that the human soul must then have had knowledge of these things prior to birth and that we need only retrieve them from our memory.
14. *a lady*: not identified.
15. *her husband*: not identified.

From Henry Bence Jones 4 August 1869 3026

Aug[ust] 4.

My dear friend

Your letter[1] from Grindelwald[2] has just come. I wish you spoke of yourself as stronger. I saw Hirst[3] last night at the Atheneum[4] & he did not look to me strong after his trip.[5] On the contrary I thought him all wrong, but perhaps he was tired. He told me a little of your laboratory (& other) sights but the country would have been better than the town for you both I suspect. I hope you will stay away a long time. Do not come back at least for Exeter.[6] Keep out of that excitement if it be possible & get yourself as well as I wish you to be

There is no news here. I went to Locker[7] on Sunday & certainly I shall not forget the beauty of Frauenhoffers lines & all that I saw altho' I am too slow to have seen all that I might have seen. It is a fine field & he does his work well tho his room & place looked as if his mind was Bunsen-like according to your description this morning. How the character comes out in little things.

I was amazed to hear Locker "pitch into" Huggins[8] & Secchi.[9] My what storms disturb the mental chromosphere. There must be an eruption soon in that direction.

I was interested to hear that you and Magnus were so smooth to each other.[10] I hope you will go to work at him when you come home & set the matter at rest for good and all & ever. I have no news for you your room looks so clean & nice. The theatre is progressing & will be a good work.[11] Lockeyer is to give six instead of three lectures[12] as Ruskin will not say yes or no & I wished to think of the list as completed by next year. Odling will give the 2nd Friday evening I have preached at his impossible abstract of his last lecture till I can say no more. The more I look at it the more utterly incomprehensible the doing & the reason for doing it appears to me. It can only answer the purpose of an impossible enigma "Give it up". However as he has shewn how far he is beyond all comprehension I hope for the future he will try to be comprehensible by every one. A much more important end for us & I think he sees it is so now.

I must finish my scrawl

Mind you stop at Folkestone[13] & sleep & eat & let me look how you are. I saw Mr H Holland this morning he is going 800 miles inland in America at 80!!! of which I & I fear you must despair

Goodbye dear friend | again & again | Ever yours | H Bence Jones | Aug 4 1869

RI MS JT/1/J/58

---

1. *Your letter*: letter 3024.
2. *Grindelwald*: see letter 3023, n. 28.
3. *Hirst*: Thomas Hirst.
4. *the Atheneum*: the Athenaeum Club (see letter 2944, n. 1).
5. *his trip*: Tyndall and Hirst had travelled together through Europe during the previous month. Hirst then continued on alone to Paris (RI MS JT/2/32d/1851).
6. *for Exeter*: see letter 3008, n. 10.
7. *Locker*: Joseph Norman Lockyer (1836–1920), a British astronomer. Lockyer was an early astrophysicist and used a spectroscope for solar observations. In 1868 he demonstrated that the protuberances seen during a solar eclipse were from the sun. He confirmed an atmosphere around the sun and came to the conclusion that it contained an as yet unknown element, which was later identified as helium, a discovery he shares with Edward Frankland. In 1869 he founded the popular science periodical *Nature* (*ODNB*; *CDSB*).

8. *Huggins*: William Huggins (1824–1910), an English astronomer known for his work, alongside his wife Margaret, in spectroscopy—specifically the emission and absorption lines produced by various celestial objects. After selling his family business in the 1850s he established a private observatory at his home where he made observations and developed methods and instruments for observing celestial spectra. In 1864 he was the first to observe the spectrum of a planetary nebula finding that they differed from stars much more than previously thought. He was elected FRS in 1865, received the RS's Royal Medal in 1866, and with William Allen Miller won the Gold Medal of the Royal Astronomical Society (1867) (*ODNB*; *CDSB*).
9. *Secchi*: Pietro Angelo Secchi (1818–78), an Italian astronomer. A Jesuit, Secchi worked at the Jesuit College in Lancashire and at Georgetown University during the expulsion of the Jesuits from France, returning in 1849 to direct the observatory at the Collegio Romano. Secchi was a gifted astronomer and did notable work on the rings of Saturn and solar radiation and made some of the earliest solar photographs during the eclipse of 1851. Throughout the 1860s he made significant spectroscopic studies of the planets, the sun, and luminous stars (*CDSB*).
10. *you and Magnus were so smooth to each other*: Tyndall and Heinrich Gustav Magnus had been involved in a dispute concerning water vapor's ability to absorb heat. In 1860 Tyndall concluded that water vapor could absorb substantial amounts of heat and thus explain temperature differences. However, the following year Magnus claimed the opposite. Despite Tyndall devoting considerable efforts to offer experimental proof of his findings, Magnus was to die in 1870 still unconvinced by Tyndall's claims. While the dispute was long-lived, it did not lead to any ill will (*Ascent of John Tyndall*, pp. 166–7).
11. *The theatre is progressing*: the South Kensington theatre and laboratory were under construction. See letter 3021.
12. *Lockeyer is to give six instead of three lectures*: Lockyer (see n. 7 above) delivered four lectures, 'On the Sun' at the RI on Saturdays from 19 March to 9 April 1870 (*The Engineer*, 29 (22 April 1870), p. 239; *Roy. Inst. Proc.*, 5 (1866–69), p. 611).
13. *Folkestone*: see letter 3021, n. 20.

To Thomas Archer Hirst [mid-August 1869][1] 3027

My dear Tom,[2]

Stegmann's nephew[3] has written to me[4] again: he says his family have sent him no money, and that he has in vain tried to earn some. He declares that he is starving, that he has no lodging and that he is compelled to walk the streets: he asks me for another sovereign. I think I ought send his letter to Stegmann. His address: 23 Dean Street Oxfordshire—though he says he does not live there nor anywhere. Would you kindly send him a sovereign having first ascertained that his address remains the same?

I remained at Grindelwald[5] for 10 or 12 days, hoping to get a little strength into my muscles, and the capacity of repose into my brain. I had one wild day upon the Wetterhorn[6] that was all, and I quitted Grindelwald less strong than when I entered it.

While there I did not neglect an wholesome practice of bathing—At first I had nothing but the ordinary baths attached to the hotel, but I But I[7] soon discovered a clear mountain stream and in the stream I found a kind of cataract, at the base of which the bed was hollowed out, so as to form a deep pool, where the clear water foamed and boiled: here I used to go even in wet weather and allow the water to boil over me. At St. Nicolas[8] also in coming up the Valley of the Visp[9] I had a plunge into the glacier torrent. There is a brisk *[1 word illeg]* below the hotel, and a jutting projection of pine trees, below which was an eddy where you were safe from the savage grasp of the river: here I had a refreshing plunge. Then I came on to the Riffelberg,[10] and the morning after my arrival we had snow: I went however to the lake below the Riffel,[11] and while the snow was falling heavily had my plunge. In fact I I have taken severe measures to harden my brain, but thus far without success. I must quit this place for a quieter one, and try what gentler methods will do.

I should not have come here at all were it not that Lady Mary Egerton had expressed so strong a desire to meet me among the mountains that it would have been absolutely uncivil on my part to back out of it. They are excessively kind, but they have but little notions of the sort of rest that I need, and the kind of life good for me. I sometimes fear that my brain has got a wrench which has damaged it permanently, but then hours of strength come to me, and hope rises once more.

Exeter[12] of course is out of the question for me. It would do me no good but rather damage to go there.
<Letter ends here>

BL Add MS 60634, ff. 53r–54r

---

1. *[mid-August 1869]*: Tyndall arrived in Grindelwald on 26 July and stayed for 10–12 days, leaving around 6 or 7 August and visiting the Eggishorn, Zermatt, and Visp, before arriving in the Riffel. He left for Brieg around 15 or 16 August and arrived at Belalp around 20 or 21 August (see letter 3030). This letter was written from the Riffel, probably around 13 August.
2. *My dear Tom*: this letter is probably an unsent draft. Much of the material is repeated in letter 3030, and this letter ends abruptly, as if unfinished.
3. *Stegmann's nephew*: see letters 2994, 3051, and 3054.
4. *written to me*: letter missing.
5. *Grindelwald*: see letter 3023, n. 28.
6. *Wetterhorn:* a peak in the Bernese Alps in west-central Switzerland located near the village of Grindelwald.

7. *but I But I*: Tyndall repeats this phrase at the end of one page and the beginning of the following page.
8. *St. Nicolas*: a village, in the Matter Valley, located in the Swiss canton of Valais some twelve miles south of the village of Visp.
9. *Valley of the Visp*: the Matter Valley is drained by the Matter Vispa river.
10. *Riffelberg*: see letter 3030, n. 15. A mountain below the Gornergrat above Zermatt, it is located in the Swiss canton of Valais. The hotel Riffelhaus, later renamed the Hotel Riffelberg, was built there in 1855.
11. *lake below the Riffel*: the Riffelsee is an alpine lake located near the town of Zermatt.
12. *Exeter*: see letter 3008, n. 10.

From Emil du Bois-Reymond 18 August 1869 3028

Berlin, 17 Victoria Str[asse]. | August 18th. '69.

Lieber Tyndall,

Ich schreibe deutsch, um sicherer klar zu sein. Wir haben uns bei Bestellung der Bussole nicht hinlänglich klar ausgesprochen. Die Bussolen sind fertig da, aber ich weiss nicht genau, welche Rollen sie wünschen. Sauerwald liefert drei Arten (immer ein Paar mit vier Drähten) 1. Hydrorollen, etwa 25000 Windungen feinsten Drahtes; 2. Thermorollen wie Sie bei mir sahen; 3. Rollen für reibungselectrische Ströme aus dickem mit Guttapercha überzogenen Drahte. Sie werden gebeten, anzugeben, was Sie zu haben wünschen, oder wenn Sie etwas Besonderes verlangen, etwa dickere Thermorollen als Sie bei mir sahen, dies näher zu bezeichnen.

2. Um das Gestell zu machen muss Sauerwald wissen wie Ihr Fernrohr beschaffen ist; ob Sie es auf dem Gestell dort in London zu befestigen beabsichtigen—ob es auf einen Theodolitenfuss steht; und alsdann wie hoch die Mitte des Objectivs über dem Tisch; oder ob Sie auch aus Deutschland ein Fernrohr miterhalten wollen u.s.w. Hierüber müssen wir durchaus genau unterrichtet sein, sonst könnte das Gestell anlangen und Ihnen sehr schlechte Dienste leisten.

3. Soll Sauerwald einen dünnen, einen dicken, Magnetspiegel, und einen Ringmagnet mit Glasspiegel schicken, oder nur letzteren?

I am going to leave Berlin for a trip in Switzerland and to the bay of Spezia, with my wife; so please write to Sauerwald 43 Kanonirstr., unless you prefer waiting till I am back, viz. till the middle of September. I send this to London, without knowing whether you are there or whether it can reach you where you are. At all events I hope it will find you in good health and spirits after your holiday excursion.

Believe me, my dear Tyndall, | Yours faithfully | E. du Bois-Reymond.

Berlin, 17 Victoria Str[asse]. | August 18th. '69.

Dear Tyndall,

I am writing in German so that I am more certain to be clear. We were not clear enough when discussing what magnetic needles to order.[1] The needles are ready but I am uncertain which coils you would like. Sauerwald[2] supplies three types (always a pair with four wires) 1. Hydro-coils, around 25000 turns of the finest wire; 2. Thermo-coils like the ones you saw here; 3. Coils made of thick wire coated with gutta-percha,[3] for static electrical currents generated by friction. You are requested to indicate what you wish to have or, if you desire something in particular, for instance thicker thermo-coils than those that you saw here, to indicate this more specifically.

2. In order to build the frame, Sauerwald needs to know the construction of your telescope; if you intend to mount it on the frame in London—whether it rests on a theodolite stand; and if so, how high the middle of the objective will be over the table; or if you also wish to receive a telescope from Germany, etc. We need quite precise information on these details; otherwise, the frame could arrive and only be of very poor service.

3. Should Sauerwald send a thin and a thick magnetic mirror, as well as a ring magnet along with the glass mirror, or only the latter?

I am going to leave Berlin for a trip to Switzerland and to the bay of Spezia,[4] with my wife;[5] so please write to Sauerwald 43 Kanonirstr., unless you prefer waiting till I am back, viz. till the middle of September. I send this to London, without knowing whether you are there or whether it can reach you where you are. At all events, I hope it will find you in good health and spirits after your holiday excursion.[6]

Believe me, my dear Tyndall, | Yours faithfully | E. du Bois-Reymond.

RI MS JT/1/D/147
RI MS JT/1/TYP/7/2434

---

1. *the magnetic needles*: du Bois-Reymond was asking about specifications for the parts of a galvanometer. For more on this, see letters 3061, 3104, 3140, and 3148.
2. *Sauerwald*: Ferdinand Sauerwald.
3. *Hydro-coils . . . thermo-coils . . . gutta percha*: the three kinds of electric coils used in a galvanometer for different measurements.
4. *Bay of Spezia*: the bay of Spezia is located on the Ligurian Sea on the north-western coast of Italy, close to the border with France.
5. *my wife*: Jeanette du Bois-Reymond (née Claude, 1833–1911). The daughter of a German merchant, Jeannette Claude grew up in Chile. Her family returned to Berlin in 1850, while she moved to England to stay with relations. She and du Bois-Reymond met in 1851 in Berlin and married in 1853 in Ambleside Church. Henry Bence Jones attended

the ceremony (G. Finkelstein, *Emil du Bois-Reymond. Neuroscience, Self, and Society in Nineteenth-Century Germany* (Cambridge, MA: MIT Press, 2013), pp. 126–8, 146–52).

6. *I am going . . . faithfully*: in English in original letter.

From Henry Bence Jones 23 August 1869 3029

My dear friend

I see by the account of Prof Phillips[1] lecture[2] that you have got to Exeter[3] safely. I hope your spirit has not been riled by Magnus on the absorption of heat[4] & by Tait on Comets.[5] When you get back write me an account of your doings & above all whether you heard anything for a Friday evening. I have seen nothing except some wonders on the growth of Quinine by Franklands[6] old assistant Mr Broughton.[7]

Pengelly[8] on Kent's Hole[9] would make a lecture & the deep sea soundings another. Pray fix any one you think worth fixing.

I suppose you had not time in Paris to go to the Rue Thorigny[10] for me?

I hope the smell of paint was gone from your study by the time you came back. I was obliged to have the little room done when I saw how bad it looked compared with the other.

I am getting lots of sea air & able to do more than I expected.

The book[11] goes on about 1/3^d^ printed

I hope you have come back very well

Ever yours truly | H. Bence Jones | Aug 23 1869

RI MS JT/1/J/59

---

1. *Prof Phillips*: John Phillips.
2. *Prof Phillips lecture*: John Phillips delivered a lecture on 'Vesuvius' at the BAAS meeting in Exeter, in Victoria Hall on 20 August 1869 (*Brit. Assoc. Rep. 1869*, p. lxxxvii).
3. *Exeter*: the BAAS meeting in Exeter (see letter 3008, n. 10). Bence Jones mistakenly thought that Tyndall attended.
4. *Magnus on the absorption of heat*: G. Magnus, 'On Emission, Absorption and Reflection of Obscure Heat', *Brit. Assoc. Rep. 1869*, pp. 214–15. See also letter 3026, n. 10.
5. *Tait on comets*: P. G. Tait, 'On Comets', *Brit. Assoc. Rep. 1869*, p. 21.
6. *Frankland*: Edward Frankland.
7. *Mr Broughton*: John Broughton (dates unknown) was Frankland's senior assistant at the RI until he resigned in November 1866 (C. A. Russell, *Edward Frankland: Chemistry, Controversy and Conspiracy in Victorian England* (Cambridge: Cambridge University Press, 1996), p. 216).
8. *Pengelly*: William Pengelly (1812–94), an English geologist and archaeologist from Cornwall. Pengelly tutored and lectured on various scientific subjects and promoted science

education. He was interested in the geology, palaeontology, and archaeology of the Devon and Cornwall region, included participating in excavations of Kent's Cavern (see n. 9) (*ODNB*).

9. *Kent's Hole*: also called Kent's Cavern, a large limestone cave near Torquay, Devonshire, England. Archaeological disputes centered around the question of whether it produced fossils demonstrating that humans and extinct animals had existed at the same time (M. J. S. Rudwick, *World's Before Adam: The Reconstruction of Geohistory in the Age of Reform* (Chicago: University of Chicago Press, 2008), pp. 227–8).
10. *Rue Thorigny*: probably a reference to the École Centrale des Arts et Manufactures, which was located at 7 Rue Thorigny.
11. *The book*: Bence Jones's *Life of Faraday*. See letter 2953, n. 1.

To Thomas Archer Hirst [29 August 1869][1] 3030

Belle Alpe Hotel Brieg | Sunday

My dear Tom

I wrote to you[2] from Grindelwald[3] and remained there a few days after I had written to you.[4] Six years ago I chanced to meet the Archdeacon of Wincester[5] and some of his family at the AEggischhorn.[6] Were you not there at the time, and did you not help me to give the young ladies their first lesson upon the mountains? I met him with an additional portion of his family at Grindelwald this year, and had some pleasant small excursions in this company. Assisting them in difficult places carried me out of myself and thus did service. It was, moreover, pleasing to gather from their conversation the innocent, active, beneficent lives which they led at home, and the manner in which their self help came into play in preparing with their own fingers all that was necessary for their Swiss journey. They knew no tedium either at home or abroad and they bubbled over with delight amid the grandeurs of the mountains.

I once made an attempt upon the Wetterhorn,[7] but not by the ordinary route. In general two days are spent upon the mountain; but it had been once accomplished in a single day by an alpine climber of high repute[8] for rapidity and endurance. Instead of rounding the precipices of the mountain he went straight up them. He had two guides. I thought I might follow his route with one guide. Well we went; climbed the precipices chasing the chamois before us, which being rarely disturbed are here exceedingly tame. We were for a long time within gunshot range of two of them. After having conquered the main difficulty we found a wild storm at work on the summit of the mountain, snow fell, and clouds wrapped the precipices all round. Up we went however into the cloud. The storm became more furious and the cold intense. We reached a col from which in ordinary weather the summit is allowed in three quarters of an hour, passed it and got upon the first snow arête. Worked up it for half an hour and more. I saw the man in front of me on his knees breaking the storm with one shoulder and trying to cut steps: so that through less than a quarter of an hour would have placed us on the summit, both of us thought the play not worth the candle and returned without reaching the summit. I have never been in a wilder scene. All my fingers had lost sensation; We lost our way for a short time among the crags as we returned. There was something savagely sublime in the aspect of the gorges and precipices, which were rendered black and of indefinable depth by the dense surrounding cloud.

While at Grindelwald I did not forget my wholesome habit of bathing. In fact the delight you seemed to take in it revived in me the pleasures of my boyhood; and nothing in the world filled me with greater delight than the practice of swimming. I found a clear brook near Grindelwald. At a certain place the water tumbled down in a cataract, and at the bottom of the cataract the bed was scooped into a foaming pool. Here I went in all weathers, and allowed every morning the water to boil over me. But notwithstanding the bathes and the glaciers, and the good company I could get nothing like health at Grindelwald, and I quitted after 10 or 12 days residence much weaker and lower than when I quitted you at Strasbourg.[9] Thence I went to Zermatt.[10] I had a fine bath in the roaring Visp[11] as I passed St. Nicholas,[12] taking advantage of an eddy formed below a projecting pile of pines to save me from the grasp of the river. I ought to have stated that I called on my way at the AEggischhorn and there picked up your brotherly letter.[13]

At St. Nicholas by the purest accident, as far as conscious volition is concerned, I met the Egertons.[14] We went in to the Riffel[15] together. At the Riffel bathing might be made a feature, but I fear they are too engaged in making money to think of any thing that does not directly promote this end. I bathed in the Riffel lake[16]—heavy snow was on the ground at the time and the thick flakes descended on my bare feet as I entered the water. All this was done to chase that terrible spell from my brain which deprived my nights of sleep and my days of enjoyment.

But the Riffel was as ineffectual as Grindelwald; so I quitted it after a time of three or four days, and came down to Brieg.[17] Here I met Layard[18] and his young wife[19] and with them ascended to the Belle Alpe.[20] You remember our arrival here together several years ago, and how disheartened you were at the time. You went to Leukerbad,[21] and I went I know not whither. The Layards remained here only a single night. She appeared to be quite unfit for mountain exercise, being seized with panting and palpitations of the heart after the most moderate exercise. They went away to Baveno,[22] and Nevile Lubbock[23] and his wife arrived. I took M^{rs}. Lubbock[24] up the Sparrenhorn.[25] We had an exceedingly pleasant day. On return we found the Egertons. They had met Layard at Brieg, and had deviated from their journey into Italy to come up here. Next day we all spent upon one of the higher glaciers. We had singing, as indeed we had upon the Sparrenhorn. M^{rs}. Nevile Lubbock sang some sweet simple things for me. I like her simple songs a hundred times better than her more ambitious productions. In fact it did my soul good to hear some of the melodies which she sang for me. I was on the whole however rendered a little thoughtful by M^{rs}. Lubbock—She appears to do what she likes with Nevile, and I could learn clearly from her account of his health that he needs a repose which I fear he is not permitted to enjoy. Still she is a hearty creature.

I have been here now about 8 or 9 days: the second day after my arrival I began to feel a health and lightness altogether new to me. The food here is good, and the air delicious. The weather since my arrival has been perfect. Every morning I saw the advance of day upon the cloudless eastern sky; and the peaks as a general rule remain clear from morning till night. The scene is glorious. Exercise taken in such air and sustained by such diet would tell upon almost any malady, and it has told most beneficially on me. In fact the only sensible benefit that I have derived from a two months holiday has been experienced here. On the 25$^{\underline{th}}$ taking with me a single guide I climbed the Aletschhorn[26] and enjoyed from the summit a view which has few parallels in my experience. The whole world of the Alps was disclosed, and indescribably magnificent it was. I considerably enriched my knowledge of atmospheric polarization.

A very genial man & his wife[27] came here two or three days ago. The lady who is considerably younger than her husband wished to go up the Sparrenhorn; she tried it in company with a young lady[28] who had been there before but on the steeper slopes her head failed her and she returned discomfited. I could see that she was greatly disappointed; and having been informed that previous to marriage she bore my name, being a niece of Chief Justice Tindal;[29] I said that I would guarantee her reaching the top, that I would take her under my own guidance. This seemed to please her very much; and it was arranged that yesterday morning we should rise at 6 and begin our march at 7 o' clock.

Since I came here I have continued to bathe copiously. Sometimes in clear mountain lakes where a good swim was possible, something permitting the clear cataracts into which the streams are broken to tumble over me. A cascade which I generally visited in the morning is not more than 3 or 4 minutes from the hotel. Yesterday, as usual, I went there; and having permitted the stream to tumble over me I was returning to my clothes dripping wet. Some slabs and blocks of granite intervening on these I stepped, and as those which were sharp to the feet I staggered and fell.[30] My shin was urged with great force against the granite, and a part of it larger than a five shilling piece was bitterly bruised & hacked: I struggled home & had to go to bed. The pain for a time was excessive; but a cold bandage diminished it; and in the afternoon

M[r] Cole the Queen's Counsel[31] & his wife[32] arrived. She had lots of Arnica oilskin & all manner of appliances, which in the kindest manner she placed at my disposal; assisting me at the same time with sound practical advise. I managed to get through the day & night well, filling the time for the most part reading novels; and this morning, though the wound looks very bad, all pain is gone. I fear it will keep me here for two or three days to come, for it is in a place difficult to heal and it will not do to irritate it.

I write to you in bed, but I hope, with the help of M[r]. Cole, to get down to the table d'hote at 5. The people, more especially my kindhearted chambermaid are very attentive to me.

I sent you a telegram[33] asking you to look at my letters & to send me a telegram if any thing required my immediate attention: no telegram has come & I therefore infer that you find nothing of the character referred to.

I must get out of this in a few days, though it would be of immense value to me to spend another fortnight here. And now the post commands me to halt.

Yours dear Tom | ever affectionately | John Tyndall

BL Add MS 63092, ff. 56–9
RI MS JT/1/HTYP/553–5

---

1. *[29 August 1869]*: the date on this letter is written in another hand, possibly Hirst's.
2. *I wrote to you*: letter 3025.
3. *Grindelwald*: see letter 3023, n. 28.
4. *remained there a few days after I had written to you*: Tyndall wrote to Hirst on 2 August (letter 3025). He intended to leave Grindelwald at the end of that week (5 or 6 August).
5. *Archdeacon of Wincester*: Philip Jacob (1802–84) was appointed Archdeacon of Winchester in 1860 and held the position until his death twenty-four years later (*Times*, 23 December 1884, p. 1).
6. *AEggischhorn*: the Eggishorn (see letter 3025, n. 11).
7. *Wetterhorn*: a mountain peak in the Swiss Alps near Grindelwald. Tyndall had attempted to climb the Wetterhorn in 1866 (*Ascent of John Tyndall*, p. 236).
8. *an alpine climber of high repute*: probably Julius Marshall Elliott (see letter 3025, n. 8).
9. *Strasbourg*: a French city located in the region of Alsace close to the German border formed by the Rhine.
10. *Zermatt*: a Swiss town in the Matter Valley, located in the canton of Valais. It is surrounded by the Pennine Alps including Switzerland's highest peak, the Monte Rosa, which Tyndall climbed for the first time in 1858.
11. *Visp*: the Matter Vispa, a tributary of the Rhône, drains the Matter Valley.
12. *St. Nicholas*: letter 3027, n. 7. St. Nicholas is a village in the Matter Valley, located in the Swiss canton of Valais some twelve miles south of Visp.
13. *your brotherly letter*: letter missing.

14. *the Egertons*: Edward Christopher Egerton (1816–69) and Lady Mary Egerton. Mary mentions them meeting Tyndall in letter 3040 (20 September 1869), see also letter 3030. Edward Christopher Egerton died a few days later from a sunstroke (See letter 3040).
15. *Riffel*: slopes below the Gornergrat near the Matterhorn, it is located in the Swiss canton of Valais. See letter 3027, n. 9.
16. *Riffel lake*: see letter 3027, n. 10.
17. *Brieg*: a city in the Swiss canton of Valais close to Zermatt ('Brig' in German).
18. *Layard*: Austen Henry Layard (1817–94), an English archaeologist and politician. In the 1840s, Layard travelled through the Middle East and Turkey, visiting Constantinople, Mosul, and Bagdad. He led excavations of the ancient Assyrian cities of Nimrud and Nineveh that uncovered several palaces, for which he received the gold medal of the Royal Geographical Society. After 1850, he entered British politics, founded the Ottoman Bank and promoted aesthetic education. In 1869 he was named ambassador to Madrid, preferring a diplomatic appointment to his continuing struggles for political reform.
19. *his young wife*: Mary Enid Evelyn (1843–1912), eighth child of Sir John Guest and his wife, Lady Charlotte Guest. Layard is rumoured to have been romantically involved with Lady Charlotte in the late 1840s.
20. *Belle Alpe*: Belalp, a village located in the Swiss canton of Valais, a 4.5 hour ascent from Brieg (K. Baedeker, *Switzerland and the Adjacent Portions of Italy, Savoy, and the Tyrol: Handbook for Travellers,* 3rd edn (Coblenz: Karl Baedeker, 1867), p. 256.
21. *Leukerbad*: a town located in the Swiss canton of Valais, west of Brieg.
22. *Baveno*: a town in the Piedmont region of northern Italy.
23. *Nevile Lubbock*: Nevile Lubbock (1839–1914), a merchant in the West Indies, who would eventually be named President of the West India Committee which aimed to promote ties with the colonies. He was the son of Sir John William, third baronet, and brother of John Lubbock (*Times*, 14 September 1914, p. 2).
24. *M^rs^. Lubbock*: Harriet Charlotte Lubbock (née Wood, d. 1878) married Nevile Lubbock on 18 June 1861.
25. *Sparrenhorn*: the Sparrhorn is a mountain in the Swiss Alps located north of Belalp.
26. *Aletschhorn*: located close to the border between the Swiss cantons of Valais and Bern, this mountain is the second highest in the Bernese Oberland.
27. *A very genial man & his wife*: not identified.
28. *a young lady*: not identified.
29. *Chief Justice Tindal*: Nicholas Conyngham Tindal (1776–1846), an English lawyer and politician. Tindal entered Parliament in 1824, became soliciter-general in 1826 and chief justice in 1829 (*ODNB*).
30. *I staggered and fell*: Tyndall contracted a serious infection after injuring his leg during this bathing accident and spent the next several months convalescing. See letters 3031, 3032, 3037, 3041, 3048, and 3049.
31. *Mr Cole the Queen's Counsel*: Henry Warwick Cole (1812–76), an English lawyer and member of the British Alpine Club.

32. *his wife*: Eliza Robinson (1819–77), an adventurous Alpinist better known under her publishing name, 'Mrs. Henry Warwick Cole', and the author of *A Lady's Tour round Monte Rosa:with Visits to the Italian Valleys of Anzasca, Mastalone, Camasco, Sesia, Lys, Challant, Aosta, and Cogne. In a Series of Excursions in the Years 1850–56–58* (London: Longman, Brown, Green, Longmans, and Roberts, 1859).
33. *Telegram*: telegram missing. Hirst mentions receiving a telegram from Tyndall on 4 September and sending a response, which is also missing (see *Hirst's Journal*, RI MS JT/z//32d/1853).

To Thomas Archer Hirst [4 September 1869][1] 3031

My dear Tom—

I was getting on exceedingly well today.[2] Kept quiet for 5 days. Thursday last I rose and moved about a little; resting in the sun in front of the hotel; but moving here also from time to time though not much. It was unwise for next day my wound was angry, red flesh & corrupt matter. I did not leave my bedroom yesterday & I am still in it. It is a pity, for I was on the way to health & strength. Even as it is I am better much than I expected to be. I do not sleep heavily, but what I get is very refreshing. I hardly think it in the least probable that I can get away from this place in less than a week: you know the state of my letters & would have time if you write promptly to & send me either them, or some account of them—I will write advice to Miss Savage[3] also by this post.

M[rs]. Ginty wishes to see me, and I told her I sh[oul][d] probably be in London on the 31st. of August: sh[oul][d] you know her address (she will probably have written to me)[4] pray explain to her the cause of my detention here.

Yours affectionately | John. | Belle Alpe Brieg | Saturday.

Professor Hirst | 18 Laughan Street | Laughan Place

BL Add MS 63092, ff. 60r–60v
RI MS JT/1/HTYP/557

---

1. *[4 September 1869]*: this letter was written on the Saturday following Tyndall's accident. See letter 3030.
2. *I was getting on exceedingly well*: Tyndall contracted a serious infection after injuring his leg during a bathing accident. He spent the next several months convalescing. See letters 3030, 3032, 3037, 3041, 3048, and 3049.
3. *Miss Savage*: Sarah Savage.
4. *she will probably have written to me*: it is unknown whether she did, since no letter has been found.

To Auguste de la Rive [8 September 1869][1] 3032

My dear Friend

Would you kindly when you receive this note do me the great favour of writing a line to me[2] to the Hotel de l'Ecu informing me of the address of an able surgeon I have hurt my leg by a fall[3] and am much in need of surgical advice.

Yours ever truly | John Tyndall

Belle Alpe Brieg Wednesday

I shall try to get to Geneva on Saturday

RI MS JT/1/TYP/1/366
LT Typescript Only

---

1. *[8 September 1869]*: the date is given by relation to letter 3031, written at Belalp on [4 September 1869], and letter 3037, written on 13 September from Geneva. The Wednesday between these two dates was 8 September. See also letter 3033 of 8 September 1869.
2. *writing a line to me*: letter 3035.
3. *I have hurt my leg by a fall*: see letters 3030, 3031, 3037, 3041, 3048, and 3049.

To Thomas Archer Hirst 8 September 1869 3033

Wednesday Sep[tember]. 8th. 1869

My dear Tom

Robson[1] has sent the proof[2] of my address[3] to the Institution. Look to it will you? I was so ill at the time I wrote it that I cannot quite trust myself for saying what I intended. Cut out that about the sky; it has no business there. Polish the introduction: I know there must be too much of the "melancholy *[*exordium*]*"[4] and look narrowly to what I have said about the emotions. I have no doubt I am right in the main, but the right may have been wrongly uttered.

How strong I was a fortnight ago. But this lying in bed has pulled me down terribly—N'importe—I must leave this for Geneva to get the advice of a surgeon. The leg is exceedingly ugly.[5]

Yours ever affect[tionately]. | John

RI MS JT/1/T/677
RI MS JT/1/HTYP/557

1. *Robson*: probably John Robson (1815–76), an English barrister, philologist, and Secretary to Council at the University of London.
2. *sent the proof*: see letter 3041.
3. *my address*: J. Tyndall, 'Address to the Students of University College, London, session 1868–69', in J. Tyndall, *Fragments of Science: A Series of Detached Essays, Addresses, and Reviews*, 6th edn, 2 vols (London: Longmans, Green, and Co., 1879), vol. 2, pp. 91–100.
4. *"melancholy exordium"*: meaning, roughly, 'sad introduction'. Exordium is Latin for 'the beginning of anything; *esp*. the introductory part of a discourse, treatise, etc.' (*OED*).
4. *The leg is exceedingly ugly*: see letters 3030, 3031, 3037, 3041, 3048, and 3049.

From Henry Bence Jones 9 September 1869 3034

just got your letter[1] | Thursday

My dear friend

I should have written to you when I got your first letter[2] but it took just as long as this and I thought mine would take as long to go back & that you would be all well before then & now I hope that may be before this reaches you & I shall not therefore say all I have to say.

But first & foremost. Do get home to this house & let your shin rest here.[3] You shall have my room and we will take such care of you. So pray come if you are not well. It would be very pleasant to me to have you here for a few days.

I think they send the foreign letters up to London & then return them here so that come without writing we are sure to have this room for you ready.

I am so sorry for your mishap. That your new experience of water (outwardly) should come to such an end is very sad, but however I dare say your thoughts on wandering will amuse thousands. These wicked satirists Field? Able? & Co[4] have been writing an essay on Punch by John Tynd-ll[5] LLD FRS. You must write one on Water inside not out. They have taken your style of writing! as closely as they could do so.

It sounds as if your wound wanted a little caustic or pressure or both. I longed to stop the arnica when you told me of Cole CB[6] giving it to you.

I saw in an Exeter paper[7] that you had assisted Phillips[8] to make an artificial geyser (which by the way gave very slight inductions when it was wanted & afterwards suddenly deluged the *[folk]*) & so I thought you were at Exeter[9] & wrote to you there[10] nothing of importance however

Some Irishman[11] I see has been working in the South of France with sunlight instead of electric light on your cloud formations & gave an account[12] of them at Exeter.

Magnus I told you had had a paper[13] on the absorption of heat by watery vapour & Tate[14] one on Comets[15] but I dare say you know all about these from Hirst or some one else. Little or nothing was in the papers about them.

I have no news to give you all at the Institution[16] seems well.

Do make up your plans to stop a day or two here that I may look at you

I am much better for the tea. Far better than last year as long as I take drugs I do well. Everybody will be glad if you will stay here for as long as you please

Good bye dear friend

I write a short letter only because I think this will reach you in London instead of the Belle Alps[17]

Ever yours | H Bence Jones | Sep 9. 1869

RI MS JT/1/J/60

1. *just got your letter*: letter missing.
2. *your first letter*: letter missing.
3. *let your shin rest here*: Tyndall was recovering from an injury to his leg. See letters 3030, 3031, 3032, 3037, 3041, 3048, and 3049.
4. *Field? Able? & Co*: not identified.
5. *an essay on Punch by John Tynd-11*: not identified..
6. *Cole CB*: Henry Warwick Cole and Eliza Robinson ('Mrs. Henry Cole'). See letter 3030, n. 31 and 32. 'CB' stands for Companions of the Order of the Bath, a British order of chivalry.
7. *In an Exeter paper*: the Exeter paper is not identified, but the experiment is described in *The Illustrated London News*, 55 (28 August 1869), p. 207.
8. *Phillips*: John Phillips.
9. *at Exeter*: meeting of the BAAS in Exeter (see letter 3008, n. 10).
10. *wrote to you there*: letter 3029.
11. *Some Irishman*: the paper was given by Jean François Auguste Morren (1840–70), a French physicist who was Dean of the science faculty of Rennes (1842), then of Marseille (1854-1870). It is unclear why Bence Jones assumed he was Irish.
12. *gave an account*: A. Morren, 'On the Chemical Reactions of Light discovered by Professor Tyndall', *Brit. Assoc. Rep. 1869*, pp. 66–72.
13. *Magnus . . . had a paper*: see letter 3029, n. 4.
14. *Tate*: Peter Guthrie Tait.
15. *Tate one on Comets*: see letter 3029, n. 5.
16. *the Institution*: RI.
17. *Belle Alps*: Belalp (see letter 3030, n. 20).

From Auguste de la Rive 10 September [1869][1] 3035

Mon cher Tyndall,

Je suis désolé d'apprendre par votre billet daté de Brigg, que vous avez eu un accident à la jambe; j'espère qu'il n'est pas grave; en tout cas il est important que vous vous fassiez bien soigner.

Je n'hésite pas à vous conseiller de faire appeler, dès que vous serez arrivé à Genève, Monsieur le Docteur Gautier. Il m'inspire toute confiance et la manière dont il m'a remis en 1865 mon bras qui avait été fracturé d'une manière grave m'a montré son habileté.

J'ai déjà prévenu Monsieur Gautier que vous le feriez chercher dès que vous seriez arrivé, et il aura soin, si par hasard il était absent, de recommander à l'un de ses collègues habiles comme lui, de le remplacer auprès de vous ; mais il est très probable que c'est lui qui ira auprès de vous.

Je mets ces lignes sous un pli à l'adresse de Monsieur Wolf le maître d'hôtel de l'Ecu en lui recommandant de vous les remettre dès que vous serez arrivé. J'écris aussi à Soret qui est à Genève de passer demain à l'Hôtel de l'Ecu pour savoir si vous êtes arrivé et comment vous êtes.

Je regrette vivement d'être dans l'impossibilité d'aller moi-même à Genève demain ; mais je ne tarderai pas à aller savoir de vos nouvelles.

Votre tout dévoué et affectionné | Aug≡ de la Rive. | Presinge—Vendredi 10 7bre

My dear Tyndall,

I am sorry to learn by your letter[2] dated from Brigg,[3] that you had a leg injury;[4] I hope that it is not serious; in any case, it is important that you seek treatment.

I do not hesitate to advise you when you arrive in Geneva, to call on, Mr Dr Gaut[h]ier. He has all my trust, and the way in 1865 that he reset my arm that was badly fractured has proven to me his skilfulness.

I have already warned Mr Gaut[h]ier that you will look for him as soon as you arrive, and he will take care, if by chance he should be absent, to recommend one of his skilful colleagues, to replace him in your care; but it is very likely that it is he who will come to you.

I put these lines in a letter to Mr Wolf[5] director of the hôtel de l'Ecu advising him to give them to you upon your arrival. I am also writing to Soret, who is in Geneva, that he pass by the hôtel de l'Ecu tomorrow to see if you arrived and how you are.

I greatly regret being unable to come myself to Geneva tomorrow; but I will not delay in having news from you.

Your very devoted and affectionate | Aug≡[6] de la Rive. | Presinge—Vendredi 10 September

RI MS JT/1/D/107
RI MS JT/1/TYP/1/368

---

1. *[1869]*: this letter is a response to letter 3032.
2. *your letter*: letter 3032.
3. *Brigg*: Brieg (see letter 3030, n. 17).
4. *a leg injury*: see letters 3030, 3031, 3032, 3037, 3041, 3048, and 3049.
5. *Mister Wolf*: Gustave Wolff, proprietor of the Hotel d'Ecu ('Murray's Handbook Advertiser. 1871', p. 4, in *A Handbook for Travellers on the Continent: being a guide to Holland, Belgium, Prussia, Northern Germany, and the Rhine from Holland to Switzerland*, 17th edn (London: John Murray, 1871).
6. *Aug≡*: de la Rive used three parallel horizontal lines to abbreviate his first name.

To Auguste de la Rive [12 September 1869][1] 3036

Sunday 13th, Sept[ember], 1869.

My dear Friend

Accept my very best thanks for the excellent and most complete arrangements that you made[2] for my comfort and well being. No man could have been better taken care of than I have been

Yours ever truly and gratefully | John Tyndall

From what I observed this morning when Dr Gauthier was dressing the wound I feel that I am on the road to recovery.

RI MS JT/1/TYP/1/369
LT Typescript Only

---

1. *[12 September 1869]*: the typescript gives the date as Sunday, 13 September, but this was actually a Monday.
2. *arrangements that you made*: see letter 3035.

To Thomas Archer Hirst 13 September 1869 3037

Hotel de l'Ecu de Genève | Geneva | 13th Sept[ember] 1869.

My dear Tom.

After 12 days waiting and being very chilled by the icy air of the Aletsch,[1] and rendered feverish by the pain and anger of my wound[2]—Having lost the most splendid of appetites and lived for two days on a lemon squeezed in water, I resolved to descend from the Belle Alpe.[3] I did so, much better than I hoped for, on an improvised stretcher. By good fortune I met an English medical man[4] on the mountain side. My foot at the time was shaped like a bladder of lard & of the colour of a live coal. He was exceedingly kind—every where I met kindness—men of all sorts ready to carry me & to do every thing possible to help me—the lotion thus obtained did me a world of good; next morning the swelling was greatly reduced. I went on to Sierre[5] & spent the night there. Stumbled against Poggendorff[6] & his kind little daughter[7] while being carried in two men's arms down the hotel steps. I had written to De la Rive asking him to send me here the address of a first rate surgeon. He is not at Geneva; but he managed matters so that Soret met me at the Railway Station; the hotel keeper here had his wits awake to find a room for me; and a second after I was carried into my room Dr. Gauthier entered it!

He immediately inspected the wound, dressed it with what I had brought with me, and wrote out another prescription. He paid me a second visit at 8 P.M. and again dressed the wound with a new cerate. I went to bed at 9 ½ or rather later. Three hours afterward the leg was in burning pain. I undid the outer bandages leaving the cerate intact, and exposing the leg & foot without any covering to the cool night air obtained great relief. This morning he came again. For several days the leg and instep have had an inflamed appearance and though the swelling had in great part subsided, I felt there was something wrong within even at a distance from the visible wound. This morning the red colour, which was very deep, ended rather sharply at the end of the shin, about 3 inches below the wound. He pressed this red part—the pain was sufficient to make a fellow sweat. He pressed his hand upwards towards the wound, and out of its ugly mouth poured a quantity of brown red slime: working with the hand, and afterwards with the sponge, continued till the sore was thoroughly cleansed. It gave me comfort to see the cause of the inner pains & soreness I had endured. The inflammation had worked downwards to the instep, and it was the bag of matter thus formed that Dr. Gauthier erupted this morning.

I have been in consequence better today than I have been for a week. Gauthier says however that a fortnight will be required to heal the wound. He lays great stress on the necessity of perfect quietness.

I fear my care was vain in sending you a telegram[8]—which cost 11 Francs—hoping that you would be able to send me an abstract of my letters; you mention neither it nor my last very long letter[9] in your recent note[10] to me.

My bedroom is a transmuted salon. It is au quatrième[11] and commands—could I stand up to enjoy it—a splendid view. I could wish it quieter; but it is by no means noisy. I hope to get well here.

And now for a little rest!

Goodbye dear Tom | Yours affectionately | John

BL Add MS 63092, f. 61
RI MS JT/1/HTYP/558

1. *Aletsch*: the Aletsch Glacier is the longest in the Alps and is located in the Swiss canton of Valais.
2. *my wound*: see letters 3030, 3031, 3032, 3041, 3048, and 3049.
3. *Belle Alpe*: see letter 3030, n. 20.
4. *an English medical man*: not identified.
5. *Sierre*: the Swiss canton of Valais is divided into districts; Sierre is one such district and its capital is also called Sierre.
6. *Poggendorff*: Johann Christian Poggendorff (1796–1877), a German physicist at the University of Berlin. His major contribution to science was as editor for fifty-two years (1824–76) of *Annalen der Physik und Chemie*, which was known as *Poggendorff's Annalen* during his editorship. He used his broad knowledge of science and scientific men in producing a *Biographisch-literarisches Handworterbuch zur Geschicte der exacten Wissenschaften* (Leipzig: Johann Ambrosius Barth), an approximate German version of the RS *Catalogue of Scientific Papers* (*NDB*; *CDSB*).
7. *his kind daughter*: Marie Poggendorff. In 1872 she married the German classics scholar, Valentin Rose (1829–1916).
8. *a telegram*: telegram missing. See letters 3030 and 3031.
9. *my last very long letter*: letter 3030.
10. *your recent note*: letter missing.
11. *au quatrième*: on the fourth floor (French).

From Henry Bence Jones 16 September 1869 3038

Folkestone Thursday | Sep[tember] 16 1869

My dear friend

I am so sorry to read the news of you this morning.[1] If I were well I should come off to Geneva but you might then have to stay & look after me. You will have first rate surgery at Geneva which you would not have here, but here you sh[oul]d have had the largest room in my house looking on the sea & we w[oul]d have taken more care of you than the Landlord of the L'Ecu[2] Do write to me often & tell me how you get on. If only a few lines. You seem to have been on the verge of Erysipelas[3] but all risk of that is past & the position in which you are will stop the matter from gravitating in the wrong direction & save you from a lower opening which a non physical sawbones would probably have made 'to let the matter out' as he w[oul]d have said. I trust all will go right now & that you will make a rapid recovery under a good surgeon as the periosteum[4] & the bone are all right. Position & place & cold water probably will soon make you heal; but I should like to have a talk with you every day for a few hours & this I sh[oul][d] have if I had you here.

All at the Institution goes on well. I have asked Miss Savage[5] to take her holiday & not to wait for your return I guess she will go off to Geneva to nurse you!!! I wrote you a line or two[6] to Exeter[7] to ask you about Magnus's paper[8] there. Is that what you allude to?[9] I saw nothing of Phillips's[10] attack[11] on any dead man in the Exeter paper.[12] What was it about? I have seen or heard nothing. One of Darwin's sons[13] was here & he told me that Darwin had been shaken by a fall riding. His horse went down & so did he & he has not gone out of a walk since. However he feels so much good from the riding that he is going at it again. Of this I am glad but I hope I shall not be the cause of his breaking his neck by telling him to ride.

Young Darwin had been in Paris & had seen some of the forgeries at Mr Chasles.[14] Yesterday's news[15] says the forger is caught in Paris. He must be a clever hand. Mr Chasles Mr. Darwin said was a simple believer void of all guile entirely. I asked him much about Clifford[16] who got a fellowship at the same time he did. He says that perhaps at the beginning of his undergraduate life Clifford might have been a Roman Catholic but that he is not now & has no tindency that way now. Mr Clifford is going to stay at Cambridge. His lectures at South Kensington[17] were much liked. Darwin is going to work at law—If his health lets him live in London of which I am not sure. Sylvester's name at Woolwich[18] is (*[1 word illeg]*). It excites his wrath horribly. I have got such a good engraving of Maul & Polyblanks[19] photo of Faraday for my book.[20] It is one of those taken with the heavy glass in his hand but with the eyes enlarged & it has all the kindness and genius I would wish in it. It is far beyond my Richmond[21] & there is also much of the body, the whole is very good. I have not seen the proofs of the laboratory & the study. We are getting on with the second volume. I have stollen from you immensely but without you the scientific part would have been very poor. The first vol is 430 pages, not very big & very large type & lots of space given to every thing. The letters as large as the text no difference in fact.

Miss Barnard[22] wrote to me from Venice I had told her to go to the Lido[23] to bathe & she says it was one of the best things in Venice as I found it when I was there. No news of the memorial it crawls on we have about £1100. Do you see for Commander Vanderbilt the Railway gambler[24] they have put up a monument. Big Statue for £50,000. Such is honor & glory. Oh my. What a farce & yet how one longs for success even at pitch & stops

Mr Scott[25] a metorology wants only to give 4 lectures so I offered J A Froude a short course & he has accepted & asks me about Calvin[26] as subject but I do not like the word & want him to take "on the influence of Fatalism of men & nations" What do you say? I must stop to day. I shall write to you again in a day or two, now I know you are fixed. Please write often

I wish I were with you or you here with me.

Good bye | dear friend

When do you go out look what a wretched state Davys monument is in at Geneva. The right hand side three parts of the way up the ground some way from the wall

RI MS JT/1/J/61

1. *the news of you this morning*: letter missing. From other letters Bence Jones had been aware of Tyndall's accident, but not yet of the dire state of his health. See letters 3030, 3031, 3032, 3034, and 3037.
2. *Landlord of L'Écu*: Gustave Wolff (see letter 3035, n. 5). Tyndall stayed at the Hotel de l'Ecu in Geneva while recuperating from his injuries.
3. *Erysipelas*: a streptococcal skin infection. It can cause high fevers, headaches, and vomiting. In the nineteenth century, before the advent of antibiotics, an erysipelas infection could be fatal.
4. *periosteum*: a layer of tissue which envelops the bones (except where they are covered by cartilage), consisting of a very dense, fibrous, and highly vascularized outer layer of areolar connective tissue and an inner layer of undifferentiated cells from which the bone substance is produced (*OED*).
5. *Miss Savage*: Sarah Savage.
6. *I wrote you a line or two*: see letter 3029.
7. *Exeter*: see letter 3008, n. 10.
8. *Magnus's paper*: see letter 3029, n. 4.
9. *what you allude to?*: letter missing.
10. *Phillips*: John Phillips.
11. *Phillips's attack*: not identified.
12. *in the Exeter paper*: not identified.
13. *One of Darwin's sons*: Darwin had five sons—William (1839–1914), George (1845–1912), Francis (1848–1925), Leonard (1850–1943), and Horace (1851–1928).
14. *some of the forgeries at Mr. Chasles*: Michel Chasles had purchased thousands of forged letters from Denis Vrain-Lucas (1818–1882), a prolific forger and con man. These included letters bearing the signatures of notable figures as Isaac Newton, Galileo, James II, and Blaise Pascal (H. Bourdier, *Prince of Forgers,* trans. Joseph Rosenblum (Newcastle, DE: Oak Knoll Press, 1998)). See also letter 3044, n. 4.
15. *Yesterday's news*: not identified.
16. *Mr. Clifford*: William Kingdon Clifford.
17. *South Kensington*: the Government School of Mines and the Royal College of Chemistry

were located in South Kensington. They were merged and expanded to form the Royal College of Science (1890), which later became part of Imperial College.

18. *Sylvester's name at Woolwich*: James Joseph Sylvester was professor of mathematics at the Royal Military Academy at Woolwich in south-east London from 1855 to 1870. Sylvester was engaged in an acrimonious dispute with the Royal Military Academy over regulation changes which forced his early retirement at the age of fifty-five without a full pension (K. H. Parshall, *James Joseph Sylvester: Jewish Mathematician in a Victorian World* (Baltimore: Johns Hopkins University Press, 2006), pp. 5, 215–7).
19. *Maul & Polyblanks*: London-based commercial photographers Maull & Polyblank (later Maull & Co.), who had studios on Gracechurch Street and on Piccadilly.
20. *my book*: *Life of Faraday* (see letter 2953, n. 1).
21. *my Richmond*: George Richmond (1809–96), an English painter and portraitist (*ODNB*). Richmond drew a chalk and charcoal portrait of Bence Jones in 1865, which is held by the RI. He would also paint a portrait of Tyndall to display at the RI (G. M. Prescott, 'Forging Identity: The Royal Institution's Visual Collections', in F. A. J. L. James (ed), *Science and Society at the Royal Institution of Great Britain* (London, New York: Routledge, 2017), p. 87).
22. *Miss Barnard*: Jane Barnard.
23. *the Lido*: the Venice Lido is a sandbar on the Adriatic in northern Italy.
24. *Commander Vanderbilt the Railway gambler*: Cornelius Vanderbilt (1794–1877), an American businessman whose extreme wealth came from his investments in railroads and shipping (*ANB*).
25. *Mr. Scott*: Robert Henry Scott (1833–1916), an Irish meteorologist. Scott studied under Heinrich Dove (1803–79) in Berlin and translated Dove's *Law of Storms* in 1862. He began teaching in Dublin in 1859 and in 1862 was named Keeper of Minerals to the Royal Dublin Society. In 1866 Edward Sabine offered him the directorship of the Royal Meteorological Office following the death-by-suicide of Captain Robert Fitzroy in April 1865; Scott accepted the position and began work on 7 February 1867 (*DIB*; *ODNB*; M. Walker, *History of the Meteorological Office* (Cambridge: Cambridge University Press, 2012), pp. 71–86).
26. *Calvin*: John (or Johannes) Calvin (1509–64), a Swiss religious reformer during the Protestant Reformation (*HLS*).

From Henry Bence Jones 18 September [1869][1] 3039

Saturday Sep 18

Dear friend

Your note of the 16[2] has come quickly. I wish it had better news of you.[3] Pray take as much brandy wine as your doctor tells you. It is no time for you to reduce your quantity when there is a discharging sore & you in a very different nerve state. I should probably give you more than your present doctor would order. A good deal more perhaps. Pray do not spare it. Good Neufchatel red wine is first rate but try the best. Some one told me of it and I got it at Neufchatel wonderful. I posted my last Brieg letter[4] at Boulogne myself a few hours after I got your letter.[5] However now you are near at hand. By the way there was no enclosure for Miss Savage[6] altho' you say there is. I hope she is gone for her holiday but I do not know where she is gone. I am so sorry to hear of your not sleeping all Wednesday night. I hope Thursday made up for it. To me the only satisfaction of a bad night is that more rest will come the next night inconsequence, but I fear that is not the case with you.

I have had a line from Froude this morning he does not like my not taking Calvin[7] as his subject.[8] It is very natural I suppose that whenever one crosses a persons idea there is an exageration of the opposition in his mind & a disposition to fight produced. However Froudes letter this morning will do to read to the managers will satisfy them that he will not willingly offend anyone's opinions. My only other piece of news is Cap[tai]n Brackenbury[9] wants to lecture on lessening the horrors of war. He calls it the highest science how tho' I can't look at in this light yet I have not refused him. Some one will want to lecture on the preservation of seagulls & sparrows. Cap. B. wants to tell what the Prussian & American ladies &c did[10] I suppose

I am very glad that you have met with kindness of all kinds in Switzerland. I wish you were in this house however instead of the hotel de l'Ecu. However you are in good hands & your account to day shews me that all chance of Erysipelus[11] is over but drink wine & eat what you can. Eggs beaten up in Brandy if you can take nothing else a dozen a day. If taken whole like oysters they are said to act on the bowels. I hope in a day or two to have a better account of you. Write often.

With kindest regards from us all Ever yours | H Bence Jones | Folkestone

RI MS JT/1/J/83

---

1. *[1869]*: the year is given by relation to letters 3030, 3031, and 3038.
2. *Your note of the 16*: letter missing. See letter 3038.
3. *better news of you*: see letters 3031 and 3038.
4. *my last Brieg letter*: presumably letter 3034 was the last sent to Tyndall at Brig/Belalp, since Tyndall left for Geneva to seek medical care a few days later.
5. *your letter*: letter missing.
6. *Miss Savage*: Sarah Savage.
7. *Calvin*: John Calvin (see letter 3038, n. 26).
8. *taking Calvin as his subject*: see letter 3038. Froude did not lecture on Calvin for the RI's Friday Evening Discourse series in 1870.
9. *Cap[tai]n Brackenbury*: Charles Booth Brackenbury (1831–90), a British army captain. After serving for several years, Brackenbury was appointed assistant instructor artillery

at the Royal Military Academy at Woolwich in 1860 and promoted in 1864 to assistant director of artillery studies. He was the military correspondent for the *Times* during the Austro-Prussian War of 1866 and the Franco-Prussian War (1870–71). He wrote several books on military and political topics (*ODNB*). Brackenbury did not lecture on war for the RI's Friday Evening Discourse series in 1870.

10. *what the Prussian & American ladies &c did*: presumably during the Austro-Prussian and American Civil Wars (1866 and 1861–65).
11. *Erisypelus*: see letter 3038, n. 3.

From Mary Egerton 20 September [1869][1] 3040

Mountfield Court | Hurst Green | Sept[ember] 20th

Dear Mr. Tyndall,

I know not whether I am wrong in yielding to the impulse that prompted me to write—I have no claim to intrude my grief[2] on you, but you are so associated with all the recollections[3] of those last bright happy days, that I cannot bear to let the icy barrier of silence grow up between us now, which will be harder and harder to break through as time goes on. We should have written to you from Baveno[4] had we known where to find you, for we could not bear that one with whom we had so lately spent such happy hours all together, should see our great sorrow only in the cold announcement of a newspaper.[5] I know we shall have had your sympathy. He was so fond of you, & I like to think that the feeling was mutual in some degree. But four short weeks ago today was that merry day on the glacier which he enjoyed so thoroughly. I seem to have lived through a lifetime since then!

Did you ever witness the awful sudden transition from the fullness of apparent health & vigour, to the cold peace of death? No gradual wearing away of the outward frame to tell that the pulse of life was beating fainter—every feature unchanged. it was so impossible to realize that the Spirit had fled! Oh! If I might not believe that a living Will is ordering all events, & that that Will is Love, my spirit would fail and faint indeed; but through all this trial there have been so many apparently accidental circumstances existing to help & support us, that I love to trace them to the tender hand of a loving Father, Who, while he alters not the working of the laws of health & disease for our sakes, can get so combine events as to spare His creatures & smooth their path of trial. I could tell you much of the boundless kindness & sympathy we experienced from utter strangers—of the comfort & help given us by the excellent clergyman there; one whom you would have thought worthy to read the words of "that glorious Bible," as you so truly called it, (he was a brother[6] of Coll. Hodson[7] of the Indian light horse;)[8] but I must not weary you with matters only interesting to myself.—The life of those few days seemed like a strange sort of trance, but here at home is the bitter awakening when every moment, every object, reminds me of him whose form now sleeps in his favorite spot in our little Church yard!

I could almost believe in presentiments when I remember how low & out of spirits I felt from the moment we left the Bel Alp,[9] & all through that weary Simplon[10] journey; how desolate the little empty inn looked after the cheerful party we had left! Italy seemed to have a cloud over it, & my daughter felt so too; but my dear Husband[11] & Hugh[12] were full of delight, & especially at the prospect of bathing in the lake! This feeling passed a little at Macugnaga,[13] where we spent a pleasant day, climbing the old moraine through your favorite little dell (?) to the further ridge, whence we looked upon the whole expanse of Monte Rosa[14] & its glaciers, traced the line of the new & old Weissthor,[15] &c. We returned that night to Ceppo Morelli[16]—the next day the end came.

May had a most kind note from M^rs^. Lubbock. In one day's acquaintance they had grown into friends whom we hoped so much to cultivate—but now!—Dear Mr Tyndall, you must not quite forget us, though the life & soul of our circle has been taken from us. We shall ever think of your kindness & patience in helping him, & all of us, along those mountain sides. Remember my dear husband's last words to you—"Report yourself to Mountfield[17] when you come back;" & let me hear some day how you are? & whether you have prolonged your holiday as you ought?

RI MS JT/1/E/49

---

1. *[1869]*: the year is given by reference to the recent death of Mary Egerton's husband in August 1869.
2. *my grief*: Mary Egerton's husband Edward Christopher Egerton (1816–69) died suddenly of a sunstroke on 27 August 1869, while hiking in the Italian Alps in Valle Anzasca, near Baveno (*Manchester Courier and Lancashire General Advertiser*, 2 September 1869, p. 8).
3. *the recollections*: Tyndall and the Egertons had met in the Swiss Alps a few days before Edward Egerton's death (see letter 3030).
4. *Baveno*: Egerton's husband died in Baveno, Provincia Verbano-Cusio-Ossola, Piemonte (or Piedmont), Italy (see n. 1 and n. 2).
5. *announcement of a newspaper*: since Edward Egerton was a British Conservative Member of Parliament, his death and the replacement discussions for his seat were reported widely in the newspapers.
6. *a brother*: G. H. Hodson (dates unknown), cleric and author of *Hodson of Hodson's Horse* (1883) which presented his brother, Colonel William Stephen Raikes Hodson, as an English Christian hero.
7. *Coll. Hodson*: William Stephen Raikes Hodson (1821–58), a British Army officer in East India Company service. He served in the first Anglo-Sikh War (1845–46) and

accompanied Henry Lawrence on an expedition to Kashmir. A difficult character, Hodson had a problematic record in the Army, and was demoted after being accused of embezzlement and mistreatment of Indian soldiers. During the Indian mutiny of 1857, he raised an irregular cavalry regiment, which became known as Hodson's Horse. His fighting tactics were both effective and brutal (*Alum. Cantab.*; 'Hodson, William Stephen Raikes', in H. Chisholm (ed), *Encyclopaedia Britannica*, 11th edn (Cambridge: Cambridge University Press, 1911), vol. 13, pp. 558–9).

8. *Indian light horse*: light horse brigade.
9. *Bel Alp*: Belalp (see letter 3030, n. 20).
10. *Simplon*: the Simplon pass lies between the Pennine and Leopontine Alps in Switzerland.
11. *my dear husband*: Edward Christopher Egerton.
12. *Hugh*: Hugh Edward Egerton (1855–1927), the younger son of Edward and Mary Egerton. He later became a historian (*ODNB*).
13. *Macugnaga*: village in the Piedmont region of Italy located at the foot of the Monte Rosa and along the Valle Anzasca.
14. *Monta Rosa*: the Monte Rosa is the highest mountain in Switzerland. Tyndall climbed it for the first time in 1858 (see M. D. Barton, J. Browne, K. Corbett, and N. McMillan (eds), *The Correspondence of John Tyndall, Volume 6: The Correspondence, November 1856–February 1859* (Pittsburgh: University of Pittsburgh Press, 2019)).
15. *new & old Weissthor*: two alpine passes near each other above Macugnaga.
16. *Ceppo Morelli*: a municipality located in the Piedmont region of northern Italy, bordering on the Pennine Alps.
17. *Mountfield*: Mountfield Court, the country estate of the Egertons.

To Thomas Archer Hirst 22 September 1869 3041

Hotel de l'Ecu de Genève[1]
22nd. Sept[embe][r]. 1869.

My dear Tom,

I am very much obliged to you indeed for the complete account you have given me[2] of my letters. It is most fortunate that neither the Trinity House nor the Board of Trades[3] has called upon me during the period of my confinement.[4]

It is weary work as you may imagine Tom, for a man like me. I have had now nearly a month of it. At the time of my accident I was in exceedingly good health, and for 10 days afterwards I was able to eat my beefsteaks every morning, and have the 5 oC[lock]. dinner of the hotel. But I then exposed myself one night for a long time to a very cold wind from the Aletsch Glacier;[5] the chilled me & produced fever and though I overcame it very soon it left me with barely a fraction of the strength that I had previously possessed.

The wound proper seems to be healing in a satisfactory manner: what troubles me most now is an irritation of the lymphatic vessels about the instep & ankle & the lower part of the leg. Nevertheless the D^r^. speaks hopefully if not confidently. He is satisfied with my progress and I therefore ought to be satisfied to.

My appetite is very poor; I feel an absolute repugnance to such food as they offer me here. I can hardly touch it—no doubt the cause resides as much in me as in the food. I shall be glad when I can put myself in a coupée lit and roll off to Paris, where I do not intend to halt. My notion is to go on to Folkstone[6] to Bence Jones.[7]

You know what rows he & I have had; but here in my hour of need he wishes to place his whole house at my disposal—offers to vacate his own room, which is the largest in the house, and which commands a full view of the sea; and pledges himself to nurse me & take care of me—In fact there is a deep stratum of good nature in the man—No brother could express himself more kindly than he has done in his letters.

I suffer very much from constipation. It makes me somnolent and apathetic.

Still notwithstanding all my ills and evils, the memory of my improvement at the Belle Alpe[8] comes back to me as a blessed influences. I felt there for some days a health and strength & *[purity]* of blood and brain such I have not experienced for the last five years. It was a joy to simply live and breathe the air of the mountains. Ennui was unknown to me. Few days more of this healthy life would have sent me back to England in a condition that would have astonished you. And although I could not have those few days, the practical demonstration that my body was capable of so great a renovation is a comfort to me now. When I get the chance I shall be able to lift myself to the same height and more.

St. Claire Deville[9] was with me here a few days ago: I cannot convey to you the force of his denunciations of Le Verrier.[10] And his picture of poor Chasles trembling under the blows of this merciless man was very touching. It is the most wonderful case of judicial blindness that history records. I wish something could be done to break the fall of Chasles,[11] by assuring him that while convinced of his error, and deploring it, no man in England entertained at any moment a doubt of his sincerity and good faith.

Let me also thank you for your care of the proof:[12] I was strongly tempted to write to Robson[13] to suppress it, and accept instead of it something that I should pledge myself to write when in a state of health. For I profoundly doubted whether any thing good could come out of a brain clouded as mine was when I wrote the little essay.[14] But then again I thought I would let it take its chance: trusting that the memory of my own experience was uttered with at all events approximate truth.

There has been a splendid fete here during the last two days—But it has passed me by like the sound of a simmering pot.

And now dear Tom I think I have exhausted all I had to say to you—it is all about myself, for illness makes us egoists. I suppose your duties commence very soon—Hope and trust that you have a stock of strength to discharge them.

Goodbye | Yours affectionately | John Tyndall

BL Add MS 63092, ff. 62–3
RI MS JT/1/HTYP/556, 559

---

1. *Hotel de l'Ecu de Genève*: Hotel de l'Ecu de Genève | Hôtel Byron près Chillon à Villeneuve | Lac de Genève | Prop[re] Gustave Wolff. Hotel headed notepaper.
2. *account you have given me*: see letter 3037.
3. *Trinity House nor the Board of Trades*: see letter 2937, n. 6.
4. *my confinement*: see letters 3030, 3031, 3032, 3037, 3048, and 3049.
5. *Aletsch Glacier*: see letter 3037, n. 1.
6. *Folkestone*: see letter 3021, n. 20.
7. *to Bence Jones*: see letters 3038, 3046, and 3050.
8. *Belle Alpe*: the Belalp (see letter 3030, n. 20).
9. *St. Claire Deville*: Henri Étienne Sainte-Claire Deville (1818–81), a French chemist at the famed École Normale Supérieure (ENS) and the Sorbonne. A talented teacher and experimentalist, Deville started as an organic chemist, and specialized in inorganic chemistry after 1849. He made numerous contributions to metallurgy, but did not become involved in industrial applications himself. His most significant research was on the disassociation and recombination of compound molecules which lead to greater understanding of the mechanisms of chemical reactions and physical chemistry (*CDSB*).
10. *Le Verrier*: Urbain Jean Le Verrier (1811–77), a French astronomer credited with predicting the existence of Neptune through mathematics. In 1854 he became François Arago's successor as Director of the Paris Observatory and Bureau of Longitude until 1870, when he was removed due to mismanagement and mass resignation of the observatory personnel. Le Verrier played a crucial role in exposing Chasles' letters as forgeries (see letter 3038, n. 14) (*CDSB*; J. Lequeux, *Le Verrier—Magnificent and Detestible Astronomer* (New York: Springer, 2013), pp. 187–93).
11. *the fall of Chasles*: see letter 3038, n. 14.
12. *the proof*: of his 'Address to the Students of University College London' (see letter 3033, n. 3).
13. *Robson*: probably John Robson (see letter 3033, n. 1).
14. *the little essay*: see n. 12.

To Auguste de la Rive 24 September 1869 3042

Hotel de L'ECU de Genève. | 24th, Sept[ember], 1869.

My dear Friend

I have been thinking since you left me to day of Mr Morren's achievements,[1] and have come to the conclusion that he is really not worthy of any attention. He will muddle himself more and more, and it will be amusing[2] to analyze his results and compare them with mine.

At the same time you could, if you would be kind enough to do so mention to Dumas privately that we have spoken together over the matter. But I should not like to make any public reclamation.

With very many thanks for your kindness | I am my dear Friend | Ever Yours | J. Tyndall[3]

RI MS JT/1/TYP/1/369
LT Typescript Only

1. *Mr. Morren's achievements*: see letter 3034, n. 11 and n. 12.
2. *amusing*: a note from LT reads: 'amusing or amazing'.
3. *J. Tyndall*: a note from LT at bottom of page reads: 'Evidently written with great effort—a shaky hand hardly legible at places'.

From Henry Bence Jones 24 September [1869][1] 3043

5 Albion Villas | Sep[tember] 24

My dear friend

Just got your letter of Tuesday[2] I wish it spoke of more rapid progress. But no appetite makes healing slow. I wish you had some Folkestone[3] mutton & some Boulogne brandy. We should soon get you up. But the weather has been very bad and Mr Penn[4] talks of leaving the end of next week & then the yacht goes also. Let me hear as soon as your plans get at all clearer for my motions in some respect depend on yours as soon as the yacht stops. There are no end of lodgings now to be had at a moments notice & first rate surgery for Simon[5] is staying here & would do all he could for you & he has more wit than any of the surgeons in London. Alas he has just been here & tells me he could have come to you in Switzerland he was close by you & would have brought you home. He leaves the end of next week to lecture at his hospital.[6] How I wish you were here under his care. If you see old Prevost[7] again I wish you w[oul][d]. ask him regarding the original statement respecting Davy & Faraday in 1814 at Geneva that Davy would not accept the Old DelaRives[8] invitation to dine with Faraday. Mr Marcet[9] has a story which doesn't do & the present DelaRive[10] I never got an account from & now I don't know from where exactly I sh[oul]d get my account exactly from. Made up of all. Is it <u>certain</u> that Davy did refuse to dine and did the old DelaRive give a second dinner to Faraday & the first one to Davy. There is a piece of work for you to make out on the spot. How I do hope you are eating better & sleeping well. What a sad time you have had. I have not any news for you. Froude is afraid of his subject & Seeley[11] who is appointed instead of Kingsley[12] at Cambridge wants to give fewer lectures. I was thinking when quite done with Froude to try Max Muller.[13]

I had hear of young Prevosts[14] state. It is the sadest of all illnesses. Bad & good both come out by intermarriage but bad only surely. Every thing is inherited & modified by circumstances.

Environments what a vile word. Surroundings would have been far more pleasant if a new word was necessary. We have got nearly 300 out of the 400 pages of the 2 vol[15] printed & all the pictures done the laboratory & the study are both too broad from the position of the drawer. But they are pleasant reminders of him to us who knew him. They will be no use to others. The grave is very good & will make the end of the book

Good bye dear friend | Ever yours | H Bence Jones

RI MS JT/1/J/84

---

1. *[1869]*: the year is given by reference to Tyndall's accident and his plans to convalesce at Folkestone.
2. *your letter of Tuesday*: letter missing.
3. *Folkestone*: see letter 3021, n. 20.
4. *Mr Penn*: possibly John Penn (1805–78), an English engineer and technician primarily of marine engines. He joined the family engineering firm after an apprenticeship with his father. The Royal Navy benefited significantly from his inventions which included a high-powered oscillating engine and the trunk engine for naval screw propeller vessels. His most important contribution to marine engineering is the co-development of the *lignum vitae* stern bearing which improved oceanic voyages for screw propeller ships (*ODNB*). See letter 3257.
5. *Simon*: probably John Simon (1816–1904), an English surgeon and public health officer. In 1848, Simon became the first elected medical officer of the Sewers Commission of the City of London, He became a strong advocate for sanitary reform and public health. In 1855 he was elected first president of the Association of Metropolitan Medical Officers. In the same year, he was named chief medical officer of the General Board of Health, after Edwin Chadwick and his associates were removed. He served as senior assistant surgeon at King's College Hospital (1840–53), lecturer in anatomical pathology at St Thomas's Hospital (1847–71), and as full surgeon there from 1863 to 1871 (*ODNB*).
6. *lecture at his hospital*: see n. 5.

7. *the old Prevost*: Alexandre Louis Prevost (1788–1876), the son of the Swiss physicist, Pierre Prevost, and member of the National Council of Geneva (*HLS*).
8. *the Old DelaRives*: Charles-Gaspard de la Rive (1770–1834), a Swiss physician, chemist, and physicist at the Academy of Geneva. After the French Revolution, he emigrated to Edinburgh to study medicine, receiving a doctorate in 1797. He returned to Geneva in 1799 and worked at the Hospice des Aliénés for thirty years. He held several honorary (unpaid) professorial positions, but had a well-equipped self-financed laboratory, at which he hosted numerous important figures such as Davy and Ampère. He was the father of Auguste de la Rive (*HLS*).
9. *Mr Marcet*: probably François Marcet. The story would, presumably, have been originally related by the elder Marcet, who visited the elder de la Rive in Geneva (J.-B. Dumas, 'Eulogy on Arthur Auguste De La Rive', in *The Miscellaneous Documents Printed by order of the House of Representatives* (Washington: Government Printing Office, 1875), pp. 184–205, on p. 186).
10. *the present DelaRive*: Auguste de la Rive.
11. *Seeley*: John Robert Seeley (1834–95), an English historian and essayist. He published on a range of topics ranging from Jesus Christ (1866) to British imperialism (1883) to Goethe (1894). His anonymous work, *Ecce homo: a Survey of the Life and Work of Jesus Christ* (1865), was controversial, appreciated by William Gladstone, but few others. Seeley succeeded Kingsley to the regius chair of modern history at Cambridge University in 1869 with Gladstone's support (*ODNB*).
12. *Kingsley*: Charles Kingsley.
13. *Max Muller*: Friedrich Max Müller (1823–1900), a German born Oxford philologist and specialist of Sanskrit. Müller visited England in 1846 to view some manuscripts and found great support for his work, returning as the turmoil of the 1848 Revolution made it difficult for him to work in Paris. In 1851 he was given a deputy professorship in modern European languages at Oxford University, and a full professorship in 1854. Müller lectured frequently on the science of language at the RI (*ODNB*).
14. *the young Prevost*: Alexandre Pierre Prevost (1821–73), a Swiss banker and member of the Society of Physics and Natural History of Geneva, who published several small papers, including on his theory binocular vision in 1843 and one on a mathematical theory of music in 1862. He married Auguste de la Rive's daughter, Augusta (M. A. de Candolle, 'Report on the Transactions of the Society of Physics and Natural History of Geneva from July 1873 to July 1874', *Annual Report of the Board of Regents of the Smithsonian Institution* (Washington: Government Printing Office, 1875), pp. 151–74, on p. 151).
15. *2 vol.*: Bence Jones was working on the second volume of his Faraday biography, *Life of Faraday*. For more correspondence, see letters 2953, 3021, 3038, 3050, 3057, 3236, and 3245. The image of Faraday's laboratory was featured as the frontispiece to the volume, his study on p. 1, and his grave on p. 479.

To Thomas Archer Hirst 26 September 1869 3044

Hotel de l'Ecu de Genève[1] | Genève | Sunday | Sep[tember]. 26. 1869

Dear Tom

Write to Lady Brodie[2] & tell her how I am circumstanced.[3] I will write to her myself when I get a little more strength. I do not know any books that I could recommend to her.

I must give up Chasles—God gave him an eye to geometric forms, but denied him wisdom and manhood. His retreat[4] was miserable; and it, if manfully conducted, might have atoned for all.

The memory just recurs to me that that man Burgess,[5] who was introduced to me by Carlyle, and who lectures at Mark Pattison's College[6] would be a capital fellow for Lady Brodie. He heard my lecture at Jermyn St.[7] & has a good knowledge of a lot of things—give him a chance.

(The only item against him in my thoughts is that he owes me 5 pounds, which as a man of spirit he ought to have paid long ago: God wot,[8] he may not have been able.)

I hope to quit this region next week—that in about 8 days from the present time—not sooner; also not later:—How wearily my time has passed!*

I take a coupée lit; and go direct to Folkestone,[9] not halting at Paris. I shall engage a body servant here to accompany me as far as Boulonge[10] or Calais.[11]

I have had my own servant, an old courier very respectable ever since I came here.

I am not equal to the writing of a longer letter at present. Yours affectionately | John.

* I have been in a state of low fever for the last three or four days and nights. It pulled me down much; but I appear to be reviving: at present even my only food is but of jelly daily.

RI MS JT/1/J/84
RI MS JT/1/HTYP/559

---

1. *Hotel de l'Ecu de Genève*: Hotel de l'Ecu de Genève | Hôtel Byron près Chillon à Villeneuve | Lac de Genève | Propre Gustave Wolff. Hotel headed notepaper.
2. *Lady Brodie*: Philothea Margaret Vincent Brodie (née Thompson, d. 1882), the wife of Benjamin Collins Brodie (1817–80), second baronet, an English chemist and FRS. They had six children (*ODNB*).
3. *how I am circumstanced*: Tyndall was bedridden recovering from serious leg wound infection. See letters 3030, 3031, 3032, 3037, 3041, 3048, and 3049.
4. *his retreat*: Michel Chasles staunchly defended the authenticity of the forged letters he

had purchased from conman Denis Vrain-Lucas (1812–88), which bore the signatures of notable figures as Isaac Newton, Galileo, James II, and Blaise Pascal (see letter 3038, n. 14, and letter 3041). For an extensive discussion of Chasles and these forgeries, see R. Higgit, '"Newton dépossédé!" The British Response to the Pascal Forgeries of 1867', *British Journal for the History of Science*, 36:4 (2003), pp. 437–53.

5. *Burgess*: Andrew Hutton Burgess (b. 1835), a Scottish journalist who in 1868 was lodging at 36 Park Walk, Chelsea (Carlyle Letters Online) (see letter 2884, n. 2, *Tyndall Corresopondence*, vol. 10).
6. *Mark Pattison's College*: Mark Pattison (1813–84), a prominent English author, priest, and intellectual (*ODNB*). Tyndall was likely referring to Lincoln College of Oxford University where Pattison served as Rector from 1861 until his death.
7. *my lecture at Jermyn St.*: the Government School of Mines and Science Applied to the Arts, established in 1851 (from 1863 known as the Royal School of Mines), was located on Jermyn Street. Tyndall held the Chair of Natural Philosophy at the Royal School of Mines from 1859–68, and in this role lectured to students during the week and in some years gave evening lectures for working men. On the Royal School of Mines, see R. M. Clary, 'The Royal School of Mines: Henry de la Bache's Convergence of Professionalization and Public Advocacy', *Earth Sciences History*, 39 (2020), pp. 291–304.
8. *God wot*: God knows (*OED*).
9. *Folkestone*: see letter 3021, n. 20. Tyndall's plan to visit Bence Jones are discussed in letters 3038 and 3041.
10. *Boulogne*: a city located south of Calais on the coast of northern France, on the English Channel.
11. *Calais*: a city and port located on the coast of northern France, on the English Channel.

From Thomas Henry Huxley — 30 September 1869 — 3045

Jermyn Street. | Sept[ember] 30th. 1869.

My dear Tyndall,

We only came back on Saturday and on Sunday I was distressed to hear from Spencer about your accident.[1] I have been inquiring at the Royal Institution, but I can't get much further than that you have damaged your leg bathing, and got an abscess. As I know what an atrociously careless old dog you are in such matters, my theory is that you have hurt your leg, neglected it and got an abscess in consequence. They tell me you have been laid up three weeks at that lively place, Brieg,[2] but that now you are at Geneva where, I am sure, De la Rive and your other friends will take care of you.

If it does not bother you to write, let me have some word of you, as we are all anxious.

After a sharp fight for Edinburgh, Liverpool was adopted as the place of meeting for the association[3] of 1870, and I am to be President; although the "Times"[4] says that my best friends tremble for me. (I hope you are not among that particular lot of my best friends).

I think we shall have a good meeting and you know you are pledged to give a lecture even if you come with your leg in a sling.

I have seen nobody but Spencer since I returned but I suppose we shall have an 'x'[5] next week. I shall live in hope you may turn up there. The chicks flourished wonderfully at Swanage[6] and they raised a howl when they heard of your being laid up the other day. The wife is not as well as I could wish she talked of writing to you but I suspect that plan is stopped for a while by the sickness of Ethel[7]—the baby-pet.

With best wishes | Ever yours | T. H. Huxley.

RI MS JT/1/TYP/9/2926
IC HP 8.75
Typed Transcript Only

---

1. *your accident*: letters 3030, 3031, 3032, 3037, 3041, 3048, and 3049.
2. *Brieg*: letter 3030, n. 17.
3. *the association*: the BAAS. Liverpool was selected as the site for its next annual meeting in 1870.
4. *"Times"*: the *Times*, a major newspaper based in London.
5. *an 'x'*: referring to a meeting of the X Club, a private London scientific dining club in which Huxley and Tyndall were members alongside George Busk, Edward Frankland, Thomas Hirst, John Lubbock, Herbert Spencer, Joseph Hooker, and William Spottiswoode. Active from 1864 to 1892, the club lacked stated objectives but the group's shared purpose was to promote science that was free from religious dogma. Over the years, its members became influential figures within British science, occupying important positions in the RS, the RI, and the Linnean Society amongst others. See letters 3047, 3048 and 3254. On the X Club, see *X Club*.
6. *Swanage*: a town on the southwest coast of England which served as an important port and holiday destination on the English Channel during the Victorian period.
7. *Ethel*: Ethel Gladys Huxley (1866–1941), the youngest daughter of Henrietta and Thomas Huxley. In 1889 she married her sister Marian's widower, the artist John Collier (1850–1934). The 1835 Marriage Act forbade marriage between a widow and his deceased wife's sister until the prohibition was removed in 1907 by the Deceased Wife's Sister's Marriage Act, so the couple wed in Christiania, Norway. Although they married with Huxley's blessing, the marriage did lead to some social ostracism (*ODNB*).

From Henry Bence Jones 30 September [1869][1] 3046

Folkestone | Sep[tember] 30

My dear friend

I have been looking for you in the tidal train hoping you might be more rapid than your letter.[2] But these dealings are slow affairs when there is any erysipelas[3] about. I shall go tomorrow to Copped Hall Epping.[4] Please write to me often there & let me know your plans and state because I may come back here to meet you if I am well. I have been bad in my heart lately & tho' we have had lovely sea weather I have not enjoyed it.

Mind you let me know your plans for my people will stay on for a week or two with a friend & even <u>when they go there is the house for you</u> and you can send for your factotum to help the woman who stays here & I can come & look at you & it is all in your road & far better than going to Sandown.[5] So whether I am here or not the house is here for you to stop in as long as you please until the sea air has mended your health if it then wants mending.

Only let me know your plans & our agent here Mr Sherwood[6] the grocer (or Dr Bowles my water doctor)[7] will help to make you comfortable. Simon[8] goes away to day. I am sorry to say we are at a temperature of 75° this morning. It is like midsummer tho we had a great storm in the night

I must end my letter once more saying Do look on this house as your own if we are away & if we are here we will make it yours as long as you like. I was going to put up a bed on the ground floor for you

Good bye. Let me hear soon if only a line how the drive suits?

Ever yours | H Bence Jones

RI MS JT/1/J/85

---

1. *[1869]*: Bence Jones invited Tyndall to convalesce at his country house after his injury in the Alps in September 1869. See letters 3038, 3041, 3049, and 3050.
2. *your letter*: letter missing.
3. *Erysipelas*: see letter 3038, n. 3.
4. *Copped Hall Epping*: country estate near Epping, Essex, somewhat north of London.
5. *Sandown*: seaside resort on the Isle of Wight.
6. *Mr. Sherwood*: John Sherwood, grocer and agent for house and lodgings, High St., Folkstone (*Visitors Guidebook for Folkestone, Sandgate, and the Neighbourhood* (Folkstone and London: J. English and Simkin, Marshall & Co., 1861), p. 37).
7. *Dr Bowles my water doctor*: Robert Leamon Bowles. See letter 3342, n. 3.
8. *Simon*: John Simon (see letter 3043, n. 5).

From Herbert Spencer 30 September 1869 3047

37 Queen's Gardens | Bayswater, | London, W. | Sep[tember] 30th, 1869.

My dear Tyndall

I called at the Royal Institution the other day anticipating the pleasure of a talk with you, and was shocked to hear of your accident[1] and the serious inconvenience and also, I should fear, pain which it has entailed. I have since been prompted, time after time, to join you, in the hope of mitigating somewhat the weariness which this long confinement to your room must cause you; and had it been at the beginning of my holiday, instead of at the end, the prompting would, I think, have resulted in action. But I have done so very little work during the whole of this year that now, after returning from a very long rustication,[2] I am eager to get something done, and cannot make up my mind to lose more time. Moreover I have the impression that you have some scientific friends in Geneva,[3] and that your life there will not be altogether so irksome as it must have been at Brieg.[4]

What a pity you did not, before going, take that long-contemplated step which we have discussed so often.[5] I suspect that, on your return, recent experiences will very decidedly accelerate matters.

Though, from what I gathered, this hurt must have been a serious one, yet, the inflammatory stage having, as I conclude, passed by, the great reparative powers of your system will, I doubt not, soon remedy the immediate mischief.

As I was returning to town I was looking forward to a lively X dinner.[6] If we have one, about which there seems some doubt, we shall feel greatly the absence of your infectious vivacity. Is it likely that if we postpone it till the 14th you will be able to join us?

Ever sincerely yours | Herbert Spencer

RI MS JT/1/S/98
RI MS JT/TYP/3/1115

---

1. *your accident*: letters 3030, 3031, 3032, 3037, 3041, 3048, and 3049.
2. *returning from a very long rustication*: Spencer had just returned from a trip to the 'rustic' Scottish countryside.
3. *some scientific friends in Geneva*: Tyndall's Geneva friends include De la Rive and Soret.
4. *Brieg*: see letter 3030, n. 17.
5. *that long-contemplated step which we have discussed so often*: with no direct evidence whatsoever, the editors suspect that Spencer is alluding here to Tyndall's bachelorhood and his unspoken love for Mary Adair. Shortly after his return to London, he proposed to her, but she declined. See letter 3059.
6. *a lively X dinner*: referring to a meeting of the X Club (see letter 3045, n. 5). See also letters 3048 and 3254.

To Thomas Henry Huxley [2 October 1869][1] 3048

Geneva, Saturday.

My dear Huxley,

I had already thought of sending you a Compte Rendu,[2] for the edification of the X[3] on Thursday, when your note[4] arrived and converted intention into act.

Be it known unto you then that for the last four years and upwards I have never known the blessing of a day's really good health. This state of things it was my desire to grapple with this year and subdue if possible—I had a four weeks scamper with Hirst in Germany,[5] and did what I could to restore myself by bathing; it did me good, but still real trustworthy health did not come back to me. I parted from Hirst and went on to Grindelwald;[6] I had some exercise here upon the mountains and glaciers, and I continued my practice of bathing every morning—but Grindelwald would by no means agree with me. The greasy chaos of a table d'hote[7] did not suit me; still I was in a measure compelled to dine at the table d'hote through the presence of friends. I finally quitted Grindelwald after a sojourn of 12 days less strong than when I entered it. An engagement took me to Zermatt,[8] I called at the AEggischhorn[9] upon the way; but that specimen of lip-service Herr Welling[10] did not please me, so I halted only a night at his hotel. On my way to Zermatt, I continued to bathe; I did not halt at Zermatt, but went up to the Riffel[11] where I stayed four days. I bathed here in the Riffel See[12] when it was snowing heavily. But the Riffel did me no good. The food was poor, and the distorted molecular force of bad beef and mutton was sufficient to neutralise all effects of air and exercise. So I quitted the Riffel unimproved. I then turned my steps towards Brieg;[13] met Layard[14] and his young wife there, and when the rain clouds cleared away we went up to the Bel Alp[15] together. The hotel stands upon a bluff of the mountain right over the lower part of the Aletsch,[16] and commands a view of the glacier from the Marjelin See[17] downwards. The air here was delicious, and the food considering the height excellent. In two days I felt a decided change for the better. The dirt in my blood seemed to vanish, my brain seemed clear and strong and my muscular power augmented enormously. Simple existence became a pleasure to me. I had my mountain walks, but still held on to my baths. The mountain streams were splendid, and the weather delightful. I used to pause by the clear pools at the bases of the cascades, strip and roll myself over and over in the delicious foam. And on quitting the water I would jump from the mountain, as from a spring board, until all the drops were shaken away, and until the sun had dried me thoroughly. There was a lovely lake about 10 minutes walk from the hotel which furnished a glorious swimming bath; five minutes further on there was a cascade about seven feet high, with a copious chute of clear water. Here one could pass to and fro through the cascade getting behind it, or in front of it, or in the middle of it. I found it very pleasant to let this cascade tumble over me. One day Mr and Mrs Neville Lubbock[18] made their appearance having brought a cavalcade of men and horses across the glacier. We climbed the Sparrenhorn[19] together and there she sang me some of her simple songs: it was perfectly delightful to me to listen to them. I do not care so much for her dramatic power as for the pathos which she throws into these simple Scotch and Irish songs and ballads. The Egerton's[20] came up from Brieg during our absence on the Sparrenhorn, and on the following day we had a most pleasant time of it on the Oberaletsch glacier.[21] (poor Egerton, then full of life and joy died suddenly[22] in the Val Ansasca[23] a few days afterwards) The Aletschhorn as you know is the second highest mountain in the Oberland.[24] I took with me one day a man from the hotel,[25] and scaled the mountain. A single guide had never attempted the thing before; but we reached the top without a flaw in the day's enjoyment. I do not think I ever saw the Alps so magnificent as from this glorious pinnacle. My object in going up was not simply to climb, but to examine the colour and polarization of the sky at this great elevation. The weather was all that could be desired for such observations. I returned, and felt as fresh after my return—though the day was one of 15 hours—as after a stroll up Primrose Hill.[26] Far fresher than I felt after spending the day with Spencer on Richmond Hill[27] where his impetuous speed almost took the breath out of me.

About five minutes below the hotel runs a watercourse; broken at one place by an interposed rock, over which the water tumbled. Here I went every morning before breakfast and permitted the water to roll over me. I had been there already seven or eight times, and the thought that an accident could possibly occur in such a place never once crossed my mind. One morning, precisely five weeks ago, I went as usual for my bath. I usually stripped at a distance of 15 or 20 yards from the cascade; and this morning on returning to my clothes I shortened my course by crossing some blocks of granite. To the present moment I do not know how or why I fell.[28] Whether the granite moved under my foot, whether I slipped, whether my foot shrank too suddenly from the contact of the sharp stone, I know not. But I staggered, tried to regain my balance, failed and fell with all my weight against the granite; which stamped three ugly wounds right through my skin, as far I believe as the bone; and bruised all the circumjacent region.

I sponged the blood away; tied a wet bandage round the wound, dressed and hobbled to the hotel. The pain was very great for two hours; but it then subsided, and as I was in overflowing health I hoped soon to get over my mishap. I added a little arnica to the water, induced to do so by the favourable opinion which I once heard a soft eyed friend of mine express regarding arnica. (Those eyes are present with me now—I hope they illumine you with that lustre which is to me one of the pleasures of memory) Well I used the arnica, though I suppose it is not quite the thing for an open wound. But I used it sparingly, and in four days during which I kept quiet in bed, the pain was so far subsided, and the wound so far healed that I felt myself able to limp down stairs on a stick and to spend seven or eight hours in the sunshine. I moved about from place to place upon the platform a good deal, thinking a little exercise would do me good. I had covered the wound with gold-beater's skin.[29] That night I found it much inflamed, the gold-beater's skin having been sucked into the deepest wound. It was evident that I had been too precipitate; so next day I remained in bed.

And now the wounds which had been previously clean began to fill with corruption.[30] They extended, and threatened to join so as to form one great pit. I tried to get rid of this corruption in all manner of ways, some of them I fear not wise ones. Still my health remained strong. I was able to consume and digest my solid beef steak in the morning, without knowing save by memory, that I had eaten it. And I was able to take my share of the good things exhibited at dinner. This satisfactory state of health continued for 8 days. On the night of the 8th., day I wished to cleanse the wound, so I took sponge, soap and water, and worked at it for nearly an hour. My window which overlooked the glacier was wide open; and I sat in my shirt letting the icy air play upon me. I cared nothing about it till I began to shiver. I then covered myself up, but all next day I was in a state of low fever. I poulticed the wound, but it did no good. My foot then began to swell and in three days resembled a bladder of lard—It glowed like a red hot coal—In fact I was in for heavy erysipelas.[31] No signs of improvement: so at length I resolved to put myself under proper surgical care at Geneva. I had an extemporary stretcher prepared, and was carried by four strong fellows down to Brieg. Mid way down I met an English medical man, and he prescribed a lotion for me which did me a world of good. I spent the night at Brieg. Next morning they sent me off in a carriage without springs to Sierre;[32] I bore the jolting as far as Visp[33] and here dismissed my springless carriage for a far more comfortable one. I spent the night at Sierre.

Next day, on entering the station of Geneva I found my friend Soret waiting for me. I went to my hotel and found everything prepared for me. I was not a second in the house before I was joined by Dr. Gauthier. In fact I had written a note to DelaRive asking him to leave at the hotel the address of some able surgeon, and he good man had everything arranged in the most admirable manner by the time of my arrival. The swelling had subsided, but an abscess had formed a little above the instep and about four inches below the wound. Happily a channel established itself between the abscess and the wound, through which the abscess was emptied without necessitating a second puncture. The abscess is now healed, but a small portion of the canal remains still unclosed, yielding a drop of matter daily. A week after I came here I had another bout of erysipelas. It brought me very low; but it is now vanished and all is in a fair way of rapid recovery. I spend three or four days on the borders of the lake, then, on Wednesday next, I proceed to Paris. I rest there Thursday and Thursday night, and proceed to Folkstone[34] on Friday. There I must remain for a few days to pick up a little strength.

I cannot tell you how kind people have been to me. A thorough American gentleman an officer in the Federal army, who himself had received a gunshot wound in the leg, brought his knowledge to bear upon me. He dressed my wound, and pressed me heartily to take his servant with me to Geneva. His servant he said was a Swiss and a capital nurse. I shall never forget the cordial brotherly kindness of this excellent fellow. His name is Howland[35]—a thorough gentleman, as I have said, in all his fibres. And then dear good Bence Jones. He would have started immediately to join me here did he not fear my being detained to take care of him. His house and all that therein is he has placed at my disposal.

For a long time—in fact all through the low fever that accompanied the erysipelas, I loathed all food; consequently have become very thin. But to day I was able to eat two mutton chops with gusto. So that I think that I am on the road to rapid improvement.

Thank your good kind wife for me—and give her and the chicks my loving remembrance.

The Times & Co.,[36] regard you as a wild bull: the Times & Co., are I fear somewhat asinine. They have no idea that a man may have the power of measuring his strength and the goodness of his cause with the opposing forces of society, and regulating his attacks with express reference to the knowledge derived from this comparison. Well I suppose old Shylock[37] you must have your pound of flesh; though my buttocks at present would hardly furnish half the weight. Tell me what <u>your</u> line is to be so that I may avoid you.

Goodbye | Yours ever | John Tyndall.

RI MS JT/1/TYP/9/2927–9
IC HP 8.72
Typed Transcript Only

---

1. *[2 October 1869]*: this letter is dated only 'Saturday', but since Tyndall left Geneva on 6 October 1869 after recovering sufficiently from the injury discussed here, it must be 2 October.
2. *Compte Rendu*: a report (French); presumably an update on his condition as Tyndall was not able to attend this meeting.
3. *the X*: referring to a meeting of the X Club (see letter 3045, n. 5). See also letters 3047 and 3254.
4. *your note*: letter 3045.

5. *scamper with Hirst in Germany*: Hirst accompanied Tyndall through the first part of his European summer tour through Belgium and Germany. See letters 3018, 3020, 3022, and 3023.
6. *Grindelwald*: see letter 3023, n. 28.
7. *table d'hote*: see letter 3020, n. 10.
8. *Zermatt*: see letter 3030, n. 10.
9. *AEggischhorn*: see letter 3025, n. 11.
10. *Herr Welling*: Alexander and Fritz Wellig ran the Hotel Jungfrau, which opened in 1865, financed in part by John Birkbeck. It was a fashionable destination, praised in John Murray's travel guides, but often crowded (R. Flückiger-Seiler, 'Das Hotel Jungfrau am Eggishorn', *Blätter aus der Walliser Geschichte*, 45 (2013), pp. 173–92).
11. *Riffel*: see letter 3030, n. 15.
12. *Riffel See*: Riffel Lake (see letter 3027, n. 10).
13. *Brieg*: see letter 3030, n. 17.
14. *Layard*: Austen Henry Layard (see letter 3030, n. 18).
15. *Bel Alp*: the Belalp (see letter 3030, n. 20).
16. *Aletsch*: see letter 3030, n. 26.
17. *Marjelin See*: the Märjelensee is a lake that, in the nineteenth century, was located on the eastern side of the Aletsch glacier.
18. *Mr and Mrs Neville Lubbock*: see letter 3030, n. 23 and n. 24.
19. *Sparrenhorn*: see letter 3030, n. 25.
20. *Egerton's*: Edward Christopher Egerton and Mary Egerton.
21. *Oberaletsch glacier*: the Oberaletsch (Upper Aletsch) Glacier is located on the south side of the Bernese Alps, in the Swiss canton of Valais.
22. *died suddenly*: see letter 3040.
23. *Val Ansasca*: the Val Anzasca, located in the Piemont (or Piedmont) region of northern Italy, is one of the seven valleys surrounding the Monte Rosa.
4. *Oberland:* the Oberland (meaning 'highland' in German) mentioned by Tyndall here is in the Swiss canton of Bern.
25. *a man from the hotel*: not identified.
26. *Primrose Hill*: Primrose Hill, measuring less than 100 m, is located at the north end of Regent's Park in London. This would not be nearly as grueling as Tyndall's expeditions in the Swiss Alps.
27. *Richmond Hill*: located in Richmond, southwest London, a hill overlooking the Thames.
28. *I fell*: Tyndall recounts this injury in letter 3030.
29. *gold-beater's skin*: made from the outside membrane of the large intestine of an ox, goldbeater's skin is known as such because it was used as an interleaf layer as goldbeaters beat the metal into gold leaf.
30. *fill with corruption*: see letter 3037.
31. *erysipelas*: see letter 3038, n. 3.
32. *Sierre*: see letter 3037, n. 5.
33. *Visp*: see letter 3030, n. 11.

34. *Folkstone*: see letter 3021, n. 20.
35. *Howland*: possibly Joseph Howland (1834–86), a Union Army Officer during the American Civil War. Howland's military career ended when he was shot in the leg. After the war, he served briefly as New York State Treasurer from 1865 to 1867. Howland later became a philanthropist involved in the foundation of Cornell University, and strongly promoted humane treatment for the mentally ill (S. R. Harlow and H. H Boone, *Life Sketches of the State Officers, Senators, and Members of the Assembly of the State of New York in 1867* (Albany: Weed, Parsons and Co., 1867), pp. 32–5.
36. *The Times & Co*: the *Times* and other newspapers. See letter 3045.
37. *Old Shylock*: Shylock is the antagonist in Shakespeare's *The Merchant of Venice*. A Jewish Venetian moneylender, Shylock agrees to lend money to Antonio, his Christian rival, setting a pound of Antonio's flesh as security interest.

To Thomas Archer Hirst 5 October 1869 3049

Hotel L'Ecu de Genève[1] | 5th Oct[obe]r 1869

My dear Tom

Your note[2] has just reached me on the 5th though written on the 18th. This comes of your not putting via France outside & paying the higher postage demanded by France. Belgium is good for the economists—good for large parcels, but bad for those who desire swift transmission. However my difficulties in this line are now at an end.

I start from Geneva tomorrow at 3.30 or thereabouts. I reach Paris at about 7 next morning. I take no courier with me, nor is it now necessary. The wound is almost healed.[3] I keep it still open in order that the last drops of matter generated in the canal which led from the wound to the abscess may be discharged and the channel completely closed. This is now almost perfectly the case: for this morning I could scarcely squeeze a single visible drop from the canal. Tomorrow I abandon ointment and let the wound skin over.

I can limp about and do so here. I have been staying for the last few days with Lady Emily Peel. She set up a bed in the library for me, so that I might not have to climb upstairs, and she has been exceedingly kind & considerate. DelaRive wanted me to go to him but his country house is too far from Geneva.

On Friday I arrive at Folkestone[4]—if you had the time to spare and would rush down on Sunday of course it would be a great delight to me to see you. But do not let me encroach upon your time.

My first aim must be to get a little strength; for though the wound has healed, two fits of erysipelas,[5] and the low fever which accompanied them have brought me very low. My appetite however has returned & I can eat a fair share of food. My sleep is also very fair. And the best of it is that it comes to me without the stimulus of brandy which I have practically given up. This is a great gain.

I wish you would look in at the Institution, and send or bring such letters as may be there for me. My address in Folkestone would be

care of D^{r}. Bence Jones | 5 Albion Villas | Folkestone

so that if you cannot come you can post the letters to me.

Yours affectionately | John Tyndall.

BL Add MS 63092, ff. 65–6
RI MS JT/1/HTYP/560

1. *Hotel de l'Ecu de Genève*: Hotel de l'Ecu de Genève | Hôtel Byron près Chillon à Villeneuve | Lac de Genève | Propre Gustave Wolff. Hotel headed notepaper.
2. *Your note*: letter missing.
3. *The wound is almost healed*: see letters 3030, 3031, 3032, 3037, 3041, and 3048.
4. *Folkestone*: see letter 3021, n. 20. Tyndall stayed in Folkestone while he recuperated from his injury. See letters 3046 and 3050.
5. *Erysipelas*: see letter 3038, n. 3.

From Henry Bence Jones 5 October [1869] 3050

Copped Hall | Epping

My dear Friend

I am so glad to hear[1] that you are able to start. I shall be still more glad to hear that you are in my room at Folkestone[2] getting sea air and as soon as I hear that you are there & I can tell how I am I will fix my plans. I have been very poorly lately & am unfit for a journey but I hope to be much better by the end of the week.

My wife will have every thing ready for you & the little Doctor Bowles[3] will do everything as he knows all about your accident[4] &c. There is lots of first rate mutton fresh air & at least one bottle of Boulogne Brandy which is the best I could get there. So go to my bed & then write to me to let me know every thing about you. You need not write to Folkestone for they will have all things ready for you on Friday & glad indeed shall we all be to help to get you up again.

I am very glad Lady Emily Peel has been taking care of you. I hope you will soon be yourself again

I shall be much interested to read DelaRives account.[5] I only hope I have stated nothing wrong. I have only said a few words but I like them to be right if possible that is. For the present good bye

Ever yours Dear friend | H Bence Jones | Oct[ober] 5

RI MS JT/1/J/87

1. *glad to hear*: letter missing.
2. *Folkestone*: see letter 3021, n. 20.
3. *Doctor Bowles*: Robert Leamon Bowles.
4. *your accident*: see letters 3030, 3031, 3032, 3037, 3041, 3048, and 3049.
5. *DelaRives account*: see letter 3043.

From Friedrich Ludwig Stegmann 6 October 1869 3051

Mein theurer Freund!

Ich war während des ganzen Monats September von hier abwesend auf einer grösseren Ferienreise, anfangs an den Süd-Bayerischen Gebirgsseen, sodann durch Tyrol in das schöne Meran, wo ich mich einige Tage aufhielt, sodann in verschiedenen Tyrolischen Gebirgsthälern, habe bedeutende Bergshöhen bestiegen, bin über Gletscher gewandert, habe hiernach an der Versammlung der Naturforscher und Aerzte in Innspruck Theil genommen, und bin dann über München und Stuttgart, wo ich sehr angenehmen Aufenthalt hatte, hierher zurückgekehrt. Unter den mittlerweile für mich eingelaufenen Sachen war ich sehr erfreut, auch von Ihnen den Brief vom 23. v. M. vorzufinden, dessen verspätete Erwiederung ich durch meine Abwesenheit zu entschuldigen bitte. Leider enthält derselbe manches Traurige. Zwar von dem Leiden, welches Sie durch einen Fall bei einer Gebirgstour zugezogen haben, und welches Sie nöthigte das Zimmer zu hüten, werden Sie hoffentlich Sich bereits erholt haben, und bei Empfang dieser Zeilen wieder ganz frisch sein werden. Dagegen was meinen unglücklichen Neffen anbelangt, so wird, wie ich fürchte, irgend eine Hülfe nicht mehr möglich sein. Ich und seine Angehörigen, wir haben ihn fortwährend nach Kräften unterstützt, aber nach den Mittheilungen, die mir in der letzten Zeit über ihn von verschiedenen Seiten zugegangen sind, muss ich leider annehmen dass er ein vollkommener Lump (a shabby fellow) geworden, was ich aber zu der Zeit, als ich ihm einen Empfehlungsbrief an Sie schrieb, unmöglich annehmen konnte. Er wird jetzt suchen müssen, entweder zu ganz gemeiner Arbeit sich zu verstehen und sich zu gewöhnen mit ganz geringem Lohne auszukommen,—oder er wird in einem Lazarethe zu Grunde gehen! Seinem Vater darf man gar nicht mehr von ihm sprechen, noch viel weniger dürfte man dem selben zumuthen, ihn nochmals und immer weiter mit Geld zu unterstützen, oder ihn gar in die Heimath zurückkommen zu lassen, wo sich ein Unterkommen für ihn durchaus nicht würde finden lassen.—Ich bin Ihnen übrigens von Herzen dankbar für die gütige Theilnahme, welche Sie ihm haben zugewendet, und bedauere in hohem Grade dass diese Unterstützung, was ich nicht habe voraussehen können, so übel angebracht gewesen ist. Ich bitte Sie jetzt nur mir gefälligst noch anzugeben den Betrag der Geldsummen, welche Sie demselben in Folge seiner schamlosen Betteleien gegeben haben; ich werde mich beeilen dieselben Ihnen mit Dank zurückzuerstatten.

Aus Ihrem lieben Briefe glaube ich entnehmen zu müssen, dass unser treuer Freund Hirst bereits wieder nach England zurückgekehrt ist. Sobald Sie ihn wiedersehen, bitte ich denselben freundlichst von mir zu grüssen. Die Hoffnung, dass ich Sie einmal in England besuche, habe ich noch keineswegs aufgegeben.

Beikommend erhalten Sie auch die versprochene Photographie. Mit der Bitte mich in freundlichem Andenken zu behalten, und mir gefälligst baldig die oben erbetene Mittheilung zu machen, unterzeichne ich in aufrichtiger Verehrung.

Ihr treuergebener | Stegmann
Marburg 6/X 69

My dear friend!

For the whole month of September I was absent from here on a long holiday journey, beginning at the alpine lakes in southern Bavaria,[1] then through Tyrol[2] to beautiful Merano[3] where I stayed for several days, then on to sundry Tyrolean valleys, I climbed significant mountain peaks, hiked across glaciers, hereafter participated in the meeting of Natural Scientists and Physicians[4] in Innsbruck, and then returned via Munich and Stuttgart, where I had a very pleasant stay. I was pleased to find your letter from the 23rd[5] of last month among the items that had arrived for me, and ask you to excuse my tardy response due to my absence. Unfortunately the letter contains much sad news. Hopefully, by the time you receive these lines, you will be entirely restored from the ailment you contracted from your fall on your alpine tour[6] and which forced you to be confined to your room. Concerning my disgraceful nephew,[7] on the other hand, I fear that any help will be in vain. His relations and I, we have continuously helped him as much as our powers allowed, but from the news that has come to me lately from different sides I must conclude, unfortunately, that he has become a perfect scoundrel (a shabby fellow),[8] something that I could not have imagined at the time I wrote a recommendation letter to you on his behalf. He will now have to look for a way either to earn a living in a quite common occupation and to accustom himself to getting by with a meagre salary, or he will perish in a military hospital! One may no longer mention him to his father and even less expect him to again continue to support him with money, let alone allow my nephew to return to his homeland, where there will be no employment to be found for him.—Incidentally, I am very grateful for the kind sympathy that you have shown him and deeply regret that your support received such an ugly reward, something that I could not have foreseen. I ask you kindly to let me know the amount of money that you gave him in response to his shameless begging; I shall hasten to reimburse you with my thanks.

From your kind letter I believe to understand that our loyal friend Hirst has already returned to England. I ask you to give him my kindest regards as soon as you see him again. I have not at all given up the hope of visiting you in England some time.

Enclosed you also receive the promised photograph.[9] With the request to remember me fondly and to inform me as soon as you can about the aforementioned matter, I sign in sincere admiration.

Your faithful | Stegmann

Marburg 6/X 69

RI MS JT/1/S/211

---

1. *Alpine lakes in southern Bavaria*: this could refer to any of the following lakes: the Ammersee, Chiemsee, Tegernsee, Eibsee, Königssee, Obersee, or Schliersee.
2. *Tyrol*: the westernmost of German territories, bordering on Bavaria and Switzerland.
3. *Merano*: is a city in northern Italy on the southern side of the Swiss Alps.
4. *Natural Scientists and Physicians*: Versammlung der Naturforscher und Aerzte. The forty-third meeting of the Assembly of German Natural Scientists and Physicians took place in Innsbruck in 1869.
5. *your letter from the 23rd*: letter missing.
6. *the fall on your alpine tour*: see letters 3030, 3031, 3032, 3037, 3041, 3048, and 3049.
7. *my disgraceful nephew*: at Stegman's behest, Tyndall used his influence to find a teaching position for the nephew earlier in the year. It apparently did not end well. See letters 2994, 3027, and 3054.
8. *a shabby fellow*: English in original.
9. *Enclosed you also receive the promised photograph*: enclosure missing.

To Thomas Archer Hirst [9 October 1869][1] 3052

5 Albion Villa

My dear Tom,

What can the matter be? I have been waiting for you here for two hours.

Yours aff[ectionte]ly | John Tyndall

RI MS JT/1/T/1002

1. *[9 October 1869]*: this letter has been dated through information about Tyndall's travel to Folkestone in letter 3049 and Hirst's journal entry for 9 October 1869 (*Hirst's Journal*, RI MS JT/2/32d/1854).

From Henry Bence Jones 10 October [1869][1] 3053

Copped Hall | Epping | Oct[ober] 10

My dear friend

How cross grained every thing has gone. I have lost a great pleasure in being with you at Folkestone[2] and at Folkestone you ought to have staid[3] until the wound[4] was well & no Birmingham[5] ought to be undertaken until all is right & quite right & I ought to be in London to look at you as you must be there and instead of that here I am a dropsical hypochondr[i]ac unable to help myself or my friend.

I wish I could write you a certificate that you ought not to go to Birmingham for another week. Send me a line here at once. Better still I have sent you the certificate in anticipation so it will be ready.

I fear you did not get the brandy I bought for you. What a sum they made you pay for your invalid carriage

Do please send me DelaRives letter[6] here tomorrow. I will soon see what should be done with it. If it contradicts anything I have said it then certainly ought to be published somehow in the volume.[7] If however it does not set me wrong then it will be a question whether it sh[oul][d] wait or be given to some reviewer but first of all (if you can put your hand on it tomorrow) let me have it. I want also to send you a sheet about the Trinity house[8] which you may have a word or two to correct.

What day shall you come back from Birmingham

Good bye dear friend | Ever yours | H Bence Jones

RI MS JT/1/J/109

1. *[1869]*: the year of this letter has been determined through its place within a series of letters from Bence Jones concerning Tyndall's convalescence at his home in Folkestone. See letters 3050, 3056, and 3057.
2. *Folkestone*: see letter 3021, n. 20.
3. *you ought to have staid*: Tyndall stayed at Folkestone while recovering from an injury sustained in the Alps. See n. 4. See letters 3038, 3041, 3046, 3049, 3050, 3057, 3060, 3061.
4. *wound*: Tyndall suffered a severe infection after wounding his leg in a fall in the Alps. See letters 3030, 3031, 3032, 3037, 3041, 3048, and 3049.
5. *Birmingham*: this is probably related to Tyndall's work with Trinity House (see letter 2937,

n. 6). In a letter to Elizabeth Dawson Steuart of spring 1867, he wrote: 'a day or two ago I received a letter from the Trinity house requesting me to meet the Deputy Master at Birmingham on Monday. I am scientific adviser to the Trinity Corporation and hence did not wish to neglect the request' (see letter 2610, *Tyndall Correspondence*, vol. 10). Birmingham was home to the Chance Brothers glassworks, who produced lighthouse lamps for Trinity House.

6. *DeLaRives letter*: letter missing.
7. *published somehow in the volume*: Bence Jones was writing his *Life of Faraday* and asked Tyndall to pass on a question to de la Rive. See letter 3043.
8. *Trinity house*: see letter 2937, n. 6.

To Thomas Archer Hirst [12 October 1869][1] 3054

Tuesday

My dear Tom

This of Stegmann's is strong,[2] but it completely coincides with the opinion which from his letters I have formed of the man.

Sylvester must be paid these two pounds & also every body else who has suffered from my introduction.[3]

Yours affectionately | John

RI MS JT/1/T/961
RI MS JT/1/HTYP/561

---

1. *[12 October 1869]*: this letter refers to letter 3051 of 6 October 1869. 12 October is the following Tuesday.
2. *This of Stegmann's is strong*: see letter 3051.
3. *my introduction*: Tyndall had helped Stegmann's nephew earlier in the year, but apparently he became rather troublesome. See letters 2994, 3027, and 3051.

To Henry Bence Jones [12 October 1869][1] 3055

Tuesday.

My dear Friend

I have now been four days and four nights at Albion Villas[2] as well fed, and as well treated in every way as a man could possibly be. The result is a great improvement in my general health.[3] The only thing I have now to complain of is the weakness of my legs. They are <u>both</u> equally weak, and it is easy to make the muscles sore and weary through exercise: but even this is slowly disappearing. Sometimes a curious stinging sensation makes itself felt in I think both legs. and a slight tendency to cramp in the bad one. But on the whole matters are going on and have gone on wonderfully well.

Hirst came down here on Saturday night.[4] He was ill poor fellow in his bedroom at the Pavilion all night long. I have made him to accept Lady Millicent's[5] invitation to luncheon on Sunday, a slice of your mutton, and a draught of your Boulogne brandy seemed to set him all right again.

He has a curious affection of the skin that attacks him at this season every year.

Yesterday, Herbert Spencer came down here in a bad condition. He had been in Scotland as usual, but it did him little good, so that when he began to work in London he immediately broke down. Now Spencer wanted to come to me in Geneva to relieve my ennui; and as one good turn deserves another, and as he suffers if left alone, I propose joining him at the West Cliff Hotel to day. But I must stay in Folkstone[6] for some days yet. The extraordinary weakness of my limbs would prevent me from doing anything in London.

I saw Dr Bowles[7] yesterday, I called more to make his acquaintance than anything else; but he looked at my wound, gave me good advice, permitted me to bathe, and refused a fee! He is a very decent man. Not only so, but he said it would do him a kindness if in case of need I called upon him. I call him a very decent man.

My sleep has been good ever since I came here, good without any stimulant stronger than Folkstone water. I think I must have had seven clear hours sleep last night. My brain needs it, for it shares the weakness of my legs. It will need a good deal of pumping of fresh blood to bring it into tone.

We shall have opportunity enough to talk about the Juveniles[8] when I return to London: what would be easier for me and certainly best for them would be *[elementary]* optics. But then I talked of light to the old folks last spring.

Harry[9] is here, and I think we shall have a header together to day.

Whenever I am in need of a really good slice of mutton I shall beg one here.

Good bye | Yours ever | John Tyndall

Dr Bowles spoke to me of the excellence of your health during your first month here: but he also spoke of your irrepressible energy and impetuosity. These <u>must</u> be toned down.

Your letter[10] has just arrived sent forward from London. Hundred of thanks. I wish I was near you.

RI MS JT/TYP/3/800
LT Typescript Only

1. *[12 October 1869]*: this letter has been dated through its place in a series of letters about Tyndall's travel to Folkestone to convalesce from an injury. See letters 3038, 3041, 3046, 3049, 3050, 3053, 3056, 3057, 3060, and 3061.
2. *Albion Villas*: Bence Jones's country home at Folkestone.
3. *improvement to my health*: Tyndall was recuperating from an injury and infection sustained while hiking in the Alps. See letters 3030, 3031, 3032, 3037, 3041, 3048, and 3049.
4. *Hirst came down here on Saturday night*: see letter 3052.
5. *Lady Millicent's*: Millicent Bence Jones (neé Acheson, 1817–87), daughter of the 2nd Earl of Gosford and a cousin of Bence Jones. She married Bence Jones in 1842, and they had seven children.
6. *Folkstone*: see letter 3021, n. 20.
7. *Dr Bowles*: Robert Leamon Bowles.
8. *the Juveniles*: the RI's annual juvenile Christmas Lectures began in 1827 and still continue. Tyndall delivered the Christmas Lectures a dozen times over his time with the RI, his first in 1861 (see F. A. J. L. James (ed), *Christmas at the Royal Institution: An Anthology of Lectures* (Hackensack, NJ: World Scientific, 2007)). His fifth time, in 1869–70, was 'On Light', and the lectures were given on 28 and 30 December 1869, and 1, 4, 6, and 8 January 1870 (*Roy. Inst. Proc.*, 5 (1866–9), p. 609).
9. *Harry*: Henry Robert Bence Jones (1844–1912), a son of Bence Jones. 'Harry' is the English short form of 'Henry'.
10. *Your letter*: possibly letter 3053, since Bence Jones assumed in that letter that Tyndall was passing through London.

From Henry Bence Jones [12 October 1869][1] 3056

Copped Hall | Epping

Dear Friend

Do pray stay on & get well. If you knew the pleasure it gives me to think that you are profiting by anything I have you would stay. My only regret is that I am not with you to talk to you & see that all is as I wish.

If you can stay on I will come let me know your plans. Eat plenty of mutton & drink plenty of brandy not sour claret.

I am all for water when you are well but it is very different now when it is medicine. I wish I had brought half a dozen bottles at Boulogne for you

Goodbye | send me DelaRives[2] letter here

Ever yours | H Bence Jones | Tuesday

RI MS JT/1/J/108

1. *[12 October 1869]*: this letter has been dated in relation to letters 3055 and 3057. In letter 3055 to Bence Jones from 12 October 1869, Tyndall refers to Spencer's visit to Folkestone,

eating mutton, and drinking brandy. In letter 3057 of 14 October 1869, Bence Jones discusses a letter from de la Rive. Since 12 October 1869 was a Tuesday, Jones must have received letter 3055 on the same day it was sent and responded immediately.

2. *send me DelaRives letter here*: most likely the letter referred to in letter 3057. See also letters 3043 and 3053.

From Henry Bence Jones 14 October 1869 3057

Oct[ober] 14. 1869.

My dear friend

I am so glad that you are staying on at Folkestone[1] but so sorry that I was not there[2] to keep you in our house.[3] In spite of Spencer or any one else you should have staid. I should have thrown your boots into the sea before you sh[oul]d have left the house. However it has all been a sad mishap from your first slip[4] to my relapse & we must make the best of it. Very many thanks for DelaRives letter[5] my account which is founded on Mr Marcets[6] statement to me is better for Davy[7] & quite creditable enough for the elder DelaRive[8] & near enough to the actual facts for the present edition[9] (the sheet is printed). The great object I had was to shew his bond to the DelaRives & as in DelaRives account I have jointed the passage out of the letter of Faraday saying that DelaRive first sustained him &c &c. I tell the story as of a single invitation to dinner instead of a stay in DelaRives house. It is strange that this stay in DelaRives house is not mentioned in Faradays journal. I think the shooting is. I must look at the Geneva time again, I may get something now I know the exact facts. I will return you DelaRives letter when we meet in London. I have just heard from Folkestone that Spencer has left you & rushed away I wish you had never left my room I am sure you w[oul]d have slept better & been better cared for than at the hotel which is by no means first rate, in anything I fear.

Bowles[10] is a good little man with a good opinion of himself which makes the world go pleasantly as long as that good opinion is not disturbed, but the disturbance which will come along now & then is not pleasant at least to most minds.

By all means take light for the juveniles[11] I have promised one lady[12] that you shall do so & therefore you must not make me out powerless in the Institution arrangements. Or my feelings of potency like poor Bowles will be hurt.

Good bye dear friend I wish I had many talks with you about many things I must send my scrawl

ever yours | H Bence Jones

RI MS JT/1/J/62

1. *Folkestone*: see letter 3021, n. 20.
2. *sorry that I was not there*: Bence Jones went to Copped Hall, in Epping, Essex, on 1 October and was away while Tyndall convalesced in his house at Folkestone. See letters 3050 and 3056.
3. *keep you in our house*: Tyndall joined Herbert Spencer at the West Cliff Hotel. See letter 3055.
4. *Your first slip*: see letters 3030, 3031, 3032, 3037, 3041, 3048, and 3049.
5. *DelaRives letter*: letter missing. Bence Jones discusses the matter in letters 3043 and 3053.
6. *Mr Marcets*: François Marcet.
7. *Davy*: Humphry Davy.
8. *the elder DelaRive*: see letter 3043, n. 8.
9. *the present edition*: Bence Jones was completing his two-volume *Life Faraday*. See letters 2953, n. 1, 3021, 3038, 3043, 3050, 3236, and 3245.
10. *Bowles*: Robert Leamon Bowles.
11. *take light for the juveniles*: see letter 3055, n. 7.
12. *one lady*: not identified.

To Thomas Archer Hirst 20 October 1869 3058

West Cliff Hotel Wednesday | Oct[ober] 20 /69

My dear Tom

I asked M[rs]. Tunstal[1] to send you a line to inform you that I return on Friday to town. I think I am quite strong enough for this, and I moreover fancy that a Turkish Bath with *[1 word illeg]*[2] as manipulator would do as much good. This will abolish all idea of your coming here which I should on no account have been a party to.

Remember me kindly to | Miss Hirst.[3] | Yours ever | John

RI MS JT/1/T/969

1. *M[rs]. Tunstal*: Mrs. Tunstall was Tyndall's housekeeper at the RI.
2. *[1 word illeg]*: not identified; presumably a person's name.
3. *Miss Hirst*: probably Hirst's niece, Lilly Hirst.

To Mary Adair 26 October 1869 3059

My dear Miss Adair.[1]

A crumpled letter, written to you in Geneva;[2] and which has lain for some time in my pocketbook, is now before me. I have written many letters to you and torn them up, but this one has remained with me so long that I will run the risk of copying it and sending it to you.

"While returning from a cascade on the Bel Alp,[3] the waters of which I permitted to tumble over me every morning, to my clothes which were placed 20 yards off, I crossed some blocks of granite with sharp edges and sharp protuberant crystals, and in doing so slipped, staggered and fell, urging my shin with all my weight against the jagged stone. I bound up the wound, or rather, wounds, for there were three of them, and hobbled up to the hotel.[4]

"I was in good health at the time, and four days of quiet brought my wound into so satisfactory a condition that I thought I might move about on the fifth day in the sun near the hotel. I did so, but the night of the day on which I permitted myself this liberty I found the wound inflamed.[5]

"I did not treat it wisely afterwards, and the consequence was that the leg and the adjacent foot swelled greatly, and glowed like a red hot coal. After 12 or 13 days at the Bel Alp, seeing no prospect of my getting better there, I came to Geneva where I have been wisely and kindly tended.

"It is now nearly five weeks since the accident occurred, and this day I quitted my bedroom for the first time. I was carried down stairs placed in a carriage and driven about for a couple of hours. And now I am again restored to my perch.

"To a quick spirit it has been a weary time; but the worst is over, and now I hope to mend rapidly.

"Many thoughts on divers subjects have passed through my mind during these five weeks of confinement. Some of these thoughts, which are indeed but a repetition and continuance of others that occupied my mind[6] long prior to the accident, relate to yourself.

"It must have been in the spring of 1863—now therefore more than six years ago—that I found myself on a Friday evening in the library of the Royal Institution in the company of two young ladies looking at a picture of the Matterhorn.[7] One of these young ladies[8] had light hair & blue eyes—unhappily she is now no more. The other had dark hair & dark eyes. I had been trying to climb the
<Hereafter LT Typescript> Matterhorn the previous Autumn,[9] and was making some remark about the difficulty of the mountain, when I noticed, or thought I noticed, in the expression of those dark eyes a depth of soul such as I have seen manifested by only one or two person in the course of my life.

"That expression, revealing as I believed it did, a nature profound, and charged with womanly emotion clings to my memory to the present hour.

"Soon afterwards I met this selfsame maiden in Cornwall Terrace.[10] After dinner she sat beside her friend, and I was near them both. Once an answer to remark made by her companion, this dark eyed maiden suddenly placed her for head against her friend's cheek. A matter-of-fact person would he smile at the importance which I attached to this apparently trivial act; but to me it seemed to shed a flood of light upon the actor. It was naturally done and it was lovingly done. I seemed to understand it so thoroughly, that I could then and there have clasped that maiden to my heart.

"I have studied her countenance from time to time since, and have always found in it those evidences of depth of soul which at first riveted my attention. The richness of her intellect had been previously discovered; but the intellect to me was less than half the attraction."

This I wrote in Geneva and a great deal more which I have not now the courage to copy. Is it necessary to tell you who the young lady here referred to is?[11]

RI MS JT/2/10/451
RI MS JT/2/13c/1359–60

---

1. *dear Miss Adair*: this letter is a draft.
2. *written to you in Geneva*: Tyndall spent much of September bedridden in Geneva because of his accident. See letter 3037.
3. *Bel Alp*: the Belalp (see letter 3030, n. 20).
4. *hobbled up to the hotel*: see letters 3030, 3031, and 3048.
5. *I found the wound inflamed*: see letters 3037 and 3048.
6. *occupied my mind*: see letter 3047, n. 5.
7. *Matterhorn*: see letter 2987, n. 17.
8. *One of these young ladies*: possibly Miss Hamilton, a good friend of Mary Adair in the early 1860s (see letter 2009, *Tyndall Correspondence*, vol. 8).
9. *Matterhorn the previous Autumn*: Tyndall first attempted the Matterhorn, one of the high peaks of the Alps located on the border between Switzerland and Italy, in August 1860 (*Hours of Exercise in the Alps*, pp. 27–52), and again, unsuccessfully, in August 1862 (*Hours of Exercise in the Alps*, pp. 153–65).
10. *Cornwall Terrace*: a block of attached mansions (rowhouses) overlooking Regent's Park (a Royal Park) in north-west London. This house appears to have belonged to Favoretta Hamilton, a distant relation of Mary Adair's.
11. *who the young lady here referred to is?*: Tyndall had secretly harbored romantic interests for Adair for some time, as this letter and his journal entries make clear. He never sent this letter, which he drafted while ill in Geneva. After returning to London, he went through considerable trouble in his attempt to find a moment to speak with her alone but had to resort to sending her a note baring his heart. In her response, Adair gently turned him down, but the two remained friends. Tyndall burned the notes and kept his heartbreak to himself (*Ascent of John Tyndall*, p. 257).

To Thomas Carlyle[1] 26 October 1869 3060

Royal Institution | 26th. October 1869

My dear Friend.

One evening towards the end of August[2] I stood upon the knob of a mountain about 7000 feet above the sea: with the largest glacier in Europe[3] just beneath me, the mountains around me, and the sky, which was still red in the West, overhead.

I was in good health. This I had sought for weeks but had not found it. It came at last. The spell of sleeplessness was broken, the food well digested and well applied and I felt as I stood there, puffing my cigar, a joy of life—a pleasure in mere existence—such as I have not felt for years. It was moreover no flash in the pan—for day after day this feeling of health made life in the hills a pleasure to me.

You crossed my thoughts and got entangled among them as I stood upon my mountain knob, and I wished intensely that I could have had you there for 10 days, to prove the effect of the mountain air upon you. The experiment of going there would be dangerous, for the hotêl, though a comparatively quiet one would not be called quiet by you. And then I schemed in thought about building a little wooden chalet apart from the hotel where you & I might sleep, and in fact the scheme seemed to me there so reasonable & so laudable that I am not sure even now whether it will not be carried out.

I was accustomed to bathe freely—to jump into the lakes to get under the cascades and allow them to tumble over me. I believe it was the day following my speculations on the knob. I allowed one of these cascades to tumble over me—as I had done a dozen times before—early in the morning. My clothes were 20 yards distant from the cascade, and in returning to there I crossed—unnecessarily—some loose blocks of granite. It was the shortest way. The thought of caution never occurred to me, for there was nothing to excite it. So I slipped and fell,[4] and stamped the sharp crystals of the granite as far as my shin bone.

I sponged off the blood, bandaged the wound, & lay four days quite still. It grew almost well. At all events it was perfectly clean & perfectly painless. Had the edges of the wounds been united, these four days—at all events a week—would have completely healed them. (I have capital healing flesh) As it was I was so well on the fifth day that the cold water bandage was abandoned, and I limped about in the neighbourhood of the hotel for some hours. That night the wounds were inflamed. I ignorantly mixed arnica with my water and this did not improve matters—the wounds filled with pus, and I took rather harsh measures to dislodge it. The leg became more & more inflamed, and finally the foot was three times its normal volume & glowing like a red hot coal. So I had myself carried down to Brieg[5] and journeyed thence to Geneva where I was wisely and kindly tended; but on the whole this paltry stagger cost me 6 weeks of bed & the loss of all my splendid health. I came over to Folkestone[6] a fortnight ago. There I halted for I could hardly walk 50 yards. Before I quitted it I was equal to 10 miles. On Friday night I came here. I have thought of writing to you daily since but could not get so far until this morning.

I saw Miss Davenport-Bromley[7] at the Station at Folkestone. She told me that Lady Ashburton was now in Scotland with a constellation of genius around her. I believe her—affinity—with genius is true & deep.

Well I have occupied you all along with my poor troubles I hope you are well, and I wish I could realize my dream of a fortnight with you upon the Bel Alp.[8] By Jove I should like to hear the trumpet note[9] that would follow.

Yours ever & always | John Tyndall

Are we ever to have our little dinner at the Athenaeum[10] together?

RI MS JT/1/T/149
RI MS JT/1/TYP/1/190–1

1. *Thomas Carlyle*: a handwritten addition on the typescript version of this letter notes that the envelope states it is docketed by Thomas Carlyle.
2. *One evening towards the end of August*: 28 August 1869 (see letter 3030).
3. *largest glacier in Europe*: the Aletsch glacier(see letter 3037, n. 1).
4. *I slipped and fell*: see letters 3030, 3031, and 3048.
5. *Brieg*: see letter 3030, n. 17.
6. *Folkestone*: see letter 3021, n. 20.
7. *Miss Davenport-Bromley*: probably Caroline Davenport Bromley (1820–91), a friend of Louisa Baring who became a close friend of Jane Carlyle's (Carlyle Letters Online).
8. *Bel Alp:* the Belalp (see letter 3030, n. 20).
9. *hear the trumpet note*: this expression is found in Virgil's *Aeneid*, in the seventh book in the prose translation from J.W. McKail, *The Aeneid of Virgil, translated into English* (London: MacMillan and Co., 1885), p. 165, lines 627–8. Since this translation was not published until the 1880s, it is unclear where Tyndall found this expression.
10. *Athenaeum*: see letter 2944, n. 1

To Emil du Bois-Reymond 26 October 1869 3061

Royal Institution | 26th. October. 1869[1]

My dear DuBois

When your note[2] reached me I was in that state of low fever which usually accompanies erysipelas.[3] I wounded myself by a fall in the Alps, treated the wound unwisely, and had to lie 6 weeks in bed.[4] I returned to London only a day or two ago.

Please ask Sauerwald[5] to send me Thermo-rollen—If I need the others I can have them subsequently. But I shall use the apparatus almost exclusively for thermo-electric experiments.

I think it will also be best to have the stand and scale and telescope complete. This will save all trouble of arrangement afterwards.

The wire of the Thermo-rollen I should like to have of such a length and thickness as to give the maximum effect with an ordinary thermo-electric pile.

I halted a fortnight at Folkestone[6] to pick up a little strength. I was greatly weakened by my confinement, and by the fever accompanying it. My strength is by no means restored even now, but it is augmenting daily. When I reached Folkestone I could scarcely walk 100 yards—when I quitted it I was equal to 10 miles.

Bence Jones is here now. I dined with him yesterday at the Athenaeum.[7] He looks exceedingly well—in fact better than I have seen him for some years.

It was a great disappointment to me to not meet Magnus here I trust they treated him well at Exeter.[8] It would have given me great pleasure to meet him there.

Let me now thank you for all the trouble you have taken in getting the galvanometer prepared for me.

Believe me dear DuBois | Yours very sincerely | John Tyndall

I ought to have gone to Switzerland before going to Germany, so as to get my head in order. It was very poorly all through my journey amongst my friends.[9]

RI MS JT/1/TYP/7/2435
LT Typescript Only

---

1. *Royal Institution | 26th. October. 1869.*: handwritten note in upper left-hand corner of typescript: 'Professor DuBois Reymond. | Perpetual Secretary to the Academy of Sciences. | Berlin'.
2. *your note*: letter 3028.
3. *erysipelas*: see letter 3038, n. 3.
4. *I wounded myself… in bed*: see letters 3030, 3031, 3032, 3037, 3041, 3048, and 3049.
5. *Sauerwald*: Ferdinand Sauerwald
6. *Folkestone*: see letter 3021, n. 20.
7. *Athenaeum*: see letter 2944, n. 1.
8. *at Exeter*: see letter 3029, n. 3.
9. *my journey amongst my friends*: Tyndall toured Germany with Hirst in July and early August 1869. See letters 3018, 3020, 3023, 3022, and 3024.

To Margaret Ginty 26 October 1869 3062

Tuesday | 26th. Oct[ober] 1869.

Dear Mrs. Ginty,

You did not name the day to Mr. Worthington[1] and he proposes Thursday. Would you arrange this and let me know the proper time and place.

I am willing to go to Mr. Worthington's office either Wednesday or Thursday or to meet him here.

I can do nothing till you decide the matter.[2]

I enclose you a cheque for £ 35 and would thank you for a receipt.

Yours very sincerely | John Tyndall

Mr. Worthington is staying at Norwood:[3] as he is ill it would be most kind to arrange for Thursday.

RI MS JT/1/TYP/11/3659
LT Typescript Only

---

1. *Mr. Worthington*: Edward Worthington, of 28 Milk Street, Cheapside, London, Agent for Solicitors (*London Gazette,* 7 January 1870, p. 164).
2. *the matter*: Tyndall had been assisting Ginty with legal and financial matters concerning her late husband William Gilbert Ginty's (*c.* 1820–66) will and served as her Trustee.
3. *Norwood*: an area on the outskirts of southeast London.

From John Morley[1] 27 October 1869 3063

My dear Mr. Tyndall,

My gratitude is boundless and fervent. You shall have the proof[2] a good fortnight in your hands. I am really very much obliged to you.

I hope cordially that you are better.

Ever yours very sincerely | John Morley.

Oct[ober]. 27. 1869.

RI MS JT/2/10/436–7/b

---

1. *John Morley*: John Morley, Viscount of Blackburn (1838–1923), an English writer, editor, and Liberal Member of Parliament (1883–95; 1896–1908). During his time as editor of the *Fortnightly Review* (1867–82) he contributed approximately 200 articles for the publication. Morley joined Tyndall as a member of the Metaphysical Society, which hoped to address rising tensions between men of science and theologians (see letter 2985, n. 3).

2. *the proof*: presumably for J. Tyndall, 'Climbing in Search of the Sky', *Fortnightly Review*, 13 (1870), pp. 1–15.

From Sarah Faraday 28 October [1869][1] 3064

12, Barnsbury Villa, | 320, Liverpool Road, N. | 28th. Oct[ober][2]

I think I must write a few lines to you my dear friend to tell you how gratified I have been in reading the translation of the preface to Helmholtz's book[3] (X) I cannot tell you the pleasure it has given me. Somehow such testimonies to my beloved husband's worth and character always seems to take me by surprise, for he was so simple, unpretending and childlike, (I might almost say), that though his character was a delight to me, I did not expect it to be appreciated by others, especially foreigners who have not the same opportunity of judging.

If you had an opportunity I could wish the author to know how he has cheered the widowed heart of your sincere friend

S. Faraday

(X) Faraday as a Discoverer. J. T.

Dr. Tyndall

RI MS JT/1/TYP/12/4184
LT Typescript Only

1. *[1869]*: the year is given by reference to the German translation of Tyndall's *Faraday as a Discoverer.*
2. *28th Oc[tober]*: handwritten note above the date line: '[this letter is copied in Tyndall's handwriting]'. Typed note above date line: 'Original sent to Helmholtz'.
3. *translation of the preface to Helmholtz's book*: Helmholtz wrote a preface to his German translation of Tyndall's *Faraday as a Discoverer,* a copy of which he sent to Tyndall in October 1869 (D. Cahan, *Helmholtz: A Life in Science* (Chicago: University of Chicago Press, 2018), p. 312). The translation itself was made by Helmholtz's wife, Anna, and published as: J. Tyndall, *Faraday und seine Entdeckungen* (Braunschweig: Friedrich Vieweg und Sohn, 1870).

From John Ruskin 1 November 1869 3065

I am sincerely grieved to hear of the time of your detention abroad[1]—I shall hope however that it will bear no evil effects—and one good result may perhaps follow—that you will stay more within reach of us all and perhaps help me a little in some plans. I have really work on the Levels as much as on the Peaks.

I ventured to send you that last book of mine (The Queen of the Air)[2] because I had been impertinent enough to use your name in it[3] but I hope not offensively to you.

RI MS JT/5/12/15
LT Typescript Only

---

1. *your detention abroad*: see letters 3030, 3031, 3032, 3037, 3041, 3048, and 3049.
2. *The Queen of the Air*: J. Ruskin, *The Queen of the Air: being a Study of the Greek Myths of Cloud and Storm* (London: Smith, Elder & Co., 1869).
3. *use your name in it*: Tyndall is thanked by name in the preface to this work (on p. v). As Roland Jackson notes, the mention was sarcastic, since this essay was 'a thinly veiled attack on Tyndall', whose work on the colour of the sky turned its deep blue colour with its many artistic allusions into a physical phenomenon describable by science (*Ascent of John Tyndall*, p. 255). Ruskin was critical of the environmental damage caused by industrialization and pollution and of the increasing status of science compared to art and artistic sensibility as sources of knowledge.

To Juliet Pollock 4 November 1869 3066

4th Nov[ember] 1869

My dear Mrs. Pollock

I shall be glad to dine with you on Tuesday.

Yours ever | John Tyndall | The unswerving fr*[iend]*

I had a splendid tipple last night with Carlyle, Forster,[1] Fitzjames Stephen[2] & Huxley.

<verso>

RI MS JT/1/T/1149
RI MS JT/1/TYP/6/2112

1. *Forster*: William Edward Forster (1818–86), an English businessman and Liberal Member of Parliament. He was elected to the Alpine Club in 1859, and he spent summers climbing in the Alps. Forster made a significant contribution to the debate on changes to the English elementary school system and was responsible for the Education Act of 1870 (*ODNB*).
2. *Fitzjames Stephen*: James Fitzjames Stephen (1829–94), an English lawyer and judge. After an undistinguished student career at Cambridge University, Stephen was called to the bar in 1854. From 1858 to 1861 he was secretary to the Royal Commission on Education. Increasingly uneasy with organised religion, Stephen wrote numerous essays expressing his skepticism and participated in the debates of the Metaphysical Society in the 1870s (*Alum. Cantab.*; *ODNB*).

To the Assistant Secretary, Harbour Dept., Board of Trade[1] 6 November 1869 3067

THE OBSERVATIONS of PROFESSOR TYNDALL upon the foregoing THIRD REPORT of MESSRS. STEVENSON are contained in the following letter:

Royal Institution, November 6, 1869.

Sir,

I beg to acknowledge the receipt of the Reports of the Messrs. Stevenson, dated May 1869 and October 1869. On these Reports you ask for my observations. I have the honour to submit the following:—

§ 1.

Regarding Burners.

As surmised by the Messrs. Stevenson, the burners employed by me[2] were different from those employed by them.[3] The lower power, for example, experimented with by Mr. Valentin and myself consisted of 28 "No. 1" burners, consuming 1¾ cubic feet of gas per jet, or for the whole burner 50 cubic feet per hour. The Messrs. Stevenson's lowest power, on the contrary, consisted of 52 jets, each consuming 3·8 cubic feet, or for the whole burner 196 cubic feet per hour.

The following table presents in a concise form the description and consumption of the respective burners. I confine myself to the first three powers of the Howth Baily[4] burner, because its second and third powers are those with which the burners employed by the Messrs. Stevenson can be most fitly compared.

Howth Baily Experiments			Granton[5] Experiments.		
Number of Jets in Burner.	Cubic Feet of Gas consumed per Jet.	Rate of Consumption per Hour.	Number of Jets in Burner.	Cubic Feet of Gas consumed per Jet.	Rate of Consumption per Hour.
28	1·78	50 Cubic Ft.	—	—	—
48	2·2	105 "	52	3·8	196 Cubic Ft.
68	2·2	148 "	72	3·3	240 "

Thus, comparing the 48-jet burner of Howth Baily with the 52-jet burner of Granton, we find the consumption of the former to be 105 cubic feet, and that of the latter 196 cubic feet.

Comparing the 68-jet burner with the 72-jet one, we find the consumption of the former to be 148 feet, and of the latter 240 feet.

The Howth Baily burners contain a smaller number of jets than the Granton ones. Adding, therefore, one twelfth to the consumption of the 48-jet burner, and one seventeenth to the consumption of the 68-jet burner, we virtually convert them into burners of 52 jets and 72 jets respectively. We thus find that with an equal number of jets the consumption at Howth Baily is to the consumption at Granton in the following rations:—

For the 52-jet burner

114 to 196.

For the 72-jet burner

157 to 240.

Thus it appears that the consumption at Granton was very much greater than that at Howth Baily, a difference due to the difference of the burners.

§ II.

Regarding cost of Production.

(A.)

In the Report of the Messrs. Stevenson (*ante* page 5) stand two tabular statements regarding the cost of gas. The first table is headed "Cost of gas from same coal as used in Experiments per annum." The "coal" which constitutes the first item of this table is *Boghead Cannel,*[6] and it is stated that to produced 740,880 cubic feet of gas, 86 tons of this coal, at a cost of 240*l.* 16*s.*, would be required.

It is, I am credibly assured, the universal experience of gas manufacturers that Boghead Cannel yields from 13, 000 to 15, 000 cubic feet of gas per ton.

Taking then the *lowest* of these figures, and calculating the quantity of coal requisite produce 740,880 cubic feet of gas, it is found to be—

		£	*s.*	*d.*
57 tons, at a cost of		159	10	0
instead of 86 tons, at a cost of		240	16	0

(B.)

The Board is aware that the coal from which gas is extracted is placed in iron retorts and subjected by the agency of external heat to destructive distillation. In the production therefore of the gas two distinct items of coal are to be taken into account; firstly, the coal from which the gas is generated, and which is *within* the retorts; secondly, the coal employed in heating, which is *without* the retorts.

In the table now under consideration, the Messrs. Stevenson assume, that to distill 86 tons of coal contained in the retorts, 232 tons of heating coal, costing 301*l.* 12*s.*, would be necessary. Taking into account every disadvantage incident to the manufacture of gas on a small scale, this seems to be an inordinately high ratio of the heating coal to the gas generating coal. The proportion of the fuel to the gas-coal at Howth Baily is stated in my Report to the Board[7] to be "exceptional." It indicates a defect in the system of manufacture. The proportions assumed by the Messrs. Stevenson would indicate a considerable exaggeration of this anomaly.

I am informed by competent authority that in our large gasworks the consumption of fuel in distilling cannel coal amounts only to 10 per cent. of the charge in the retort. The fuel demanded by the Messrs. Stevenson in relation to the charge in their retorts, would be exactly twenty-seven times the quantity found sufficient when the gas is manufactured on a large scale. This point needs reconsideration.

The estimate in the second table in Messrs. Stevenson's Report (ante page 5) seems to be a much closer approach to the normal conditions of gas production than that in the first. In the second table we find that to carbonize 67 1/3 tons of the best gas coal, 10 tons of coal, in addition to the coke left as a residue by the charge itself, are assumed to be sufficient. Supposing the residue of coke to amount to 34 tons, or to one half of the charge, and adding 34 to 10 we should still have only 44 tons of coal and coke, devoted as fuel to the distillation of 67 1/3 tons. I am unable to reconcile this with the first table, where, to distil 86 tons, fuel to the amount of 232 tons is stated to be necessary.

It is true that the coal of the first table is Boghead Cannel, while the coal of the second table is not; but I am credibly informed that Boghead Cannel does not require more heat for its distillation than the best gas coal mentioned by Messrs. Stevenson in the second table of page 5.

(C.)

As regards the *purification* of the gas, there is also a great discrepancy between the two tables referred to. In the first the cost of the lime for purifiers is stated to be 38*l.* 8*s.*; in the second the cost of purification of exactly the same amount of gas is only 7*l.* 8*s.* The former sum is stated to be calculated on the scale of Howth. If so it would prove that the gas markers at Howth were not at the time this item reached so high a figure by any means masters of the art of gas purification.

A similar remark would apply to the last paragraph of the Messrs. Stevenson's postscript. It is there stated that 120,430 cubic feet of gas were manufactured at Howth Baily at a cost of 135*l.* 10*s.* 4*d.*, and from this it is inferred that the cost of the gas per 1,000 feet is 22*s.* 6*d.*

Referring to the report of Captain Roberts,[8] dated February 8th, 1867 (page 8 of the Papers presented to Parliament by Command in 1869), we find the amount of shale (Boghead Cannel) employed in the generation of 120,430 cubic feet of gas to be 14 tons. Dividing 120,430 by 14 we obtain as a quotient 8,600, which represents the number of cubic feet per ton alleged to be the yield of Boghead Cannel; but the yield is really from 13,000 to 15,000 feet per ton, which proves the number 120,430 to be wrong. This is of course the error of the Howth Baily observers or reporters, who might be fitly called upon to explain how the mistake arose.*

§ III.

What is here stated will doubles leave upon the Board the general impression that the data on which the Messrs. Stevenson have formed their calculations are very different from those which form the basis of mine. Fortunately the table already referred to, namely, the second one in Messrs. Stevenson's Report (*ante* page 5), enables me to place the question before the Board in a more definite light.

There are *four* items in the table, each of which is directly proportional to the quantity of gas generated. Firstly, the coal; secondly, the carbonization of the coal; thirdly, the purification of the gas; and fourthly, the wear and tear of retorts. These four items, I say, would vary as the quantity of gas. Adding them together we have a sum of 127*l.* 16*s.* The gas marker's wages, and the interest on the outlay for buildings are here purposely excluded.

The quantity of gas assumed as necessary by the Messrs. Stevenson is 740,880 feet per annum. But with the Howth Baily burner No. 1, yielding a light which I have strong grounds for believing to be more powerful than that operated on by the Messrs. Stevenson, the consumption per annum is only 189,000 cubic feet. Reducing the foregoing sum of 137*l.* 16*s.* in this proportion, we have-

	189,000 x 137*l.* 16*s.* 4*d* / 740,800.	=	35	3	0
Add this to wages			91	5	0
and interest			32	18	4
and we have a total of			159	6	4

Thus, taking the prices of materials given by the Messrs. Stevenson; taking the full rate of wages, and of interest which they deem necessary; taking also the quantity of gas which they assume to be yielded per ton of coal, and altering nothing save the consumption of the gas, *which is diminished by the use of a better burner than they employed*, we reduce their estimate from 261*l.* 19*s.* 8*d.* to 159*l.* 6*s.* 4*d.* The Board, I trust, will bear in mind that if the Messrs. Stevenson's prices be correct this reduction *is experimentally demonstrable.*

If we take the Howth Baily rate of wages, viz., 36*l.* 10*s.* per annum, instead of 91*l.* 5*s.*, the annual expenditure falls from 159*l.* 6*s.* 4*d.* to 104*l.* 11*s.* 4*d.*

This is less than I made it in my Report; and the reason is that while the Messrs. Stevenson have assumed the yield of gas to be 11,000 cubic feet per ton, I, to keep myself within a safe margin, assumed the yield to be only 10,000 cubic feet.

§ IV.

Regarding Cost of Oil.

In my Report the price of oil with which the gas was compared is assumed to be 3*s.* 10½*d.* per gallon. This, I was assured, is the average of the sums paid for oil during the last five years at Howth Baily. In Messrs. Stevenson's Report the price of oil is stated to be 2*s.* 9*d.* per gallon. This serious difference in the cost of so important an item needs an explanation which ought to be demanded.

A similar remark applies to the cost of wicks, lamp glasses, and burners in Ireland as compared with their cost in Scotland. For the information of the Board I here give the two statements of cost:-

IRELAND.

	£	*s.*	*d.*				
800 gallons of oil at 3*s.* 10½*d.* per gallon -	155	0	0				
Wicks, lampglasses, and burners -	-	-			13	0	0
					168	0	0

SCOTLAND.

	£	*s.*	*d.*				
800 gallons of oil at 2*s.* 9*d.* per gallon -	110	0	0				
Wicks, lampglasses, and burners -	-	-			4	12	6
					114	12	6

§ V.

A perfectly clear summary of the comparative cost of the gas and oil-lights is now possible.

Assuming Messrs. Stevenson's estimates of prices in all particulars, the relation is:—

Gas.	Oil.					
	£159 6 4	to	£114	12	6	

Assuming Messers. Stevenson's estimates in all particulars, excepting the item oil, the relation is:-

Gas.	Oil.				
	£159 6 4	to	£168	0	0

Assuming the smaller wages paid at Howth Baily, but preserving Messrs. Stevenson's estimates in other respects, the relation is:—

Gas					Oil			
£104	11	4	to	{	£168	0	0	Capt. Roberts
					£114	12	6	Messrs. Stevenson.

In the one case then, there is in Ireland a saving by the gas of 53*l.* 8*s.* 8*d.*, and in the other a saving of 10*l.* 1*s.* 2*d.* Should the Howth Baily burner be still further improved, or should a cheaper Cannel than Stockwell Cannel be found, to give as good a light (which is quite possible), the saving would be greater.

I therefore adhere to my recommendation to the Board of "the further encouragement of this system of illumination in Ireland."

§ VI.

Regarding Illuminating Power.

On this point, as far as regards the Reports of Messrs. Stevenson and myself, it would be useless to make any remarks at present; for it is plain that the Messrs. Stevenson and I have operated on different lights, and that our results are not comparable with each other. On one subject, however, I should like to obtain a little information.

The "liquid photometer"[9] of Mr. Thomas Stevenson, referred to in the Report of the Messrs. Stevenson, measures the power of a light by the depth to which its rays can penetrate a darkened liquid. It may prove a highly valuable instrument.

As a reference to this photometer is likely frequently to occur, I should feel exceedingly obliged if the Messrs. Stevenson would give me, as soon as it is convenient to them, answers to the following three questions:—

What is the matter employed to darken the liquid in the photometer?

What is the proof that the absorption of light by this matter follows the same law as the absorption of light by the atmosphere in clear and in foggy weather?

Have any discrepancies been established between the indications of the liquid photometer and those of Bunsen's photometer?[10]

Remark.

In the third paragraph of their Postscript the Messrs. Stevenson express themselves thus:—"We observe that it is not now gas made from shale, or Boghead coal, that is proposed to be used, but ordinary coal-gas, and this Professor Tyndall assumes will cost 10*s*. per 1,000 feet." In my Report it is stated that the gas actually used, not "proposed" to be used, at Howth Baily is extracted from Stockwell Cannel. I am not aware of any proposal to employ ordinary coal gas, nor have I made any estimate for such gas.

I have, &c. | JOHN TYNDALL.

The Assistant Secretary, | Harbour Department, Board of Trade.

Postscript.

I think it very desirable that experimenters who have already proved their competence in the matter of fixed lights in so satisfactory a manner, should, if they see a prospect of success, be also encouraged, within reasonable limits, to apply their skill and ingenuity to revolving lights. The sum requested by the Irish Light Commissioners is not greater than I am prepared to recommend the Board to grant. I think it desirable then to grant a sum of 54*l*. for the purchase of an annular lens of a first order revolving apparatus, for the purpose of making gas experiments for revolving lights which have not yet been tried.

Throughout these inquiries I have had to congratulate myself on the able co-operation of Mr. Valentin, Principal Assistant in the Royal College of Chemistry, Oxford Street.

(Signed) JOHN TYNDALL.

*This error was detected by Mr. Valentin immediately after his arrival in Dublin on the 5th of June. The explanation offered by Capt. Roberts and Mr. Wigham was that for some time no station meter had been at work to register the gas made or consumed.

'Gas in lighthouses: Further papers relative to a proposal to substitute gas for oil as an illuminating power in lighthouses', C 282, Parliamentary Papers (1871), pp. 7–11

---

1. *Assistant Secretary, Harbour Dept., Board of Trade*: C. Cecil Trevor.
2. *The burners employed by me*: Tyndall used the series of gas burners developed by John Wigham, which gradually increased in power. See letter 3019. For Tyndall's work on lighthouses, see letter 2937, n. 6.
3. *employed by them*: the Stevenson brothers, engineers for Scottish lighthouses, tested Wigham's setup at the lighthouse in Granton in 1868 (G. Dodd, 'The Lighthouse System of the British Islands', *Companion to the Almanac; or, Year-Book of General Information for 1873* (London: Stationers' Company, 1872), pp. 33–59, on p. 53).
4. *Howth Baily*: see letter 3019, n. 3.
5. *Granton*: a part of North Edinburgh. The Northern Board of Lighthouses ran an experimental lighthouse at Granton, where the Stevenson brothers conducted their experiment. See n. 3.

6. *Boghead Cannel*: Boghead Cannel (coal), also known as Torbanite, is a mineral substance with properties intermediate between oil shale and coal. Torbanite is so named because it mainly occurs near Torbane Hill, around Bathgate, Scotland. It resembles cannel coal but contains microscopic remains of algae (*OED*).
7. *my Report to the Board*: letter 3019, section 4.
8. *Captain Roberts*: see letter 3019, n. 9.
9. *"liquid photometer"*: Stevenson's 'liquid photometer' consisted of an India rubber bag filled with diluted ink. Light passed through the bag, which could be compressed to alter the length of fluid that the light passed through (T. Stevenson, *Lighthouse Construction and Illumination* (London and New York: E & F N Spon, 1881), pp. 244–5).
10. *Bunsen's photometer*: see letter 3019, n. 8.

To the Editor of *the Times* 6 November 1869 3068

Sir—A tidy little book,[1] recently published by Messrs. Cassell and Co., bears upon its cover in prominent gild letters the title *Tyndall's Natural Philosophy*, and I am informed that an ably-conducted scientific journal,[2] under the impression that the book is a product of this year of grace 1869, has congratulated the Messrs. Cassell on the wisdom they have shown in selecting its author.

The impression just referred to is general, and it is natural, for neither upon the cover nor upon the title-page is there a word or date to inform the public that the book, so far as I am related to it, is a reprint of some short articles written 14 or 15 years ago for *Hughes' Book of Lessons*,[3] and published in that book side by side with contributions from various other scientific men.

I use the words 'as far as I am related to it' because the essays on "Light," and on "Heat and Chemical Power," with which the volume ends, are not mine, though I am at the present moment reaping unmerited credit for them. They are from the pen of Mr. Robert Hunt,[4] and bear in a very subdued form their author's name.

It is by no means my wish to injure the sale of this little volume, which I hold to be cheap at its price; but it is right that the public who purchase it should know its real genesis,[5] and that the articles here bound together, and which I tried to render good in their way, could never stand, and were never intended to stand, as a separate work on natural philosophy.

I am, Sir, your obedient servant, | JOHN TYNDALL. | Royal Institution, Nov[ember]. 6

Times, 8 November 1869, p. 4
RI MS JT/5/12/17–8

1. *little book*: J. Tyndall, *Natural Philosophy, in easy Lessons* (London: Cassell, Petter, and Galpin, 1869). Note that while the cover of the book showed 'Tyndall's Natural Philosophy', on the title page it is just 'Natural Philosophy'.
2. *scientific journal*: not identified.
3. *Hughes' Book of Lessons*: Tyndall contributed the chapter on natural philosophy to four volumes of Edward Hughes' *Reading Lessons,* published by Longmans from 1855 to 1858.
4. *Mr Robert Hunt*: Robert Hunt (1807–87), an English mineralogist, chemist, and photographer. In 1845 he was appointed keeper of the mining record office, and began compiling statistics. When the Government School of Mines opened in 1851, he was appointed Professor of Mechanical Science. From 1860 onward, he edited *Ures Dictionary of Arts, Manufacture and Mines* (*ODNB*).
5. *its real Genesis*: for further correspondence on this dispute, see letters 3072 and 3075.

From Emily Peel 7 November 1869 3069

Whitehall | Nov[ember] 7/69

Dear Professor Tyndall,

I shall be very glad to know that you are now all well again,[1] & that you got over y[ou]r journey without too great fatigue.

I have just come over, & am off on Tuesday to Drayton,[2] but am anxious to have news of you before leaving.

Yours sincerely| Emily Peel

RI MS JT/1/P/50

1. *well again*: see letter 3071.
2. *Drayton*: Drayton Bassett is a village in Staffordshire, England. Lady Emily, her husband Robert, Member of Parliament and son of the British Prime Minister Sir Robert Peel, and their children lived in Drayton Manor.

To Robin Allen 8 November 1869 3070

CORRESPONDENCE:—LIGHTHOUSE LANTERNS. | Royal Institution, 8 November 1869.

Sir,

I AM glad to find the discussion between the Messrs. Stevenson and Mr. Douglass regarding lighthouse lanterns narrowed down to the consideration of what becomes of so small a fraction as 6 ½ per cent. of the total light. It is also gratifying to observe the amicable spirit with which the discussion closes.[1]

Mr. Douglass stated in his report that more than three-fourths of the foregoing fraction was utilized by his cylindrical glass panes. Messrs. Stevenson contend that only about one-fourth of this reflected light is utilized.

The truth, as Mr. Douglass states in his last remarks, lies, I believe, between these estimates.

The Messrs. Stevenson, if I understand them aright, have treated the quantities of light reflected back through the cylindrical refractor, the upper prisms and the lower prisms respectively, as if they were in the proportions of the angles subtended by these three parts of the apparatus, that is to say, in the proportions of 56, 46, 21. The loss of light which they prove to occur has reference to the upper and lower prisms mainly, consequently the greater the amount of light apportioned to those prisms the greater would be the loss.

For the sake of more easy comparison, I will express the above proportions by the figures—

70 57 26 (A.)

but, according to the experiments of both M. Reynaud[2] and Mr. James Chance,[3] the proportions of the light emergent from the cylindrical refractor, the upper prisms and the lower prisms, and consequently the proportion *reflected back* from the cylindrical lantern, are expressed by the numbers—

70 20 10 (B.)

Such rough observations as I have made myself give the cylindrical refractor a still greater predominance.

The loss of light calculated from the three last figures (B.) would be less than that calculated from the three first (A.), and, therefore, less than that which Messrs. Stevenson make it.

Supposing his panes to be perfectly cylindrical, Mr. Douglass's principal gain would accrue from the light reflected back through the cylindrical refractor. The data from which Messrs. Stevenson conclude that this is reduced to 0·410 of its original value are not given with sufficient precision to enable me to discuss them. The reduction is greater than I should have anticipated.

To sum up, the quantity of light here in question is at best but small, and the portion of it utilized lies, in my opinion, between that claimed in the first report of Mr. Douglass, and that conceded in the last report of the Messrs. Stevenson.

I have, &c, | (signed) *John Tyndall.*

Robin Allen, Esq., &c., &c., Trinity House.[4]

'Correspondence relative to Comparative Advantages of Lighthouse Lanterns adopted by Trinity House and Coms. of Northern Lighthouses,' C 4159, Parliamentary Papers (1870), pp. 31–2

1. *the discussion closes*: see letters 2937, 2969, 2984, 2989, and 3067. For Tyndall's work on lighthouses, see letter 2937, n. 6.

2. *M. Reynaud*: Léonce Reynaud (see letter 2969, n. 2).
3. *Mr. James Chance*: see letter 2969, n. 12.
4. *Trinity House*: see letter 2937, n. 6.

To Emily Peel 8 November 1869 3071

Royal Institution. | 8th November, 1869.

My dear Lady Emily,

I knew your full, bold hand, such as Minerva[1] might have written had she been given to penmanship, before I opened your letter.

I had a good passage from Geneva to Paris; halted there a day and night, and on the Friday reached Folkestone.[2] I remained there a fortnight,[3] for it was quite useless to come to London. Each day made a distinct addition to my strength; so that before I quitted the seaside I was able to walk from Dover[4] to the South Foreland Lighthouse[5] and back again.

Still I am very weak in the legs—that is, weak for me, who previous to my accident[6] had a good stock of muscular whipcord. I weigh nearly a stone less than when I quitted London, and remember this in my case is clear loss. I have made arrangements with a friend[7] to spend 8 or 10 days at Clifton[8] and this I hope will set all things right.

Your note was a pleasant phenomenon to me. I had thought of writing to you to Geneva. Though the impulse to do so naturally runs to the nib of my pen I will not thank you; and considering what I told you regarding my disbelief in the Nicene Creed,[9] and my detestation of the Athenasian,[10] you would only laugh at me if I blessed you.

Yours ever and always, | John Tyndall.

RI MS JT/1/TYP/3/958
LT Typescript Only

---

1. *Minerva*: Minerva (for Ancient Greeks, Athena) is the Roman goddess of medicine, commerce, handicrafts, poetry, and the arts generally, but was primarily worshipped as the goddess of wisdom and warfare.
2. *Folkestone*: see letter 3021, n. 20.
3. *I remained there a fortnight*: Tyndall convalesced from a serious injury at Bence Jones' house in Folkestone. See n. 6 and letter 3061.
4. *Dover*: a town and significant port in Kent, in southeastern England on the English Channel.
5. *South Foreland Lighthouse*: a lighthouse on the White Cliffs of Dover, just under 6 miles from the town of Dover. It had been refurbished in 1846.
6. *my accident*: see letters 3030, 3031, 3032, 3037, 3041, 3048, and 3049.

7. *a friend*: probably Heinrich Debus who was Science Master at Clifton College.
8. *Clifton*: see letter 2996, n. 7 and n. 8.
9. *Nicene Creed*: the Nicene Creed is a profession of Christian Faith used in the Roman Catholic Church, Eastern Orthodox Churches, and in most Protestant denominations. It was first attested at the First Council of Nicaea (an Ancient Greek city in Anatolia, modern Turkey) in 325 AD and approved in amplified form at the Council of Constantinople in 381 AD (*Catholic Encyclopedia*, vol. 11, online at http://www.newadvent.org/cathen/11049a.htm [Accessed 24 February 2020]).
10. *Athenasian*: the Athanasian creed is a short, clear exposition of the doctrines of the Trinity and the Incarnation. It was probably written in the 4th century AD (*Catholic Encyclopedia*, vol. 2, online at http://www.newadvent.org/cathen/02033b.htm [accessed 24 February 2020]).

To the Editor of the *Times* 10 November 1869 3072

Sir,—I am very sorry to perceive that the pointing out of their error[1] to Messrs. Cassell and Co. instead of exciting repentance has invoked excuse.[2]

It appears, then, that because I, in the civilest manner, declined to correct their proofs, or to write an introduction to those 15-year-old "Lessons,"[3] they deem themselves justified in giving their reprint a form and title calculated to mislead the public, in suppressing a date calculated to enlighten the public, and in supplementing my shortcomings by two unannounced articles from another pen.

Messrs. Cassell and Co. I know, will reply that the articles referred to are announced, but how? I am reminded of the exhortation of Luther, "Give nobly,"[4] not in this way.

Had this been a purely personal question between Messrs. Cassell and myself, I should have taken a course different from that of resorting to your columns. It is, however a matter of more than personal moment that those who publish books should know their duty towards those who write them as well as to the public; and in cases where the positive and inward sense of what is right and honourable is weak, there is no course open but to employ such outward props and influences as you, Sir, have permitted me to use, to secure at least a negative observance of good morals.

I am, Sir, your obedient and obliged servant, | JOHN TYNDALL. | Royal Institution, Nov[ember]. 10.

Times, 11 November 1869, p. 9
RI MS JT/5/12/18–19

---

1. *pointing out of their error*: see letter 3068.

2. *has invoked excuse*: for Cassell and Co.'s response to this letter of Tyndall's, see *Times*, 10 November 1869, p. 11.
3. *those 15-year-old "Lessons"*: see letter 3068, n. 3.
4. *"Give nobly"*: see letter 2073.

To Thomas Archer Hirst 11 November 1869 3073

11th Nov[ember] / 69

My dear Tom.

The Times has quoted Luther "give nobly" [1]—Luther's words, & mine, were sin nobly[2]—will you correct this in conversation when opportunity occurs!

Do not mention that subject to L.L.[3] for the present—It may appear more gracious if I mention it myself.

Yours affect[tionately] | John.

RI MS JT/1/T/678

1. *The Times has quoted Luther "give nobly"*: see Tyndall's letter to *the Times* in letter 3072.
2. *sin nobly*: the standard translation is 'sin boldly!'.
3. *L.L.*: possibly Lady (Rosina) Lytton (see letters 3074, 3079, and 3080).

From Rosina Lytton 11 November 1869 3074

Sir

Allow me to thank you in the interest of all the other beasts of burden called Authors for your salutary little exposée[1] of Messieurs Cassell in this days Times. Sydney Smith[2] foresaw the necessity of a Bishop being immolated, in a railway accident,[3] before the safety of the general public could be insured; and while astonished and sincerely regretting that the "Publishers" should have ventured upon so illustrious a victim as yourself, the publicity you have given to the audacious fact cannot be otherwise than satisfactory to us minnows; at all events, I, for one, never should have dared to make my particular Publishers aggressions known to you but for it; as

"Nisi dextro tempore, Flacci verba per attentam non ibunt Caesaris aurem."[4]—

But before I do so;

I must say, that in calling the Publisher "Inward sense of what is right, and honourable weak";—[5] you have taken Luther's advice to "give nobly"[6] as it has never been patent that even the weakest "positive inward sense of what is right, and honourable" existed in those "fellows"; I use the word advisedly since

"manners make the man, want of them the fellow."—now it so happens, that I wrote to a "fellow" alias Publisher of the name of Hotten[7] last July, a most scandalously civil letter to have addressed to any of the tribe by whom in order to be well treated one should "grasp like man of metal"[8] well, in this letter, I confidentially informed M^r^. Hotten, of the peculiarities of my painful and exceptional position for which reason, in order to insure something like fair play I wished to publish the Historical Tale I had ready under a nom de plume; at the same time, sending him a book (with the stamps to return it) to show him the old style in which I wished my work to be published; and urgently requesting a speedy answer "yea", or "nay". Two months past and no answer came. I wrote again and received a fiction in three lines from one of his myrmidons[9] saying that Mr Hotten was out of Town! A palpable falsehood, for at that time he was engaged in a paper war, with the publisher Norton[10] via "The Times".—Bref—from that day in July last, to this present 11th of November; Mr Hotten has "made" no sign—even of returning the Book I sent him,—*[proof],* though we all know, that: "Barrabas"[11] as Lord Byron[12] applied it to Murray[13] is the proper generic term for Publishers; yet, I was not before aware that book-stealing was included in their peculiar mode of Book-keeping. With many apologies for so long an intrusion upon your time; I have the Honor to be Sir,

With great respect | your obedient Servant | Rosina Lytton | Brenton Villa | Taunton | Somerset | November the 11th 1869 | To Professor Tyndall

RI MS JT/1/L/81

---

1. *your salutary little exposée*: see letters 3068 and 3072.
2. *Sidney Smith*: Sidney Smith (1771–1845), a British author and cleric. Born to a poor family, Smith attended New College at Oxford University, becoming a priest in 1776. He was one of the founders of the *Edinburgh Review*, a journal of literature and culture, in 1801. Smith was a prolific author for the review, which paved his way into educated cultural circles. In 1804, he lectured at the RI on moral philosophy. Concurrently, he began actively preaching a social gospel and became more involved in religious and political matters as his influence increased until his death (*ODNB*).
3. *in a railway accident*: see S. Smith, '"Locking in" on Railways', *The Works of the Rev. Sidney Smith*, 4th edn, 3 vols (London: Longman, Brown, Green, and Longmans, 1848), vol. 3, pp. 457–60, on p. 460.
4. *"Nisi dextro . . . aurem."*: 'Only at an auspicious moment will the words of a Flaccus find with Caesar entrance to an attentive ear' (Latin), from Horace, *Satires II,* LCL 194, pp. 19, 128–9.

5. *"Inward sense of what is right, and honourable weak"*: slight misquote from Tyndall's letter 3074.
6. *Give nobly*: see letter 3072. Tyndall claims he was misquoted, but then misquotes Luther himself (letter 3073).
7. *Hotten*: James Camden Hotten (1832–73), an English publisher and writer. He had a dubious reputation. His catalogue of publications was diverse, ranging from literature and classics to how-to books, local history, and pornography. In 1873 his firm was taken over by the publisher Andrew Chatto (S. Eliot, 'Hotten: Rotten: Forgotten? An Apologia for a General Publisher', *Book History*, 3 (2000), pp. 61–93).
8. *"grasp like man of metal"*: 'Grasp it like a man of mettle'; from the poem 'A Useful Hint' by Aaron Hill (1685–1750).
9. *myrmidons*: faithful followers, group or team of attendants, servants, or assistants (*OED*).
10. *Norton*: Caroline Elizabeth Sarah Norton, Lady Stirling Maxwell (1808–77), an author and social reformer. Norton campaigned especially for the rights of women to divorce, their own property and earnings, and custody of their children, after all were denied to her when she left her controlling and abusive husband (*ODNB*; J. G. Perkins, *The Life of Mrs. Norton* (London: John Murray, 1909). She published a letter criticising Hotten in the *Times* of 28 October 1869.
11. *Barrabas*: Barrabas is a figure from the New Testament. He was an insurrectionist held prisoner and destined to be crucified at the same time as Jesus but was freed at Passover by Pontius Pilate.
12. *Lord Byron*: George Gordon Byron (1788–1824), Sixth Baron Byron, an English Romantic poet. Byron was one of the most famous and influential poets of the nineteenth century, even though the suggestive nature of his poems often ran counter to Victorian high moralism (*ODNB*). Tyndall was deeply influenced by Byron's poetry and by the figure of the independent, defiant Byronic hero.
13. *Murray*: the elder John Murray. In 1824, Murray burned Lord Byron's autobiographical manuscript, considering it too salacious to publish (see letter 2963, n. 9).

To the Editor of the *Times* 12 November 1869 3075

Sir,—In *The Times* of yesterday,[1] which I had the pleasure of seeing late last night, Messrs. Cassell represent themselves as "repelling" my insinuations.

To my mind the case of Messrs. Cassell lacks the moral elasticity out of which a repellent force could be generated. The facts adduced go right into them without a tendency to rebound.

The Messrs. Cassell did send me their proofs neatly made up, with blank leaves for correction and emendation. I, point blank, but civilly, declined having anything whatever to do with them. At the same time I ventured the opinion that if the lessons were published their date ought to be made known. It was partly my unwillingness to take up an unnecessary fragment of your space and partly my tenderness to Messrs. Cassell that caused me to "omit" these points in my first letter

Here is a question for wiser heads than mine. I wrote those articles for Hughes' *Book of Lessons*[2] and for nothing else. They were written at the earnest request of the editor, and because good men who thought such lessons likely to be a public benefit gave Mr. Hughes the aid of their pens. Have Messrs. Cassell a right to isolate my articles from their environment, and to publish them in a form the least mention of which would have prevented me from writing them?

Again thanking you, I make my last bow and retire.

Your obedient servant, | JOHN TYNDALL. | Clifton, Down Hotel, Nov[ember]. 12

Times, 15 November 1869, p. 6
RI MS JT/5/12/20–1

1. *the Times of yesterday*: see Cassell, Petter, and Galpin, 'To the Editor of the Times', *Times*, 12 November 1869, p. 8. See also letters 3068 and 3072.
2. *Hughes' Book of Lessons*: see letter 3068, n. 3.

From Mary Egerton 12 November [1869][1] 3076

Mountfield Court | Hurst Green | Nov[ember]. 12th

Dear Mr. Tyndall,

I was very glad to hear of you;[2] I have wondered often how you were getting on, & thought of you especially on that Sunday morning. It was a day to remember all one's friends, for it was one on which we celebrated the simple memorials of that one friend whose love embraces all.—Your account of yourself is on the whole satisfactory I think—muscle must take longer to recover than flesh and till that is done, I suppose, strength of limbs cannot completely return. Fever takes so much out of the Constitution, and truly you had nothing external for it to pull upon! I hope your lecture work will not be of a heavy nature—will you ask my friend Mr Brown[3] to send me a Programme as soon as they are printing. I like to know what is going on.

Does your evening lecture always open the series? I wish it happened to be at the end of the winter season instead of the beginning! After being in town a little while I might summon more courage than I feel now[4] to face a crowd, much as I should like to see the Comet, or the Light Clouds.[5] Your party at Mr Forster's[6] sounds the perfection of society! I am very fond of Mr Venables;[7] if you come across him speak to him of us—I sh[oul]d like him not to forget us.

I always think of his great nature acting charades over here with the children, till he got into such fits of laughing he c[oul][d] not speak! An unusual phase in which to see his fine half melancholy countenance. That grand old Carlyle! O he knows his scorn of the selfish systems of Religion.—I meant only just to thank you for writing & it has come to this! Adieu

Y[ou][r] most loving | M.F. Egerton

I rejoice to hear of your old papers being republished[8] though I fear they will be too deep for one. I wonder what you are experimenting upon!

RI MS JT/1/E/60

1. *[1869]*: the year is given by reference to Tyndall's accident and injury of 1869.
2. *hear of you*: letter missing.
3. *M[r] Brown*: possibly William Brown, a porter at the RI.
4. *more courage than I feel now*: Egerton's husband, Edward died suddenly in August 1869 (see letter 3040).
5. *the Comet or the Light Clouds*: the lectures Egerton was referring to took place earlier in 1869. She and Tyndall discussed his cometary theory earlier in the year. See letters 2981 and 3007.
6. *M[r] Forster's*: see letter 3066, n. 1.
7. *M[r] Venables*: probably George Stovin Venables (1810–88), an English barrister and journalist. Venables was called to the bar in 1836 and became a queen's counsel in 1863. He had an anonymous parallel career as author of numerous essays for journals such as the *Saturday Review*, the *Literary Review*, and *MacMillan's Magazine* (*ODNB*). Tyndall and Venables had many acquaintances in common.
8. *Your old papers being republished*: in 1870, two edited collections of Tyndall's previous publications were published: *Essays on the Use and Limit of the Imagination in Science* (London: Longmans, Green, and Co., 1870) and *Diamagnetism*. The following year, 1871, another such edited collection appeared: *Fragments of Science*. Egerton was likely referring to the collection on diamagnetism.

From Mary Egerton [15 November 1869][1] 3077

<Letter starts here>

Mountfield Court | Hurst Green

Do read again that last part from chapter 13th or 14th; it is full of such expressions, and so beautiful. It is the other book that I have of yours[2] you know—(or rather of Mr. Hirst's, according to the inscriptions; I have always wondered how you came to have it back again!).

But what a digression your dinner party[3] has led me into! Forgive me:

these "super sensual" subjects come before me the more forcibly just now, from the longing I cannot help feeling that I might lift, if but for a moment, the veil that hides one so dearly loved from my sight. Life seems so strange without him![4] A shadow continually hangs over it, though with so many young ones *[round]*, it is not one of unmitigated gloom. Except my children, I rarely see anybody to speak to, (much less any soul,) and there is some unaccountable satisfaction in letting out some of what is in one's mind, only unfortunately thoughts flow much more readily upward than downward in the mental scale! I only wish I knew verily & truly whether it bored you very much? I hope this may follow you to Clifton[5] & to not interrupt your busy time.

Your little friend Hugh[6] is a great happiness to me; he is doing so well at Rugby[7] & writes me long letters full of discussions on Greek philosophers, Roman history or Poetry and geology. The latter he is very much interested in, they are reading Sir Charles Lyell's "Principles."[8] By the by Mr Hutchinson[9] called him up the other day to enquire about your accident[10] he had heard some how of our having met you.

May[11] & I are reading a little Chymistry but we get sadly puzzled among the ates & etes & ides & ics! I do not think Professor Roscoe's book[12] gives the general principles of the nomenclature very clearly—and I cannot find anywhere an explanation of the change of the numbers representing the combining weights; 0 from 8 to 16 &c. The books all adopt one or the other system, but do not explain the alteration.

<Letter ends here>

RI MS JT/1/E/99q

---

1. *[15 November 1869]*: the date of this letter has been estimated based on following information: Egerton's husband died 27 August 1869 (see letter 3040); Tyndall did not return to Britain after his accident until 8 October (see letter 3049); and he was in Clifton on 16 November (see letter 3081).
2. *the other book that I have of yours*: not identified.
3. *your dinner party*: possibly the dinner party referred to in letter 3677.
4. *Life seems so strange without him!*: Egerton's husband Edward passed away in August 1869 (see letter 3040).
5. *Clifton*: see letter 2996, n. 7 and n. 8.
6. *your little friend Hugh*: Hugh Edward Egerton (see letter 3040, n. 12).
7. *Rugby*: Rugby School, an established independent school in Rugby, Warwickshire, England.
8. *Sir Charles Lyell's "Principles"*: C. Lyell, *Principles of Geography: being an Attempt to Explain the Former Changes of the Earth's Surface, by Reference to Causes now in Operation*, 3 vols (London: John Murray, 1830–33).
9. *Mr Hutchison*: Thomas Hutchison (1826–99), master at Rugby School.

10. *about your accident*: see letters 3030, 3031, 3032, 3037, 3041, 3048, and 3049.
11. *May*: Mary Alice Egerton.
12. *Professor Roscoe's book*: Henry Roscoe (1833–1915), an English chemist who studied at University College, London before earning his PhD under Robert Bunsen at Heidelberg. After Edward Frankland left for London, in 1857 Roscoe replaced him as professor of chemistry at Owens College, Manchester where he remained until 1886. Roscoe's research included collaborations with Bunsen on photochemistry between 1855 and 1862, as well as his own work on vanadium of which he was the first to prepare a pure sample. He wrote a number of popular works on chemistry including *Lessons in Elementary Chemistry* (1866) and *Spectrum Analysis* (1869). Lady Egerton is probably referring to the former. Roscoe was knighted in 1884 (*ODNB*; *CDSB*).

From John Frederick William Herschel — 15 November 1869 — 3078

Collingwood | Nov[ember] 15th 1869.

My dear Sir,

Seeing a letter of yours in a Newspaper[1] I gather that you are returned to London and I hope quite recovered from the effects of your accident[2] which I was very much grieved to hear of.

Looking over a number of Comic Pamphlets[3] of all degrees of absurdity from mere raving to some that have a semblance of plausibility—I find two of the latter class which perhaps you may like to look at as (one of them at least) seems to bear a distant analogy to your views—you can keep them as long as you like only ultimately let me have them again as I keep them to bind up with other absurd speculations.

Yours very truly | Herschel.

RI MS JT/1/TYP/2/568
LT Typescript Only

---

1. *a letter of yours in a Newspaper*: Tyndall wrote letters 3068 and 3072 to the editor of the *Times* in the days preceding this letter.
2. *your accident*: see letters 3030, 3031, 3032, 3037, 3041, 3048, and 3049.
3. *Comic Pamphlets*: enclosure missing.

From Rosina Lytton 15 November 1869 3079

<Page missing>
from soiling my fingers by again writing to him; for you Sir, as a man, and moreover a celebrity, can have no idea of the crushing process brought to bear upon defenceless women in this country; where might is right, and where the power, and the impunity of vice, is so mathematically graduated; for nothing can, exceed the unimpeded vice of the haute Volée of English Society; unless it be, the *[immodest]* corruption of English political life, which in its turn, is at once capped, and guarded by the venal mendacity of our Newspaper Press.—And alas, for us poor legal victims of bad men,[1] I fear our position, is not likely to be either bettered or strengthened, by Becker,[2] having become a synonyme for that omnipotent masculine garment, whose initial letter is also B.[3]—I am disappointed, but by no means surprised;—at not seeing your second letter in todays Times:[4] I don't know about there being "honor among thieves"? as I have never yet found it, long as I have "fallen among thieves," but there can be no doubt of the proverbial "thickness", which exists between the journalistic Press, and the Publishers.

I am Sir Yours, with much esteem | Rosina Lytton

Brenton Villa | Taunton Monday | November | the 15th 1869

RI MS JT/1/L/83

1. *us poor legal victims of bad men*: see letters 3068, 3072, and 3075 for Tyndall's dispute with a publisher and letters 3074 and 3080 for Lytton's publisher grievances.
2. *Becker*: not identified.
3. *B.*: not identified.
4. *your second letter in todays Times*: letter 3075. See also letter 3080.

From Rosina Lytton 15 November 1869 3080

Sir

Pray pardon my boring you with a second letter, but when I wrote my first[1] not being well, I trusted to another, to look over The Times for me; Upon taking it up myself just now, the first thing I saw, was your admirable second letter,[2] *[and of]* also Mr Hunts[3] letter exonerating himself from any participation in the Cassell *[cansillerie]*.[4] I hope the Messieurs Cassell will not die of the shock, as the straining at gnats, Camel–swallowing Anglo

Saxons, are always terribly shocked, at the disencumbered estate, in which Truth appears, upon the rare occasions on which she issues from her Well.—However, it is to be hoped, that your two letters, will in some degree clear the public eye-sight. I remember being much disgusted, a few years ago at a paragraph in "Cassells Paper"[5] which professes to be an Educational paper, it was à propos of the present Emperor of the French, it said "In making a Romance of their mothers, the National heir of France, Louis Napolèon has pandered to the Sentimentality of the French"[6] yea verily, an affectionate!! Respect for a mother's memory is a "sentimentality" which all Frenchmen from the highest, to the lowest, are guilty; of course an unpardonable juvenile absurdity, in the eyes of self-seeking Mammon worshipping Englishmen, who are not guilty of that, or any other "sentimentality" inculpating human affection.

Wishing you great speed in your Cassell crusade—believe me Sir yours with great respect

Rosinda Lytton | Benton Villa

To Professor Tyndall | November 15th 1869

RI MS JT/1/L/82

1. *my first*: letter 3079.
2. *your admirable second letter*: see letter 3075.
3. *Mr Hunts letter*: in the *Times*, 15 November 1869.
4. *the Cassell [cansillerie]*: see letters 3068, 3072, and 3075.
5. *"Cassells Paper"*: *Cassell's Illustrated Family Paper.* It became *Cassell's Magazine* in 1867 and later *Cassell's Family Magazine*.
6. *"In making a Romance . . . the French"*: source not identified.

To John Frederick William Herschel — 16 November 1869 — 3081

Clifton Down Hotel | Clifton | 16th Nov[ember] 1869

My dear Sir John,

I thank you very much for your kind note,[1] with added thanks for the papers[2] which I shall read with pleasure.

I shall find them on my return home.

I brought down here with me your book entitled "Familiar Lectures on Scientific Subjects";[3] and this day before your note reached me I thought of writing to you to express to you the delight with which I read your expositions regarding Light. I never read anything to exceed or equal your exposition of the colours of thin plates and of the measurement of a wave of light. I intended to say something upon these subjects at a future day, but I shall be positively afraid or ashamed to approach the subject after your treatment of it. Perhaps I may ask your permission to quote bodily, with due acknowledgement, this particular exposition.

I suppose all the other parts of the book are equally luminous, but them I have not yet read.

I am now engaged chastening into some kind of order a chaos of experiments—which when "reduced" I mean to present to the Royal Society.[4]

Believe me | Yours very sincerely | John Tyndall.

RS HS 408
RI MS JT/1/TYP/2/569

1. *your kind note*: letter 3078.
2. *the papers*: Herschel sent Tyndall two 'Comic Pamphlets' (see letter 3078).
3. *"Familiar Lectures on Scientific Subjects"*: J. Herschel, *Familiar Lectures on Scientific Subjects* (London: A. Strahan & Co., 1866).
4. *present to the Royal Society*: probably 'On the Action of Rays of High Refrangibility upon Gaseous Matter', which Tyndall read to the RS on 27 January 1870, and which was published later that year in *Phil. Trans.*, 160 (1870), pp. 333–65.

From John Frederick William Herschel — 19 November 1869 — 3082

Collingwood | Nov[ember] 19th 1869.

My dear Sir,

I am very glad you like my explanation of the colours of thin plates and that part of my lecture which refers to the wave lengths &c.[1] It appeared to me quite indispensable to make perfectly plain "to the meanest capacity" that seemingly gratuitous postulate about the loss of a half undulation at reflexion—knowing from my own experience—how very arbitrary the assumption seems at first sight (and has the effect of casting quite a doubt in the learner's mind as to the real validity of the undulatory explanation of the phenomenon)

Of course I should only feel too much flattered by your quoting this or any other part of those lectures which may seem to you calculated to make the subject more intelligible to any of your audience.

I wish you or any one could give a satisfactory explanation of the actual mechanical process by which, on the undulatory theory, the dispersion of light is effected—It is all very well to say that in the integration of the partial diff., equation an arbitrary function is involved which may contain among its constant parameters a term or terms depending on the No. per second of the vibrations made—What I want to see is, how this sort of parameter gets in? What sort of modification of the acting forces, or their incidence, this greater or less frequency produces?, and how it produces it?—In a word—the modus operandi. . . . failing which, the X S and Y S and arbitrary constants are more like a fog obscuring than a light illuminating the subject. Challis[2] has just made the matter more obscure than ever!

Believe me dear Sir | yours very truly | J. F. W. Herschel.

RI MS JT/1/TYP/2/570
LT Typescript Only

1. *my explanation . . . wave lengths &c.*: see letter 3081.
2. *Challis*: James Challis (1803–82), an English physicist and astronomer. In 1836 he succeeded Airy as Plumian Professor of Astronomy and Experimental Philosophy at Cambridge University and was also named director of the Cambridge Observatory. He lectured and published on astronomy, mathematics, physics, and religion (*ODNB*). Herschel may be referring to Challis' recently published *Notes on the Principles of Pure and Applied Calculation; and Applications of Mathematical Principles to Theories of the Physical Forces* (Cambridge: Deighton, Bell and Co.; London: Bell and Daldy, 1869).

To [John Tyndall][1] 20 November [1869][2] 3083

20th Nov[ember]

My dear Friend

I can truly say that my health and strength are greatly improved. I came down here to Clifton[3] about 10 days ago[4] and have derived great benefit from the excursion. Indeed I feel myself almost completely restored.

I return to London on Tuesday next.

It was my privilege to make yesterday one of the most interesting excursions I ever made in my life. The good people of Gloucestershire[5] erected a couple of years ago upon an eminence of the Cotswold Hills[6] a high tower as a monument to William Tyndall the martyr. The little town of Nibley[7] stands at the foot of the hill on which the tower is built; and close at hand is the spot where the martyr was born. The people are full of his history. It is a most delightful spot—so charming indeed as to excite in me the desire to build a little house there where I might end my work and life.

Yours ever | John Tyndall

Very kindest regards to all at Gorey.[8] I am very much rejoiced to hear that Emma[9] is so well.

RI MS JT/1/TYP/10/3317
RI MS JT/5/14
LT Typescript Only

---

1. *[John Tyndall]*: the recipient of this letter has been identified as Tyndall's cousin, Dr. John Tyndall, who lived in Gorey and was caring for Tyndall's sister, Emily. See letter 2987, n. 7.
2. *[1869]*: the letter has been dated to 1869, since Tyndall's sister Emily stayed with Dr. Tyndall in Gorey in that year.
3. *Clifton*: see letter 2996, n. 7 and n. 8.
4. *about 10 days ago*: Tyndall arrived in Clifton on 16 November. The date LT provided for this letter (20 November) may be incorrect.
5. *Gloucestershire*: in the Cotswold Hills in southwestern England, a county bordering Bristol.
6. *Cotswold Hills*: also known as Cotswolds, a ridge of limestone hills, part of the Jurassic uplands, that stretch across south-central England.
7. *Nibley*: is a village in the southern portion of Gloucestershire, England.
8. *Gorey*: see letter 2987, n. 7.
9. *Emma:* Emily Tyndall, Tyndall's sister, currently lodging with Dr. John Tyndall at Gorey. See n. 1.

To William Hughes[1] 26 November 1869 3084

26th November 1869.

My dear Sir

The subject of the Christmas lectures[2] this year is to be Light. They are intended as you know for a juvenile audience.

Yours Truly | John Tyndall | Mr. William Hughes | &c. &c. &c.

Private Collection #4

---

1. *William Hughes*: since the letter is addressed to *Mr.* William Hughes, it is probably William Hughes, Clerk of Accounts and Collector at the RI. Possibly this is William Hughes (1818–76), Professor of Geography at King's College and Queen's Colleges, London, who was interested in juvenile education, primarily in the field of geography.
2. *Christmas Lectures this year*: see letter 3055, n. 7.

To Francis Fox Tuckett[1] 27 November [1869][2] 3085

Royal Institution of Great Britain | 27th Nov[ember].

My dear Mr. Tuckett.

Your cheque has arrived & been thankfully accepted by those interested in the Faraday Memorial.[3] Kind regards to your sisters.

Yours faithfully | John Tyndall

Wellcome Collection PP/HO/G/A441

1. *Francis Fox Tuckett*: Francis Fox Tuckett (1834–1913), a British mountaineer and early member of the Alpine Club. Tuckett explored and collected observations of the High Alps. In the 1860s, he explored the Dauphiné and Orteler group, correcting incomplete data from the geographical survey. Tuckett published reports of his climbs in the *Alpine Journal* as well as in *Peaks, Passes, and Glaciers*, a series of essay collections edited by John Ball (D. W. F., 'Obituary: Francis Fox Tuckett', *Geographical Journal*, 42:2 (1913), pp. 206–7).
2. *[1869]*: the year of this letter was determined through the postmark and the mention of the Faraday Memorial. Tyndall collected funds for it in 1869. See letter 3098.
3. *Faraday Memorial*: in June 1869 a public meeting was held at the RI to explore a memorial to Faraday. It was attended by numerous well-known figures from politics and science, including Tyndall, and the Faraday Memorial Fund was established to collect subscriptions for this purpose. It did not come to fruition (Anon, 'Proposed Memorial to Faraday', *Chemical News*, 25 June 1869, pp. 308–9; *Ascent of John Tyndall*, p. 209). See letter 3098, n. 3.

To George Biddell Airy [28 November 1869][1] 3086

Private. | Royal Institution of Great Britain

My dear Sir,

Would you allow me purely on my own responsibility to ask you the following question.

Would you accept the office of President[2] to the Royal Society?

It appears to me that the society is in a difficulty as to finding a man of real eminence to preside over it. And in this one day it would be unworthy of us to aim at any thing less than the highest eminence.

Yours very faithfully | John Tyndall

RGO MS 6/393

1. *[28 November 1869]*: this letter has been dated through its place in the correspondence between Tyndall and Airy on whether Airy would be willing to take on the Presidency of the Royal Society. See letters 3088, 3090, and 3089.
2. *would you accept the office of President*: see letters 3088, 3089, 3090, 3094, and 3095. Edward Sabine had been President of the RS since 1861, and feeling pressure from younger members of the RS's Council, such as Tyndall, Huxley, and Hooker, that he was too old (in his 80s) and had served too long, he did not seek re-election in 1871. Airy was proposed and elected, serving from 1871–3 (*Ascent of John Tyndall*, pp. 257–8; *X Club*, pp. 269, 271–2).

To [Juliet Pollock][1] 29 November 1869 3087

29th Nov[ember] 1869

Thanks & many thanks for that kind note—But am I right in calling it so? Is it kind to paint the Crimson snows and purple of air of the Alps to one doomed to the twilight of Seven-dials?[2] But I must not be selfish. Your letter[3] was an esthetic picture, and the love you bear your friend very beautiful to me. I will not be jealous because I expect you to make me the other pole of the magnet of your affections. Thus your husband is to occupy the centre or meridian—Your Augusta[4] is to be your North pole and I am to be your South. Thus do we form an organic group pleasant to the vision of Gods and men.[5]

Yours ever | The South Pole. | true.

RI MS JT/1/T/1150
RI MS JT/1/TYP/6/2114

1. *[Juliet Pollock]*: the recipient has been identified as Juliet Pollock, since Tyndall occasionally signed letters to her as 'The South Pole'.
2. *the twilight of Seven-dials*: a road junction located just west of Covent Garden in London, Seven Dials in the eighteenth century was synonymous with dereliction, with many instances of mob violence.
3. *Your letter*: letter missing.
4. *Your Augusta*: possibly Augusta Huth (née von Westenholz), wife of Henry Huth (1815–78); or Augusta Elizabeth Frederica Stanley (née Bruce, 1822–76), a courtier who served as lady in waiting to Queen Victoria before marrying Arthur Stanley (Dean of Westminster) in 1863 (*ODNB*); or Augusta Ritchie (1847–1911), who married Douglas Freshfield in 1869. The family was related through marriage to the family of author William Makepeace Thackray (J. Aplin, *Memory and Legacy. A Thackeray Family Biography 1876–1919* (Cambridge, UK: Lutterworth Press, 2011), pp. 278–9).
5. *pleasant to the vision of Gods and men*: allusion not identified.

From George Biddell Airy 29 November 1869 3088

Royal Observatory, Greenwich. | London, S.E. | 1869 November 29[1]

My dear Sir

In haste—

I presume that your question about Presidency[2] can have no relation to proceedings tomorrow.[3] This allows me a little time for consideration.

My present impression is, that I cannot undertake the presidency. I would, if I could, diminish my cares: and I live an hour from London. Otherwise, I should be very proud indeed of the position.

I will write again before long.

I am, my dear Sir, | Yours very truly | GB Airy

Professor Tyndall

RI MS JT/1/A/50
RGO MS 6/393

1. *Royal Observatory, Greenwich, | London, S.E.*: Royal Observatory headed notepaper.
2. *your question about Presidency*: see letter 3086. For further discussion see also letters 3089, 3090, 3094, and 3095.
3. *proceedings tomorrow*: the annual Anniversary Meeting of the RS took place on 30 November 1869. These meetings dealt with the formal business of the Society including reports on accounts and from the Society's various committees, a presidential address, and the election of officers. In 1869 Edward Sabine was returned as President, while Tyndall was elected to the Council ('Anniversary Meeting', *Roy. Soc. Proc.*, 18 (1870), pp 101–15).

To George Biddell Airy 30 November 1869 3089

Private. | Royal Institution of Great Britain | 30th Nov[ember]

My dear Sir,

If you could let me have a line holding out even a hope that you would be able to accept the presidency[1] by Thursday morning it would enable me to do something in reference to the matter on that day.

I would beg of you to remember that I am acting purely on my own responsibility—doing what I as a fellow deem in my private mind to be for the advantage of the Royal Society.

Yours very faithfully | John Tyndall

RGO MS 6/393

1. *accept the Presidency*: Tyndall approached Airy to serve as President of the RS, shortly before he became a member of the Council of the RS. See letters 3086, 3088, 3090, 3094, and 3095.

From George Biddell Airy 30 November 1869 3090

Royal Observatory, Greenwich, | London, S.E. | 1869 Nov[ember] 30[1]

My dear Sir

In thinking further of your Presidential question,[2] my thoughts still incline to the negative.

I could, I believe, manage my various business fairly with the assistance of a good Private Secretary: but such a one is not to be obtained without expense. And that is the point on which very much bears. The tenure of the office would cause expense in various ways. And on the whole it would press heavily on my limited income. This is not the first occasion, by two or three, in which I have felt it my duty, for the same reason, to decline offers of an honorable character. There is no disguising the law, that money must govern every thing.

It would also be inconvenient to me and to my family, to attempt regular attendance at the weekly meetings.

On the whole, I do not see that it is possible to remove the difficulties. Meantime, I am much flattered by your privately proposing the question to me.

I am, my dear Sir, | Yours very truly | GB Airy
Professor Tyndall

RI MS JT/1/A/51
RI MS JT/1/TYP/1/42
RGO MS 6/393

1. *Royal Observatory, Greenwich, | London, S.E.*: Royal Observatory headed notepaper.
2. *your Presidential question*: see letter 3086 and for a continuation of this discussion letters 3088, 3089, 3094, and 3095.

From Emily Peel 30 November 1869 3091

Drayton Manor, Tamworth. | Nov[ember] 30/69.

My dear Professor,

"Minerva"[1] is anxious to judge for herself whether you are really sound & whole again[2]—Will you come & spend a couple of days with us next week—Tuesday 7th to Friday 10th? Sir Robert[3] will be charmed to see you, & I shall be happy to renew our pleasant conversations of Geneva memory.[4]

Yours sincerely | Emily Peel

Have you worked out your theory about the blueness of the Lake Water?[5]—

RI MS JT/1/P/51

---

1. *"Minerva"*: in letter 3071 Tyndall likened Peel's handwriting to that of the Roman goddess Minerva.
2. *whether you are really sound & whole again*: for Tyndall's mountaineering accident and recovery see letters 3030, 3031, 3032, 3037, 3041, 3048, and 3049.
3. *Sir Robert*: Emily Peel's husband, Sir Robert Peel (1822–95), a Liberal politician, eldest son of the prime minister Sir Robert Peel, and Member of Parliament (1850–86). He was often critical of Gladstone. He married Lady Emily in 1856, but she later left him to live in Geneva (*ODNB*).
4. *of Geneva memory*: Emily Peel hosted Tyndall in Geneva when he was recovering from his mountaineering injuries. See letters 3049 and 3050.
5. *your theory of the blueness of the Lake Water*: Tyndall's experiments on polarization helped him determine that the blue colour reflected was caused by the reflection of the waves of shorter blue light. He presented this theory in a paper titled 'On the Blue Colour of the Sky' (see letter 2953, n. 4). Emily Peel later acquired water from Lake Geneva for his experiments. See letter 3156.

From Jacques-Louis Soret 30 November 1869 3092

Genève 30 Nov[ember] 1869

Mon cher Tyndall

Je vous envoie ci joint le brouillon d'une petite communication que j'ai *[*prié*]* M. Dumas de présenter à l'Académie, en réponse aux notes de M. Lellamand. Le brouillon est fort mal écrit cependant j'espère que vous pourrez le lire; en tout cas je pense que la chose paraîtra bientôt dans les Ctes rendus.

J'aurais voulu pouvoir vous expliquer aujourd'hui comment je crois que le phénomène de la polarisation des matières *[1 word illeg]* peut *[*s'interpréter*]* en *[*admettant*]* que sur des particules très petites les lois ordinaires de la réflexion cessent de *[*se vérifier*]*. Mais c'est un peu long pour aujourd'hui où je suis pressé. Je tacherai de vous écrire prochainement à ce sujet ainsi que sur la coloration de l'eau.

Comment *[*allez*]* vous maintenant? Etes-vous tout à fait *[*remis*]*?

Milles compliments affectueux | L. Soret

Geneva 30 November 1869

My dear Tyndall,

I send you enclosed the draft[1] of a brief paper[2] that I asked Mr. Dumas to present to the Academy,[3] in response to Mr. Lallemand's[4] notes. The draft is very badly written but I hope that you will be able to read it. In any case, I think that the thing will soon appear in the Comptes rendus.[5]

I would have liked to be able to explain to you today how I believe the phenomenon of the polarization of *[1 word illeg]* matter can */*be explained*/* by assuming that the ordinary laws of reflection cease to */*hold true*/* on very small particles. But it is a bit too long for today since I am in a hurry. I will try to write to you soon about this subject,[6] as well as about the colouring of water.

How are you now? Have you recovered?[7]

A thousand affectionate regards | L. Soret

RI MS JT/1/S/90

1. *enclosed the draft*: enclosure missing.
2. *a brief paper*: J.-L. Soret, 'Sur l'illumination des corps transparents', *Paris, Comptes Rendus*, 69 (1869), pp. 1192–6.
3. *the Academy*: Parisian Academy of Science.
4. *Mr. Lallemand*: Alexandre Lallemand (1816–86), the first Professor of Physics at the University of Poitiers and later Dean of the Faculty of Science. Although professor of physics, most of his publications were in chemistry (M. Durrande, 'Nécrologie', *Bulletin de la Faculté des lettres de Poitiers, Année 1886* (Paris and Poitiers: Leroux, Blanchier, Druinaud, 1887), pp. 175–6).
5. *Comptes rendus*: *Paris, Comptes Rendus*, the proceedings of the French Academy of Sciences. See n. 2.
6. *Write to you soon about this subject*: letter 3119.
7. *Have you recovered?*: contacted by de la Rive about Tyndall's injury and travel to Geneva for medical care, Soret picked him up at the train station and brought him to his hotel. See letters 3035 and 3048.

To [Emily Peel][1] 1 December 1869 3093

1st. December, 1869.

That goddess[2] was the dream of my youth and has become the idol of my mature years.

In my hour of need and trial she spread her gracious AEgis[3] over me. It would therefore ill become me to hesitate when she commands.

I shall therefore report myself at Drayton Manor[4] on Wednesday the 8th,
John Tyndall

RI MS JT/1/TYP/3/948
LT Typescript Only

1. *[Emily Peel]*: Drayton Manor was the home of Emily Peel, who invited Tyndall for a visit in letter 3093.
2. *That goddess*: Minerva. Tyndall and Lady Peel were sharing an inside joke. See letters 3071 and 3091.
3. *AEgis*: Though the material nature of Minerva's Aegis (a shield or cloak) is uncertain, it was used for protection. Tyndall was alluding to the care Emily Peel provided when he was convalescing in Geneva after his mountaineering injury. See letters 3049 and 3050.
4. *Drayton Manor*: Lady Emily lived with her husband, Sir Robert Peel, and their children at Drayton Manor. Drayton Bassett is a village in Staffordshire, England.

From George Biddell Airy 1 December 1869 3094

Royal Observatory, Greenwich, | London, S.E. | 1869 December 1[1]

My dear Sir

I think that I can now answer you[2] definitively.

1. I cannot undertake presidency of the R[oyal] Society as from the present time or from any very near time.[3] This is a matter on which I am not free to choose: I have some troublesome engagements, and have not the physical power to undertake more at present.

2. If arrangements could be deferred till next June or July, I could undertake it, and would if necessary:—nothing unforeseen preventing.

3. Even then, I should be better pleased not to be called upon to undertake it: but this only expresses my wish, not contradicting the engagement of last paragraph.

4. I think it right to mention that, with other infirmities of approaching age, I have a slight deafness creeping upon me, which is a sad disablement for a Chairman.

I am, my dear Sir, | Yours very truly | GB Airy
Professor Tyndall

RGO MS 6/393
RI MS JT/1/A/52
RI MS JT/1/TYP/1/43

1. *Royal Observatory, Greenwich, | London, S.E.*: Royal Observatory headed notepaper.
2. *I can now answer you*: see letters 3086, 3088, 3089, 3090, and 3095.
3. *any very near time*: Airy changed his mind a few years later and succeeded Sabine as president in 1871 (see letter 3086, n. 2).

To George Biddell Airy 5 December [1869][1] 3095

Royal Institution of Great Britain | 5th Dec[ember]

My dear Sir,

There is no immediate call for action.[2]

Thereafter stands thus:—I find many of the best men in the Society asking what is to come! They are looking out for the future and need a rallying point. I have heard many conversations on the subject. My conviction after all I have heard is that you are the best man & that the Society is likely to think you so.[3] Your difficulties[4] I think can be ready met—Meanwhile accept all this as before as very private interpretation of things, and the direction of my own convictions.

Yours very truly | John Tyndall

RGO MS 6/393

1. *[1869]*: this letter has been dated through its place in an ongoing discussion between Tyndall and Airy. See letters 3086, 3088, 3089, 3090, and 3094.
2. *no immediate call for action*: see letter 3094.
3. *the Society is likely to think you so*: Tyndall has been trying to convince Airy to serve as President of the RS. See letters 3086, 3088, 3089, 3090, and 3094.
4. *your difficulties*: see letter 3094.

To Auguste de la Rive 5 December 1869 3096

5th. December, 1869.

My dear Friend

I have been for some time thinking of writing to you and now at length I convert my intention into an act.

I quitted you with my wound[1] quite healed, but I was still very weak. I halted for a fortnight at Folkestone[2] where I obtained some strength and then came on to London. Not feeling myself strong enough to continue in London I went down to Clifton[3] near Bristol and remained there ten days. The improvement in my health and strength has been considerable.

And now I am preparing my Christmas Lecture[4] and I wish you and yours every happiness which the season can bring.

Bence Jones had gone too far with the printing of his book[5] to be able to make use of your letter[6] in the first edition. The work has just made its appearance. I think you will be pleased at the manner in which he speaks of Faraday's relationship to your father.[7]

While at Clifton I worked hard to put a few of my experiments into a condition fit for presentation to the Royal Society. I have just handed in a paper[8] which constitutes the first part of these researches on the action of the short waves of the spectrum on gaseous matter.

Will you convey my kindest regards and remembrances to Madame DelaRive[9] and to your two sons.[10] I shall not soon forget the kindness which all of you shewed me during my time of trial in Geneva.

I see M. Morren[11] has been objecting to the use of Sulphuric acid in the drying of gases, and has found his sulphuric acid impregnated with sulphurous acid. I can only say he must have worked with a very impure acid. In fact it requires a considerable quantity of sulphurous acid to produce any visible action at all.

And now with best wishes | believe me | Yours always | John Tyndall

RI MS JT/1/TYP/1/372
LT Typescript Only

---

1. *my wound*: Tyndall had written to De La Rive for assistance after suffering a serious injury while mountaineering. See letters 3030, 3031, 3032, 3037, 3041, 3048, and 3049.
2. *Folkestone*: see letter 3021, n. 20. Tyndall recuperated further at Bence Jones' home at Folkestone. See letters 3038, 3041, 3046, 3049, 3050, 3053, 3056, 3057, 3060, and 3061.
3. *Clifton near Bristol*: Tyndall continued his recovery with a visit to Heinrich Debus in Clifton. See letter 3071.
4. *Christmas Lecture*: Tyndall planned to lecture on Light at this year's RI Christmas lecture (see letter 3055, n. 7).
5. *his book*: *Life of Faraday* (see letter 2953, n. 1).
6. *make use of your letter*: see letters 3043 and 3057.
7. *your father*: Charles-Gaspard de la Rive (see letter 3043, n. 8).
8. *a paper*: likely J. Tyndall, 'On the Action of Rays of High Refrangibility upon Gaseous Matter', *Phil. Trans.*, 160 (1870), pp. 333–65.
9. *Madame DelaRive*: a widower, de la Rive married for a second time in 1855 to Louise-Victoire-Marie de la Rive (née Fatio, 1808–74).
10. *two sons*: de la Rive had six children from his first wife, Jeanne-Mathilde Duppa (1808–50; married in 1826): Eugène de la Rive (b. *c.* 1827), Jeanne Adèle de la Rive (1829–95), William de la Rive (b. *c.* 1831), Lucien Charles de la Rive (1834–1924), Augusta de la Rive (b. *c.* 1836), and Alice de la Rive (b. *c.* 1838).
11. *M. Morren*: Jean François Auguste Morren.

To Miss Moore[1] 5 December 1869 3097

London 5th Dec[ember] 1869.

My Dear Miss Moore

I received your kind note[2] a moment ago and reply to it on the spot. I am right glad to hear that you like the skies of Cannes. I shall never forget the drive with Carlyle over the heights between Nice and Mentone,[3] nor the splendour of the moon reflected as an unbroken pillar of light from the slightly rippled surface of the Mediterranean. It was heavenly. But to ease my repining for not now enjoying such a chill you doubtless put in that remark about the mosquitos at the end. What shocking beasts they are! They set me quite in a rage, and I would prefer smoky London without them to the fairest clime on earth with them.

I was long weak in the legs[4] so ran down to Clifton[5] for 10 days a short time ago and came back immensely improved. I am now preparing for the juveniles and in connexion with them am putting together some pages about Light,[6] with a view of making a real boy's book.[7] The lectures begin on the 28th I also open the Friday evenings, when I propose making the dust of London my principal theme.[8]

Dr Bence Jones has just brought out his Life of Faraday.[9] You have got an image of the man all through his life from his own letters. I have not yet read the book, nor do I know how the public are likely to receive it. It has only just made its appearance.

I did not quite relish those remarks of Murchison[10] though they are sure to increase our funds. I am persuaded that if people really know what is needed we should get in a moment all we need; but the difficulty is to make people know without appearance of solicitation. My notion that a series of advertisements would have been the most impersonal and the least objectionable form of making the thing known. Numbers of people have written to myself not knowing where to send their money.

I have not yet seen Mrs Faraday, but I am going up to her on Tuesday next. I am also going down to Drayton Manor[11] for a day or two. Lady Emily Peel was so exceedingly kind to me in Geneva[12] that *[I]* could not say "no" to her invitation.[13]

The Pollocks are all well. Mrs Pollock lost a young friend of hers whom she loved dearly a few days ago, Miss Ritchie.[14] She lost her in the sense that a husband has snapped her up, and Mrs Pollock is inconsolable for her loss.

Now goodbye. I wish you a prosperous time and give my best regards to your excellent sister.[15]

Ever yours | John Tyndall.

RI MS JT/1/TYP/3/864b–65a
Transcript Only

---

1. *Miss Moore*: probably Harriet Jane Moore (1801–84), a British watercolour artist who painted scenes of Faraday's apartment, study, and laboratory at the RI during the 1850s. She was close to many of the leading figures of the RI, especially Faraday and Tyndall. Moore was the eldest child of surgeon James Carrick-Moore and Harriet Moore (née Henderson). She joined the RI in 1852. Her brother, John Carrick Moore, was a field geologist and was elected member at the RI in 1836 (F. A. J. L. James, 'Harriet Jane Moore, Michael Faraday, and Moore's mid-nineteenth century watercolours of the interior of the Royal Institution', in James Hamilton (ed), *Fields of Influence: Conjunctions of Artists and Scientists, 1815–1860* (Birmingham: University of Birmingham Press, 2001), pp. 111–28).
2. *your kind note*: letter missing.
3. *Nice and Mentone*: in December 1866, Carlyle and Tyndall visited Louisa Baring's house in the south of France at Mentone, some twenty miles east of Nice, on the Côte d'Azur. See letter 2535 (from Tyndall to Ellen Busk), *Tyndall Correspondence*, vol. 10.
4. *weak in the leg*: Tyndall spent the autumn recovering from a leg injury. See letters 3030, 3031, 3032, 3037, 3041, 3048, and 3049.
5. *Clifton*: Tyndall continued his recovery in Clifton, where his friend, Henrich Debus lived. See letters 2996, n. 7 and n. 8; and 3071.
6. *some pages about Light*: 'Light' was the topic for Tyndall's RI Christmas lectures in 1869 (see letter 3055, n. 7).
7. *a real boy's book*: J. Tyndall, *Notes of a Course of Nine Lectures on Light Delivered at the Royal Institution of Great Britain, April 8—June 3, 1869* (London: Longmans, Green, and Co., 1870). According to the preface this volume was published in a cheap form following requests from teachers and students.
8. *open the Friday evenings . . . the dust of London my principal theme*: Tyndall delivered a Friday Evening Discouse at the RI on 'Dust and Disease' on 21 January 1870 ('Dust and Disease', *Roy. Inst. Proc.*, 6 (1870–72), pp. 1–14).
9. *Life of Faraday*: Bence Jones's Life of Faraday consisted of two volumes.
10. *those remarks of Murchison*: Roderick Murchison had made a statement on 22 November 1869 at a meeting of the Royal Geographical Society about the committee for a Michael Faraday memorial having not yet raised sufficient funds ('The Faraday Memorial', *The Lancet*, 27 November 1869, p. 748).
11. *Drayton Manor*: see letter 3093, n. 4.
12. *in Geneva*: Emily Peel attended to Tyndall during his recovery from a serious injury. See letters 3091, n. 4, 3049, and 3050.
13. *her invitation*: see letters 3091 and 3093.
14. *Miss Ritchie*: probably Augusta Ritchie (see letter 3087, n. 4).
15. *your excellent sister*: Julia Moore (1803–1904), the youngest of the three daughters of Harriet and James Carrick-Moore. Their other daughter was Louisa (1802–53).

To Arthur Stanley[1] 6 December 1869 3098

6th Dec[ember] 1869

My dear Dean.

Would you kindly inform me whether you think a suitable place could be found in the Abbey[2] for a statue to Faraday?[3]

Would you also mention what the costs of permitting the statue to be placed in the Abbey will be?

Ever yours sinc[ere]ly | John Tyndall

Library of Congress, Papers of Robert Browning

1. *Arthur Stanley*: Arthur Penrhyn Stanley (1815–81), Dean of Westminster (1864–81) and a supporter of religious tolerance. Having served as a tutor at the University of Oxford, in 1851 he was offered a canonry at Canterbury. In 1856, he was appointed Regius professor of ecclesiastical history at Oxford and installed as a canon of Christ Church in 1858. He wrote many books on church history over the course of his life. Following his marriage to (Lady) Augusta Bruce (1822–76) in 1863 he was a confident of Queen Victoria on matters of church and state (*ODNB*).
2. *in the Abbey*: Westminster Abbey. The Abbey contained tombs and memorials of numerous illustrious individuals, including Isaac Newton and James Watt.
3. *a statue to Faraday*: this particular initiative did not come to fruition, due to disagreement about what would constitute a sufficiently worthy memorial, what it would cost, and how it would be funded (*Ascent of John Tyndall*, p. 209).

From Alfred M. Mayer[1] 7 December 1869 3099

The Lehigh University | South Bethlehem | Pennsylvania | United States of America | 7. December 1869.
To | Professor John Tyndall D.C.L. F.R.S. | London

Respected Sir:

In writing your name I do not feel that this is the first time I have had the honour of addressing you,—so closely have you been brought to us, in this side the Atlantic, by the individuality of your universally known works.

I write to tell you that I have, to-day, taken the liberty of sending to your address, at the Royal Institution, several writings of mine[2] which I will hope to learn have safely reached you—

The report on the Total Solar Eclipse[3] of Aug[ust] last cost me much labour and I can speak quite confidently of its accuracy.

The determination of the time of exposure, and the method of measuring and delineating the details of the photographs by means of the camera-lucida are, as far as I know, the first attempts in these directions.

Last summer I made a series of experiments at the Falls of Niagara[4] and at Trenton Falls,[5] the results of which are embodied in a paper entitled "The Thermo-Dynamics of Waterfalls"[6] read before our American Association for the Advancement of Science.[7] I will send a copy, when published, for the thermometric facts obtained cannot fail to interest you.

In conclusion, I will venture to confess a wish (which I have often heard repeated) that one of these days we may see you in the United States and personally and socially know one who has, intellectually, our fullest sympathy and regard. If you should come you will find a dearly welcome from us all—

I remain, my dear Sir, respectfully | Alfred M. Mayer

P.S. I am so bold as to send you my photograph, may I not expect yours in return?

RI MS JT/1/M/71

---

1. *Alfred M. Mayer*: Alfred Marshall Mayer (1836–97), an American physicist. He became Professor of Physics and Astronomy at the newly established Lehigh University in 1867, where he worked extensively on electro-magnetism and astronomy. His ground-breaking study of the 1869 solar eclipse, funded by the U.S. Nautical Almanac Office, utilized photographs he had taken of the phenomenon (*ANB*; A. G. Mayer and R. S. Woodward, *Biographical Memoir of Alfred Marshall Mayer 1836–1897* (Washington, DC: National Academy of Sciences, 1916).
2. *writings of mine*: although which publications Mayer sent Tyndall are not identified, some likely possibilities are: A. M. Mayer. 'The Solar Eclipse of August 7. 1869', *Journal of the Franklin Institute*, 88 (October 1869), pp. 249–75; and A. M. Mayer, 'An abstract of some of the results of measurements and examinations of the photographs of the total solar eclipse of August 7, 1869', *Amer. Phil. Soc. Proc.*, 2 (1869), pp. 205–9.
3. *Report on the Total Solar Eclipse*: A. M. Mayer, *Reports of observations of the total eclipse of the sun, August 7, 1869, made under the general direction of Professor J. H. C. Coffin, U. S. N., Nautical Almanac Office* (Washington: Government Printing Office, 1869), pp. 129–49.
4. *Falls of Niagara*: Mayer's first observations at Niagara Falls, located on the Niagara River in southern Ontario, Canada, were made on 7 July 1869.
5. *Trenton Falls*: Mayer's observations at Trenton Falls in Trenton (central New York state) were made on 4 July 1869.
6. *the Thermo-Dynamics of Waterfalls*: A. M. Mayer, 'On the thermo-dynamics of waterfalls', *Amer. Assoc. Proc.*, 18 (1869), pp. 64–9.

7. *American Association for the Advancement of Science*: the American Association for the Advancement of Science (AAAS) was founded in 1848 to pursue the advancement of science in the United States, as well as to establish a national scientific community and a national identity as American scientists (K. Benson and J. Maienchein, 'Introduction. AAAS narrative history', in S. Gregory Kohlstedt, M. M. Sokal, and B. V. Lewenstein (eds), *The Establishment of Science in America: 150 Years of the American Association for the Advancement of Science* (New Brunswick, NJ/London: Rutgers University Press, 1999), pp. 1–6, on p. 2).

To Thomas Archer Hirst 8 December 1869 3100

Dear Tom.

In place of your ordinary light reading cast your eye over this.[1]

Yours aff[ectionate]ly | John Tyndall

8th. Dec[ember] 69

I am going out to the country today, but shall be back on Saturday[2]

RI MS JT/1/T/947

1. *this*: not identified.
2. *back on Saturday*: according to Hirst's journal, Tyndall returned on Saturday, 11 December 1869 from a visit to Robert Peel and Lady Wrottesley (1798–1867), widow of John Wrottesley, astronomer and second Baron of Wrottesley (*Hirst's journal*, RI MS JT/2/32d/1856).

From Louisa Baring [9 December 1869][1] 3101

Latimer | Chesham | Thursday night.

Dear Professor Tyndal.

I am to be in town tomorrow—Lady Marian Alford[2] dines with me at 6¼ to go and see Hamlet[3] after—Do come and meet her, and go with us—I leave the next morning Sat[urda]y for Melchet[4]—& I shall be so glad to see you again—don't disappoint me. But come.

Ever most truly Yours | L. Ashburton

Elton Hall, Ashburton Collection
RI MS JT/TYP/1/132

---

1. *[9 December 1869]*: the date of this letter has been determined through Tyndall's travels to visit Lady Emily Peel from 8–10 December (letters 3091 and 3093) and his response to this letter on 13 December 1869 (letter 3103).
2. *Lady Marian Alford*: Marianne Margaret Egerton (née Compton), Viscountess Alford (1817–88), was an art patron and writer on needlework. Born in Naples, she lived with her parents in Rome from 1820 to 1830. She married John Hume Cust (afterwards Egerton), Viscount Alford (1812–51), heir to John Cust, first Earl Brownlow, in 1841. After her husband's death, she lived the life of a wealthy upper-class widow: hosting gatherings, socializing, and wintering in Rome. She was instrumental in the formation of the Royal School of Needlework, where she served as vice-president (*ODNB*).
3. *Hamlet*: play by William Shakespeare. Several versions of the play ran in London in 1869, including an opera version from the French composer Ambroise Thomas, which opened in London at the Royal Italian Opera, Covent Garden on 19 June and was performed on at least one occasion in December 1869 (on the 7th). However Baring and Alford would more likely have attended a performance at the Princess's Theatre, where it was played on several dates in December, including the 10th. The Princess's Theatre was a popular venue on Oxford Street, known for the 'scenic excesses' and 'lavish productions' of its Shakespeare performances (G. Marshall (ed), *Shakespeare in the Nineteenth Century* (Cambridge: Cambridge University Press, 2012), p. 180).
4. *Melchet*: see letter 2971, n. 5.

To Thomas Archer Hirst [10 December 1869][1] 3102

Drayton Manor Friday

My dear Tom.

There is a pheasant at Albermarle Street[2]—I wish you would take it and get it nicely roasted for the Lily[3] and yourself tomorrow or Sunday.

Yours af[fectionate]ly | John

I sweep round by Wrottesley[4] today & return tomorrow.

RI MS JT/1/T/976

---

1. *[10 December 1869]*: the date was written in a second hand but has been confirmed: Tyndall was at Drayton Manor visiting Emily Peel on Friday, 10 December 1869. See letters 3091, 3093, and 3097.
2. *Albermarle Street*: address of the RI.
3. *Lily*: Lilly Hirst.
4. *Wrottesley*: Wrottesley Hall is the family home of the Wrottesley family. Tyndall visited Lady Wrottesley there on Saturday, 11 December 1869. See letter 3100.

To Louisa Baring 13 December 1869 3103

13th. Dec[ember]. 1869

My dear Lady Ashburton.

Nothing could have given me greater pleasure than to meet you & Lady Marian Alford[1] in the terms of your most kind invitation:[2] but when you wrote to me I was in Staffordshire,[3] and I did not get your note until 8 oC[lock]. on Saturday evening.

It give[s] me great pleasure to see once more your well-known hand.

Yours always | John Tyndall

A very merry Xmas & a happy new year to you and Miss Mazy[4]—I am overwhelmed with the juveniles[5]

National Library of Scotland Ashburton Papers Acc.11388 N0.101

---

1. *Lady Marian Alford*: see letter 3101, n. 2.
2. *your most kind invitation*: in letter 3101 Baring (Lady Ashburton) invited Tyndall to join her and Lady Alford for dinner and then to see a theatrical production of *Hamlet*.
3. *in Staffordshire*: Tyndall was visiting the Peels at Drayton Manor in Staffordshire, England. See letters 3091 and 3093.
4. *Miss Mazy*: Mary Florence Baring.
5. *the juveniles*: Tyndall delivered the annual Christmas lecture at the RI in 1869 (see letter 3055, n. 7).

From Emil du Bois-Reymond 13 December 1869 3104

Berlin, 17 Victoria Str[asse]. | December 13th, '69.

My dear Tyndall,

Your apparatus are steadily approaching completion at Sauerwald's hands.[1] The telescope is from Steinheil[2] at Munich. I have contrived the stand which supports the division, the lamps and the telescope so that all these things may be cleared away and the stand be used for any other purpose. In the course of January Sauerwald will be able to send you the whole and he should like to know whether you have some particular directions to give him relating to the mode of transport etc.

I have been extremely sorry to learn that and how your holiday trip was spoiled by your accident.[3] I also had a narrow escape this time, and I am still suffering of the evil consequences. I was descending a steep gully or couloir down from Mont Perse[4] to the Diavolezza Lake[5] near Pontresina.[6] The couloir was filled with frozen snow in the very slippery surface of which my guide was cutting steps. While at work he did not notice the blunder I made by following him a great deal too closely. I was standing quietly, and as I thought very securely supported upon my Alpenstock,[7] when suddenly I found myself rolling down the snow together with my guide. I experienced the full sensation of something terrible going to happen, when we safely landed together on a stony slope, some fifty feet beneath the place where either I slipped or the snow gave way under my weight. Had I been at some height above the man, he might have steadied himself on seeing me fall, and have stopped me easily; but so I went with my legs between his, as he told me afterwards, and threw him off his legs irretrievably. We were very much bruised both of us; at the time however, I did not care about it. But the same night my right ankle began to hurt me and to swell, and also my left shoulder; and the next day I was fairly laid up, and climbing for this season had come to an end. By a singularly lucky chance, my wife going over to Enderlin[8] to pay a visit, met my friend Bardeleben,[9] one of our best surgeons, who had just arrived, and he cured my sprained ankle most skilfully, without leeches and any other messes, simply by firmly bandaging the joint and making cold cataplasms. In this state I went over the Bernina-Pass[10] to le Prese,[11] the lovely little watering-place on the lake of Poschiavo,[12] where I stayed till I could again limp about. The sprain of my ankle is all but forgotten now, but my shoulder is still stiff, and will remain so, I am afraid, for a long time. This will not prevent me from trying the Tödi[13] the next time.

You will shortly get my paper[14] on the aperiodic motion of the magnetic needle, and I should be greatly obliged to you if you would have it translated in the Phil. Mag.[15]

Believe me, my dear Tyndall, | Yours faithfully | E du Bois-Reymond

I forgot wishing you the compliments of the season, and asking you to remember me to Dr. Hirst.[16]

RI MS JT/1/D/148
RI MS JT/1/TYP/7/2436–7

---

1. *Sauerwald's hands*: Tyndall had commissioned Ferdinand Sauerwald to construct some instruments for him. See letters 3028, 3061, 3140, and 3148.
2. *Steinheil*: Carl August Steinheil (1801–70), a German physicist and astronomer who was professor of mathematics and physics at the University of Munich from 1832 to 1870. In 1854 he opened an optical workshop in Munich that became renowned for its instruments and telescopes for astronomy and telegraphy. In 1865, he left the company C. A. Steinheil

and Sons in the hands of his sons, Adolph (1832–93) and Eduard (1830–78) (*CDSB*; *NDB*).

3. *your accident*: Tyndall spent the autumn of 1869 recovering from a mountaineering injury. See letters 3030, 3031, 3032, 3037, 3041, 3048, and 3049.
4. *Mont Perse*: Mont Perse (also called Munt Pers) is a mountain in the Bernina Range of the Alps in the Swiss canton of Graubünden.
5. *Diabolezza Lake*: the Diabolezza Lake is located in Graubünden, in the south east of Switzerland.
6. *Pontresina*: a village in the Swiss canton of Graubünden.
7. *Alpenstock*: long, metal-tipped wooden pole used in mountain climbing (*OED*).
8. *Enderlin*: Leonard Enderlin was proprietor of the Weisses Kreuz Hotel in Pontresina and a well-known mountain guide (G. Rosenbusch and A. de Knecht-van Eekelen, *Wilhelm Conrad Röntgen: The Birth of Radiology* (Switzerland: Springer Nature Switzerland, 2019), p. 66).
9. *Bardeleben*: Heinrich Adolf von Bardeleben (1819–95), a German surgeon who worked at the Charité-Universitätsmedizin, an important university health center (*NDB*).
10. *Bernina-Pass*: the Bernina Pass is a mountain pass connecting St. Moritz in the Swiss canton of Graubünden to the northern Italian town of Tirano.
11. *Le Prese*: a town in Val Poschiavo, Switzerland, close to the terminus of the Bernina Pass.
12. *Lake of Poschiavo*: a lake in the Swiss canton of Grisons which borders the southern limit of Le Prese.
13. *Tödi*: the highest mountain in the Glarus Alps of Switzerland, on the border to Graubünden.
14. *my paper*: E. du Bois-Reymond, 'Ueber die aperiodische Bewegung gedämpfter Magnete', *Berlin, Monatsber. Akad.*, (1869), pp. 807–52.
15. *translated in the Phil. Mag.*: this article was not translated and published in the *Phil. Mag.*
16. *Dr. Hirst*: Thomas Hirst.

From Frederick Guthrie[1] 17 December 1869 3105

ROYAL SCHOOL OF MINES | JERMYN STREET[2] |
17th Dec[embe]r 1869

My dear Sir,

I cannot deny myself the pleasure of thanking you for the very great kindness with which you spoke of the account of my experiments at the Royal Society[3] yesterday.

You judged rightly in thinking that a word of approbation from you would encourage me greatly.

If science should ever be considerably enriched through my instrumentality, it will in no small degree be due to you who have on this and on former occasions cheered me with your approval of my attempts.

Faithfully yours | Fred[eric][k] Guthrie

Prof[fesso]r Tyndall.

RI MS JT/2/10/454–5

---

1. *Frederick Guthrie*: Frederick Guthrie (1833–86), a British chemist and physicist. He worked with Tyndall on the thermal resistance of liquids and on 16 October 1868 Tyndall communicated Guthrie's paper 'On the thermal resistance of liquids' to the RS (F. Guthrie, 'On the thermal resistance of liquids (abstract)', *Roy Soc. Proc.*, 17 (1868–69), pp. 234–6). Guthrie was Tyndall's successor as lecturer in physics at the Department of Science and Art's School of Mines, where Huxley was also a lecturer (*ODNB*). His son Frederick Bickell Guthrie (1861–1927) was an agricultural chemist (*ODNB*).
2. *ROYAL SCHOOL OF MiINES | JERMYN STREET*: on the Royal School of Mines, see letter 3044, n. 7.
3. *account of my experiments at the Royal Society*: Guthrie's experiments were communicated to the RS by George Gabriel Stokes on 18 November 1869. F. Guthrie, 'On Approach by Vibration (abstract)', *Roy. Soc. Proc.*, 19 (1869–70), pp. 93–4.

From George Wheatley Bennett[1] — 18 December 1869 — 3106

London Dialectical Society | 1 Adam Street |
Adelphi N.C. | 18th Dec[ember] 1869

Sir,

The Committee appointed by the Council of the above Society[2] to 'investigate the phenomena alleged to be spiritual manifestations' have instructed me to bring the following facts under your notice;—

Since the commencement of their enquiry (in January last) they have carefully avoided all discussion as to causes, all search for theories, and have confined their attention mainly to the collection of facts. Subcommittees, consisting of such men as Mr Sergeant Cox,[3] Mr A. Russell Wallace, Mr W[illia]m Crookes (Editor "Chemical News")[4] have sat at regular intervals, and have succeeded in recording a number of phenomena, which, while they do not, in the opinion of the majority of our members, point to any spiritual origin, send at least to demonstrate the existence of some power (magnetic or otherwise) which has not yet been generally recognised by men of science. The members have sought, by every means that could be suggested, to guard themselves against all possibility of deception or trickery. None but members of the Committee or gentlemen of undoubted honour and respectability have been present; the full blaze of gas-light has been rigidly insisted upon, and the séances have always been held at the residences of members of the Committee who were not themselves Spiritualists. Yet even with these conditions phenomena have occurred of so surprising & unmistakable a character as to have carried the conviction of their reality to all present.

The Committee have therefore instructed me to request either that you will attend two or three of these sittings, or that you will at least assist them in suggesting some further tests by which the nature & character of these phenomena may be satisfactorily & conclusively shown.

I have the honour to be, | Sir, Your obed[ien]t Servant | G Wheatley Bennett | Hon[ourary] Sec[retary]

RI MS JT/2/10/458–9/b
RI MS JT/2/13c/1362–3
Transcript Only

---

1. *George Wheatley Bennett*: George Wheatley Bennett (dates unknown), was honorary secretary for the committee, appointed by the London Dialectical Society, to investigate the 'Phenomena alleged to be Spiritual Manifestations' (*Report on Spiritualism, of The Committee of the London Dialectical Society* (London: Longmans, Green, Reader and Dyer, 1871)).
2. *the above Society*: founded in 1867, the London Dialectical Society investigated alleged manifestations of Spiritual phenomena. It is considered to be the predecessor to the better-known Society for Psychical Research. While these groups were dedicated to revealing the tricks of charlatans, fake mediums and cons, members often believed strongly in the existence of psychical phenomena and sought to separate false claims from real spiritual phenomena.
3. *Mr Sergeant Cox*: Edward William Cox (known as Serjeant Cox) (1809–79), an English lawyer and publisher. In 1868 he was named Serjeant-at-Law. He was best known for his publishing expertise. In addition to publishing the popular weekly periodical *Exchange and Mart* (1868), Cox was also the proprietor of *The Country: a Journal of Rural Pursuits* (1873), *The Critic* (1843–63), a prolific publisher of legal texts, and the founder and editor of *Law Times* (1843) (*ODNB*).
4. *"Chemical News"*: originally the *Chemical Gazette*, published by William Francis, it was purchased by William Crookes in 1859 and renamed the *Chemical News.* Crookes turned it into a weekly research and trade journal with original papers, reports of scientific meetings, correspondence, information on new books, patents and other information relevant to the chemical community. It was tremendously successful (H. Cooke, 'A Historical Review of the Chemistry Periodical Literature until 1950', *Learned Publishing*, 17 (2004), pp. 125–34, on p. 127).

From Mary E. Lyell[1] 20 December [1869][2] 3107

73, Harley St | London. W. | 20 Dec[embe]r

Dear Dr Tyndall—

Rosamond[3] is looking forward to the Christmas Lectures,[4] but will I am sure be proud to receive a ticket from you.

Ever dear Dr Tyndall | Sincerely y[our]s | Mary E. Lyell

RI MS JT/1/L/63
RI MS JT/1/TYP/3/854

---

1. *Mary E. Lyell*: Mary Elizabeth Lyell (née Horner, 1808–73), daughter of Whig reformer and geologist Leonard Horner (1785–64). She married geologist Charles Lyell in 1832. Fluent in French and German, she helped her husband with translation, but she was a geologist and conchologist in her own right. There were no children from their marriage (M. Ogilvie and J. Harvey (eds), *Biographical Dictionary of Women in Science*, 2 vols (London: Routledge, 2000), vol. 2, pp. 813–14)).
2. *[1869]*: the year of this letter has been added to the typescript by hand, and has been confirmed in reference to letter 3108.
3. *Rosamond*: Rosamond Lyell.
4. *Christmas Lectures*: Tyndall delivered the annual Christmas lectures at the RI in 1869 (see letter 3055, n. 7)

From Rosamond Lyell 21 December 1869 3108

21 Dec[ember] 1869. | 42. Regents Park Road. N.W.

My dear Dr. Tyndall:

I thank you very much for the ticket[1] which I have just received from my aunt.[2]

Mamma[3] has been for long intending to take Arthur[4] and myself to your Christmas course,[5] and he will have much pleasure in going also.

It was most kind of you to think of me.

I remain dear Dr. Tyndall | Your young friend | Rosamond.

RI MS JT/1/L/64
RI MS JT/1/TYP/3/854

---

1. *the ticket*: see letter 3107.
2. *my aunt*: Mary Lyell (see letter 3107, n. 1).
3. *Mamma*: Katherine (or Katharine) Murray Lyell (1817–1915), an English botanist and an

editor of important volumes on the correspondence of several notable scientists, including Charles Lyell, Leonard Horner, and Charles Bunbury (R. Desmond, *Dictionary of British and Irish Botanists and Horticulturalists* (London: BMNH, 1994); M. Ogilvie and J. Harvey (eds), *Biographical Dictionary of Women in Science*, 2 vols (London: Routledge, 2000), vol. 2, p. 813).

4. *Arthur*: Arthur Henry Lyell (1853–1925), Rosamond's brother. At the time of this letter, Rosamond was thirteen.
5. *your Christmas course*: Tyndall delivered the annual Christmas lectures at the RI in 1869 (see letter 3055, n. 7).

From Ellen Busk[1] 22 December [1869] 3109

My dear M^r^. Tyndall

Pray accept our best thanks for the Tickets[2] it is very pleasant to be so kindly remembered & I need not tell you with what delight as well as instruction I know we shall listen to your "thrice told tale"—I am very glad to hear from M^r^. Hirst how completely you have recovered from the effects of y[ou]r accident[3] which made your friends so anxious—& that you are remarkably well, but as you do not dine out & never come to see us in any other way I fear we are not likely to meet. With our united kind regards—wishing you the best wishes of the season.

Believe me | Yours v[er]y truly | Ellen Busk
32 Harley St. | Dec[em]b[e]^r^ 22^d^

RI MS JT/1/B/157
RI MS JT/1/TYP/1/173

---

1. *Ellen Busk*: Ellen Busk (1816–90), the daughter of Jacob Hans Busk. In 1843 she married her cousin George Busk with whom she had four children.
2. *the Tickets*: tickets for Tyndall's RI Christmas lectures on 'Light' (see letter3055, n. 7).
3. *y[ou]r accident*: Tyndall spent September and October 1869 recovering from a mountaineering injury. See letters 3030, 3031, 3032, 3037, 3041, 3048, and 3049.

To George Wheatley Bennett[1] 22 December 1869 3110

22nd Dec[ember]. 1869

Sir,—You mention in your note[2] to me three gentlemen, two of whom are personally known to me, and for both of whom I entertain a sincere esteem.

The house of one of these, namely Mr. Wallace, I have already visited, and made there the acquaintance of the lady who was the reputed medium[3] between Mr. Wallace and the supernatural.

And if earnestly invited by Mr. Crookes (the editor of the 'Chemical News')[4] to witness phenomena which in his opinion 'tend to demonstrate the existence of some power (magnetic or otherwise) which has not yet been recognized by men of science,' I should pay due respect to his invitation.

But understand my position: more than a year ago Mr. Cromwell Varley,[5] who is, I believe, one of the greatest modern spiritualists, did me the favour to pay me a visit, and he then employed a comparison which, though flattering to my spiritual strength, seems to mark me out as unfit for spiritual investigation. He said that my presence at a *séance* resembled that of a great magnet[6] among a number of small ones. I throw all into confusion. Still he expressed a hope that arrangements might be made to show me the phenomena, and I expressed my willingness to witness such things as Mr. Varley might think worth showing to me. I have not since been favoured by a visit from Mr. Varley.

I am now perfectly willing to accept the personal invitation of Mr. Crookes, should he consider that he can show me phenomena of the character you describe.

I am sir, your obedient servant, | John Tyndall

G. W. Bennett, Esq.

Report on Spiritualism of the Committee of the London Dialectical Society (London: Longmans, Green, Reader and Dyer, 1871), p. 265
RI MS JT/2/10/458–9
RI MS JT/2/13c/1363

---

1. *George Wheatley Bennett*: see letter 3106, n. 1.
2. *your note*: letter 3106.
3. *the lady who was the reputed medium*: probably Agnes Elisabeth Nicholl [Guppy; Guppy-Volckman] (née White, 1838–1917), an English medium. After having visions as a child Nicholl was convinced of her powers by Fanny Sims, sister of Alfred Russel Wallace. Nicholl began holding séances in 1866 at first in the home of Wallace (*ODNB*). See letter 2817, *Tyndall Correspondence*, vol. 10. Tyndall had some interaction with spiritualistically-inclined scientists in 1867 and 1868. (See also letters 2558, 2560, 2819, and 2822, *Tyndall Correspondence*, vol. 10.)
4. *'Chemical News'*: see letter 3106, n. 4.
5. *Mr. Cromwell Varley*: Cromwell Fleetwood Varley (1828–83), a telegraph engineer who worked for the Electric Telegraph Company and the Atlantic Telegraph Company. He was responsible for the development of the 'Varley loop test', 'Varley's multiplier', and a type of harmonic telegraph called the 'cymaphen' (1870). A founding member of the Society

of Telegraph Engineers in 1871, Varley was made FRS that same year. Varley was a believer in spiritualism and conducted investigations of his own; he introduced chemist William Crookes to the subject (*ODNB*).

6. *a great magnet*: Crookes teased Tyndall about this in letter 3111. See also *Ascent of John Tyndall*, p. 217.

From William Crookes 22 December 1869 3111

Dec[em]b[e]r. 22nd 1869. | 20, Mornington Road, | London, N.W.

Dear Dr Tyndall,

I feel thankful for the opportunity you have afforded me of reading your letter,[1] of this date, to the Secretary of the Dialectical Society,[2] and especially for the flattering words in which you speak of me. So highly do I value your good opinion that I would rather not at present join in the invitation which it appears you have received.

This is how I stand with regard to Spiritualism. Some six months ago, having no knowledge of the subject beyond what I had derived from sundry published letters and replies, a friend told me of some very extraordinary occurrences which had happened within his own experience. I do not feel at liberty to mention my friend's name, but he is an F.R.S., and deservedly stands in the foremost rank of experimental philosophers. He assured me that he had witnessed phenomena, alleged to be spiritual, which he was unable to explain by any known physical force, and advised me to take the first opportunity of witnessing such things for myself and forming my own judgment upon them.

A short time afterwards I happened to make the acquaintance of some leading members of the Dialectical Society, and was invited by them to join their investigation committee. This I declined. I occasionally attend as a visitor, but only on conditions that I should not be asked to give any opinion, and that my name should not be mentioned in connection with the subject. Mr Bennett[3] forgot this when he wrote to you.[4]

At some of these meetings nothing took place which could not easily be accounted for; at others I have seen phenomena which could only be explained on the almost impossible supposition of gross fraud and collusion on the part of many ladies & gentlemen present; but at one or two I have certainly witnessed occurrences which appear beyond the domain of any known physical force.

Something new and worthy of the notice of the man of science I am tolerably certain we are getting glimpses of, and I fancy my thoughts are shaping themselves in the direction of a power in some way connected with gravitation. I have seen just enough to raise my curiosity and urge me to investigate; but I must do so in my own laboratory and at my own time, and test everything more calmly than is possible when I am in a party of enthusiasts.

Most of the observed facts are so encrusted over with superstition that it is hard to separate one from the other, and certainly the bulk of spiritualists are amongst the most credulous of mortals, although there are some honorable exceptions, Mr Varley[5] & Mr Wallace for instance. I see nothing unphilosophical in the idea that there exist in nature unseen entities of a higher order, or possessing different powers, than the human race, but as yet I have witnessed nothing which satisfies me that such intelligences are instrumental in producing these phenomena.

I have lately devised some test experiments, which, whether they succeed or not, can hardly fail to give me some information. I wish however to keep quiet in the matter, and not to have my name mixed up with spiritualism until I feel justified in bringing the results of my investigation before the Royal Society, or in showing them experimentally at the Royal Institution table.

If in the mean time I should meet with anything in my laboratory which I think worthy of your notice, would you mind the trouble of coming here, as a friend, and in your private capacity only, and seeing what I can show?—taking your chance of the "small magnets" being thrown out of gear[6] by your presence?.

Believe me | very sincerely yours | William Crookes.

RI MS JT/2/10/458–9
RI MS JT/5/12/31

1. *your letter*: letter 3110.
2. *Dialectical Society*: see letter 3106, n. 2.
3. *Mr Bennett*: see letter 3106, n. 1.
4. *he wrote to you*: letter 3106.
5. *Mr Varley*: see letter 3110, n. 5.
6. *the "small magnets" being thrown out of gear*: see letter 3110.

From Jane Barnard 23 December 1869 3112

320 Liverpool R[oa][d] | Barnsbury Villas, | Islington. | 23 Dec[ember] 1869

My dear Dr Tyndall

You are very kind in thinking of us, in sending the tickets[1] & my Aunt[2] & I feel much gratified. I hope to have the pleasure of hearing you & of taking my little nephews & nieces. I am afraid they are rather young but I do not like them to lose this opportunity which they may remember with great interest in after life. I am yours

very sincerely | Jane Barnard

RI MS JT/1/B/44

1. *the tickets*: tickets to Tyndall's RI Christmas lectures on 'Light' (see letter 3055, n. 7).
2. *my aunt*: Sarah Faraday.

To Margaret Ginty 24 December 1869 3113

Royal Institution | 24th. Dec[ember] 1869.

My dear Mrs Ginty,

It is very natural that you should cling to a place connected with so many loving memories;[1] and depend upon it it is better to have such memories though sadness is diffused through them, than none at all. I am glad you have listened to the reasoning of Mr James,[2] because whatever you may think, there is a future for yourself and your children provision for which must not be neglected. I will not tease you further—I am pleased with your resolution, for you know in general I give you very little credit for resolution. I wish you a happy quiet Christmas and a great many of them to come.

Yours very sincerely | John Tyndall.

RI MS JT/1/TYP/11/3660
LT Typescript Only

1. *so many loving memories*: Tyndall assisted Ginty in legal and financial matters after the death of her husband William Ginty in 1866 (see letter 3062). He is perhaps referring to considerations about selling the family home or property in Brazil, where Ginty had lived prior to the death of her husband.
2. *Mr. James*: Bartlett James (dates unknown), was manager, or 'Gerente', of the Imperial Gas Works Company in Rio de Janiero in Brazil, established by Baron de Maua. He worked there with Tyndall's friend William Ginty, who was the engineer, or 'Engenheiro' (*Almanak Administrativo, Mercantil e Industrial da Corte e Provincia do Rio de Janeiro para o anno de 1865* (Rio de Janeiro: Eduardo & Henrique Laemmert, 1865), pp. 414–5). According to a 5 December 1882 letter from Ginty's daughter Fanny Unwin (née Ginty) to Tyndall, James had taken on the role of executor to her late father's will (RI MS JT/1/TYP/11/3682).

To Edward Livingston Youmans 24 December 1869 3114

MY DEAR YOUMANS: It is a long time since I have heard from you, and the reason, no doubt, is that you wrote to me last.[1] Well, I must not allow you to fall utterly away from me, so by this day's post I send you the copy of an article[2] which is to appear in the next number of the Fortnightly Review.[3]

Your last letter[4] made me smile. I know you imagine me to be a screw in money matters, and therefore you thought it would please me to know that I should be well paid for that short scrap[5] from Macmillan.[6] Well, if you feel an interest in the matter you may ask my friends whether I am a screw or not. Sometimes I certainly wish to put the screw upon publishers; for they sometimes need it much. Let me say now that you may do just what you please with any article of mine, and feel not a thought on the money side of it, as far as I am concerned.

I am trying very hard on a boy's book on optics.[7] Ostensibly for boys, but equally for teachers; for boys thus far do not know how to learn and teachers do not know how to teach. I am so treating the subject that boys and teachers may make the experiments for themselves. My aim is to teach them both to experiment and to reason upon experiment. I suppose a boy to be alongside, and that we are working together. I try to overcome the apathy and the repugnance arising from awkwardness in the first stages of experiment. I speak, therefore, not only to the boy's *brain,* but to his *blood*—stirring him to action.

I had a fall with ugly consequences[8] in the Alps this year. One morning, after allowing a mountain cascade to tumble over me, I was returning across some blocks of granite naked to my clothes when I staggered and fell all my weight against the sharp crystals. Three of them stamped themselves into the fiber of my shin, and the shin was generally much bruised. But four days of perfect quiet destroyed all pain, and there was no inflammation. So I came down stairs, moved about, excited inflammation, had erysipelas[9] twice over, and was six weeks in bed. It required three months to set me right. I am now well, and just on the point of beginning the Christmas lectures.[10]

I wish much you would tell me what kind of lectures (scientific) you are accustomed to in New York.

Yours ever, | JOHN TYNDALL.

E. A. Youmans, 'Tyndall's American Visit', *Popular Science Monthly*, 44 (February 1894), pp. 502–13, on pp. 506–7

---

1. *you wrote to me last*: letter missing.
2. *an article*: 'Climbing in Search of the sky'. See letter 3063, n. 2.

3. *Fortnightly Review*: a Victorian periodical founded in 1865 by Anthony Trollope and Frederick Chapman, the *Fortnightly Review* was diffusely liberal, but open to a variety of positions. It published essays on topics as wide-ranging as philosophy, politics, religion, literature, and science (*Wellesley's Index*). It was edited by John Morley from 1867 to 1882. See letter 3062, n. 1.
4. *Your last letter*: letter missing.
5. *that short scrap*: see letter 2943.
6. *Macmillan*: see letter 2943, n. 1.
7. *a boy's book on optics*: see letter 3097, n. 7.
8. *a fall with ugly consequences*: Tyndall injured himself while mountaineering in September 1869. See letters 3030, 3031, 3032, 3037, 3041, 3048, 3049, and 3060.
9. *erysipelas*: see letter 3038, n. 3.
10. *Christmas lectures*: Tyndall's RI Christmas lectures on 'Light' (see letter 3055, n. 7).

1870

From Antonin Bergeret[1] 11 January 1870 3115

Monsieur Tyndall

C'est un de vos lecteurs assidus, un de vos nombreux admirateurs qui vous adresse un livre sur la biologie médicale positive. C'est un ouvrage de vulgarisation scientifique; puisse-t-il avoir un petit succès! La page 301 est un *[*emprunt*]* textuel à votre livre sur les Montagnes.

L'idée de la nouvelle publication—La Nature—ne peut pas moins faire que de réussir; elle fera honneur aux courageux pionniers de l'ère scientifique qui doit infailliblement succéder à l'ère métaphysique. Elle aura le même succès que la Revue des cours scientifiques—*[*La*] [1 word illeg] [*aimé*]*. Si je connaissais la langue anglaise, je m'y abonnerais, mais je dois me *[*contenter*]* des comptes-rendus des journaux Français.

Veuillez agréer, Monsieur Tyndall, mes meilleurs compliments.

Dr Bergeret | 11 janvier 1870.

Dr Bergeret, de Saint Léger Chalon S/S. Saône et Loire.

Mr Tyndall,

This is one of your regular readers, one of your numerous admirers, who is sending you a book[2] on positive medical biology. It is a book of popular science; could it have a bit of success! Page 301 is a textual *[*borrowing*]*[3] from your book on Mountains.[4]

The idea behind the new publication—Nature[5]—can be nothing less than successful; it will honour the courageous pioneers of the scientific era, which inevitably must follow the metaphysical era. It will have the same success as the la Revue des cours scientifiques[6]—*[*The*] [1 word illeg] [*liked*]*. If I understood English, I would have subscribed to it, but I must content myself with the reports in French newspapers.

Please accept, Mr Tyndall, my best regards.

Dr Bergeret | 11 January 1870.

Dr Bergeret, from de Saint Léger Chalon S/S. Saône et Loire.

RI MS JT/1/B/90

---

1. *Antonin Bergeret*: Antonin Bergeret (1829–?), a French physician and member of the Royal Medical Society of Lyon. Bergeret's work focused on metabolism, microscopy, and public health, but he also wrote on cosmology and the vine fungus *Oidium tuckeri.*
2. *a book*: A. Bergeret, *Petit manuel pratique de la santé; nutrition, alimentation, hygiène* (Paris: Germer-Baillière, 1870).
3. *a textual [borrowing]*: 'Le corps entier, quoique plus lentement que le sang, se consume peu à peu. Comment le sentiment de l'identité personnelle peut-il se conserver malgré cette fuite de molécules? La matière est nécessaire à l'homme tel que nous le connaissons, pour qu'il ait conscience de lui-même; mais la matière peut être entièrement renouvelée dans un temps donné, sans que la conscience de soi-même éprouve aucune altération. L'oxygène qui le remplace, de telle sorte que le *non moi* s'altère et se renouvelle, pendant que le *moi* demeure intact. C'est la persistance de la *forme* dans le groupement des molécules, et non la persistance de ces molécules elles-mêmes qui est la cause de la croyance à l'identité. La vie est semblable à une vague qui, à deux instants successifs, n'est jamais composée des mêmes particules' (Bergeret, *Petit manuel,* p. 301).
4. *your book on mountains*: J. Tyndall, *Dans les montagnes,* 2nd edn, trans. Louis Lortet (Paris: J. Hetzel, 1869).
5. *Nature*: the popular science journal *Nature* was founded in 1869 by Norman Lockyer (see letter 3026, n. 7) with the aim of bringing scientific knowledge to a general public. Many Victorian scientists published articles in *Nature.*
6. *la Revue des cours scientifiques*: *Review of Scientific Progress*, a French science journal edited by Odysse Barot (1830–1907). It mainly presented the work of French scientists, physicians, institutions, and societies, but also included summaries of the work of foreign scientists.

To Henry Morton[1] 13 January 1870 3116

Royal Institution of Great Britain, 13th Jan[uary], 1870.

My Dear Sir: I am most heartily obliged both to Mr. Dickson[2] and yourself for the exceedingly interesting series of photographs[3] which you have been good enough to send me.[4]

The marks of kindly feeling which I have received, as exhibited by invitations from the United States, have been particularly agreeable to me, and I propose some day, and that not a very distant one, paying your country a visit. Whether I shall lecture or not must for the present remain an open question. In connection with lecturing I fear two things; one of these, strange to say, is your proverbial American hospitality; the other is that, however lucky I may be in striking, when well and strong, the keynote which suits the people here, I might not be equally successful among your people. They are accustomed to things on a very large scale, and I, if I dealt in experiments, might not be able to come up to their desires in this respect. I have sometimes thought of preparing a few lectures that I might carry in my pocket, and thus quit experimenting altogether.

I see that as regards light you have already cut the ground away from me.

Whatever the decision may be, I am truly grateful to you for your offer of co-operation.

Yours faithfully | John Tyndall.

C. Sellers, *Biographical Notice of Pres't Henry Morton* (New York: Engineering Press, 1892), pp. 57–8

1. *Henry Morton*: Henry Morton (1836–1902), a college administrator and popular science instructor. He was known for the use of magic lanterns in his theatrically performed lectures. This led to more than sixteen publications on the use of magic lanterns in teaching. In 1870 he was appointed the first president of Stevens Institute of Technology in Hoboken, New Jersey (*ANB*).
2. *Mr. Dickson*: not identified.
3. *interesting series of photographs*: possibly of a solar eclipse. Morton participated in an eclipse viewing in 1869 and made striking photographs (H. Morton, 'Solar Eclipse—August 7, 1869. The Photographic Expedition', *Journal of the Franklin Institute*, 88, pp. 200–16).
4. *to send me*: letter missing.

From Sarah Faraday 15 January 1870 3117

Barnsbury Villa, | 320, Liverpool Road, N. | 15th. Jan[uary] 1870.

You are very kind to think of me dear Dr. Tyndall, and I must say I have thought much of you, since you were here, for I saw that you were so unfit and disinclined for the juvenile lectures,[1] I wish I could see you and be of any use or pleasure to you, but what is the use of saying so our paths lie in different directions and I can only give you my good wishes and thanks, as to the books,[2] if you really have one or two to spare I shall be glad to have them, for we are surrounded by a rising generation to whom the name of Faraday is of great interest.

I trust you will get a little relaxation before you enter upon your spring duties.

I am well and believe me | Most sincerely yours | S[arah] Faraday

Jane[3] begs me to thank you for the tickets.[4]

RI MS JT/1/TYP/12/4185
LT Typescript Only

1. *the juvenile lectures*: Tyndall delivered the annual Christmas lectures at the RI in 1869 (see letter 3055, n. 7).
2. *the books*: probably Tyndall's *Faraday as a Discoverer* and possibly also Henry Bence Jones's *Life of Faraday* (see letters 2951, n. 3, and 2953, n. 1).
3. *Jane*: Jane Barnard.
4. *the tickets*: see letter 3112.

To Frederick Arthur Paget[1] 23 January 1870 3118

Royal Institution of Great Britain | 23rd. Jan[uary]. 1870

My Dear Sir,

If I could in any degree forward you I would willingly do so, but I have entirely discharged the subject of Compasses from my mind, and filled it for the present with something else. Indeed the practical construction of compasses has never engaged my attention. It would therefore be a useless expenditure of time to bring you here to show me a compass. My opinion about it without investigation would be of no value, and an investigation out of the question at present.

I am yours faithfully | John Tyndall

F. A. Paget Esq[ui]r[e]

The Waller Manuscript Collection, Uppsala University, alb–40:187

1. *Frederick Arthur Paget*: Frederick Arthur Paget (d. 1903), a British inventor and civil and mechanical engineer. He worked, among other things, on magnets and compasses in iron ships (F. A. Paget, 'On a New Form of Permanent Magnet', *English Mechanic and the Mirror of Science*, 9 (12 February 1869), p. 458). He received a patent for a magnetic compass card in 1873 (1 Jan. 1873, No. 5, English Patents of Inventions, Specifications: 1873, 1–72).

From Jacques-Louis Soret 24 January 1870 3119

*On Illumination and Polarization in Transparent Sub*stances. By L. SORET.[1]

Geneva, Jan[uary] 24, 1870.

MY DEAR TYNDALL,

I HAVE been somewhat long in replying to your kind letter of the 5th of December;[2] numerous occupations have prevented me from communicating to you my idea of how, by the presence of particles diffused in a transparent medium, we may theoretically explain the various phenomena of illumination and of polarization which are observed when a pencil of light passes through a transparent substance. I am far from believing that my theory is not open to objection, and I should be at no loss to criticise it myself and to raise doubts in respect of it. You know, moreover, how many delicate points are involved in all that concerns the theory of reflection. I think, nevertheless, that my views on this subject will not be uninteresting to you.

Let us see, then, how we can account for the action exerted by these particles.

I. The first hypothesis would be to assume that the vibratory motion of aether is communicated to those minute particles which occur in the path of the pencil of light pretty much as sound-vibrations in air are communicated to solids, such as strings or membranes. The corpuscles would thus become centres of disturbance, sending in all directions vibrations of the same kind as those which constitute the pencil of light; if, more especially, this pencil were polarized, the waves emitted by the particles would consist of rectilinear vibrations parallel to the incident vibrations. The explanation of the phenomenon would be the same as that given by M. Lallemand,[3] except in what concerns the origin of these waves.

I confine myself here to mentioning this first hypothesis: as it has not hitherto been supported by any experimental proof, nor even by any analogy with other phenomena of light, it would be useless to discuss it at present.

II. A second hypothesis would assume that a phenomenon of diffraction is produced analogous to that of gratings. Let us imagine the transparent medium divided into layers perpendicular to the pencil of light; in each of these layers there are particles which should act like a very fine powder deposited on a transparent plate; that is to say, part of the light is no longer propagated in a right line, but diffuses itself in various directions. The vibrations producing this scattered light retain the form and the direction of the vibrations of the illuminating pencil; hence, if this is polarized, the scattered light will consist of rectilinear vibrations parallel to those of the pencil.

I am inclined to think that this diffraction really takes place, and contributes to the production of the phenomena observed, especially in directions which differ little from the direction of the illuminating pencil; but in any case the particles should also reflect light, and this effect ought to predominate over that which has been mentioned.

III. Let us, then, consider the reflected light. Let *a b c d* (fig. 1) be a body illuminated by a pencil of parallel rays, *s' b*, *s a*, *s" c*.

Fig. 1.

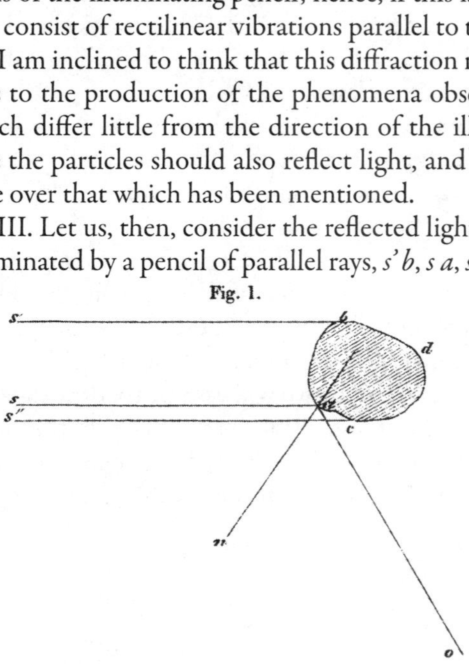

Suppose the observer's eye placed in any position *o*, looking at the body *a b c d*. For shortness' sake let us call *the angle of vision* the angle of *o a s* formed by the visual ray and the incident pencil. Let the *plane of vision* be the plane *o a s* comprising the pencil of light and the visual ray.

If the dimensions of the body *a b c d* are large as compared with the wave-lengths of the light, according to the ordinary laws of reflection the eye in the position *o* will only receive the rays reflected by a point *a*, such that the normal *a n* to the surface of the body at this point is situated in the plane of vision, and that the angle of incidence *s a n* is equal to the angle of reflection *n a o*. The reflected light will in general be partially polarized; and if the eye is in such a position that the visual angle is double the angle of polarization, the ray *a o* will be totally polarized, the plane of polarization coinciding with the plane of vision.

But we know that if the reflecting surfaces are not very great as compared with the wave-lengths of the light, the ordinary laws of reflection are no longer applicable, and each point of the reflecting surface may be considered as a centre of disturbance from which waves are really emitted—or, in other words, that the motion of the aether in all directions other than that of the reflected ray in the ordinary case of reflection is no longer annulled by interference. This is what results from the reasonings and experiments of Fresnel.[4] If, then, the body *a b c d* is very small and has dimensions comparable to the wave-lengths, the eye will receive more or less intense reflected rays proceeding from all points of the *illuminated surface* of the body. Thus, for instance, the visual angle being double the angle of polarization, the eye no longer receives merely the light totally polarized in the plane of vision reflected from the point *a*, but also non-polarized light, or light partially polarized, emitted by the other points of the surface illuminated by the incident pencil. There is no longer any reason why the maximum polarization should be exhibited when the angle of vision is twice the polarization angle.

It is clear that this reasoning applies to all the corpuscles which are illuminated by the luminous pencil; thus a first source of difficulty is removed, that, namely, which you mentioned in your communication to the Royal Society[5] on the 16th of December 1868.

IV. Let us now suppose that the pencil of light *s a*, *s′c* (fig. 2) is polarized in a horizontal plane perpendicular to the plane of the figure; it illuminates the very small particles, which first of all we will suppose to be spherical. Let us now examine what takes place in the case of one of these particles, *a b c d*. The plane of polarization being horizontal, we will assume that the incident vibrations of the aether are rectilinear and vertical, and therefore parallel to the plane of the figure. The vibrations will be reflected by each point of the illuminated part of the sphere; and as the dimensions of this sphere are very small, each

Fig. 2.

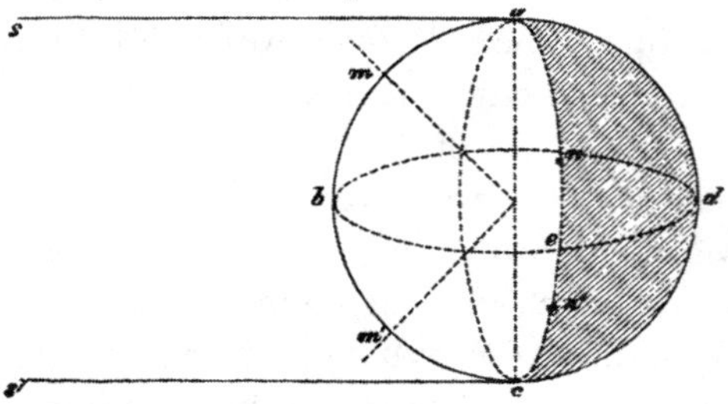

point will become a centre of disturbance or of waves. The reflected vibrations forming the system of waves which proceed from one of the points of the spherical surface will be rectilinear, and their direction will be given by the laws of reflection of polarized light. (You see that I assume the *direction of the reflected rays* in these conditions to be the same as that which Fresnel assumes in the case of light reflected from large surfaces. The tenability of this hypothesis might in certain cases be contested; but I think that even if it ought to be modified, that would make no change in the considerations I am about to develope, except for a part of those laid down in § VIII.)

Each reflected vibration may be regarded as resulting from three component vibrations—one vertical, the second horizontal and parallel to the luminous pencil, the third horizontal and perpendicular to the pencil.

Suppose the eye placed in the horizontal plane passing through the centre of the sphere. There is no reason why the vertical components of the various reflected waves should not reach the eye, which thus receives light polarized in a horizontal plane and proceeding in greater or less quantity from all the illuminated points of the reflecting corpuscle.

Let us pass to the horizontal components parallel to the pencil. Consider, for instance, the point *m* on the vertical great circle of the sphere parallel to the pencil, the normal at this point making an angle of 45° with the direction of the pencil. The incident vertical vibrations give rise at the point *a* to reflected waves entirely formed of vibrations horizontal and parallel to the pencil; that is to say, the other components are null for this point. These vibrations are propagated in all directions, and therefore also in the direction of the eye placed in the horizontal plane passing through the centre of the sphere. But, on the other hand, the point *m'*, which is symmetrical with the point *m,* also gives rise to waves of equal intensity formed of horizontal vibrations parallel to the pencil; only there will be a difference of phase of half a wavelength between the waves arising from the point *m'* and those from the point *m.* In fact the incident wave formed of vertical vibrations reaches the points *m* and *m'* in the same phase. Hence, if the molecules of aether in the incident vibration approach the reflecting surface at *m,* they are at the same time moving away from this surface at *m'*; hence the motion will be from right to left in the vibration reflected by the point *m*, while it will be from left to right in the vibration reflected by the point *m'*. The reflected waves will therefore be in opposite phases, and will destroy each other in all points at equal distance from *m* and *m'*—that is to say, in all points of the horizontal plane passing through the centre of the sphere: thus the eye placed in this plane will not be affected by the horizontal vibrations proceeding from *m* and *m'*. In other corresponding points the effect will be the same (excepting that it will not be to the whole of the reflected vibrations, but merely to the horizontal component parallel to the beam that the reasoning applies). In fine, the horizontal components parallel to the luminous pencil proceeding from the illuminated part of the sphere situated above the plane of vision will be destroyed by interference with the same components proceeding from the part of the sphere which is below the plane of vision.

The same reasoning applies to the third components, horizontal and perpendicular to the pencil of light. Thus the effect produced by the point *n* will be destroyed by that of the corresponding point *n'* for the eye placed in the horizontal plane passing through the centre of the sphere.

In short, the eye in the position in question only receives light formed by the first components—that is, composed of vertical vibrations. In other words, the light is polarized in a horizontal plane.

It is easy to understand that, if the spherical corpuscles are small enough, this destruction by interference will be produced even outside the horizontal plane which passes through the centre of the sphere, provided the observer is not too far off. Thus when the plane of vision coincides with the plane of polarization of the pencil of light (which I suppose to be very thin), all the illuminated corpuscles will transmit to the eye light consisting only of vertical vibrations. This is confirmed by experiment.

V. Always assuming that the corpuscles are spherical, let us examine what ensues when the horizontal pencil of light is polarized in a vertical plane—that is to say, when the vibrations of the incident light are horizontal.

For the eye placed in the horizontal plane which passes through the centre of the sphere, the first vertical components will destroy each other by interference. Let us, in fact, take two symmetrical points, *n* and *n';* the incident horizontal vibrations will give rise to vertical reflected vibrations (or at least to vertical components); but, the inclination of the reflecting surface in reference to the horizontal vibrations being in opposite directions for the points *n* and *n',* the reflected motion must also be in the opposite direction.

The second components, which are horizontal and parallel to the pencil, may reach the eye; there is no reason why they should interfere. The intensity of the light due to these components must be very feeble, as we shall subsequently see.

Nor will the third, horizontal components perpendicular to the pencil neutralize each other.

Thus the eye placed in the horizontal plane passing through the centre of the sphere will receive light from the second and third components. If the visual angle is 90°, the third components will produce no effect upon the eye, since they act longitudinally. Thus the eye in this position will only receive the impression of the horizontal components parallel to the pencil, which, as we have already said, are very feeble, and often quite imperceptible.

We readily deduce from this the explanation of the experiment which was suggested to you by Mr. Stokes.*

If the visual angle is greater or less than 90°, the eye will receive the combined effect of the second and third components; but the direction of the resulting vibrations will be always parallel to the horizontal plane passing through the incident pencil. This result is also quite in accordance with experiment.

VI. If the illuminating pencil consists of natural light, we shall easily find what takes place by regarding it as composed of two luminous pencils polarized at right angles, one in the plane of vision (which I suppose to be always horizontal), the other in a perpendicular plane. We have the maximum polarization when the angle of vision is 90°; for in this position the eye receives vertical components in large quantities, while from the horizontal components it only receives the impression of the second components of the reflected vibrations proceeding from the pencil polarized perpendicularly to the plane of vision. Now, as we have already said and as we shall show, these components are very feeble.

If the angle of vision is less or more than 90°, then the horizontal components perpendicular to the pencil, proceeding from the pencil polarized perpendicularly to the plane, begin to act, and the polarization ceases to be at its maximum.

VII. Let us still suppose the plane of vision to be horizontal, and that the illuminating pencil is formed of natural light—that is to say, of two pencils polarized at right angles, one in the horizontal plane, the other in the vertical plane. Let us inquire why in the reflected light the components parallel to the pencil are very feeble as compared with the vertical components.

In the first place, even if the sphere had an absolute reflecting power, there would be a considerable predominance of the vertical components. In fact, on the vertical great circle *a e c*, the pencil polarized in a horizontal plane will give rise to reflected vertical components at all points, except the two points at 45° from *e*. The pencil polarized in the vertical plane, on the contrary, will not produce at any part of the great circle components parallel to the pencil; the same is the case with the great circle *a b c*. As regards the horizontal great

circle *b e*, the first pencil will everywhere produce only vertical components, while the second pencil will produce no components parallel to itself at *b,* nor will it at *e;* it is only between these points that it will produce any.
But, further, this preponderance of the vertical components will be greatly increased by the fact that the reflecting power of the sphere is not perfect. Thus on the horizontal great circle *b e* (which in the case of perfect reflecting power would produce most components parallel to the pencil), the intensity of the vertical components will be least at *b*, where its value will be proportional to $(\frac{n-1}{n+1})^2$, *n* being the refractive index of the corpuscles as compared with the transparent medium; from *b* this intensity will continually increase, up to *e*, where its value is proportional to 1. On the other hand, the intensity of the components parallel to the pencil will be zero at *b*, zero again at the point for which the angle of incidence is equal to the angle of polarization, zero again at the point *e.* Passing three times through zero, the intensity of these components must nowhere be considerable on the great circle *b e*, &c.

The exact calculation of these intensities is very complicated. M. Cellérier[6] has been kind enough to do it at my request for several values of the refractive index, and he has found that the vertical components greatly predominate over those parallel to the pencil.

VIII. I have hitherto always supposed that the corpuscles were spherical. Is this condition necessary for the explanation which I have given of the phenomena of illumination? I think not. For, on the one hand, whatever be the shape of the particles, the surface is always tarnished, and the phenomena of interference which I have pointed out will always be partially produced. On the other hand, we have not to consider a single particle. In a layer perpendicular to the luminous pencil there is a large number of particles; there must, then, in general be in this layer as many surfaces in a certain position as in the position symmetrical with it. These symmetrical surfaces will produce the interferences of certain components of the reflected vibrations, as if they belonged to the same particle.

This is the summary of my theory, or rather my *attempt* at a theory.

Although this letter is already long, will you permit me to add a few words? You will have seen that M. Lallemand has answered[7] (*Comptes Rendus*, Dec. 20) the note I communicated[8] to the Academy (*Comptes Rendus*, Dec. 6), and that he still thinks the illumination of liquids and of solids should be attributed to a lateral propagation of the motion of the aether, and not to particles diffused in the transparent medium. The reasons which he adduces in support of his opinion do not seem to me conclusive; but I will not dwell upon this subject. I will merely mention that I have recently made a series of experiments by taking water and producing light precipitates in it by means of certain reagents. Whenever the precipitate was formed of particles so fine that they were not visible to the naked eye, and in such small quantity that the transparency of the water was not sensibly diminished in diffused light, I observed a great increase in the illuminating power of the line of light, with the same phenomena of polarization.

Thus I take water from the Lake of Geneva[9] which has quite settled, and which has very little illuminating power; it contains some traces of chlorides in solution; I add to it, by means of a glass rod, a drop of an extremely dilute solution of nitrate of silver. The line of light becomes far more visible, and the polarization is complete. I add a larger quantity of nitrate of silver, and the water becomes a little dull; the line is still more distinctly visible, of a bluish tint; but if we examine with a Nicol's prism,[10] in a suitable position, the extinction is no longer complete; the line is still seen, and is of a beautiful deep blue (almost the tint of the indigo of the spectrum).

It appears to me probable that this incomplete polarization is due, at least partially, to multiple reflections when the particles in suspension become very numerous. If water containing chloride of silver is exposed to the sun's light, the precipitate changes its nature, the line is extremely distinct, but the phenomena of polarization are far less definite than before.

I do not see how these facts can be due to any other cause than that of particles in suspension.

Believe me, | My dear Tyndall, | Yours very sincerely, | L. SORET.

* In a private note Professor Stokes suggested to me the employment of polarized light to illuminate the actinic cloud, and I feel pleasure in giving him the credit of the suggestion. Had the suggestion not come to me, however, the result would have been substantially the same.—J. T.

Phil. Mag., 39 (1870), pp. 221–9

---

1. *On illumination . . . Soret*: see letter 3092, n. 2.
2. *your kind letter of the 5th of December*: letter missing. This letter was also a reply to letter 3092 of 30 November.
3. *M. Lallemand*: Alexandre Lallemand (see letter 3092, n. 4).
4. *Fresnel*: Augustin-Jean Fresnel (1788–1827), a French engineer and physicist who contributed significantly to the establishment of the theory of wave optics. Fresnel studied the behaviour of light both theoretically and experimentally. He is the inventor of the Fresnel lens, first adopted in lighthouses while he was a French commissioner of lighthouses (*CDSB*).
5. *your communication to the Royal Society*: 'On the Blue Colour of the Sky' (see letter 2940, n. 3).
6. *M. Cellérier:* Gustave Cellérier (1855–1914), assistant astronomer to director Colonel Gautier at the Geneva Observatory ('Gustave Cellérier (1855–1914)', *Memoires de la Société de physique et d'histoire naturelle de Genève*, 38 (1914–16), pp. 203–4). Cellérier co-authored with Swiss physicist Raoul Pictet (1846–1929; *HLS*), *Méthode générale*

d'intégration continue d'une fonction numériqe quelconcque: à propos de quelques théorèmes fournis par l'analyse mathématique appliquée au calcul des courbes d' une nouveau thermographe (Geneva: H. Georg, 1879).

7. *M. Lallemand has answered*: M. A. Lallemand, 'Observations sur la Communication de M. Soret, relative à l'illumination des corps transparents', *Paris, Comptes Rendus*, 69 (1869), pp. 1294–6.
8. *the note I communicated*: letter 3092.
9. *Lake of Geneva*: a lake on the north side of the Alps located between Switzerland and France. The city of Geneva is located along the southwestern edge of the lake.
10. *Nicol's prism*: see letter 2966, n. 4.

To Thomas Archer Hirst [25 January 1870][1] 3120

Royal Institution | Jan[uary] 25th 1870] | Tuesday

Dear Tom

Mrs. Tunstal[2] says that she can cook the hare admirably here—I take the liberty therefore of saying "come here" at half past 6.

John

RI MS JT/1/T/960
RI MS JT/1/HTYP/563

1. *[25 January 1870]*: the date was added by a second hand.
2. *Mrs. Tunstal*: see letter 3058, n. 1.

To the Editor[1] of the *Pall Mall Gazette* 26 January 1870 3121

To the EDITOR *of the* PALL MALL GAZETTE.[2]

SIR,—Would the senior physician[3] of Colchester Hospital, whose letter you published[4] in yesterday's *Pall Mall Gazette*, permit me to take him to the top of the Matterhorn[5] and to show him the sky of Italy overhead? If not, will he accept my assurance that the sky is blue?

Will he believe me when I say that this deep colour is, or may be, produced by particles suspended in the air which are not only invisible to the naked eye but irreducible by the highest powers of the microscope? That hence, without seeing the individual particles, we may have indubitable evidence of their existence.

This, indeed, is the point wherein my method differs from preceding ones, and is, I think, destined powerfully to supplement them. The microscope seeks for single particles; I take them *en masse*, and can demonstrate their existence by the light which they scatter after they have passed utterly beyond the range of the microscope.

Were the "dark spaces" referred to by Dr. Bree "full of germs," I should, in my opinion, have there the blue of the sky instead of "the darkness of stellar space."

"It is doubtful," says Dr. Bree, "whether the germs of disease have been ever seen by the more powerful microscope; it is therefore clear that they would pass through cotton wool with the atmosphere, and pass into the lungs as though the wool were absent."

My experiments render the very reverse of this "quite clear" to me.

I have been favoured by Dr. Angus Smith[6] with a copy of his fifth annual report,[7] from which I glean some interesting facts regarding the air of Manchester. To catch in water the floating matter of the air, Dr. Smith places a small quantity of the liquid in a bottle and shakes it up with successive charges of air. In one instance, he did this 500 times and then handed over his bottle to an able microscopist, Mr. J. B. Dancer,[8] for examination. The bottle has been shaken in the open air, through which, however, Dr. Smith could not see any dust blowing; at all events, if there were dust, it was only such as people are called upon to breathe. Here are some of the revelations of Mr. Dancer:—

"*Fungoid Matter*.—Spores or Sporidic appeared in numbers, and, to ascertain as nearly as possible the numerical proportion of these bodies a single drop of the liquid, the contents of the bottle were well shaken, and then one drop was taken with a pipette. This was spread out by compressing to a circle half an inch in diameter. A magnifying power was then employed, which gave a field of vision of an area exactly 100th of an inch in diameter and it was found that more than 100 spores were contained in this space. Consequently the average number of spores in a single drop would be 250,000. These spores varied from 10,000th to 50,000th of an inch in diameter.

"For the purpose of obtaining a rough approximation to the number of spores or germs of organic matter contained in the entire fluid received from Dr. Smith, I measured a quantity by the pipette and found it contained 150 drops of the size used in each examination. Now I have previously seen that in each drop there were about 250,000 of the spores, and as there were 150 drops, the sum then reaches the startling number of 37½ millions; as these, exclusive of other substances, were collected from 2,495 litres of the air of this city—a quantity which would be respired in about ten hours by a man of ordinary size when actively employed. I may add that there was a marked absence of particles of carbon among the collected matter."

Apart from their other effects, the mere medicinal irritation produced by the disposition of these particles in tender lungs must go for something. They may be entirely withheld by a cotton wool respirator. In various dusty trades and occupations the respirator will also be found a comfort and protection. I may add, in reference, to inquiries addressed to me, that the respirator is placed unconditionally at the disposal of medical men.

Since writing these lines, I have received from Wigan[9] a letter, which I enclose;[10] it shows that in this case, as in so many others, the needs of men have prompted remedies in advance of the suggestions of science.—I am, Sir., Your obedient servant,

JOHN TYNDALL. | Royal Institution, Jan[uary]. 26.

J. Tyndall, 'Dust and Disease', *Pall Mall Gazette*, 1547 (27 January1870), p. 6

---

1. *Editor*: Frederick Greenwood (1830–1909), an English journalist and editor. He edited the *Cornhill Magazine* (1864) and served as the first editor of the *Pall Mall Gazette* (1865–80) and the *St James Gazette* (1880–88). Under his direction the *Pall Mall Gazette* forcefully defended Tory (Conservative) policies. Greenwood played an important but unacknowledged role in the British government efforts to purchase part of the Suez Canal in 1875 by passing on intelligence of the French government's plans to buy the canal (*ODNB*).
2. *Paul Mall Gazette:* the *Paul Mall Gazette* was established in 1865 by George Murray Smith (1824–1901). It tended toward the conservative end of the political spectrum, albeit not exclusively.
3. *senior physician:* Charles Robert Bree (1811–86), Senior Physician to the Colchester Hospital, an anti-Darwinian enthusiast zoologist who researched and published on the birds of Europe ('Charles Robert Bree, M.D. Edin., M.R.C.S. Eng., L.S.A., &c.', *British Medical Journal*, 2 (13 November 1886), p. 954; J. R. Moore, *The Post-Darwinian Controversies: A Study of the Protestant Struggle to Come to Terms with Darwin in Great Britain and America, 1870–1900* (Cambridge: Cambridge University Press, 1979), pp. 199–200).
4. *whose letter you published*: Bree argued against Tyndall's findings in his lecture on dust, stating that the particles that Tyndall observed could not represent germs which were invisible to the naked eye. Consequently, while Tyndall's dust particles might or might not contain germs, it would be impossible for the observer to know. C. Bree, 'Occasional Notes', *Pall Mall Gazette*, 25 January 1870, p. 3.
5. *Matterhorn*: see letter 2987, n. 17.
6. *Angus Smith*: Robert Angus Smith (1817–84), an applied chemist, environment scientist, and FRS (1857). He became fellow of the Chemical Society (1845), council member (1870–72) and vice-president (1878–80), although he was much more active in Manchester-based societies. Smith was appointed alkali inspector under the Alkali Act of 1863 and later appointed inspector under the Prevention of Rivers Pollution Act (1876) (*ODNB*).
7. *fifth annual report*: 'Alkali Act, 1863. Fifth annual report to the lords of the committee of Privy Council for Trade, by the inspector, of his proceedings during the year 1868', C 4167, Parliamentary Papers (1868–69). On the Alkali Acts, see R. M. MacLeod, 'The Alkali Acts Administration, 1863–84: The Emergence of the Civil Scientist', *Victorian Studies*, 9:2 (December 1965), pp. 85–112.

8. *Mr. Dancer*: John Benjamin Dancer (1812–87), a British optical instrument maker. From 1841 Dancer operated out of Manchester and was a member in both the Manchester Literary and Philosophical Society and the Manchester Photographic Society. He made several contributions to magic lantern technology and lighting, and developed microscopes and other optical instruments. His skills as a microscopist and photographer were combined in his development of techniques and apparatus for micro-photography (*ODNB*).
9. *Wigan*: a town near Manchester.
10. *which I enclose*: enclosure missing.

To the Editor of *The Times* 26 January 1870 3122

Sir—Would you allow me to supplement the report of my lecture[1] which you have done me the honour of publishing, by stating that, according to an analysis communicated to me by Dr. Percy,[2] the dust collected *from the walls* of the British Museum[3] contains fully 50 per cent of inorganic matter? I have every confidence in the results obtained by this distinguished chymist.[4] They show that the floating dust of our rooms is, as it were, winnowed from the heavier matter. As bearing directly on this point, I may quote the following passage from *Pasteur*:[5]—

"Mais ici se présente une remarque: la poussière que l'on trouve à la surface de tous les corps est soumise constamment à des courants d'air, qui doivent soulever ses particules les plus légères, au nombre desquelles se trouvent, sans doute, de préférence les corpuscules organisés, œufs ou spores, moins lourds généralement que les particules minérales."[6]

I am, Sir, your obedient servant, | JOHN TYNDALL. | Royal Institution, Jan[uary]. 26.

Times, 27 January 1870, p. 8
RI MS JT/6/5/1/28

---

1. *report of my lecture*: 'Professor Tyndall on dust and disease', *Times*, 22 January 1870, p. 6. Tyndall delivered a Friday Evening Discourse at the RI, 'On Dust and Disease', on 21 January 1870 (*Roy. Inst. Proc.*, 6 (1870–72), pp. 1–14).
2. *Dr. Percy*: John Percy (1817–89), an English metallurgist and art collector. Percy was elected fellow of the Geological Society (1851) and appointed lecturer and later professor of metallurgy to the museum at the Metropolitan School of Science in London. The first volume of his *Treatise on Metallurgy* was published in 1861 and though it contains over 3500 pages of careful research, it was never completed. Percy collected prints and drawings, which he intended to bequeath to the British Museum. His metallurgical specimens were bequeathed to the South Kensington museum (*ODNB*).
3. *British Museum*: established in 1753, the British Museum was the primary museum in

London dedicated to human history, art, and culture. Several institutions, such as the Natural History Museum and the British Library, were formed from its collections.

4. *chymist*: an antiquated spelling of chemist (often interchangeable with alchemist), its use became less common in the nineteenth century.
5. *the following passage from Pasteur*: from L. Pasteur, 'Mémoire sur les corpuscule organisés qui existent dans l'atmosphère. Examen de la doctrine des générations spontanées', *Annales des sciences naturelles*, 16 (1861), pp. 5–98.
6. *"Mais ici se... les particules minérales"*: "But here is a comment: the dust that we find on the surface of all bodies is constantly subjected to streams of air, which must lift the lightest particles, among which we undoubtedly find organized particles, eggs or spores, which are generally lighter than mineral particles" (French).

From Mr. Higginbotham[1] *c.* 29 January 1870 3123

Your theory[2] is confirmed by a practice very generally adopted for many years past by persons working in the preparatory process of the cotton-spinning trade—viz., to hold in their teeth a ball of cotton sufficiently large to protect, not only the mouth, but also the apertures of the nose, so as to prevent the inhalation of dust and small particles of cotton.

Sheffield Daily Telegraph, 29 January 1870, p. 7

---

1. *Mr. Higginbotham*: resident of Preston, not further identified.
2. *Your theory*: Higginbotham is presumably referring to the report of Tyndall's RI discourse 'On Dust and Disease' (see letter 3122, n. 1).

From Elliot H. Hawes[1] 1 February 1870 3124

[2]Office of Irish Lights | Dublin 1st February 1870.

Dear Sir,

Since writing to you last[3] I have gone a good deal more into the matter of the use of Gas for annular lenses and have requested Mr Wigham to make many experiments on the subject. We had just succeeded in reducing the small secondary flash, to which your last note to me alluded, to an exceedingly trifling twinkling Light when we devised an improvement upon the method of saving Gas which we showed to you when you were here and have succeeded in producing an exceedingly beautiful and distinguishing light, it consists in causing the Gas to produce while each lens passes the eye, three or four rapid flashes, caused by the extinction and reignition of the Gas. The first had this light fixed on the roof of Messr Edmondson's Factory[4] and I observed it from the upper Windows of this House and confess I was very much delighted with its performance and I think I never saw any Light that seemed to me more suitable for catching the eye of the mariner. Wishing to ascertain whether the same appearance would be seen by an observer at a greater distance, I had the lens removed to the experimental House at Howth Bailey[5] and for the last two evenings have been witnessing the effect at Kingstown.[6]

My previous favourable impressions are more than confirmed by these observations, there are as you know a great many Lights about Kingstown Harbour *[&c]* but whenever the flashes of this new Light were displayed they caught the eye as a flash of a Gun would, even though not exactly turned in the direction of the Lighthouse.

I look upon this matter as one of exceeding importance and would be glad to hear from you on the subject. You will observe from the foregoing that while the saving of the Gas by its extinction is not accomplished exactly in the same mode as you described in your last report to the Board of Trade yet that the main gist of your report, Viz, the saving of the Gas is effected to the same extent as you stated whether you would like to send a supplemental note to the Board of Trade to that effect or whether you would like first to come and see the Light, of course you know best, but I should very much like to see Gas introduced for Rockabill Lighthouse[7] and am persuaded by this new system of rapid flashes, many most desirable things, for example connected with signaling which we now perhaps hardly think of might be accomplished.

I am | Dear Sir, | Your obedient Servant | E H Hawes | Inspector of Lights

Doctor John Tyndall L.L.D. F.R.S. &c. | Royal Institution of Great Britain | 21. Albermarle Street London W

National Archives MT 10/31 H64B 2314–16

---

1. *Elliot H. Hawes*: see letter 3019, n. 12.
2. A stamp in the upper left-hand corner reads: 'It is requested that any further communication on this subject may be addressed to— | "The Secretary," | Office of Irish Lights, | Dublin. | *Quoting the official number and noting on left hand corner of Envelope.* | FOR INSPECTOR OF LIGHTS'.
3. *Since writing to you last*: letter missing. For Tyndall's work on lighthouses, see letter 2937, n. 6.
4. *Messrs Edmondson's Factory*: Joshua Edmundson & Co. was a small Dublin hardware and manufacturing business. John Wigham began work there as an apprentice to his brother-in-law in 1843 and eventually took control of the business.
5. *Howth Bailey*: see letter 3019, n. 3.
6. *Kingstown*: now known as Dún Laoghaire, Kingstown is a town and port located on the east coast of Ireland in County Dublin. It was constructed as a port to serve Dublin.
7. *Rockabill Lighthouse*: Rockabill is a group of two islands located in the Irish sea off the coast of Skerries, County Skerries. Construction of a lighthouse began there in 1855 and a gas power light was installed in 1872.

To George Gabriel Stokes 7 February [1870][1] 3125

Royal Institution of Great Britain | 7th Feb[ruary][2]

Dear Stokes.

I have promised to make a few of the experiments recorded in my paper[3] now in your hands before the Chemical Society[4] of next week. Would you do me the favour to lend me the paper so that I may refresh my memory for the occasion? An early answer will oblige me.

yours faithfully | John Tyndall

Wellcome Library MS. 7777/40

1. *[1870]*: Tyndall demonstrated some experiments on light and vapours to the Chemical Society on 17 February 1870 ('Proceedings at the Meetings of the Chemical Society', *Journal of the Chemical Society*, 23 (1870), pp. 286–322, on p. 289).
2. *7th Feb[ruary]*: a note in the upper left-hand corner in Stokes' hand reads: 'paper sent to author Feb 8'.
3. *my paper*: J. Tyndall, 'On the Action of Rays of High Refrangibility upon Gaseous Matter', *Phil. Trans.*, 160 (1870), pp. 333–65.
4. *before the Chemical Society*: see n. 1.

To Auguste de la Rive 16 February 1870 3126

16th, Feb[ruary], 1870.

My dear Friend

I am exceedingly obliged to you for your letter,[1] and for the frank friendliness with which you speak of my lecture.[2]

That lecture was got up in ten days. It was given at the urgent recommendation of Dr Bence Jones. I was so pressed for time that I had to ask Dr Bence Jones and Sir Henry Holland to help me to get up the literature of the subject. I was employed wholly upon the experiments, intending to return to the subject and deal fully with its history. I thought meantime that it might be of practical use to medical men and therefore printed it in the form in which it has reached you.

I am still engaged upon the subject, and will publish the historic account of it before I begin my next course of lectures.[3]

In the lecture I spoke in the very highest terms of Pasteur. So high indeed that I have been gravely censured by his opponents. But I had to pull down a portion of the type of the lecture through inability to finish it aright, or to compress it within the hour.

I should be exceedingly obliged to you for any further remarks upon the subject which would enable me to complete my knowledge of the literature. My only desire is to do faithfully what is just and right

Yours ever truly | John Tyndall

RI MS JT/1/TYP/1/373
LT Typescript Only

1. *your letter*: letter missing.
2. *my lecture*: 'On Dust and Disease' (see letter 3122, n. 1).
3. *my next course of lectures*: after Easter, Tyndall delivered a series of seven lectures at the RI on 'Electrical Phenomena and Theories', on Thursdays from 28 April to 9 June 1870 (*Ascent of John Tyndall*, p. 266; RI Guard Book vol. III, p. 89; *Roy. Inst. Proc.*, 5 (1866 – 69, p. 611). See letter 3169, n. 4.

To Auguste de la Rive 17 February 1870 3127

17th. Feb[ruary], 1870.

My dear Friend

I find on looking once more over your letter[1] that I had omitted the reading of one page when I wrote to you yesterday.[2] Indeed I am not yet able to make out the whole of it; but the points of importance seem clear.

You desire to write me a letter about your own researches on the impurities of the atmosphere. By all means do so my dear friend. I will translate the letter myself and send it to the Phil[osophical] Mag[azine] or if you prefer it it shall be sent in the original. This way is open to you or else if you wish you can send me the article you are writing for the Archives, and leave it to me[3] to give the history of your labours. Do what you please in the matter; I shall be glad to second all your wishes to the best of my ability.

In fact it was Bence Jones instigated me to give that lecture;[4] and I consented to do so on condition that he would get up the literature for me. I wished him to give the lecture, and I was willing to place the experiments at his disposal—so that he might speak upon the subject with the authority of a Medical man.

But I really do not see that I claim an iota in the lecture that could be objected to. It is perfectly manifest from my words that I make not the slightest claim to the germ theory. Indeed my mind is in suspense about it. That germs are in the atmosphere has been known for centuries, but what was new to me, and what surprised me was that all the floating matter of our London rooms is of organic origin.

I have been attacked in a Manchester newspaper[5] for not giving due credit to Dr Angus Smith.[6] So that the lecture has got me into hot water. Happily it is not hotter than I can bear.

I know full well that you will say nothing of me or of my work unworthy of a gentleman and a friend. Your friendship has been always one of my highest prizes, and I have no doubt it will continue to be so.

I was somewhat surprised at your complaint regarding the English Journals: for by some means or other you have attained a position in the estimation of Englishmen which few have reached.

For my own part I was obliged to withdraw myself long ago from the editorship[7] of the Phil[osophical] Mag[azine]. At the present time I should say I have less influence with that Journal than many men of my acquaintance. The reason is natural—I do not write for it, and it is simply by the favour of Mr Francis[8] that I get anything inserted in it.

You draw my attention to a photometer exhibited at Geneva in 1865. Did you shew it to me? If so I have entirely forgotten it. This however is not an unusual thing with me. I shall set about a historic sketch of this subject immediately, and should any thing occur to you that you would wish me to know, pray let me know, and every thing right and proper shall be done.

I am my dear Friend | Yours as ever | John Tyndall

RI MS JT/1/TYP/1/374
LT Typescript Only

1. *your letter*: letter missing.
2. *wrote to you yesterday*: letter 3126.
3. *article . . . to me*: A. de la Rive, 'De la poussière qui flotte dans l'air atmosphérique', *Archives des Sciences*, 37 (1870), pp. 229–42 ; and 'On the Organic Dust of the Air', *Phil. Mag.*, 39 (1870), pp. 229–30.
4. *that lecture*: 'On Dust and Disease' (see letter 3122, n. 1).
5. *a Manchester newspaper*: not identified.
6. *Dr. Angus Smith*: see letter 3121, n. 6.
7. *withdraw myself . . . from the editorship*: Tyndall ceased being an editor of the *Phil. Mag.* in 1863.
8. *Mr. Francis*: William Francis.

To Emil du Bois-Reymond 18 February 1870 3128

18th Feb[ruary] 1870

My dear DuBois,

I read your last letter to me[1] with extreme interest. It is an odd coincidence that both of us in the same year should be put hors de combat in the Alps.[2] I should dearly like to join you in some excursion as your performance on the Chalk cliffs of Scratchell's Bay[3] gave me a high idea both of your head and feet. You never, I dare say, told Madame DuBois[4] of that dangerous escapade.

Your papers[5] have come to me—I enclosed one of them immediately to Francis[6] with a letter[7] in which I mentioned the very extraordinary results you showed me in Berlin. I told him that so much was I struck by their importance in a practical as well as purely scientific point of view that I ordered the whole apparatus. I hope and think he will print the memoir;[8] but I have not so much influence with the Phil[osophical] Mag[azine] as I used to have. For the good reason that I am unable to contribute to it.[9]

I send you by this post an abstract[10] of a lecture[11] which I gave here sometime ago. Bence Jones was the instigator of it. I wished him to give it, and thus speak on the subject with the authority of a medical man. You know everything and if you could help me in any way by pointing out to me the German literature which immediately bears upon the subject you would oblige me by doing so. The lecture has caused a surprising row among the newspapers,[12] and some murmuring, on account of my omitting to go into the history of the subject; but this within the time at my disposal was simply impossible. I am still at work upon the subject, and will return to its history by and by.

Yours ever truly | John Tyndall

RI MS JT/1/T/402
RI MS JT/1/TYP/7/2437

---

1. *your last letter to me*: letter 3104.
2. *hors de combat*: 'out of combat' (French). Like Tyndall, du Bois-Reymond had suffered a mountaineering injury in autumn 1869 (see letter 3104). For Tyndall's injury, see letters 3030, 3031, 3032, 3037, 3041, 3048, 3049, and 3060.
3. *Scratchell's Bay*: Scratchell's Bay is located on the south-west coast of the Isle of Wight, England.
4. *Madame DuBois*: Jeanette du Bois-Reymond (née Claude, 1833–1911). She and Emil du Bois-Reymond had nine children (R. Tobies, *Iris Runge: A Life at the Crossroads of Mathematics, Science, and Industry* (Basel: Sprinter, 2012), p. 26).
5. *Your papers*: one of the papers is that which is cited in letter 3104, n. 14. See also letter 3140.
6. *Francis*: William Francis.
7. *with a letter*: letter missing.
8. *print the memoir*: no memoir from du Bois-Reymond was published in the *Phil. Mag.* in 1870.
9. *unable to contribute to it*: Tyndall explained this in more detail in letter 3127 to de la Rive.
10. *I send you by this post an abstract*: enclosure missing.
11. *abstract of a lecture*: 'On Dust and Disease' (see letters 3122, n. 1, and 3127).
12. *a surprising row among the newspapers*: see letters 3121, 3122, 3126, and 3127.

To George Gabriel Stokes 18 February 1870 3129

Royal Institution | 18th February 1870.

Dear Stokes,

In connexion with the subject of thermometer stands brought yesterday before the Government Grant Committee[1] would you permit me to suggest through you to the society that the real difficulty of obtaining the true temperature of the air is due to the interaction as regards radiation & absorption between the bulb of the thermometer and surrounding basis. Destroy this interaction & you can obtain the true temperature of the air. I would therefore so propose that in experiments conducted under the auspices of the Royal Society thermometer bulbs silvered on the outside be subjected to heat. It appears to me that such thermometers might save the cost and labour now expended upon stands.

I am | yours very faithfully | John Tyndall

Prof Stokes Sec. R.S.

CUL SC—add 7656 RS729

1. *Government Grant Committee*: from 1849 the British government supplied an annual grant to the RS for the promotion of scientific research. These funds were allocated by a Committee within the Society (known from 1853 as the 'Government Grant Committee') to individual scientists to aid in their research. Between 1849 and 1914 the Grant provided around £179,000 for such projects. Tyndall was a member of the Council and attended the meeting on 17 February (RS Minutes of Council 1870–1877 4, 37; R. M. MacLeod, 'The Royal Society and the Government Grant: Notes on the Administration of Scientific Research, 1849–1911', *The Historical Journal*, 14 (1971), pp. 323–58).

From William Thompson[1] 18 February 1870 3130

To PROFESSOR TYNDALL[2]

I have been engaged since September last in making a survey for the Intercolonial railway,[3] up the valley of the Matapediac,[4] a stream about two hundred feet in width and four or five feet in depth, which discharges into the Restijinche river,[5] about twenty miles above the head of the Baie des Chaleures.[6]

The Matapediac, which is fed by large fresh water springs, runs over a rocky bottom covered with loose stones, ranging in size from coarse gravel to boulders as large as a hogshead,[7] and the average current is about four miles an hour.

Early in November last the temperature went down in one night to 12°F., and on going out of camp the following morning I noticed large quantities of what appeared to be snow saturated with water floating down the stream, but not a particle of snow had fallen near us or for many miles round, as far as I could see by the mountain tops, nor had any ice formed on the surface of the river.

The water opposite where I stood was about six feet deep, and perfectly clear, so that I could see every stone on the bottom, and with the exception of the floating slush, the river was as it had been the previous day when the temperature was about 50°F.; I got into a canoe and paddled with the current for half a mile or so, and in shooting some small rapids, where the water in places was not more than two or three feet deep, I noticed on the bottom, masses of the slush clustered round and between the boulders, and a slight touch with the paddle was sufficient to free these clusters, when they rose to the surface, and were carried away by the current. I continued down the stream for three or four miles, and noticed the same thing in every rapid, where the water was shallow and *ruffled* by stones at the bottom.

The buoyancy of this slush was such that when detached from the bottom it rose so rapidly as to force itself well out of the water, and then floated off about half submerged.

I watched this forming of slush for many days, and in several cases found small stones imbedded in the floating slush, having been torn from the bottom when the buoyancy of the slush, aided by the running water, caused it to rise.

The temperature continued getting lower daily, and the slush in the rapids formed more rapidly than it was carried away, so much so that a bar or dam was formed across the river at each rapid, backing up the water in some cases five and six feet, when it generally found an outlet over the adjoining land, and into its natural bed again, or the head of water became sufficient to tear away the obstruction, which by this time had become a solid frozen mass.

All this time, no properly crystallised ice had formed on the surface of the river, the current being too rapid, but the slush or "anchor ice" as the trappers call it, was forming in deeper water than it had formed in before, indeed all over the river bottom, and was rising and floating away as I have already described, Eventually the temperature got down to two and three degrees below zero, when the river *surface* began to freeze in the eddies and along the edges, and the open water space became narrower every day, and was filled with floating "anchor ice" and detached masses of solid ice, which here and there became jammed and frozen together, so as to form ice-bridges on which we could cross.

These ice-bridges served as booms to stop much of the floating ice, which froze solid the moment it came to rest; and in this manner the river at last became completely frozen over for about forty miles of its length, but not until after we had experienced five weeks of steady cold, with the thermometer never above +12°F., and frequently down to -16°F.

It is just possible that what I have endeavoured to describe may suggest something to you, or it may be an old story; if so, please pardon the intrusion.

W. G. THOMPSON | Dalhousie, N.B., February 18

Nature 1 (31 March 1870), p. 555

---

1. *William Thompson*: William George Macneill Thompson (1833–1903), an engineer, was first Resident Engineer for the railway construction from 1868–72 (S. Fleming, *The Intercolonial: a Historical Sketch of the Inception, Location, Construction and Completion of the Line of Railway Uniting the Inland and Atlantic Provinces of the Dominion* (Montreal: Dawson Brothers, 1876), p. 162; 'Obituary', *Minutes of Proceedings of the Institution of Civil Engineers*, 154 (1903), pp. 384–5).
2. *To Professor Tyndall*: this letter was published in the 31 March 1870 issue of *Nature* along with letter 3160.
3. *intercolonial Railway*: construction on the Intercolonial Railway of Canada began in 1867 and was largely completed by the mid-1870s. It connected Nova Scotia, New Brunswick, Quebec, and Ontario.
4. *Matapediac*: a valley in the Chi-Choc Mountains, eastern Quebec.
5. *Restijinche River*: Restigouche River in northwestern New Brunswick.
6. *Baie des Chaleueres*: Chaleur Bay is at the mouth of Restigouche River in the Gulf of St. Lawrence between New Brunswick and Quebec.
7. *hogshead*: a large cask used as a unit of measurement. Equal to about 64 US gallons or 245 liters (*OED*).

To Elinor Colvile 19 February 1870 3131

<Envelope> Lady Colvile | 8 Rutland Gate | S.W.
19th. Feb[ruary]. 1870 | Royal Institution

My dear Lady Colvile.

It would give me exceeding pleasure to join you on Thursday, for I feel somehow that you are a kind & warm friend of mine; but to my horror I learn on consulting my little note book that I am engaged for Thursday & for several other days next week.

Yours always brief | John Tyndall

I am very glad to see Sir James[1] looking so well—I hope you also are well.

BL Add MS 60634, ff. 126–7

---

1. *Sir James*: James William Colvile (1810–80), an English lawyer and civil servant. Educated at Eton College and at Trinity College, Cambridge, Colvile was called to the bar in 1835. He became advocate general of Bengal in 1845 and puisne judge of the supreme court of Bengal in 1848 and was knighted that same year. He was appointed chief justice in 1855. In 1859 he retired to London (*ODNB*).

To George Gabriel Stokes 21 February 1870 3132

Royal Institution of Great Britain | 21st. Feb[ruary]. 1870

My dear Stokes.

I must have expressed myself very imperfectly—however if the thing has been already tried pray throw my note[1] in your waste-paper basket.

Yours faithfully | John Tyndall

CUL SC—add 7656 RS732

---

1. *my note*: possibly letter 3129. If so, Stokes's response to letter 3129 is missing.

From Auguste de La Rive 22 February 1870 3133

Geneva, February 22, 1870.

MY DEAR TYNDALL,[1]

I have read with great interest the analysis of your lecture[2] at the Royal Institution on the 21st of January, and admired the experiments by which you succeed in demonstrating that all the dust floating in the air of London is of an organic nature. Your conclusions as regards the probable cause of epidemic diseases remind me that I had announced the same consequences in an inquiry[3] in which, by a process different from yours, I sought to prove the presence of organic dust in atmospheric air. Permit me to state in a few words the results of this inquiry.

I had followed with care the labours of Pasteur, who, like you, had deprived the air of its organic matter, sometimes by passing it through a red-hot tube, and sometimes by the employment of cotton wool. I had been struck by the fact that the air on high mountains, and more particularly on the summits covered with snow, does not contain any such matter. I was thus led to believe that the remarkable transparency of atmospheric air when it is charged with humidity, both before and after rain, arises from the fact that the organic particles which float constantly in the air, at least over plains, become transparent by absorbing water, or fall to the earth on becoming heavier by the fact of that absorption. Finally, I found in the observations of De Saussure,[4] and those of Humboldt,[5] facts relative to the transparency of air which seemed to indicate that the clearness always exists in those places where organic matters are absent, or where the humidity annuls their influence.

These considerations led me to believe, as stated in the memoir communicated to the Academy of Sciences in Paris (*Comptes Rendus*, vol. lxiv. p. 1221, and *Ann. De Chim. et de Phys.* vol. xii. p. 243),[6] that it would be interesting to embrace the transparency of the atmosphere in the number of meteorological elements which are subjected to regular determination, and thus to connect it in a precise manner with the other elements, such as the pressure, the temperature, the degree of humidity, the direction of the wind, and, above all, the hour of the day, the period of the year, or the seasons. This class of observations, I added, possessed interest, not only for science properly so called, but perhaps also for medicine, *as regards the hygiene of epidemic diseases*. It is very probable indeed, I said, that miasmas, the hydrogenous nature of which was demonstrated in 1834 by M. Boussingault,[7] are due to these organic germs, whose presence would be thus established in a manner tolerably exact by the degree of transparency, more or less, of the air.

I expressed myself thus in 1867; but for the purpose of verifying my conclusions, I had previously sought to appreciate the variations of transparency by a photometer of a new construction, the description of which was given at the Meeting of the Société Helvétique[8] in Geneva in 1865. (See my discourse[9] on opening the Society, August 23, 1865, *Ann. de Chim. et de Phys.* vol. xxiv. p. 55.)[10] This instrument, the detailed description of which is to be found in the memoir above referred to, has since served me in my regular observations. It confirms the idea which I entertained regarding the influence on the transparency of the air of organic matter, which floats in greater or less quantity according to the hour of the day, the period of the year, &c., and the effect of which is more or less abolished by the humidity of the air.

I intend to publish immediately the result of these observations, which continue to be regularly made at the Observatory of Geneva. Perhaps the Editors of the Philosophical Magazine would be good enough to make them known. They might be preceded by a description of the instrument which I name an *atmospheric photometer*,[11] and by means of which the observations were made.

AUGUSTE DE LA RIVE.

A. de la Rive, 'On the Organic Dust of the Air', *Phil. Mag.*, 39 (1870), pp. 229–30

---

1. *My dear Tyndall*: this letter was published in the *Phil. Mag.* with an accompanying note from Tyndall. See letter 3134.

2. *analysis of your lecture*: 'On Dust and Disease' (see letter 3122, n. 1).
3. *an inquiry*: probably discussed in his papers: A. de la Rive, 'Note sur un photomètre destiné à mesurer la transparence de l'air', *Comptes Rendus*, 64 (1867), pp. 1221–5 ; and *Annal. Chim. et Phys.*, 12 (1867), pp. 243–9.
4. *De Saussure*: Horace Bénédict de Saussure (1740–99), a Swiss meteorologist, physicist, geologist, and mountaineer. De Saussure was an early promoter of Alpine exploration and popularized the term *geology* in his *Voyages dans les Alpes* (1779–96), a work which contains more than thirty years of geological research. He was particularly interested in the mechanism that led to the formation of the Alps, arguing that they were formed by horizontal thrusting in opposite directions. In 1783, he built the first hygrometer, a device used to measure humidity (*CDSB*; D. W. Freshfield, *The Life of Horace Benedict de Saussure* (London: Edward Arnold, 1920); *HLS*).
5. *Humboldt*: Alexander von Humboldt (1769–1859), the world-renowned Prussian naturalist and explorer, and by the 1850s one of the most notable scientific figures in Europe. From 1799 to 1804 Humboldt journeyed through South America with French botanist Aimé Bonpland (1773–1858), publishing thirty volumes upon his return. His meticulous attention to climate, temperature, and the ecology of his discoveries were made possible by the many scientific instruments that he took with him and he defined a new era of accuracy in science. His wide-ranging *Kosmos* appeared in five volumes between 1845 and 1862. Financial constraints stemming from the publication of his works forced him to spend the last thirty years of his life in Berlin working for the Prussian government (*CDSB*; *NDB*).
6. *the memoir . . . 243)*: see n. 3.
7. *Boussingault*: Jean-Baptiste Joseph Dieudonné Boussingault (1802–87), a French agricultural scientist and chemist also known for important contributions to agriculture research, metallurgy, and petroleum science. He published in seven volumes, *Agronomie, agricole, et Physiologie* from 1860 to 1884. In 1878, he received the Copley Medal from the RS for his contributions to agricultural science. His primary work in agricultural science was on nitrogen fixation and the chemistry of nitrification (*CDSB*; 'M. Boussinault', *Nature*, 36:919 (9 June 1887), pp. 134–5). See J.-B. Boussingault, 'Sur la possibilité de constater l'existence des miasmes', *Annal. Chim. et Phys.*, 57 (September 1834), pp. 148–61.
8. *Société Helvétique*: Société Helvétique des Sciences Naturelles (Swiss Society for Natural Sciences).
9. *my discourse*: A. de la Rive, 'Discours prononcé le 21 Aout 1865 à l'ouverture de la quarante-neuvième Session de la Sociéte Helvétque des Sciences Naturelles réunie à Genève (la question des glaciers en géologie)', *Archives Sci. Phys. Nat.*, 24 (1865), pp. 48–75.
10. *August 23 . . . 55*: de la Rive seems to have misremembered both the exact date of his discourse and the journal in which it was published. See n. 9.
11. *atmospheric photometer*: an apparatus for measuring the transparency of the atmosphere. It allowed for the simultaneous observation of two similar objects at different distances with one eye with the differences in brightness and colour between the two objects caused by the intervening air.

To the Editors of the *Phil. Mag.* 24 February 1870 3134

Gentlemen,

I received last night and translated rapidly this morning the following letter[1] from M. De la Rive, which I trust will reach you in time for your forthcoming Number. I may add that the current Number of Fraser's Magazine contains a brief historic summary[2] of this subject, to which I would refer the reader. I look forward with extreme interest to the publication of M. De la Rive's photometric observations.

John Tyndall. | Royal Institution, | February 24, 1870.

[. . .][3]

A. de la Rive, 'On the Organic Dust of the Air', *Phil. Mag.*, 39:260 (1870), pp. 229–30

1. *the following letter*: see letter 3133.
2. *a brief historic summary*: J. Tyndall, 'On Dust and Disease', *Fraser's Magazine*, 1 (1870), pp. 302–10.
3. *[. . .]*: letter 3133 inserted here.

To Auguste de la Rive [24 February 1870][1] 3135

Royal Institution, Thursday

My dear Friend

Your letter[2] reached me late last night. I rose early this morning—translated it and took it myself to the office of the Phil[osophical] Mag[azine]. The proof has already reached me and I have corrected it and sent it back: so that no delay has occurred.

Soret's letter[3] will appear at the same time.

Ever Yours truly | John Tyndall

RI MS JT/1/TYP/1/351
LT Typescript Only

1. *[24 February 1870]*: Tyndall submitted de la Rive's letter of 22 February (letter 3133) to the *Phil. Mag.* on 24 February (letter 3134) and sent this letter to de la Rive immediately after.
2. *Your letter*: letter 3133.
3. *Soret's letter*: letter 3119.

To John Lubbock 26 February [1870][1] 3136

26th. Feb[ruary].

My dear Lubbock

Let me add my small drop to the tide of congratulation[2] which must now be flowing in upon you.

Yours faithfully | John Tyndall

BL Add MS 49643, f. 59

1. *[1870]*: the year of this letter is presumably 1870 (it is written on the letter by another hand], since Tyndall is likely to be congratulating Lubbock on his recent election as MP for Maidstone. See n. 2.
2. *tide of congratulation*: Lubbock was elected MP for Maidstone on 25 February 1870. He served in Maidstone until 1880, then as MP for London University for twenty years (M. Patton, *Science, Politics and Business in the Work of Sir John Lubbock: A Man of Universal Mind* (Burlington, VT: Ashgate, 2013).

To Charles Rogers[1] 26 February 1870 3137

26th Feb[ruary] 1870

Sir,

I like the name of your club,[2] for it suggests mountain air, and also brings to my mind, the first scrap of dramatic poetry that ever I committed to memory. I also like ruins, but I am not in the least an antiquary. Would it be therefore right in me to accept an honour[3] to which I have no shadow of claim? I hardly think so.

Your obedient servant | John Tyndall | Charles Rogers LLD.

Wellcome Library MS. 7777/12

1. *Charles Rogers*: Charles Rogers (1825–90), a Church of Scotland minister and historian. Born in Dunino, Fife, Rogers attended the University of St Andrews and was licensed as a minister in 1846. Appointed a garrison chaplain in Stirling Castle (1855–63) he left Scotland due to accusations of financial irregularities in a project to build a monument to William Wallace. In November 1868 Rogers established the Grampian Club (which published his autobiography) and the Royal Historical Society.
2. *the name of your club*: the Grampian Club references the Grampian Mountains, an extensive mountain range in the Scottish Highlands.

3. *accept an honour*: Rogers offered Tyndall membership to the Grampian Club, the object of which was 'the editing and printing of works illustrative of Scottish literature, history, and antiquities'. See C. Rogers, *Leaves from My Autobiography* (London: Grampian Club, 1876), pp. 332–3.

From Joseph Plateau 1 March 1870 3138

Gand, 1er Mars 1870

Mon Cher Monsieur Tyndall,

J'ai eu l'honneur de vous adresser il y a près d'un an, des exemplaires de mes dernières séries pour vous, pour l'Institution Royale et pour la Société Royale; j'espère que vous les avez reçues. Dans ma lettre d'envoi, je vous disais que je ne voulais pas bouder la Société Royale, et que, malgré son silence à mon égard, je lui faisais cependant hommage de ces mémoires. Depuis, l'espoir que vous aviez bien voulu me donner a été encore déçu: de nouvelles nominations ont eu lieu, dans lesquelles je ne suis pas compris; *[aussi]* est-ce pour la dernière fois que je vous parle de cette affaire; si la Société Royale ne veut pas de moi, je me résignerai à son indifférence, comme je dois me résigner à bien autre chose. Je prends néanmoins la liberté de vous rappeler que j'appartiens depuis un grand nombre d'années à l'Institut de France, ainsi qu'à l'Institution Royale de Londres; j'ajouterai que, sans que j'aie fait à cet égard la moindre démarche, l'Académie de Berlin m'a admis récemment, à l'unanimité, au nombre de ses correspondants.

Agréez, Mon Cher Monsieur Tyndall, mes compliments affectueux, et l'assurance de ma haute considération.

J. Plateau | Professeur à l'Université | Place du Casino, *[15]* | Gand.

Ghent, 1 March 1870

My dear Mr Tyndall,

I was honoured to send copies of my latest series[1] almost a year ago to you, to the Royal Institution and to the Royal Society. I hope that you received them. In the accompanying letter,[2] I told you that I did not wish to sulk over the Royal Society, and that despite their snubbing me, I was still willing to send them those essays. Since then, the hope that you have kindly given me was again dashed: new appointments were made, in which I am not included. Thus, it is the last time that I will speak to you about this matter. If the Royal Society does not want me,[3] then I will resign myself to its indifference, as I must resign myself to many other things. I will nevertheless take the liberty to remind you that I have long been a member of the Institute of France,[4] as well as the Royal Institution of London.[5] I will add that without my having pursued them, the Academy of Berlin[6] has recently admitted me, unanimously, among its members.

Please accept, my dear Mr Tyndall, my affectionate compliments and the assurance of my highest regards.

J. Plateau | Professor at the University | *[15]* Place du Casino | Ghent.

RI MS JT/1/P/105

---

1. *my latest series*: see letter 2975, n. 2.
2. *the accompanying letter*: letter 2975.
3. *If the Royal Society does not want me*: after Tyndall's considerable efforts, Plateau was made ForMemRS in November 1870 (see letter 2975, n. 4).
4. *Institute of France*: the Institute de France is a French learned society which was founded in 1795 and is comprised of five academies including an Academy of Sciences which was formed in 1666.
5. *Royal Institution of London*: Plateau errs in the full name of the RI (of Great Britain).
6. *Academy of Berlin*: the Akademie der Wissenschaften zu Berlin (or Prussian Academy of Science) was founded in Berlin in 1700. Fredrick II reorganised the society attempting to rival the Academy in Paris and its activities and institutional infrastructure were expanded considerably throughout the eighteenth century.

From John Tyndall 2 March 1870 3139

The Lodge | Gorey | 2nd March 1870.

My dear Friend,

Your letter[1] with cheque for £30 enclosed I have just received, which, indeed, surprises me not a little, Your liberality and kindness surpasses, which truly is more than I could have expected, all I can say is I feel deeply grateful to you for this and all your former kindness.

I send you a receipt.

Emma,[2] I am happy to inform you, is now quite recovered[3] and out every fine day, and she desires me to tell you that "she never felt better in health, more happy and contented in her life, and that she remembers you continually in her prayers" This is the fact and I am happy to add that she is "growing in Grace, and in the Knowledge of the Lord Jesus Christ". She is continually searching the Scriptures and asking most profitable questions on them.

I have already, as you requested, that I should, supplied her with any money she requires, indeed she is exceedingly judicious in selecting and distributing her alms to poor creatures such as are in need.

You may rely on it this I should be most happy to do if you never reminded me.

All my family, I include Emma, kindly and cordially unite with me in kindest love and regards.

ever yours very sincerely | John Tyndall
Professor Tyndall
Kindly bear John[4] in your mind if you can do anything for him.

RI MS JT/1/TYP/10/3385
LT Typescript Only

1. *Your letter*: letter missing.
2. *Emma*: Emily Tyndall.
3. *quite recovered*: see letter 2987, n. 7.
4. *John:* son of John Tyndall of Gorey. Tyndall of Gorey had three sons. One was a surgeon and member of the Royal College of Surgeons of Ireland, while two were intended for mercantile professions. Tyndall (of the RI) had already been asked to help secure positions for the latter in 1863, when they spent a few weeks of the summer with him. See letter 2019, n. 2, and letter 2032, n. 10, *Tyndall Correspondence*, vol. 8.

From Emil du Bois-Reymond 6 March 1870 3140

Berlin 17 Victoria Str[asse], | March 6th, 1870.

My dear Tyndall,

I called at Sauerwald's[1] yesterday to look after your apparatus[2] and he told me that he would be ready to send them at the end of next week, (meaning next Saturday)

Many thanks for your endeavours to get my paper translated[3] in the Phil[osophical] Mag[azine].

I read your lecture on dust and disease[4] with great interest and I think your experiments are of very great importance in as much as they furnish us with a method of controling the presence of small bodies in the air. All the experiments of Pasteur's and of so many others on spontaneous generation ought to be repeated now with this control. I know of no observations in this line of which I cannot but suppose that they are known to you as well as to me. For instance Pouchet's Micrographie atmosphérique, in his Hétérogénie ou Traité de la génération spontanée Paris '59. p. 432–459;[5] Bennett's[6] paper: On the atmospheric germ theory and Origin of Infusoria. Edinburgh 1868.[7] Both Pouchet and Bennett in their experiments have extensively used the method of washing the air with sulfuric acid or potassa which by your experiments is now shown to be utterly worthless.

You will be extremely sorry to learn that our friend Magnus[8] is very dangerously ill,[9] so that there is only a faint hope of his recovery. He is affected of a disease of the intestinal tube, the nature of which is not perfectly clear, but very suspicious, and he suffers fearfully: you would hardly to be able to recognise him.

As you have sent Francis[10] the copy of my paper on the aperiodic motion[11] etc. I send you another copy together with this letter.

Goodbye and believe me | yours faithfully | E du Bois-Reymond

RI MS JT/1/D/149
RI MS JT/1/TYP/7/2438

1. *Sauerwald*: Ferdinand Sauerwald.
2. *your apparatus*: see letters 3028, 3061, 3104, and 3148.
3. *my paper translated*: see letter 3128.
4. *your lecture on dust and disease*: see letter 3122, n. 1.
5. *Pouchet . . . 459*: F.-A. Pouchet, *Hétérogenie, ou Traité de la génération spontanée* (Paris: J. B. Baillière et Fils, 1859).
6. *Bennet*: John Hughes Bennett (1812–75), an English physician and physiologist. A prolific writer, he published 105 papers, and a *Textbook of Physiology* (1871). He was well known for his skills as a lecturer and his instructional methods were modelled at other teaching institutions. He is credited for encouraging the use of the microscope in the clinical investigation of disease, the reintroduction of the use of cod-liver oil as a beneficial therapeutic agent, and for putting an end to the common practice of bloodletting (*ODNB*).
7. *On the atmospheric . . . 1868*: J. H. Bennett, *On the Atmospheric Germ Theory and the Origin of Infusoria* (Edinburgh: Adam & Charles Black, 1868).
8. *our friend Magnus*: Gustav Magnus.
9. *dangerously ill*: Magnus passed away 4 April 1870. See letter 3174.
10. *Francis*: William Francis.
11. *my paper on the aperiodic motion*: see letter 3128, n. 5.

To [Juliet Pollock][1] 7 March 1870 3141

The "sweet South" presents his Compliments to the "bank of violets"[2] and regrets exceedingly that on Thursday he has to breathe to the west of Montagu Square.[3] | 7th March 1870

RI MS JT/1/T/1151
RI MS JT/1/TYP/6/2116

1. *[Juliet Pollock]*: in this letter Tyndall states that he can't visit Montagu Square, which was the home of the Pollocks. That and the tone of the letter indicate that it was sent to Juliet Pollock.

2. *"sweet South" . . ."bank of violets"*: allusion to 'O, it came o'er my ear like the sweet sound | That breathes upon a bank of violets', from Shakespeare, *Twelfth Night*, Act 1, Scene 1, Lines 5–6.
3. *Montagu Square*: the Pollocks' city home was at Montagu Square in Marylebone, a fashionable London neighbourhood.

To Elizabeth Dawson Steuart 7 March [1870][1] 3142

Royal Institution. | 7th. March

My dear Mrs Steuart,

I have just time to send you this horrible portrait[2] by this post. I was very glad to get your letter[3] I thank you for your offer of hospitality. I wish I could accept it. But alas in April and May and part of June I shall be bound by chains of iron to London.

Your ever | John Tyndall.

RI MS JT/1/TYP/10/3386
LT Typescript Only

1. *[1870]*: the letter has been dated with reference to letter 3144.
2. *this horrible portrait*: see letter 3144. Portrait missing.
3. *your letter*: letter missing.

From George Wynne 7 March 1870 3143

7 Mountjoy Square | Dublin | 7 March 70

My Dear Tyndall

Have you time to write me a short line—Frank[1] is in Constantinople carrying on some work for Mr Siemens.[2] I do not exactly know what it is, but he has written to me for some books one of which Percy's Metallurgy[3] is not yet published tho' advertised—I enclose you the advertisement[4] of another book on the same subject and I want you to tell me if you know anything about it and whether it is one that you would recommend that I send out to him, the 1st. vol. contains the subject which he requires.

Harry and Edward[5] are in India so that Lucy[6] and I are all alone, I am very near my promotion to M. General[7] and when I get that step my military career will be closed and I shall be free to go where I please—I suppose you are very busy and working your brain as usual, I should like to know the special subject you are engaged on—

If Lucy were in I am sure she would like to be particularly remembered to you—farewell for the present

from your old friend | Most truly y[ou]rs | Geo[rge] Wynne

RI MS JT/1/W/97
RI MS JT/1/TYP/5/1863

---

1. *Frank*: Frank Wynne.
2. *Mr Siemens*: Charles William Siemens (formerly Karl Wilhelm) (1823–83), was a German-British electrical engineer and metallurgist. His company advised the British government on marine telegraph cables. Siemens served as president of several professional societies, including the Society of Telegraph Engineers (1871 and 1878), Institution of Mechanical Engineers (1872), and BAAS (1882). He received several medals and foreign orders and was knighted seven months before his death. His collected works, in three volumes, were edited by E. F. Bamber in 1889 (*ODNB*).
3. *Percy's Metallurgy:* J. Percy, *Metallurgy: the art of extracting metals from their ores, and adapting them to various purposes of manufacture* (London: J. Murray, 1861–80). Vol. 3 on 'Lead, including Desilvarization and Cuppellation' was published in 1870; vol. 4 on 'Gold, Silver, and Mercury' in 1871. John Percy (1817–89), a metallurgist and professor at the Royal School of Mines (*ONDB*).
4. *I [enclose] you the advertisement*: enclosure missing.
5. *Harry and Edward*: Henry and Edward Wynne.
6. *Lucy*: Lucy Wynne.
7. *M. General*: Wynne was promoted to Major General, the third highest rank in the British Army, on 4 June 1870, with the promotion antedated to 6 March 1868 (T. Walker, *Bulletins and other State Intelligence for the Year 1870 in Two Parts, part II, July to December* (London: London Gazette Office, 1871), p. 625).

To Elizabeth Dawson Steuart [8 March 1870][1] 3144

My dear Mrs Steuart,

You asked me[2] after Emma:[3] the enclosed letter[4] will reply for me. The accounts of her are exceedingly gratifying.

Yours ever | John Tyndall.

I chanced to have a copy of the bad portrait by me I send it to you at once. It has been in my drawer since it was executed so I do not know whether it has suffered If so please tell me. I did not open it.

RI MS JT/1/TYP/10/3387
LT Typescript Only

1. *[8 March 1870]*: this letter has been dated through reference to letter 3139 (2 March 1870) and letter 3142 (7 March 1870).
2. *you asked me*: letter missing.
3. *Emma*: Emily Tyndall.
4. *enclosed letter*: letter 3139.

To George Gabriel Stokes 10 March 1870 3145

10th March 1870

My dear Stokes

Sir Edward Sabine has had the kindness to show me one of the thermometers referred to in your last note[1] to me.* It is a very beautiful differential instrument, one of whose bulbs is bare, the other being placed in the centre of a hollow silver sphere. The space between the globe and its envelope was a vacuum. Thus the two halves, joined together along a great circle resembled in some degree two little Magdeburg hemispheres. At the time that the thermometer was employed (now fifty years ago) for experiments on radiation a similarly silvered bulb was used in determining the temperature of the air.

Now what I would propose is this. Do away with the space between the envelope and bulb and bring the silvering down upon the bulb itself. This would enable the bulb to take up rapidly the temperature of the air and be at the same time a powerful shield against gain or loss by radiation. I think I proposed this some years ago, and the expedient is so obvious that it has probably been tried though I have not heard of any such trial.

Yours faithfully | John Tyndall

*And employed fifty years ago.

RI MS JT/1/T/1401
RI MS JT/1/TYP/4/1479

1. *your last note*: letter missing. Tyndall and Stokes are having an ongoing conversation about experimental thermometers. See letter 3129.

To Rudolf Clausius 11 March 1870 3146

London 11th. March 1870

Liebster Clausius.[1]

Warum schreibst Du mir nicht?[2] Aber ich veilleicht bin der Schuldner.[3] When I saw you last in Bonn[4] I was in very bad health, and I continued so for a good while afterwards. Nevertheless it was a great pleasure to me to see M^rs^. Clausius[5] and yourself and the children. I have not yet forgotten their bright faces, and I feel proud in having Johnny for my godson.[6] I went through Marburg[7] and the Hartz[8] to Berlin, and then through Halle[9] to Heidelberg.[10] Afterwards I called upon Wiedemann[11] at Carslruhe, and then went on to Switzerland. Ich war aber stark angegriffen[12] and I was many weeks in Switzerland before I became any thing like well. And a great misfortune occurred to [me].[13]

I had recovered an extraordinary amount of strength when one day at the Bel Alp[14] I went to have a shower bath under a cascade—it was my daily custom to do so. But this day in returning to my clothes which I had placed at a distance of about twenty or thirty yards from the cascade, in crossing small blocks of granite my foot slipped, and I fell with all my weight,[15] bringing my shin into collision with the sharp rock. Three wounds passed through to the bone, and the parts around were much bruised. Well I sponged the blood away, tied a cold water bandage round the wound and went to bed. In four or five days every thing looked so well that I abandoned my cold water bandage and limped down stairs. I inflamed the wound, and then pus (Eiter)[16] began to appear. This I did not like, and tried various unwise means to remove it. But after nearly a fortnight spent in bed the leg and foot were greatly inflamed, erysipelas had set in and I saw clearly that I might very easily lose my leg, if not my life. So I went to Geneva; and there was well cared for. But even under the best surgical attendance I lay in bed nearly a month in Geneva. When I came home I was very weak. In fact this accident cost me three-months of work; and altogether fully five months were taken out my time last year.

I have now got over all my ailments, and indeed feel in better health than I have felt for a long time at this season of the year. Curiously enough Du Bois[17] had an accident[18] in the Engadin[19] about the same time. And he has not yet quite recovered from the effects of it.

I have very often thought of writing to you, but as you may imagine I have had to work hard to make up for lost time. How are you getting on with your house—how is M^rs^. Clausius, how are the children?[20] Write to me and tell me all.

I will send you in a day or two a little paper[21] which will tell you what I have been doing of late.

Yours ever | John Tyndall

Have your read Maxwell's[22] paper on the Electromagnetic Field,[23] where he shows the velocity of light & of magnetic propagation to be approximately the same. (Phil. Trans. Dec 1864) What do you think of it? I wish he had given some physical image of the field.[24]

RI MS JT/1/T/204

---

1. *Liebster Clausius*: Dearest Clausius (original in German).
2. *Warum schreibst Du mir nicht?*: Why do you not write to me? (original in German).
3. *Aber ich veilleicht bin der Schuldner*: But perhaps I am the guilty party (original in incorrect German).
4. *When I saw you last in Bonn*: Tyndall and Hirst visited Clausius during their European travels in the summer of 1869. They met with Clausius on 7 July. See letter 3020.
5. *M[rs]. Clausius*: Adelheid Clausius.
6. *Johnny for my godson:* Tyndall was delighted when he was asked to be the godfather of Clausius' son, Rudolf John Clausius, who was named after him. He took his responsibilities seriously. See letter 2147, *Tyndall Correspondence*, vol. 8.
7. *Marburg*: see letter 3018, n. 3.
8. *Harz*: see letter 3022, n. 10.
9. *Halle*: see letter 3023, n. 12.
10. *Heidelberg*: see letter 3023, n. 24.
11. *Wiedemann*: Gustav Heinrich Wiedemann (see letter 2960, n. 13).
12. *Ich war aber stark angegriffen*: I was however quite ill (original in German).
13. *[me]*: Tyndall seems to have forgotten a word, as he began a new page of the letter.
14. *Bel Alp:* the Belalp (see letter 3030, n. 20).
15. *I fell with all my weight*: Tyndall spent the autumn of 1869 recovering from this injury. See letters 3030, 3031, 3032, 3037, 3041, 3048, 3049, and 3060.
16. *Eiter*: pus (original in German).
17. *DuBois*: Emil du Bois-Reymond.
18. *had an accident*: see letter 3104.
19. *Engadin*: a region in the eastern Swiss Alps.
20. *children*: at this time, Rudolf and Adelheid Clausius had five children: Mathilde (1861–1907), Helene (1863–1919), Rudolf John (1864–1915?), Hedwig (1866–1946), and Alfred (1868–1950). A sixth child was born two years later, Else (1872–1954).
21. *a little paper*: probably 'On the Action of Rays of High Refrangibility upon Gaseous Matter'. See letter 3125, n. 3.
22. *Maxwell*: James Clerk Maxwell.
23. *paper on the Electromagnetic Field*: J. C. Maxwell, 'A Dynamical Theory of the Electromagnetic Field', *Roy. Soc. Proc.*, 13 (1864), pp. 531–6.
24. *I wish he had given some physical image of the field*: Tyndall expressed this desire for a visual representation of Maxwell's mathematically complex theory in letter 3147 as well and points to his more matter-based approach to physics, in contrast to Maxwell's conceptual one.

To Hermann Helmholtz 15 March 1870 3147

Royal Institution | 15th. March 1870.

My dear Helmholtz,

Your last little paper[1] has reached me, and I thank you for it.

I sent you a day or two ago[2] a scrap written on a subject on which you have worked and in which your name is mentioned.

I wish you or Clerk Maxwell,[3] or somebody with the requisite force of imagination would give the world some physical image of an electric current.[4] Without some such image there is a certain emptiness[5] in that remarkable paper of Maxwell's on the Electromagnetic Field.[6]

When I saw you last I was far from well. I remained ill for nearly three weeks in Switzerland, but then shook all my ailments away and became very strong. Then however I had a fall[7]—I had been bathing and was returning naked to my clothes when in crossing some blocks of granite I slipped and stamped the sharp crystals into my shin. I lay quiet for four or five days and was going on exceedingly well, but unfortunately cast my cold bandage away too soon, and my leg became fearfully inflamed. I was carried down in a Sessel[8] from the Bel Alp[9]—and went on to Geneva, where I had good surgical care. But I was 6 weeks in bed and three months were necessary to make me well again.

Curiously enough Du Bois[10] had a severe fall in the Engadin[11] about the same time.

I was well pleased to see the way in which the Academy of Science did its duty.[12] My only sorrow was that Joule,[13] who so eminently deserves the honour, was not among those elected.

Give my kind regards to Mrs Helmholtz[14]—to Bunsen and to Kirchhoff.

Yours ever faithfully | John Tyndall.

RI MS JT/5/15b/189
LT Typescript Only

---

1. *Your last little paper*: possibly H. Helmholtz, 'Ueber die Bewegungsgleichungen der Elektricität für ruhende leitende Körpe', *Crelles Journal der Mathematik*, 72 (1870), pp. 57–129.
2. *I sent you a day or two ago*: letter missing.
3. *Clerk Maxwell*: James Clerk Maxwell.
4. *some physical image of an electric current*: Tyndall expressed the same desire to Clausius in letter 3146.

5. *Without some such image there is a certain emptiness*: see letter 3146, n. 24.
6. *Maxwell's on the Electromagnetic Field*: see letter 3146, n. 23.
7. *I had a fall*: Tyndall recounts for Helmholtz the story of his injury in September 1869. See letters 3030, 3031, 3032, 3037, 3041, 3048, 3049, and 3060.
8. *Sessel*: large chair. In this case, probably Sedan chair (original German).
9. *Bel Alp*: the Belalp (see letter 3030, n. 20).
10. *DuBois*: Emil du Bois-Reymond.
11. *Engadin*: see letter 3146, n. 19.
12. *the Academy of Science did its duty*: in January 1870 Robert Mayer was elected corresponding member of the Parisian Academy of Science. Joule was elected shortly after, in May of the same year.
13. *Joule*: James Joule.
14. *Mrs Helmholtz*: Anna Helmholtz.

From Ferdinand Sauerwald 15 March 1870 3148

Berlin 15 März 1870

Verehrter Herr Professor!

Hierdurch erlaube ich mir Ihnen anzuzeigen, daß ich heut eine Kiste H.P.T N° 606 London an Sie abgesandt. Verschiedene dringende Aufträge vom vorigen Jahr, haben verursacht daß dieser Apparat jetzt erst fertig geworden, und bitte deshalb recht sehr um Entschuldigung. Herr Prof. du Bois hatte mir ein Rollenpaar zur Boussole bestellt, ich habe den Draht ausnahmsweise für Sie mit weisser Seide bespinnen lassen, obgleich ich jetzt trotzdem mit grüner Seide (des eisenfreien Kupfers wegen) keine freiwillige Ablenkung des Nadelpaars bei der grösten Astasie bemerken kann, und schon seit langer Zeit keine Corrections Vorrichtung an den Galvanometern anbringe. Auf jeder Rolle sind 2 Mal 50 Windungen zugleich nebeneinander gewunden. Anzahl Fuß und Grammes sind ebenfalls angegeben. Ausser dem Magnet mit drehbarem Glasplanspiegel, in allen Richtungen zu beobachten legte ich noch 2 magnetische Stahlspiegel nach der alten Manier zu beobachten bei, von denen der dünne allerdings der grossen Schwierigkeit des Planmachens wegen, nicht ganz so vollkommen ist. Beiläufig will ich bemerken daß die Spiegel gut magnetisirt sind, bei etwaiger späteren Magnetisirung darauf zu achten ist daß nicht zu große Streichmagnete angewandt werden, da dadurch die Spiegel sich verziehen und total unbrauchbar werden. Der Astasirungsstab wird unter das Boussolenbrett mittelst der 4 beiliegender Schrauben befestigt und durchs Consol ein Loch gebohrt, den Holzstab mit Theilung durchzulassen. Das Fernrohr (Steinheil'sches) wird so in sein Lager gelegt und befestigt, daß das Objectiv in der Ebene des getheilten Lineals kommt, das Unterbrett wird mit 3 Flügelschrauben, (welche gezeichnet sind) auf den

Statiftisch geschraubt. Die Befestigung der Spiegel ergiebt sich von selbst. Ich erlaube mir nun hier zugleich die Rechnung einzulegen und um gütige Einsendung des Betrags o.Z. zu bitten.

Hochachtungsvoll und ergebenst | F. Sauerwald

Mechaniker, Kanonier St 43.

P.S. Die 3 Fußschrauben von der Boussole sind des besseren Transports wegen, herausgeschraubt und mit Zahlen gezeichnet, ein Coconfaden ist im Glasrohr und 2 sind zur Reserve bei einem Stahlspiegel eingelegt. Die Magnete liegen zerstreut in der Kiste. | FS.

Berlin 15 March 1870

Dear Professor!

I take the liberty of informing you that I have today sent you a crate[1] H.P.T N° 606 London.[2] Due to a number of urgent orders from last year the apparatus has just now been completed, and offer sincere apologies. Prof. Du Bois had ordered a pair of coils for the magnetic compass, I made an exception for you and had the wire wound in white silk, although I have yet to notice when using green silk (due to the iron-free copper), any voluntary deflection of the needle pair at the point of maximum astasis, and have not attached any corrective device to the galvanometers for quite a while now. Each coil has 2 times 50 turns, wound next to one another. The numbers are given likewise in feet and grammes. Beside the magnet with rotatory plane glass mirror, I included 2 additional magnetic steel mirrors for observations in all directions in the old way. Of those, the thin one is not quite perfect given the great difficulties in smoothing it. Incidentally, I want to mention that the mirrors are well-magnetised and for later magnetisation one must be take care not to use too large magnetic sticks, as the mirrors can distort and become completely useless. The rod used to create astasis is fixed underneath the board of the magnetic compass using the 4 accompanying screws, and a hole must be drilled through the console in order to introduce the wooden bar with the subdivisions. The telescope (after Steinheil[3]) can thereby be positioned and fixed so that the objective falls in the plane of the subdivided straight edge, the base board is attached to the mount using 3 wing screws (which are marked). The mirrors will then fall into place on their own. I take the liberty of including here the invoice and ask you kindly to send the amount without fee.

Respectfully and truly | F. Sauerwald

Mechanic, Kanonier Str. 43.

P.S. For the purpose transportation, the 3 screws for the base of the magnetic compass were unscrewed and marked with numbers, a cocoon thread is in the glass tube and 2 more for replacement were included with one of the steel mirrors. The magnets are all scattered in the box. | FS.

RI MS JT/1/S/55

1. *sent you a crate*: see the earlier correpondence between Tyndall and Emil du Bois-Reymond on these instruments in letters 3028, 3061, 3104, and 3140.
2. *H.P.T N° 606 London*: H.P.T. was British telegraph code for 'post office'. Sauerwald is shipping the crate to a London post office.
3. *Steinheil*: see letter 3104, n. 2.

From Cornil Woestyn[1] 16 March 1870 3149

Société de la Médaille D'Or | Orlovetz—District de Tcherkasse | Gouvernement de Kieff | Russie | Épuration des Produits Sucrés | Sans Emploi de Sang
16 Mars 1870 | Paris | 98 Rue Mirosménil

Monsieur John Tyndall

La lecture de vos intéressants travaux sur les germes contenus dans l'air m'a suggéré l'idée d'en faire l'application à la destruction des miasmes contagieux qui rendent si souvent mortelle l'atmosphère des hôpitaux.

Vos belles expériences ont montré qu'un courant d'une certaine vitesse à travers une flamme suffit pour carboniser ces particules organiques, et qu'en diminuant encore la vitesse de ce courant on peut obtenir leur destruction complète par la transformation de ce charbon en gaz carbonique.

Or, pour résoudre la question dont j'ai abordé la solution il n'y a même pas besoin de carboniser les particules pour paralyser leur action; il faut une température relativement beaucoup plus faible pour arrêter l'action de tous les ferments connus—J'ajouterai encore que les procédés usités pour la conservation des substances alimentaires en vase clos reposent sur le fait qu'une température de 100 degrés suffit pour empêcher les germes de l'air d'engendrer la putréfaction.

Telles sont, Monsieur, les considérations que j'ose soumettre à votre appréciation si compétente en pareille matière; votre approbation serait pour moi un honneur que j'ose à peine espérer.

Agréez, Monsieur, les salutations empressées de votre tout dévoué serviteur | Cornill Woestyn

Gold Medal Society[2] |
16 March 1870 | Paris | 98 Rue Mirosménil

Mr John Tyndall,

Reading your interesting work[3] on the germs present in the air brought to my mind the idea of putting this into practice to eliminate the contagious miasma that so often renders the atmosphere of hospitals fatal.

Your fine experiments demonstrated that a current of air *[travelling at]* a certain speed through a flame is sufficient to carbonize these organic particles, and that by further reducing the speed of this current, it is possible to destroy them completely through the transformation of this coal into carbon dioxide.

Yet, to solve the question whose solution I broached, carbonizing the particles to paralyse their movements is not even required; a comparatively much lower temperature is required in order to stop the action of all known catalysts. I will also add that the processes used for the preservation of food substances in airtight jars rest upon the fact that a temperature of 100 degrees is sufficient to prevent germs in the air from leading to putrefaction.

Such are, Sir, the thoughts that I dare submit for your consideration, given your expertise in such matters; your approval would be an honour that I hardly dare to hope for.

Please accept, Sir, the eager greetings of your much devoted servant | Cornill Woestyn

RI MS JT/1/W/1

---

1. *Cornill Woestyn*: Cornill [Cornell] Alphonse Woestyn (dates unknown) studied at the École Polytechnique and became director of a sugar factory in Russia. His research on methods of treating sugar and molasses can be found in the first three volumes of *The Sugar Cane: A Monthly Magazine Devoted to the Interests of the Sugar Cane Industry* (1869–71) (J. R. Partington, *History of Chemistry,* 4 vols (London: MacMillan, 1964), vol. 4, p. 202).
2. *Gold Medal Society*: Gold Medal Society | Orlovetz—Cherkessia District | Government of Kiev | Russia | Refining of Sugar Products | Blood-free process. Gold Medal Society headed notepaper. Blood was used in the refining process to clean the impurities in raw sugar. Mixed with the sugar, the blood was boiled to coagulation, and, would have carried all foreign matters, to settle in its final phase separated state floating atop the sugar.
3. *your interesting work*: Woestyn may have read about Tyndall's discourse 'On Dust and Disease' of 21 January 1870 in the newspaper (see letter 3122, n. 1) or been familiar with his paper 'On Haze and Dust', *Nature,* 1 (27 January 1870), pp. 339–42. See letters 3121, 3122, n. 1, 3126, 3127, and 3128.

From Lucy Wynne 17 March [1870][1] 3150

Mar[ch] 17 | 7 Mountjoy Square

My dear M^r^. Tyndall

I must write you a few lines in Papa's[2] letter[3] to tell you how extremely sorry I am to hear of your having been so ill, & to beg you to try & force yourself to take a little <u>much</u>-needed repose—you are just overworking yourself dear M^r^. Tyndall & we do wish you would give your brain a little necessary relaxation & rest—you ought <u>now</u> to go to Italy for a few weeks & come in the Summer & pay us a visit in our little fishing Lodge in Connemara[4]—do try & do this it would be real pleasure to see you there. We shall be there in July & give you such a cordial welcome.

Did you see Harry's wife?[5] We have such pleasant bright letters f[ro]^{m} Harry & her f[ro]$^{\underline{m}}$ Calcutta. India so far agrees well with her & they expect to move to the hills in April.

Frank[6] writes us most amusing letters f[ro]$^{\underline{m}}$ Constantinople he makes himself very happy wherever he goes. Edward[7] too enjoys India & is now quite well.

We are in the midst of packing up to leave this house. Our time is so nearly up here in Ireland that we think it prudent to sell the house while we can. Now dear M^{r}. Tyndall try & follow out my advice & give yourself Entire rest for a few weeks. You have been ruling yourself with work.

I am | y[ou]^{r} aff[ectiona]^{te} friend | Lucy H Wynne

RI MS JT/1/W/110

1. *17 March [1870]*: this letter was enclosed with letter 3157 from her father, George Wynne, which is dated 25 March 1870. It is unknown why there is a week between Lucy's letter and her father's.
2. *Papa*: George Wynne.
3. *Papa's letter*: letter 3157.
4. *Connemara*: a region in the West of Ireland.
5. *Harry's wife*: the wife of Lucy Wynne's brother, Henry Wynne, has not been identified.
6. *Frank*: Frank Wynne.
7. *Edward*: Edward Wynne.

To Josiah Dwight Whitney[1] 18 March 1870 3151

Royal Institution of Great Britain | 18th March 1870

My dear Sir,

I am truly obliged to you for your kindness in writing to me as you have written. Indeed it is a pleasure greater than I can well describe to find such a generally warm feeling towards me and my work prevalent in your country. I shall thank you for the photograph of Mount Tyndall[2] should it ever come; but these thanks can hardly be warmer than those I now return you for your letter.

Perhaps some of this feeling arises from the devotion to science which your letter[3] reveals. It must be a deep true love that prompted you to make the sacrifices that you have made, and which causes you to contemplate still further sacrifices. The public, both in America and England, is an inert mass, and it is only by effort long continued that we can hope to influence it. I wish you good speed, or better still a steadfast heart to follow up your work.

Were it not that I am so entangled in London I should make an attempt to see those Californian wilds myself. Your papers[4] about them are full of interest.

The guide book has not yet come to hand,[5] but no doubt it will reach me in good time.

I have been deflected for the moment from my usual course of investigation to consider the question of "Spontaneous generation". As regards the evidence my mind is already made up. So I shall make a few experiments & then have done with the subject.

Believe me my dear Sir | Most faithfully yours | John Tyndall

Yale MS 555, Box 15, 1–4

---

1. *Josiah Dwight Whitney*: Josiah Dwight Whitney (1819–96), an American geologist. Whitney was inspired to become a geologist after hearing a lecture by Charles Lyell. After working as a surveyor and mining consultant, in 1855 he became professor of geology at Iowa State University. In 1860 he was appointed state geologist in California and led its geological survey, expanding it to include botanical, zoological, geographical, and palaeontological data. Mount Whitney and the Whitney glacier are named after him (*ANB*; *CDSB*).
2. *Mount Tyndall*: a peak in the Sierra Nevada near Mount Whitney (the highest peak). Part of the California Geological Survey's explorations in the Sierra Nevada included an ascent and measurement of a peak (on 6 July 1864) that Clarence King (1842–1901), an American geologist who would later go on to help establish the U.S. Geological Survey in 1879 (*CDSB*), named Mount Tyndall (it rises to just over 14,000 feet). This peak is part of the Mount Whitney range, named after Josiah Dwight Whitney. King recounts the ascent and descent of Mount Tyndall in chapters 3 and 4 in his *Mountaineering in the Sierra Nevada* (Boston: James Osgood & Co., 1872), pp. 49–93.
3. *your letter*: letter missing.
4. *your papers*: Whitney published numerous papers on fossils and geology in the late 1860s. The paper that would have interested Tyndall the most was likely J. D. Whitney, 'Notice of Explorations in the Rocky Mountains', *Proceedings of the California Academy of Science*, 4 (1868), pp. 90–1 and perhaps also J. D. Whitney, 'Remarks on the Mineral Species Occurring in California', *Proceedings of the California Academy of Sciences*, 3 (1868), pp. 741–6.
5. *Guide book has not yet come to hand*: J. D. Whitney, *The Yosemite Guide-Book, a Description of the Yosemite Valley and the Adjacent Region of the Sierra Nevada, and of the Big Trees of California* (Geological Survey of California, 1870). See letter 3228.

From John Lucas Tupper[1] 18 March [1870][2] 3152

On an Optical Illusion. By J. L. TUPPER*. | *To Professor Tyndall, F.R.S.*

DEAR SIR,

AS the following fallacious "Solution"[3] that appeared in No. 251 of 'The English Mechanic' has not found correction, and as there are points of real interest connected with a true solution of this not new, but, so far as I know, yet unexplained optical illusion, you will perhaps favour me with your attention to the question.

"[898.] *Optical Illusion.*—I beg to forward the following solution:—It is known that if an object be placed at the focus F (fig.1) of a converging glass, the rays will be so refracted by the glass that, after going out of it, they will become parallel; if the object be *further* than the focus (fig. 2), the rays will meet in a point after they have passed the lens; and if *nearer*, they will diverge (fig. 3); hence results that the rays of light proceeding

Fig. 1.

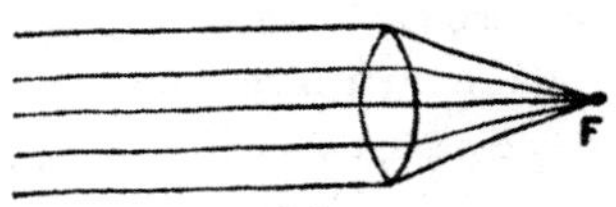

Fig. 2.

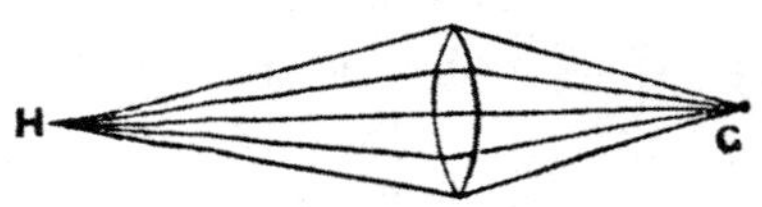

Fig. 3.

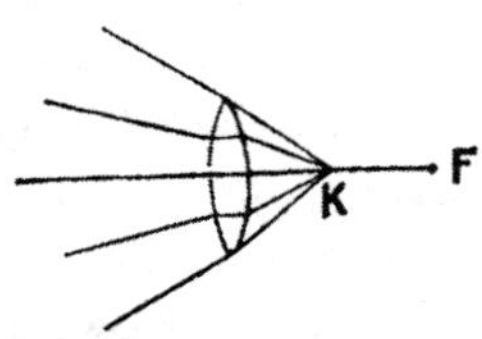

from certain objects, refracted through the crystalline of the eye, must converge to a point, or there would be no distinct picture of them on the retina. An object, to become visible, must then be placed further than the principal focus of the eye; if an object be placed between that focus and the eye, it becomes directly invisible (*p. ex.* the eyelashes); but that object may intercept the rays emitted by another visible object situated further—in a word, it may act as a *screen*. These particulars understood, let A (fig. 4) be the eye, M N a white illuminated surface, P Q the

Fig. 4.

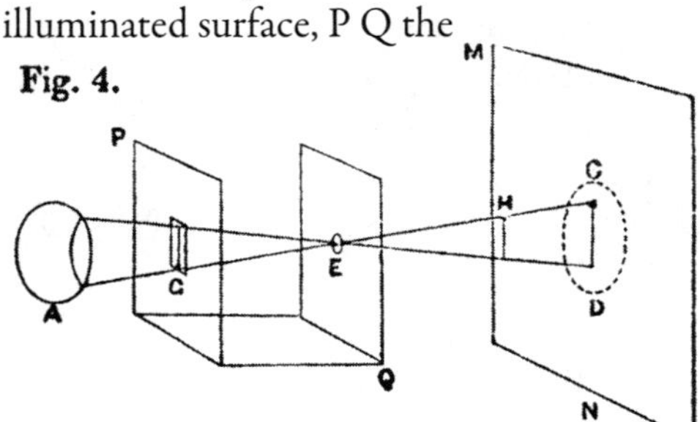

bent card. It is easy to understand that of all the white rays emitted by the surface M N, those alone that are situated on the circle C D can enter into the eye, crossing at the aperture E. If, now, an object be placed on the cone E C D, this object, acting as screen, will intercept some rays of the circle C D, and the eye will perceive in black the image of this object H on the circle. The same must take place when the object is in G, provided it be reversed; then, as all the rays cross at E, the object G will intercept the same rays as H (supposed E G—E H), with that difference, that at G the object, being directly invisible, can only act as a screen, while at H it could at the same time act as a screen for the surface M N and be visible, if it is in the conditions of distance above mentioned. This will explain, then, why the object G is projected in M N in an inverse position.—

MICHAEL."[4]

Here the writer explains the erect appearance of the inverted pin G by identifying it with the erect appearance of the erect pin H, as seen by crossing rays—quite forgetting that, if objects ordinarily appear in their true position, a special arrangement which inverts them before they reach the eye will make them appear reversed, that H therefore ought to appear reversed; and that "why it does not" demands explanation as much as "why G does." He has, by employing two cones of rays, brought the pin on to the lens in an inverted position, and yet expects that we should see it erect, in defiance of hourly experience, that just as objects stand upon the lens so they are seen. What has to be shown is, how the position of an object can appear the reverse of what it is upon the lens: this he has not attempted. Another singular fallacy is where he postulates G too near for vision (because its rays do not converge upon the retina), and then supposes H (because of proper seeing-distance) made visible by these self-same rays. If ray A G cannot do the work of vision for G because of non-convergence on the retina, how can it do it for H? His second cone is useless unless we suppose a sensorium at E.

Again, he supposes the light which comes through the hole E reflected from a surface M N beyond the hole, and hence derives his second cone and his means of solution. But it is not necessary to the success of the illusion that light proceed from any such surface, or from any distance beyond the hole; neither is such hole required, any radiant point (such as a spark) being in all respects sufficient. Thus the basis of solution is cut away.

I will now explain the phenomenon by simply following the rays to the retina, first alluding to normal vision, fig. 5, where,

Fig. 5.

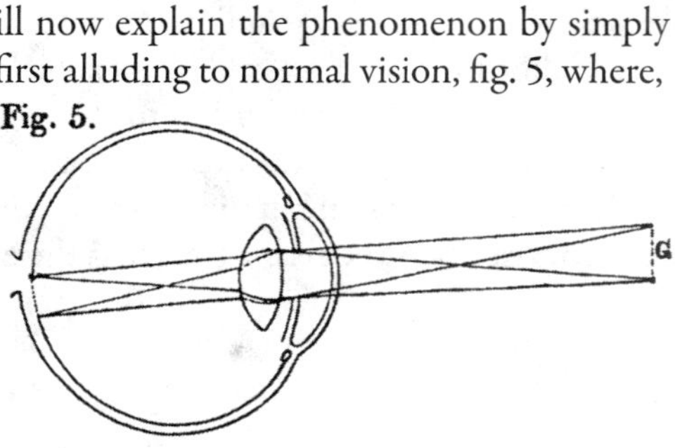

the eye seeing the pin G under ordinary conditions, every point in the object delivers its cone of rays (whose base is just within the pupil) to the crystalline, whence, after refraction, the rays pass in another cone whose apex is in the retina (only two cones, depicting the point and head of the pin, are figured). Here the impress of the object on the retina is not only the reverse of the object's real position (or of its position on the lens), but it is moreover the strongest retinal impress that the object can produce,

Fig. 6.

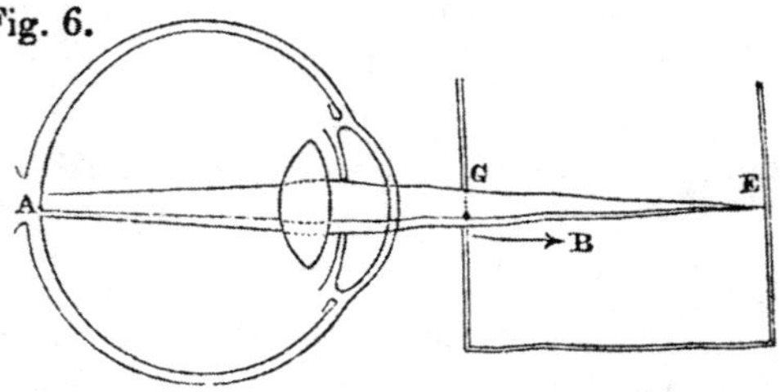

its *every point* being depicted by the convergence of an entire cone of rays. Now (fig. 6) trace the rays as, under the conditions of the experiment, they pass in a cone from the radiant point E, where they surround the pin, and proceed to the lens, whence, after refraction, they slightly converge and end on the retina in a circle of light containing a dark space corresponding to a section of the pin. But observe here, the impress is not erect, as in fig. 5; neither is every point in its outline depicted by the convergence of a cone of rays (as in fig. 5), but, on the contrary, it is obvious that no more than one ray can go to the depicting of any one point of this impress, which is consequently the faintest that can be produced.

Now, whilst it is manifest that two diametrically opposed positions of the pin on the sensitive retina will produce perceptions of diametrically opposite position, the erect pin on retina fig. 5 giving perception of an inverted pin, and the inverted pin on retina fig. 6 giving perception of an erect pin (and so explaining the illusion), it is equally clear that in fig. 5 there will be the same faint impact of rays on the retina that we find in fig. 6, seeing that there is nothing to hinder a cone of rays from any radiant *point* impinging on retina fig. 5, inasmuch as fig. 6 differs only from fig. 5 in point of deficiency; *i.e.* retina fig. 5 has all the light of retina fig. 6, and a great deal more. Now the experiment is here interesting because, by a special arrangement, it makes this faint impact of light appreciable; and this it does by excluding the more vigorous impact of light; *i.e.* it negatives delineation by cones and admits of it by single rays only; and it shows once more, supposing perception to the *appreciated nervous differentiation*, how such differentiation may exist unappreciated.

For it may, as I shall now show, be proved that both impressions, the stronger and the fainter, do indeed coexist upon the retina, and that the exclusive appreciation of the former is owing solely to its greater intensity. For by introducing the means of delineation by cones of *faint light*, both images (the one erect, the other inverted) may be made visible at once. At E, fig. 6, have a hole (1/8 inch diameter) in the card and cover it with transparent paper. The faint light from this enlarged source will depict all points of the pin by *cones of rays*, which cross a little behind the crystalline, and consequently give a tracing of the pin head upward on the retina, head downward to perception. Now let a small hole be pierced in the transparent paper to do the office of radiant E, *i.e.* to delineate by single rays; and a little nice adjustment, somewhat increasing the distance of G from the eye, will bring this single-ray tracing of the pin head downward on the retina, and head upward to perception. So that we have on the retina at the same time two impressions in contrary positions, and consequently perceive two pins in contrary positions, each retinal impression being respectively reversed to the mind. To procure this result we first fix the eye upon the radiant point E till we see the pin head upward, as in the illusion; then steadily increasing the distance, as stated, we presently get sight of the actual pin as it is, head downward, with now a little light circle in its head, and a little upright pin within that circle. Remove the transparent paper, and the bright cones obliterate the smaller image; or cut off the cones by opaque paper, still retaining the radiant point, and the larger image vanishes. It is only by reducing the intensity of the cone-depicted image that we make the other image appreciable.

The deductions from these induced facts are, first, concerning erect vision with an inverted image.

However conclusive the reasoning against there being any direct deliverance to the retina, or other neurine, as to the position of the object, however consistent the theory that experience is the sole interpreter of sensory excitement, and that there never is "an inverted retinal image," inasmuch as there is *no image* till experience has made it (made it in accordance with muscular evidence, and so erect), however conclusive all this to reason, by various analogies, direct experimental evidence has been wanting. It is here supplied. For we have here one object before us and two opposite perceptions of its position at the same moment, a present voucher that there is no direct deliverance from the object as to its position. If there were (as still contended by a high authority) "a law" by which intuitively we interpret an effect on the retina in behoof of the object's position, we should here see both images in one and the true position. But we do not. And what is it we do? We fall back upon hitherto experience, and interpret the retinal effect accordingly. It is of no avail that the pin falls on the retina in its true position; we must see it as we have hitherto found it.

A second deduction regards "the law of visible direction," by which, it is said, an object is seen in the direction of a perpendicular to the retina at the object's point of incidence; and it is cited by its promulgator to prove the sense of vision, primary and unacquired; therefore its operation should be absolutely efficient and unindebted to experience, unconditionally indicating the direction of the object. But (fig. 6) the law proves fallacious, the object not being in the direction of the perpendicular A B: it is seen where it is not. That it is seen under novel and unexperienced conditions cannot be urged without admitting experience as a prerequisite to sense of direction.

Thirdly, there are some points of interest in regard to focus, which I omit for want of space, but would observe that the pin neither is, nor I presented as in focus, its outline being everywhere jagged with rays of light. I have shown the attempt to increase the focal distance by supposed "projection" on the plane M N illusory, and will only add that, while all similar vague theories of "projection" or out-seeing are begging the question when the question is of vision, in this case it is obvious that as the plane M N was removed further off, the image thus projected would increase in size; and if the projection represented the appearance as proposed, the pin would appear larger: but it proves otherwise.

It may be noticed that this analysis demonstrates the peculiar effect of what I call single-ray delineation on the retina, which will not only happen when the radiant point is behind the object, showing it as a shadow, but also when an object, itself reflecting rays, is seen through a pinhole. The effect in both cases is faint vision, and an *uninverted* image on the retina, which would result in both cases in *inverted* vision, save that in the latter there has been a previous inversion which sets things right; so that if, while we wonder at the result of this "illusion," we are not surprised at an object seen through a pinhole looking upright, it is not from knowing more of what takes place.

Your obliged Servant, | J. L. TUPPER. | Rugby School, | March 18, 1869.

Note.—A spark at the end of a smouldering thread I have found to exhibit the phenomenon most perfectly. Multiply these sparks, and you multiply the images of the pin; and the same if you multiply the pinholes in the card E, their local relations appearing the same as in nature.

* Communicated by Professor Tyndall

'On an Optical Illusion', *Phil. Mag.*, 39 (June 1870), pp. 423–8

---

1. *John Lucas Tupper*: John Lucas Tupper (1826?–79), a pre-Raphaelite artist and sculptor. He studied at the Royal Academy Schools from about 1844 and was an anatomical draftsman at Guy's Hospital, London until 1863. Tupper later became master of drawing at Rugby School. He published a number of literary, scientific, and educational articles ('John Lucas Tupper', *Mapping the Practice and Profession of Sculpture in Britain and Ireland 1851–1951*,

University of Glasgow History of Art and HATII (2011), at http://sculpture.gla.ac.uk/view/person.php?id=msib7_1220373335 [accessed 25 January 2020]).

2. *[1870]*: although the letter is dated at the signature line with 1869, the "Solution" Tupper discusses was not published until January 1870. It is not clear whether this error comes from Tyndall, Tupper, or the Editor of the *Phil. Mag.*
3. *"Solution"*: Michael, 'Optical Illusion', *The English Mechanic and World of Science*, 10:251 (14 January 1870), p. 439.
4. *MICHAEL*: not identified.

From Wilhelm Weber[1] 18 March 1870 3153

Göttingen, 18 März 1870.

Mein lieber Professor Tyndall

An den schönen und scharfsinnigen Untersuchungen des Herrn Maxwell nehme ich gleiches Interesse wie Sie, und knüpfe es besonders an die Elektrodynamische Theorie des Lichtes, die Herr Maxwell entwickelt hat.

Die Nachweisung eines Mediums, durch dessen Molecularkräfte die Wirkungen genau bestimmt werden könnten, welche elektrische Ströme und elektrische Ladungen aus der Ferne auf einander ausüben, wäre an sich sehr interessant. Die Annahme eines solchen Mediums, welches dies wirklich leistete, halte ich für ebenso zulässig, wie die Annahme fernwirkender Kräfte, aus denen diese Wirkungen bisher bestimmt worden sind. Würde aber ferner bewiesen, dass aus der Annahme dieses Mediums zugleich auch die vom Lichte in die Ferne ausgeübten Wirkungen bestimmt werden könnten, so würde die Alternative zwischen beiden Annahmen dadurch meines Erachtens entschieden sein.—Die Hoffnung einer solchen Combination ist durch Faradays Entdeckung der Drehung der Polarisationsebene durch magnetische Directionskräfte zuerst, und sodann durch die Übereinstimmung derjenigen Geschwindigkeit, welche das Verhältnis der elektrodynamischen Einheit zur elektrostatischen ausdrückt, mit der Geschwindigkeit des Lichts angeregt worden, und mir scheint von allen Versuchen, welche zur Verwirklichung dieser Hoffnung gemacht worden sind, der von Herrn Maxwell gemachte am erfolgreichsten. Ein Urtheil darüber von mathematischer Seite steht mir zwar nicht zu; die durch meist sehr einfache Combinationen gewonnenen Resultate sind aber sehr überraschend und lenken die Aufmerksamkeit auf die von weiterer Durchführung noch zu erwartenden Resultate. Was das Medium selbst betrifft und die Bestimmung der darin wirksamem Molecularkräfte, so erweckt die Übereinstimmung der analytischen Ausdrücke mit den Resultaten von Faradays experimentellen Forschungen grosses Vertrauen, wenn es dabei auch noch, wie mir scheint, an deutlicher Einsicht in den inneren

Zusammenhang zwischen molecularen Kräften und Verhältnissen fehlt, was im Grunde aber überall der Fall ist, wo die Forschung auf die innere moleculare Constitution des Körpers bisher geführt hat. Ich möchte glauben, dass eine Übertragung der Gesetze der Fernwirkung auf Molecularverhältnisse, wie C. Neumann sie versucht hat, weiter führen könnte.

Entschuldigen Sie diese unmaßgeblichen Aeusserungen über einen so wichtigen Gegenstand, worüber Ihnen selbst und Anderen ein weit competenteres Urtheil zusteht als mir. Ich habe dadurch nur das Interesse, welches ich an Herrn Maxwells Untersuchungen nehme, ausdrücken wollen in der Hoffnung mit Ihnen darin wesentlich übereinzustimmen, insbesondere auch darin, dass die Gesetze der Fernwirkung, die auch Gegenstand meiner Untersuchungen gewesen, ihre Bedeutung in der Wissenschaft ebenso wenig verlieren würden, wie die Lehre vom Magnetismus, wenn sich auch zuletzt ergeben sollte, dass jene fernwirkenden Kräfte, ebenso wie die magnetischen Fluida, nur ideale Vorstellungen, aber in weitem Gebiete den wahren äquivalent wären.

Ihr | treu ergebenster | Wilhelm Weber

Göttingen, 18 März 1870.

My dear Professor Tyndall

I take the same interest in Maxwell's[2] beautiful and penetrating investigations[3] as yourself, and it is tied particularly to the electrodynamic theory of light developed by Mr. Maxwell.

It would be very interesting in itself to provide evidence of a medium with molecular forces that would allow one to precisely determine the effects that electric currents and electric charges exert on each other at a distance. I consider it equally acceptable to assume a medium that really acts in this way, as to assume forces acting at a distance, to which these effects had until now been attributed. If it were further proven that with the assumption of this medium the effects of light at a distance could also be determined, the alternative between the two assumptions would, in my opinion, be decided.—The hope of such a combination was first encouraged through Faraday's[4] discovery that the plane of polarization rotates through directional magnetic forces, and then through the agreement between the velocity that expresses the relation of the electrodynamic unit to the electrostatic one and the velocity of light,[5] and it appears to me that of all experiments which have been made to realise this hope, those of Maxwell are the most successful. Admittedly, I am not entitled to judge it from a mathematical perspective; the results, which are obtained through combinations that are for the most part very simple are very astonishing and draw attention to the results that can be expected from continuing development. As far as the medium itself is concerned, and the determination of the molecular forces effective within it, it inspires considerable confidence that the analytical expressions agree with the results of Faraday's experimental researches, even if we lack, as it appears to me, clear insight into the inner relationship between molecular forces and properties, which is at bottom the case everywhere, where research into the inner molecular constitution of matter has led so far. I would like to believe that applying the laws of action at a distance to molecular interactions, as C. Neumann[6] has attempted, could lead further.

Do pardon these humble remarks on such an important matter, over which you yourself and others are able to make a far more competent judgement. I desired only to express the interest I take in Maxwell's investigations,[7] in the hope of agreeing with you in most aspects, and in particular in that the laws of action at a distance, which have also been the object of my research, would no more lose their significance for science than the theory of magnetism, if in the end it should be shown that these forces acting at a distance, and equally magnetic fluids, were only ideal concepts, but for many areas were equivalent to the real ones.

Your | most devoted | Wilhelm Weber

RI MS JT/1/W/17

---

1. *Wilhelm Weber*: Wilhelm Eduard Weber (1804–91), a German physicist noted for his work on electrodynamics. A graduate of the University of Halle, Weber began a distinguished career as professor of physics there in 1828, the year he first met Gauss. From 1831 until his resignation in political protest in 1837, he served as professor of physics at the University of Göttingen, where he collaborated with Gauss on magnetic and telegraphic researches. In 1843 he was called to the chair of physics at the University of Leipzig, where he engaged in research on electrodynamics. He returned to the chair at Göttingen in 1848 and resumed his partnership with Gauss (*CDSB*; *NDB*). Tyndall arranged for several of his papers to be translated in the *Phil. Mag.*
2. *Maxwell*: James Clerk Maxwell.
3. *penetrating investigations*: Maxwell is probably best remembered for his work on electromagnetic radiation.
4. *Faraday's*: Michael Faraday.
5. *discovery that the plane of polarization rotates . . . the velocity of light*: Faraday discovered in 1845 that light and electromagnetism are related.
6. *C. Neumann*: Carl Neumann (1832–1925), a German mathematician. He taught at the University of Königsberg from 1831, became a professor at Halle in 1856 and was appointed to a chair at the University of Leipzig. His work used applied mathematics in a range of areas including mathematical physics and electrodynamics (*NDB*).
7. *Maxwell's investigations*: see n. 3.

To Frederic W. Farrar[1] 21 March [1870][2] 3154

Royal Institution of Great Britain | 21st. March.

My dear Mr. Farrar.

You must have deemed me dwelling beyond the pale[3] of courtesy, to account for my not sooner thanking you for your kindness in sending me the "Families of Speech".[4] Let me now do so. and let me at the same time ask you to accept a little volume from me[5] in return.

I had a horrid accident[6] in the Alps which threw my work much back, but I am overtaking it by degrees.

Kind regards to Mrs. Farrar. | yours ever faithfully | John Tyndall

From the private collection of Ryan Tweney

1. *Frederic W. Farrar*: Frederic Farrar (1831–1903), dean of Canterbury, a novelist and a philologist, was elected FRS in 1866. His argument for the evolutionary development of language was associated with Darwinian evolution (*ODNB*).
2. *[1870]*: this letter refers to Tyndall's accident of September 1869, mentions a publication from 1870, and is accompanied by a book with a dedication dated March 1870.
3. *the pale*: the Pale was the area of Ireland that continued under English control prior to the Tudor reconquest of Ireland in the sixteenth century. Those within the Pale were the English in Ireland, those beyond the native Irish. Hence the expression 'beyond the Pale' is used to describe something considered beyond acceptable civil behaviour.
4. *"Families of Speech"*: F. W. Farrar, *Families of speech: Four Lectures Delivered Before the Royal Institution of Great Britain in March 1869* (London: Longmans, Green, and Co., 1870).
5. *a little volume from me*: *Faraday as a Discoverer* (see letter 2951, n. 3). Included with this letter is a scan of title page of the book, presumably from the same private collection as the letter. 'Michael Faraday' is printed on the title page, and it bears the following inscription written at the top right: 'The Rev. F. W. Farrar. | With the author's kind regards | March 1870'.
6. *horrid accident*: for more on Tyndall's injury, due to an accident in the Alps, see letters 3030, 3031, 3032, 3037, 3041, 3048, 3049, and 3060.

From Joseph Plateau 22 March 1870 3155

Gand, 22 Mars 1870

Mon Cher Monsieur Tyndall

J'ai l'honneur de vous adresser ci-joint le résumé que vous avez bien voulu me demander. Ce résumé est celui de travaux de 40 années, et vous voulez que je le condense en quatre petites pages; je l'ai fait, mais je doute qu'il suffise pour que la Société Royale puisse juger un peu sainement de mes recherches.

Je sais depuis longtemps que ces mêmes <reche>rches, surtout celles qui concernent les <liqu>ides supposés sans pesanteur, n'ont pas <gr>ande faveur en Angleterre; mais je ne crois pas que ce soit pour la raison que vous m'indiquez; la véritable raison est, je pense, que personne, dans votre pays, ne s'occupe de sujets analogues.

Je remets entre vos mains les intérêts de ma nomination, et je vous remercie d'avance des peines que vous vous donnerez pour la faire réussir; quel que soit le succès de l'affaire, agréez l'assurance de mes sentiments affectueux et de ma haute considération.

J. Plateau

Ghent, 22 March 1870

My dear Mr. Tyndall,

I have the honour of sending you the enclosed summary[1] that you kindly requested of me. This summary is one of 40 years work, and you ask that I condense it into four short pages. I did so, but I doubt that it will be sufficient for the Royal Society[2] to evaluate my research in a sensible enough way.

I have known for a long time that these same <studies>,[3] especially those concerning the purportedly weightless <liquids>, have not won <many> favours in England; but I do not believe that this is the result of the reason you gave me; the real reason is, I think, that no one, in your country, deals with comparable subjects.

I commend my nomination[4] into your hands, and I thank you in advance for the trouble that you will be taking to ensure its success. Regardless of the success of this matter,[5] accept the assurance of my warmest affections and highest regards.

J. Plateau

RI MS JT/1/P/106

---

1. *the enclosed summary*: enclosure missing.
2. *I doubt . . . for the Royal Society*: Tyndall had been involved in ongoing efforts to have Plateau elected ForMemRS (see letter 2975, n. 4).
3. *<studies>*: part of this page is torn.
4. *my nomination*: see n. 2.
5. *Regardless of the success of this matter*: Plateau was elected ForMemRS on 24 November 1870 (see letters 3184 and 3192).

From Emily Peel 24 March 1870 3156

Whitehall Gardens | March 24/70

Dear Professor Tyndall,

I have received only today the bottles containing water from the Mediterranean which Baroness Rothschild[1] had promised to send for some time since. It accompanies this note, & I hope some day to hear you relate the result of y[ou]r investigation both into the reason of its lovely colour as of the beautiful blue of my Swiss Lake.[2]—

You must excuse this pencil scrawl for I am not yet able to sit up and write & pen & ink are therefore more difficult to handle.—

Yours very sin[cerel]y | Emily Peel

RI MS JT/1/P/52

1. *Baroness Rothschild*: Charlotte von Rothschild.
2. *beautiful blue of my Swiss Lake*: Tyndall was carrying out research into the blueness of water and air. See J. Tyndall, 'On the Colour of the Lake of Geneva and the Mediterranean Sea', *Nature*, 2 (20 October 1870), pp. 489–90; J. Tyndall, 'On the Colour of Water and on the Scattering of Light in Later and Air', *Roy. Inst. Proc.*, 6 (1871), pp. 189–99. See letter 3091.

From George Wynne 25 March 1870[1] 3157

7 Mountjoy Squ[ar]e | 25 March 70

My dear Old Friend

I was truly grieved to hear of your illness.[2] I am afraid if you do not give that overworked frame a little more rest that some important part of the machinery will give way and that the repair will then take longer than you are prepared for. Whereas if you had treated that delicate machinery with that consideration which it had a right to expect from you from the good service it has already done you, it would stand you in stead for many a year to come, work your 7 or 8 hours a day and then take a good walk or ride and probably if you take this course your brain will become more prolific than it now is, cease to burn the midnight oil and apply it to the lubrication of the machine in the shape of sleep. I thank you very much for your book about Faraday[3] which at present I have only been able to glance at. What a wonderful field of inquiry those experiments of yours of the chemical reactions produced by light seem to open.

Continue to hear good accounts from my scattered sons,[4] a letter from Frank today from Constantinople says he is getting on well there tho' the Feast of Barram[5] is rather interfering with his progress, his work there as well as I can make it out is putting up one of Siemen's sleet furnaces.[6] I enclose you a few lines from your old friend Lucy.[7]

Ever affectionately | Geo[rge] Wynne

RI MS JT/1/W/98
RI MS JT/1/TYP/5/1865

---

1. *25 March 1870*: the dating of this letter raises some questions. Lousia has dated it as 25 March 1870, and the barely decipherable date on the original letter seems to confirm this, but letter 3150 from Lucy Wynne was sent with this one and is dated 17 March.
2. *your illness*: for more on Tyndall's injury, due to an accident in the Alps, see letters 3030, 3031, 3032, 3037, 3041, 3048, 3049, and 3060.
3. *your book about Faraday*: *Faraday as a Discoverer* (see letter 2951, n. 3).
4. *my scattered sons*: Frank, Edward, and Henry Wynne.
5. *Feast of Barram:* Eid, the end of Ramadan, the Muslim month of fasting and prayer.
6. *Siemen's sleet furnaces:* an open hearth furnace invented by Charles William Siemens (see letter 3143, n. 2).
7. *a few lines from your old friend Lucy*: letter 3150 from Lucy Wynne, George Wynne's daughter. It is unclear why Lucy's letter was written over a week before this one.

To Emily Peel 26 March 1870 3158

Royal Institution. | 26th March, 1870.

My dear Lady Emily,

A thousand thanks to you for your great kindness. The pencilled words[1] are ten times more precious to me than if they had been written in ink. I hope that you will soon be quite strong again.

Ever yours, | John Tyndall.

Pray offer the Baroness[2] my very best thanks.

RI MS JT/1/TYP/3/959
LT Typescript Only

---

1. *The pencilled words*: see letter 3156.
2. *the Baroness*: Charlotte von Rothschild.

From Sarah Faraday [26 March 1870][1] 3159

26 March, 1870. | Barnsbury Villa, | 320, Liverpool Road.

Dear Dr. Tyndall,

I should, as you have suggested, have sent these slips back without any comment, being quite satisfied and pleased with what you say of the letters to me;[2] but the notice of the experiments at the latter end of the article has brought back so vividly to my mind the occurrence, that I cannot refrain from telling you my feelings.

In making the experiments, he told me that he had just succeeded in accomplishing certain <*1 word excised*> when Sir Humphrey[3] came in. It being his own private investigation, he did not feel obliged to tell him all he was doing, but as a compliment did inform him how far he had gone; and he felt aggrieved when Davy[4] suggested he should proceed in a certain way, which he considered an interference, for it was the thing he would most likely have done of his own accord.

I had no wish to bring these things forward, nor should I have done so, but your notice has made me feel that you had not understood all the circumstances. I am in too much pain to sit up to the table to write, so have borrowed my niece's hand.[5]

With regard to the letters to me, as you seem to appreciate them so highly, I may say I have one which Dr. Bence Jones seemed to be more pleased with than any, which though I could not wish to have published, I would willingly show to you, as a warmly attached friend, some time when you come, if you would like to see it. These are my impressions as far as I can recollect without looking into any records. Farewell.

Yours most truly, | S Faraday. | Dr. Tyndall.

RI MS JT/1/TYP/12/4190
LT Typescript Only

---

1. *[26 March 1870]*: it is possible that the date on this typescript letter is incorrect. It may have been written after the letter from Jane Barnard on the 31 March 1870. See letter 3162.
2. *letters to me*: Tyndall notes the warmth in Faraday's letters to his wife Sarah in his review of Bence Jones's *Life of Faraday* (J. Tyndall, 'Life and Letters of Faraday, by Dr. Bence Jones', *The Academy, a Monthly Record of Literature, Learning, Science, and Art*, 1 (14 May 1870), pp. 204–7, and 1 (11 June 1870), pp. 232–4, at p. 205). It is apparent that Sarah was reviewing Tyndall's review. The review was also published in *Fragments of Science*, pp. 349–72.
3. *Sir Humphrey*: Humphry Davy.
4. *Davy*: Humphry Davy.
5. *my niece's hand*: Jane Barnard.

To the Editor of *Nature*[1] 27 March 1870 3160

I THINK the enclosed letter[2] contains as good a description of the formation of ground ice as I have hitherto seen; I therefore send it to you, thinking it of sufficient interest to appear in NATURE.

JOHN TYNDALL | Royal Institution, March 27

[. . .][3]

Nature, 1 (31 March 1870), p. 555.

1. Editor of *Nature*: Norman Lockyer (see letter 3026, n. 7).
2. *the enclosed letter*: letter 3130.
3. *[. . .]*: letter 3130 inserted here.

To Thomas Stevenson 27 March 1870 3161

Royal Institution of Great Britain | 27th March.

Will Mr. Thomas Stevenson accept my best thanks for the present of his paper on "Illumination of Beacons and Buoys"?[1] I wish Mr. Stevenson every success in his important experiments.

John Tyndall

Thomas Stevenson Esq[ui]r[e] F.R.S.E.[2] | &c. &c. &c. | Edinburgh.

National Library of Scotland, MS.785, f.55 r, f.57 r

1. *"Illumination of Beacons and Buoys"*: T. Stevenson, *Proposals for the Illumination of Beacons and Buoys* (Edinburgh: A. and C. Black, 1870).
2. *F.R.S.E.*: Fellow of the Royal Society of Edinburgh.

From Jane Barnard 31 March 1870 3162

Barnsbury Villa | 31 March 1870 | 11 P.M.

Dear Dr Tyndall

Your packet[1] came in by the last post this evening, too late for my Aunt[2] to venture to look at, without running the risk of a disturbed night's rest; but by her leave & your permission I have read it through, & as it impresses me very pleasantly, I run the risk of appearing forward & write down some of my feelings at once.

Some facts strike me as exceedingly happy, so fresh & to the point. Among them the end of the second paragraph beginning, "Faraday notes his own impetuosity"[3]—

Again, the paragraph beginning, "Over & over again."[4]—the last sentence in this is beautiful. Then the concluding piece of the paper. I cannot tell whether it is yours, or quoted from the book, or elsewhere as the commas could seem to denote, but I know that I like it exceedingly. Indeed all your remarks on his character & his powers of mind &c are, as you may well imagine, very delightful & truthful to me.

There are facts which I cannot judge of not knowing all the circumstances, but it seems to me from your point of view to be a just & clear account.

I hope you will excuse me dear D[r] Tyndall for intruding these lines on you, but I longed to thank you for the pleasure your words have given me.

Yours ever sincerely | Jane Barnard

RI MS JT/1/B/45

---

1. *Your packet*: probably Tyndall's review of Henry Bence Jones's *Life of Faraday* and his own *Faraday as a Discoverer*. See letter 3164.
2. *my Aunt*: Sarah Faraday.
3. *"Faraday notes his own impetuosity"*: from Tyndall's review of Henry Bence Jones's *Life of Faraday* (p. 204). See letter 3159, n. 2.
4. *"Over & over again"*: see n. 3, at p. 205.

From Mary Egerton [March 1870][1] 3163

<Letter begins here>

I could not resist sending it,[2] I thought much of it so beautiful, (of course I do not agree with him[3] in everything, some passages I confess to having marked because I thought them <u>hard</u> <u>for</u> <u>you</u> <u>to</u> *[answer]*. If you wish to know why I ventured to do so, without fearing to make you angry, (which I sh[oul][d]. not like to do,) I refer you to the Phil[osophical]: Magazine, for 1859,[4] (I think!) (I have the passage & reference at home!) but I forget who the paper is by at this moment!

I was almost thinking of writing, to ask you to tell me about M[rs] Spottiswoode.[5] I thought M[r] S[pottiswoode]. spoke in a tone of much anxiety about her that evening, though she seemed cheerful when I saw her & noted herself much better. If you had leisure to send me a line, I should be truly glad to hear she was improving & had had no further relapses one dreads inflammation of the lungs all the more after this last sad event!

My address will be Thoresby Park Nework Nott[ingham]s[hire]

I am so sorry to miss your first lecture! If you have any printed notes keep me & May two copies. I shall try to come in for the second, but our plans are somewhat uncertain owing to sick servants.

I have changed daughters, & have the two others[6] with me now.

Ever yours | M F Egerton

RI MS JT/1/E/99k

1. *[March 1870]*: in 1870 Egerton began dividing her time between her Mountfield home and Thoresby Park. Thoresby Park was the home of her brother, Sydney Pierrepont (1826–1900), the 3rd Earl Manvers, who had it rebuilt during the late 1860s and early 1870s. Since Egerton mentions missing Tyndall's first lecture, and the RI lecture season began in April, this letter has been given the tentative date of March 1870.
2. *sending it*: enclosure missing.
3. *him*: not identified
4. *Phil[osophical]. Magazine, for 1859*: it is unclear what paper she is referring to.
5. *Mrs Spottiswoode*: Eliza Spottiswoode.
6. *the two others*: Charlotte and Georgina. Her usual companion was her daughter Mary.

From Jane Barnard 2 April 1870 3164

Barnsbury Villa | 320 Liverpool Rd. | N

Dear Dr Tyndall

Your note[1] came to me this afternoon, but I am sorry that I have had no opportunity of shewing it to my Aunt;[2] I am obliged to be careful of her just now.

As I do not wish to leave it on my mind tomorrow, I answer it tonight. You are very kind & flattering in your desire to hear any remarks or suggestions. I felt sorry that I had left an impression last night[3] that there might be parts of your paper not so agreeable to us—but how could it be otherwise?

This article,[4] as well as your book on the same subject,[5] seems to me like a fresh & sparkling draught & it is *[but]* natural that some parts being especially striking & delightful to us, others cannot be equally consonant with our feelings.

If you still determine to send the second part & afterwards come as you proposed, you may if you wish it draw out some remarks. But really the whole has left as pleasant an impression that I cannot criticize it now. x

I am yours | very sincerely | Jane Barnard

2nd April. 1870 | John Tyndall Esq
x & besides it would be most miserable to alter & cut out your work to others patterns.

RI MS JT/1/B/46

1. *Your note*: letter missing.
2. *my Aunt*: Sarah Faraday.
3. *left an impression last night*: see letter 3162.
4. *This article*: Tyndall's review of Henry Bence Jones's *Life of Faraday*. Barnard and Tyndall are discussing the proofs, which were also being read by Sarah Faraday. See letter 3159, n. 2, and letter 3165.
5. *your book on the same subject: Faraday as a Discoverer*. See letter 2951, n. 3.

To Jane Barnard 2 April [1870][1] 3165

Royal Institution | 2nd April

My dear Miss Barnard.

Your letter[2] has given me very great satisfaction; and this is to say to you that the proof[3] is likely to remain in my hands for nearly a month, and that any alteration or suggestion, or any remark which you think likely to modify or render more correct my point of view; or any alteration as to style or expression—in short any thing that could bring the article still more into harmony with the views of your aunt[4] and yourself shall, if mentioned to me, receive my best attention. I have such implicit faith in your aunt's rightmindedness, and I know so well my own strong desire to say & do only what is right, that there can I am persuaded be no real difficulty in rendering the article all that she & you would wish it to be.

Yours ever sincerely | John Tyndall

RI MS JT/1/T/118
RI MS JT/1/TYP/12/4200

1. *[1870]*: a note from LT in the typescript indicates that the year was added to the original letter in pencil later. Since the letter is part of an ongoing exchange between Tyndall and Barnard, the dating is credible.
2. *your letter*: letter 3164.
3. *the proof*: Barnard and Tyndall are discussing the proofs for his review of Henry Bence Jones's, *Life of Faraday*. See letters 3159, 3162, and 3164.
4. *your aunt*: Sarah Faraday.

To William Thomson 3 April 1870 3166

Royal Institution[1] 3rd April 1870

My dear Thomson.

I observed with immense pleasure the sweep of your scimitar through the atoms[2] and molecules. To this we must come at last. There can be no permanent peace of mind until we know something definitif of the size and other physical characters of atoms and molecules, and to this you have made a contribution marked as might be expected by extraordinary penetration. I have recently been engaged in certain experiments which have a general bearing upon this question and which I should therefore like briefly to communicate to you.

One feature of the clouds produced by the decomposition of vapours by light is that you can regulate, at will, the growth of their particles, making it either slow or rapid by varying the quantity of matter acted on by the lights. When the vapour is very attenuated the cloud first appears of a beautiful blue colour, and afterwards, in the great majority of cases, passes from blue through whitish blue to white. Did the particles first formed merely grow uniformly without the accession of new particles we should have a succession of colours, and it is (I suppose) because side by side with the growth of the larger particles new small ones are introduced, thus abolishing the predominance of any particular magnitude, that the passage is, in most cases, that which I have indicated—variety from blue through whitish blue to white.

The depth of the colour is a rough measure of the relative size of the particles. If for example two nebulas be formed under precisely the same conditions with the sole difference that one has been acted upon for three minutes and the other for four minutes, then the relative grossness of the latter is marked by its whiter hue. You must not imagine that the particles are seen as floating notes. The light shed by the nebula is as continuous as the spectrum blue of incandescent carbon.

You know Brücke's receipt for precipitating mastic in water,[3] so as to imitate the blue of the sky. Now when you find water containing such a precipitate whiter than an actinic cloud is it not fair to infer that its particles are coarser than those which constitute the cloud? I have prepared various precipitates which yield blue light and have placed them in the hands of eminent microscopists, but they can make nothing of the precipitated particles. The blue nebula is irresolvable by the highest microscopic powers. Huxley and Bence Jones have both sought for the particles, but the liquid in which they are suspended appears as free from particles as pure water. The nebulises may appear perfectly homogenous.

Now let us return to the clouds. Imagine an attenuated vapour acted on by the light. For a time all show the darkness of pure transparency, but in two or three minutes a delicate blue appears at the most concentrated portion of the beam. At right angles to the beam the light discharged is perfectly polarized: the cloud can be quenched utterly by a Nicol prism.[4] Let the light continue to act. During its action the molecules fall in showers upon the particles. It is, on a small scale, your cannonade of meteors against the sun; and yet, after twenty minutes or half an hour of rich action the light scattered by the nebula in the direction of the normal recurrences perfectly polarized—the cloud can be utterly quenched by the Nicol. This, of itself indicates the amazing smallness of the particles. But we can also compare the colour of the nebula with that of the mastic precipitate, and infer that after the twenty minutes action the particles of the cloud are still smaller than the particles of the mastic. That is to say smaller than particles which demonstrably defy the highest powers of the microscope. Now if this be the state of matters at the end of the 20 minutes, what must be the size of the particles when they first appear? And yet I suppose we must conclude that even then each particle is a heap of molecules & each molecule a heap of atoms. We may in time find out how many of the molecules thus grouped together are competent, under the conditions referred to, to send a sensible amount of light to the eye.

The escape of the normally scattered light from perfect polarization as the particles grow larger is beautifully marked. The short waves are those which may be expected to first feel the influence of size and accordingly we find the first colour which reaches the eye when the Nicol is in its position of minimum transmission to be a perfectly gorgeous blue. Neither in the skies of the Alps nor *[those of]* Italy have I seen any thing to compete with this "residual blue" in depth and lustre. This also grows under your eyes to whiteness.

I have been deflected for a moment from my usual work to look into the question of "spontaneous generation". It is amusing to observe how much the notion of littleness has had to do with the diffusion of this doctrine. It was all but dead when the invention of the compound microscope revived it. The amazing smallness of the creatures revealed by this instrument gave the doctrine a new foothold. I say it is amusing because the observer seems not to have the least idea that in their vibrios and bacteriums, they are dealing with perfect behemoths as compared with the particles of active clouds, not to say any thing about your world of molecules which lies behind those particles.

I was glad to hear from M[rs]. *[King]*[5] that you are well. The atoms indeed show that your mind, at all events, retains a first rate edge and temper.

Believe me | yours very sincerely | John Tyndall

Sir W[illia][m] Thomson.

How *[do]* you figure the contents of your elastic molecular bags? Does not the power of restitution depend on the interaction of their parts? What are those parts?

RI MS JT/4/7b/291–2

---

1. *Royal Institution*: while written from the RI, this letter was written on stationary with letterhead for the Athenaeum Club.
2. *scimitar through the atoms*: Thomson was a proponent of the vortex theory of atoms which arose in the late 1860s and was mostly abandoned by 1890. Tyndall could be referring to one of the following publications: W. Thomson, 'On Vortex Atoms', *Phil. Mag.*, 32 (1867), pp. 15–34; W. Thomson, 'On Vortex Motion', *Edin. Roy. Soc. Trans.*, 25 (1869), pp. 217–60. See H. Kragh, 'The Vortex Atom: A Victorian Theory of Everything', *Centaurus*, 44: 1–2 (2002), pp. 32–114.
3. *receipt for precipitating mastic in water*: probably in E. Brücke, 'Ueber die Farben, welche trübe Medien im auffallenden und durchfallenden Lichte zeigen', *Annal. Phys. und Chem.*, 88 (1853), pp. 363–85.
4. *Nicol prism*: see letter 2966, n. 4.
5. *M^rs^. [King]*: probably Elizabeth King (née Thomson, 1819–96), Thomson's sister who married David King (1806–83), a minister of the United Presbyterian church.

To Heinrich Debus [5 April 1870][1] 3167

Royal Institution | 5th March[2]

My dear Heins

It grieves me much to hear of the bad state of your head. I wished to see you connected with the Kensington papers;[3] but your health is the paramount object, and you must not undertake anything to overburden it too much. I think even at a little loss in the first instance, if you find these symptoms continue, you will be obliged to look forward to releasing yourself from Clifton.[4] The life is more than meat and the body than raiment. I hope and think there is nothing seriously wrong with your head. But I fear you have not for many years past given it sufficient rest; and it now begins to rebel.

Matthessen,[5] Guthrie,[6] Hooker, Barrett,[7] and Huxley are to be the examiners this year. I intend to ask Goodeve[8] today to take the place which I had hoped to offer to you. So that I shall have a very good staff. I think Matthessen got the medal of the Royal Society[9] mainly for his physical work; but if I recollect aright some chemical investigations were also mentioned as a ground for the award. Did he not work at opium—he and another?

I am still in the fangs of the printers. The fourth edition of my work on

Heat[10] is just printed and I am thankful to have it off my hands. Longman prints 3000 copies. The book has sold remarkably. If I mistake not 6 thousand have been already sold. So that this new edition would bring it up to 9000 or thereabouts. Roscoe[11] told me a few days ago that 21000 of his Chemistry[12] had been sold.

I now almost wish that instead of going to Clifton you had sat down & written a good school book—but then on the other hand the experience you have had at Clifton must be invaluable. It is one of the model modern Schools.

I hope in a few days to have completed my work on Diamagnetism & magnecrystallic action.[13] It will be a handsome volume; and you if you are a good boy & promise to look at it—shall have a copy. My lectures[14] are now looming upon me. And I have also some experiments going on regarding Spontaneous Generation; which subject I hope to settle conclusively as far as my own convictions are concerned. Prof. Lister of Edinburgh a very able man—writes to tell me[15] that the dust lecture[16] has borne good fruit, and has suggested to him a most simple and most effective new method for the treatment of wounds.

I hoped to get off to the Eifel[17] for 10 days. & indeed promised Huxley yesterday to go with him: but I forgot my date of lecturing, & find that I must forego the pleasure of a scamper in that delightful country. I have to give only seven lectures, and therefore must try to introduce something new.

What are you going to do with yourself at Easter?

I run down on Thursday to spend a day with M^{r}. & M^{rs}. Grote[18] in the Country—you know Grote—He is a finely cultured & most courteous gentleman. A gentleman I say to the core. He never reveals by even a chance movement or utterance that he could be anything else.

Yours ever dear Heins | John Tyndall

RI MS JT/1/T/306
RI MS JT/1/TYP/7/2371

---

1. *[5 April 1870]*: the year of this letter has been determined through its mention of the fourth edition of *Heat*, which was published in 1870. Although dated with 5 March, the editors have determined that it was written on 5 April through its place in relation to letters 3171 and 3169.
2. *5 March*: this letter has been misdated. See n. 1.
3. *Kensington papers*: examination papers of the Science Schools, South Kensington Museum, London.
4. *Clifton*: Clifton College (see letter 2996, n. 8).
5. *Matthessen*: Augustus Matthiessen (1831–70), a British chemist and physicist. Matthiessen held a position as lecturer in chemistry at St Mary's Hospital from 1862 to 1868 and

worked with the BAAS committee on electrical standards from 1862 to 1865. In 1869 he was awarded the Royal Medal of the RS for his work on metal alloys. He was an editor at the *Phil. Mag.* from 1869 to 1870 and joined the University of London Board of Examiners in 1870. He took his own life later that year (*ODNB*).

6. *Guthrie*: Frederick Guthrie (see letter 3105, n. 1).
7. *Barrett*: William Fletcher Barrett (1844–1925), was a physicist and psychical researcher. He became assistant to John Tyndall at the RI in 1863. Following this post, he was science master at the International College, London (1867–69), physics lecturer at the Royal School of Naval Architecture (1869–73), and professor of physics at the Royal College of Science, Dublin (1873–1910). He is primarily remembered for his pivotal role in the foundation of the Society for Psychical Research (SPR) (*ODNB*). Barrett's endeavours in spiritual phenomena put him in conflict with Tyndall (see letter 2906, *Tyndall Correspondence*, vol. 10; and R. Noakes, 'The "Bridge Which Is Between Physical and Psychical Research": William Fletcher Barrett, Sensitive Flames, and Spiritualism', *History of Science*, 42 (2004), pp. 419–64).
8. *Goodeve*: Thomas Minchin Goodeve (1821–1902), lecturer of Mathematics (1846–56), professor of Natural Philosophy (1854–60) and Manufacturing Art and Machinery (1852–60) at King's College, London. In 1869, he was professor of mathematics and applied mechanics at the Royal School of Mines, South Kensington. His publications include *The Elements of Mechanism* (1860) and *Text-Books of Science, Adapted for the use of Artisans* (1870) (*Alum. Cantab.*; 'Obituary', *The Eagle: a Magazine supported by Members of St John's College*, 34 (March 1903), p. 246).
9. *Matthessen got the medal of the Royal Society*: see n. 5.
10. *fourth edition of my work on Heat*: J. Tyndall, *Heat Considered as a Mode of Motion* (London: Longman, Green, and Co., 1870).
11. *Roscoe*: Henry Enfield Roscoe (see letter 3077, n. 12).
12. *his Chemistry*: H. E. Roscoe, *Lessons in Elementary Chemistry* (London: MacMillan and Co., 1866).
13. *my work on Diamagnetism & magnecrystallic action*: *Diamagnetism.*
14. *My lectures*: see letter 3126, n. 3.
15. *writes to tell me*: letter missing.
16. *the dust lecture*: see letter 3126, n. 2.
17. *the Eifel*: mountain range in western Germany and eastern Belgium.
18. *M^r^. and M^rs^. Grote*: George Grote (1794–1871), was an English historian and politician (*ODNB*). In 1820, he married Harriett (née Lewin, 1792–1878), who helped him achieve his academic and political goals. Harriet was a woman of letters, a suffragist and helped to found the Society of Female Artists (1857) (*Ascent of John Tyndall*, p. 374; *ODNB*).

From Joseph Plateau 5 April 1870 3168

Gand 5 Avril 1870

Mon cher Monsieur Tyndall,

Je vous remercie de vos bonnes intentions à mon égard, et j'ai l'honneur de vous adresser les renseignements que vous voulez bien me demander.

Le premier des deux mémoires où Beer applique le calcul aux résultats de mes expériences, mémoire intitulé: Ueber die Oberflächen rotirender Flüssigkeiten im algemeinen, insbesondere über den Plateau'schen Rotationsversuch, est inséré dans les Ann<ales> de M. Poggendorff, t.XCVI, pages *[7]* et 210. Le second, intitulé: Tractatus de theoria mathematica phenomenorum in liquidis actioni gravitatis detractis observatorum, a été publié isolément à Bonn, en 1857. Vous trouverez, du reste, tout ce qui concerne ces deux mémoires dans les §§ 2 à 26 de ma 10ème Série.

En France, ma théorie des phénomènes subjectifs de la vision a été exposée d'une manière parfaitement exacte dans la 2me édition du traité de physique de M. Lamé et dans le traité de physiologie de M. Longet, de plus M. Paul Bert, professeur à la Faculté des Sciences de Paris, l'a soutenue et défendue, l'année dernière contre les attaques des Allemands dans son cours sur la physiologie des sensations. Quant à mes recherches sur les figures d'équilibre, Mr. Lamarle qui est français quoique professeur en Belgique, a publié, sur la théorie de mes système laminaires un grand mémoire intitulé: Sur la Stabilité des systèmes liquides en lames minces. Enfin, vous pouvez voir dans les Comptes rendus de 1861, tome 53, page 463, l'opinion énoncée par M. Faye en ce qui me concerne.

En Amérique, l'Institut Smithsonian publie des traductions anglaises in extenso de toutes mes Séries; la traduction de la première est précédée d'une préface de M. Henry.

En Russie, mes recherches paraissent avoir été l'occasion d'un mémoire de M. Lindelöf intitulé: Théorie des surfaces de révélation à courbure moyenne constante, travail publié dans les mémoires de la Société des Sciences de Finlande.

En Angleterre, Faraday m'a fait l'honneur de répéter, à l'Institution Royale, les expériences de ma première Série, et mes systèmes laminaires ont donné lieu à un mémoire de Sir David Brewster, intitulé: On the figures of equilibrium of liquid films.

Voilà, mon cher Monsieur Tyndall, les principaux témoignages en ma faveur. Vous désirez une lettre d'un savant étranger pour appuyer ma candidature; je joins ici celle par laquelle M. Magnus m'annonça ma nomination à l'Académie de Berlin. Enfin, vous me demandez l'indication des principales sociétés savantes auxquelles j'appartiens; les voici: l'Académie royale de Belgique, l'Institut de France, la Société Philomatique de Paris, l'Institution Royale de Londres, l'Académie de Berlin, la Société de Physique et d'histoire naturelle de Genève, plus une suite d'autres sociétés moins importantes en France, en Allemagne, en Suisse, en Hollande et en Ecosse.

Et maintenant, laissez moi vous dire qu'il est triste, après avoir tant travaillé et tant lutté, d'être obligé d'énumérer tous ces témoignages pour tâcher de persuader à la Société Royale qu'en me nommant dans son sein, elle ne fera pas un choix indigne d'elle; c'est pour moi une leçon d'humilité.

Recevez, mon cher Monsieur Tyndall, mes compliments affectueux, et agréez l'assurance de tous mes sentiments de haute considération.

J. Plateau

Dans le résumé de mes recherches, j'ai oublié un point important; en étudiant les lames liquides, je suis arrivé à une limite supérieure bien *[petite]* du rayon d'activité sensible de l'attraction moléculaire; j'ai trouvé que du moins dans mon liquide glycémique, ce rayon devait être moindre que 1/17000 de millimètre. C'est la première fois, je pense, qu'on ait tenté une semblable évaluation.

Ghent | 5 April 1870

My dear Mr. Tyndall,

I thank you for all your good intentions towards me, and I am honoured to address to you the information that you had the kindness to request of me.[1]

The first of the two essays where Beer[2] applies calculus to the results of my experiments, an essay entitled: Ueber die Oberflächen rotirender Flüssigkeiten im algemeinen, insbesondere über den Plateau'schen Rotationsversuch,[3] is included in the Annals of Mr. Poggendorff, t.XCVI, pages *[7]* and 210. The second, entitled Tractatus de theoria mathematica phenomenorum in liquidis actioni gravitatis detractis observatorum,[4] was published separately in Bonn, in 1857. Indeed, you will find everything concerning these two essays in the §§ 2 to 26 of my Tenth Series.[5]

In France, my theory of subjective visual phenomena was presented in an accurate manner in the 2nd edition of the *Traité de Physique*[6] by Mr. Lamé[7] and in the *Traité de Physiologie*[8] de Mr. Longet;[9] moreover, last year, Paul Bert,[10] professor at the Faculty of Sciences in Paris, supported and defended it, last year, against the attacks from the Germans in his class on the physiology of sensations.

As for my research on the figures of equilibrium, Mr. Lamarle,[11] who is French teaching in Belgium, published, a large essay entitled Sur la Stabilité des systèmes liquides en lames minces[12] regarding the theory of my laminar systems. Finally, you can see in the Comptes rendus[13] from 1861, tome 53, page 463, the opinion stated by M. Faye[14] concerning myself.

In America, the Smithsonian Institute[15] is publishing *in extenso* English translations of all my series; the translation of the first one is preceded by a preface from Mr. Henry.

In Russia, my research seems to have brought about an essay from Mr, Lindelöf[16] entitled: Théorie des surfaces de révélation à courbure moyenne constante,[17] a work published in the memoirs of the Finnish Society of Science.[18]

In England, Faraday[19] honoured me in recreating, in the Royal Institution, the experiments of my first Series, and my laminar systems have led to an essay by Sir David Brewster,[20] entitled: On the figures of equilibrium of liquid films.[21]

These are, my dear Tyndall, the principal testimonies in my favour. You desire a letter from a foreign scientist to support my candidacy; I attach here the one by which Mr. Magnus[22] announced my nomination to the Academy in Berlin. Finally, you ask me to indicate the principal scientific societies I belong to. Here they are: the Royal Academy of Belgium,[23] the French Institute,[24] the Philomatic Society of Paris,[25] the Royal Institution in London, the Academy of Berlin,[26] the Society of Physics and Natural History of Geneva,[27] in addition to other societies of lesser importance in France, in Germany, in Switzerland, in Holland and in Scotland.

And now, let me tell you that it is unfortunate, after having worked and fought so hard, to be compelled to enumerate all of these testimonies in an attempt to persuade the Royal Society that my nomination as a member would not be a choice beneath its dignity; this is for me a lesson in humility.

Please accept, my dear Mr. Tyndall, my sincere compliments, and the assurance of my highest consideration.

J. Plateau

I forgot an important point in the summary of my research; while studying the liquid *[strips]*, I reached a very small superior limit of the *[sensitive activity ray]* of molecular attraction; I have found that, at least in my glycemic liquid, this ray should be less than 1/17000 mm. This is the first time, I believe, that someone has attempted a similar evaluation.

RI MS JT/1/P/107

---

1. *information that you had the kindness to request of me*: see letter 3155. This letter is part of a series of correspondence related to Tyndall's efforts to have Plateau elected ForMemRS (see letter 2975, n. 4).
2. *Beer*: August Beer (1825–63), a German physicist and mathematician. An extra-ordinary professor of mathematics in Bonn, Beer specialized in optics. He is best remembered for Beer's Law, also called the Lambert-Beer Law, regarding the absorption of radiant energy by an absorbing medium (*NDB*).

3. *Ueber die Oberflächen . . . Rotationsversuch*: A. Beer, 'Ueber die Oberflächen rotirender Flüssigkeiten im allgemeinen, insbesondere über den Plateau' schen Rotationsversuch', *Annal. Phys. und Chem.*, 96 (1855), pp. 1–18, 210–35.
4. *Tractatus de theoria . . . observatorum*: A. Beer, *Tractatus de theoria mathematica phenomenorum in liquidis actioni gravitatis detractis observatorum* (Bonn: C. Georgii, 1857).
5. *of my Tenth Series*: presumably J. Plateau, *Recherches expérimentales et théoriques sur les figures d'équilibre d'une masse liquide sans pesanteur. Serie 8–10* (Brussels: Hayez, 1868).
6. *Traité de Physique*: probably M. Lamé, *Cours de physique de l'École Polytechnique* (Paris: Bachelier, 1836–40).
7. *M. Lamé*: Gabriel Lamé (1795–1870), a French mathematician. After studying at the École Polytechnique (1813–17) and the School of Mines (graduating 1820), he travelled to Russia, where he directed the School of Highways and Transportation in St. Peterburg. In 1832, he accepted a chair in physics the École Polytechnique, where he remained until 1844. He became graduate examiner (1844) and professor (1851) of mathematical physics and probability for the University of Paris. His most significant contribution to mathematics was in the area of curvilinear coordinates in theoretical and applied mathematics (*CDSB*).
8. *Traité de Physiologie*: F. A. Longet, *Traité de Physiologie* (Paris: Germer-Baillière, 1868), vol. 1.
9. *M. Longet*: François Achille Longet (1811–71), a French physiologist and anatomist. Longet was physician at the hospitals of St Denis and Ecouen and considered to be one of the best experimental physiologists in France. He specialized in the function of the nervous system (C. Knight (ed.), *Biography, or Third Division of "The English Cyclopedia". Supplement* (London: Bradbury, Evans & Co., 1872), p. 822).
10. *Paul Bert*: Paul Bert (1833–86), a French zoologist and physiologist. He was awarded a doctorate in natural science in 1866 for his thesis on animal grafting. He was professor of zoology and physiology at the University of Bordeaux from 1866 to 1867 before being appointed the chair of comparative physiology at the Muséum d'Histoire Naturelle in 1868. Bert was elected to the Chamber of Deputies in 1872 and was heavily involved in efforts to reform education. His most important publication, *La pression barométrique* (1878), focused on the impact of high altitudes on individuals and spurred systematic research on the topic in the Alps (*CDSB*).
11. *Mr. Lamarle*: Anatole Henri Ernest Lamarle (1806–75), a Belgian engineer and professor at the University of Ghent ('Lamarle', *Mémoires de l'Académie des sciences, lettres et arts d'Arras*, 14 (1883), pp. 273–81).
12. *Sur la Stabilité des systèmes liquides en lames minces*: E. Lamarle, 'Sur la Stabilité des systèmes liquides en lames minces', *Mémoires de l'Académie Royale des Sciences, des Lettres et des Beaux-Arts de Belgique*, 35 (1865), pp. 1–104.
13. *Comptes rendus*: *Comptes Rendus de l'Académie des sciences,* the proceedings of the French Academy of Sciences, is a scientific journal that has been published since 1666.

14. *the opinion stated by M. Faye*: Hervé-Auguste-Étienne-Albans Faye (1814–1902), a French astronomer. In 1843, while at the Paris Observatory, Faye discovered the comet that was given his name. He was elected member of the Parisian Academy of Sciences in 1846 and taught geodesy at the École Polytechnique from 1848 to 1854, before becoming professor of astronomy at the University of Nancy. Faye contributed to astronomical measurement techniques and methodology and physical astronomy (*Catholic Encyclopedia*, vol. 6, online at http://www.newadvent.org/cathen/06020a.htm [accessed 23 March 2020]). For his opinion, see M. Faye, 'Remarques sur la Note de M. Plateau', *Paris, Comptes Rendus* (1861), pp. 463–5.
15. *Smithsonian Institute*: the Smithsonian Institution was founded in 1846 and is a group of museums and research centers located in Washington, DC.
16. *Lindelöf*: Lorenz Leonard Lindelöf (1827–1908), a Finnish mathematician and astronomer. In the 1850s, he worked at Polkovo Observatory. He was professor of mathematics and astronomy at the University of Helsinki from 1857 to 1874 and permanent secretary of the Finnish Society of Science and Letters from 1867. In 1860, Lindelöf joined the British Eclipse Expedition to Spain at Airy's invitation. He resigned his professorship in 1874 to become minister of state education (R. Lehti and T. Markkanen, *History of Astronomy in Finland, 1828–1918* (Helsinki: Societas Scientiarum Fennica, 2010), pp. 139–41).
17. *Théorie . . . constante*: L. L. Lindelöf, 'Théorie des surfaces de révélation à courbure moyenne constante', *Acta Societatis Scientarium Fennicae, Helsingfors*, 7 (1863), pp. 559–63.
18. *Finnish Society of Science*: the Finnish Society of Sciences and Letters was founded in 1838. It is the national academy for the promotion of science and scholarship.
19. *Faraday*: Michael Faraday.
20. *David Brewster*: see letter 3010, n. 8.
21. *On the figures of equilibrium of liquid films*: D. Brewster, 'On the Figures of Equilibrium in Liquid Films', *Edin. Roy. Soc. Trans.*, 24:3 (1867), pp. 505–13.
22. *Mr. Magnus*: Gustav Magnus.
23. *Royal Academy of Belgium*: the Royal Academy of Science, Letters and Fine Arts of Belgium was founded in the eighteenth century.
24. *French Institute*: see letter 3138, n. 4.
25. *Philomatic Society of Paris*: la Société Philomatique de Paris was founded in 1788 which aimed to provide an association for scientists.
26. *Academy of Berlin*: the Prussian Academy of Science was founded in Berlin in 1700. Fredrick II reorganised the society attempting to rival the Academy in Paris and its activities and institutional infrastructure were expanded considerably throughout the eighteenth century.
27. *Society of Physics and Natural History of Geneva*: founded as the Geneva Society of Natural History in 1791 it was soon opened up to all scientific disciplines and published a journal, *Archives des Sciences*.

To Thomas Henry Huxley [6 April 1870][1] 3169

Royal Institution. | 6th.

My dear Huxley,

I was rendered drunk by the excess of prospective pleasure when you mentioned the Eifel[2] yesterday,[3] and took no account of my lectures.[x] They begin on the 28th,[4] and I have studiously to this hour excluded them from my thoughts. I have made arrangements to see various experiments involving the practical application of electricity before the lectures begin; I find myself in short cut off from the expedition. My regret on this score is commensurate with the pleasures I promised myself. Confound the lectures!

And yours on Friday[5] is creating a pretty hubbub already. I am torn to pieces by women in search of tickets. Any thing that touches progenitorship interests them. You will have a crammed house I doubt not.

Yours ever | John Tyndall.

[x] On Electrical Phenomena.[6]

IC HP 8.76
Typed Transcript Only

1. *[6 April 1870]*: this letter has been dated with reference to letters 3167 and 3171, which mention Huxley's upcoming (8 April) lecture at the RI.
2. *the Eifel*: see letter 3167, n. 17.
3. *when you mentioned the Eifel yesterday*: if Huxley wrote a letter, it is now missing, but Tyndall mentioned the plans to Debus in letter 3167.
4. *They begin on the 28th*: see letter 3126, n. 3.
5. *yours on Friday*: Huxley delivered a Friday Evening Discourse at the RI, 'On the Pedigree of the Horse', on 8 April 1870 (*Roy. Inst. Proc.*, 6 (1870–71), p. 129).
6. *On Electrical Phenomena*: added in pencil to the typescript. See letter 3126, n. 3.

To the Editor of the *Times* 6 April 1870 3170

PROFESSOR TYNDALL ON FILTERED AIR. |
TO THE EDITOR OF THE TIMES

Sir,—The theory of disease was never discussed with more earnestness, or with greater precision, than at the present time. The exact methods pursued in physics and chymistry,[1] both as regards reasoning and experiment, are making their influence felt in medicine and surgery; and they promise, while assigning narrow limits to our present accurate knowledge, to insure its healthy growth. It is, I think, of capital importance to mark each successive step by which that knowledge is surely and certainly augmented; to detach from the domain of vagueness and uncertainty each successive fragment of demonstrated truth. Now, if the published *data* be correct, it seems to me that such a step has been recently taken with reference to the germ-theory of the putrefaction of wounds, and that the evidence in favour of that theory amounts to a physical demonstration of its truth. This result and its basis I propose here to describe and define.

The entrance of air into a wound is the dread of the surgeon. When an abscess is opened he must prevent the air from mingling with the blood-clots if he would avoid putrefaction and its teeming accompaniment of animalcule[2] life. Some eminent London surgeons inform me that they never squeeze an abscess, lest when the pressure is relaxed the air should be sucked in. Now, whence this dreaded power? Is it the air itself that causes putrefaction, or is it something carried mechanically by the air? A follower of Gay-Lussac[3] would affirm the former;[4] a heterogenist[5] would refer the animalcules to "spontaneous generation;" a holder of the germ-theory[6] would ascribe the putrefaction to seeds or eggs floating in the atmosphere, and which, when sown upon the wound, sprout into this crop of minute organisms. Do any *data* exist which will enable us to say, with certainty, which party is right? I think so.

It would be very difficult to reduce the putrefying power of pure air, even if it existed, to absolute demonstration; for, however cleansed in appearance, a stubborn objector might still urge that the air was not cleansed in reality; that germs exist, though they baffle our attempts to reveal them. But this difficulty does not hamper the other side; for if, notwithstanding the risk of these residual germs, *visibly pure air* can be proved incompetent to produce the phenomena of putrefaction there is no escape from the inference that, as regards the point to be decided, such air is perfectly filtered; and its proved impotence would be demonstration of the truth of the germ-theory. By "visibly pure air" I mean air which where traversed by a powerful and intensely concentrated beam of light, in a space not otherwise illuminated, reveals no trace of floating matter to the eye.

How, then, are we to obtain our filtered air, and, having obtained it, how are we to apply it to a wound and mix it effectually with the blood? Two or three years ago an observation and an inference, which, taken together, reflect the highest credit on his sagacity, were made and drawn by Professor Joseph Lister, of Edinburgh. He found, and I believe it is the universal experience of surgery to find, that when the lung is wounded by the spike of a broken rib, air from the pleural cavity may mingle freely with the blood, but that putrefaction never ensues. Here is the statement of Professor Lister, abbreviated, but in his own words:—

"I have explained to my own mind the remarkable fact that in simple fracture of the ribs, if the lung be penetrated by a fragment, the blood effused into the pleural cavity, through freely mixed with air, undergoes no decomposition. The air is sometimes pumped into the pleural cavity in such abundance that making its way through the wound, it inflates the cellular tissue of the whole body. Yet this occasions (as regards putrefaction) no alarm to the surgeon. Why air introduced into the pleural cavity through a wounded lung should have such wholly different effects from that entering through a permanently open wound penetrating from without, was to me a complete mystery till I heard of the germ theory of putrefaction, when it at once occurred to me that it was only natural that the air should be filtered of germs by the air passages, one of whose offices is to arrest inhaled particles of dust and prevent them from entering the air-cells. In truth this fact in practical surgery, when duly considered, affords as good evidence in support of the germ theory of putrefaction as any experiment that can be performed artificially."*

Here is a surmise which bears upon it the mark of genius, but which nevertheless needs verification. If in the place of the words "it is only natural," we were authorized to write "it is perfectly certain," the demonstration would be complete. Now, this is exactly what experiments with a beam of light enable us to do. One evening towards the close of last year, while pouring various gases across the dust track of a beam in the laboratory of the Royal Institution, the thought occurred to me of displacing by my breath the illuminated dust. I then noticed, for the first time, the extraordinary darkness produced by the air expired towards the end of an expiration. By an intentional effort of expulsion the lungs may be far more effectually emptied of air than by ordinary respiration; and by such an effort, which discharges the air from the interior portion of the lungs into the beam, the darkness is changed to absolute blackness. There is no speck or mote of any kind in such air. It is a true elastic fluid, without a trace of cloud or floating matter.

Thus, by ocular evidence we prove the filtering power of the lungs, and by the experience of surgery we prove the incompetence of air so filtered to produce putrefaction. The germs removed by the process of filtration are therefore the cause of the putrefaction, and its associated phenomena of animalcule life, which was to be demonstrated.

As a guide to the practical surgeon the establishment of this fact is plainly of the very highest importance. Professor Lister now avails himself of the filtering power of cotton wool, and by a proper lotion he kills those which may be scattered on the flesh. The cleansed wool placed upon the wound permits of a free diffusion of the air, but entirely intercepts the germs, and thus keeps the blood perfectly sweet. It is essential that no matter from the wound should reach the outside air, for such matter would open a highway for the animalcules. I may add that when the foregoing observations on the filtering power of the lungs were made I had no thought, and but little knowledge, of the germ theory. Their value as evidence is enhanced by the consideration that they are absolutely independent of all theoretic bias.†

I am, Sir, your obedient servant, | JOHN TYNDALL. | Royal Institution, April 6.

* *British Medical Journal*, 1868, p. 56.[7]

† The black wreaths produced by placing the flame of a spirit lamp underneath the track of a sunbeam may now be clearly though imperfectly seen in every drawing-room in London. The light, save that passing through a single aperture, ought, as far as possible, to be excluded. A candle flame also shows the effect, but very imperfectly.

Times, 7 April 1870, p. 5
RI MS JT/6/5/1/31–4

---

1. *chymstry*: an antiquated spelling of chemistry (often interchangeable with alchemy), its use became less common in the nineteenth century.
2. *animalcule*: 'little animals'; a term for microscopic organisms used from the sixteenth into the nineteenth century.
3. *Gay-Lussac*: Joseph Louis Gay-Lussac (1778–1850), a French chemist. He graduated from the Ecole Polytechnique in 1800 and entered the civil engineering school, the Ecole Nationale des Ponts et Chaussées. Gay-Lussac became professor of practical chemistry (1809) and professor of chemistry (1810) at the Ecole Polytechnique. From 1808 to 1832 he was also professor of physics at the Sorbonne, when he became chair of general chemistry at the Museum National d'Histoire Naturelle. He best known for his discovery of the law of volumes for gases, and his resultant paper on the subject (1802). It is now known as Gay-Lussac's Law (*CDSB*).
4. *the former*: Gay-Lussac did several studies of fermentation and asserted that oxygen was necessary for the appearance of microorganisms in fermentation. (J. L. Gay-Lussac, 'Sur le fermentation', *Annales de Chimie*, 76 (1810), pp. 245–60).
5. *heterogenist*: in debates around spontaneous generation in the nineteenth century, heterogenesis was the theory that living organisms could arise from organic matter alone, without parent organisms (*OED*).
6. *the germ theory*: in the late 1850s and early 1860s, Pasteur conducted experiments that challenged theories of spontaneous generation and demonstrated that microorganisms were necessary for fermentation, against the theories of Justus Liebig and Henry Charlton Bastian. He later proposed that microorganisms were also the agents of contagion. See J. Strick, *Sparks of Life: Darwinism and the Victorian Debates over Spontaneous Generation* (Cambridge, MA: Harvard University Press, 2000).
7. *British Medical Journal, 1869, p. 56*: J. Lister, 'An Address on the Antiseptic System of Treatment in Surgery', *British Medical Journal*, 2:394 (1868), pp. 53–6.

From Thomas Henry Huxley [6 April 1870][1] 3171

Geological Survey of England and Wales[2]

My dear Tyndall,

DAMN

the

L
e
c
t
u
r
e
s

T. H. H.

That's a practical application of Electricity for you.[3]

RI MS JT/1/TYP/9/2931
IC HP 8.77
Typed Transcript Only

1. *[6 April 1870]*: this letter is a response to letter 3169 of April 1870. See also letter 3167.
2. *Geological Survey of England and Wales*: Huxley was naturalist with the Geological Survey, which had been founded in 1835 under Henry de La Beche.
3. *for you*: Huxley was referring to his single lecture at the RI (see letter 3169, n. 5) and Tyndall's lecture course at the RI about electrical phenomena (see letter 3126, n. 3).

From W. Palmer[1] 8 April 1870 3172

German Gallery | 168 New Bond-street, | April 8, 1870

Dr. Tyndall:

Sir

I beg to call your attention to an opening made in the wall of your Institution[2] which opening overlooks these premises, the continuance of which if allowed to exist without some arrangement may grow to my detriment and injury: also to that of a steam-pipe made to discharge itself into the gutter, whereby damage and injury is done. An early attention to this will oblige[4]

Your obedient servant, | W. F. Palmer | for F. H. Evans,[3] Proprietor.

RI Manager's Minutes (1853–1874), XII, p. 322
Transcript Only

1. *W. Palmer*: not identified, but presumably a solicitor representing the German Gallery.
2. *your Institution*: the RI.
3. *F. H. Evans*: F. H. Evans, proprietor of the German Gallery, 168 New Bond Street, London.
4. *An early attention to this will oblige*: in the Manager's Meeting minutes, directly below the transcript of this letter, is: 'Resolved That Dr. Tyndall and Mr. Pole be appointed <to> a Committee to arrange this matter with Mr. Palmer'. 'Mr. Pole' is William Pole (1814–1900), an engineer and musician (*ODNB*), a member of the RI who was often a Manager, though not at this time. He would serve as a Manager beginning the following month (*Roy. Inst. Proc.*, 6 (1870–72), p. 142).

To Jane Barnard 11 April 1870 3173

Royal Institution | 11th. April 1870[1]

My dear Miss Barnard.

I send you the first notice of the Life and Letters.[2]

I have given it my very best attention, and I cannot note an error in statement or in style. I think it is above all things necessary that an atmosphere of freedom should surround this matter—and especially breathe through my sayings and doings in regard to it.

I shall be only too happy if any mistake be pointed out to me to correct it; or if any improvement be suggested to me to adopt it.

I will walk up tomorrow afternoon: I need a walk and will take it then; reaching you at about 4 p.m. I dine in this region.

Yours ever truly | John Tyndall.

I asked Mr. Gassiot's[3] opinion of the article, and I enclose his reply.[4]

RI MS JT/1/T/112
RI MS JT/1/TYP/12/4200

1. *11th. April 1870*: Tyndall wrote, and then struck through '10th' before writing '11th. April 1870'.
2. *Life and Letters*: H. B. Jones, *The Life and Letters of Michael Faraday*, 2nd edn, 2 vols (London: Longmans, Green, and Co., 1870). Note that the first edition, abbreviated in this volume as *Life of Faraday*, was also published in 1870.
3. *Mr. Gassiot*: John Peter Gassiot.
4. *his reply*: letter missing.

To the Editor of *Nature*[1] 11 April 1870 3174

DEATH OF PROFESSOR MAGNUS

ON the 4th of April, 1870, at a quarter-past 10 p.m., died peacefully, after a long illness, Dr. Gustav Magnus, Professor of Physics, and Director of the Physical Cabinet in the University of Berlin. He was an experimental philosopher of great and varied excellence, executing his work with the choicest apparatus and with the most conscientious care. His numerous labours are known to all students of physics, and they are such as to secure for him an enduring fame. On the 28th of April, 1851, I first saw Professor Magnus on his own doorstep in Berlin.[2] His aspect won my immediate regard, which was strengthened to affection by our subsequent intercourse. He gave me a working place in his laboratory, and it was there I carried out the investigation of Diamagnetism and Magne-crystallic Action, which is published in the *Philosophical Magazine* for Sept. 1851.[3] In 1853 I was again in Berlin,[4] and found under his roof the same ready help and sympathy. Professor Hirst and myself paid him a visit last summer; and he afterwards attended the Exeter Meeting of the British Association,[5] where his frank, genial, and gentlemanly demeanour were conspicuous to all. Over and above his direct contributions to Science, Prof. Magnus exercised a powerful indirect influence, through the kindly aid and countenance which he lent to young inquirers. When I bade him good-bye in 1851 his last words to me were, "If you should meet any really able young fellow, willing to work, and to whom such assistance as I can render would be valuable, send him to me." There are many such, now no longer young, who, like myself, will mingle a grateful memory of his goodness with their grief for his loss.

JOHN TYNDALL | Royal Institution, 11th April

Nature, 1 (14 April 1870), p. 607

---

1. *Editor of* Nature: Norman Lockyer (see letter 3026, n. 7).
2. *doorstep in Berlin*: Tyndall visited Marburg in October 1850, leaving there on 6 April 1851 on his way to Berlin. He experimented on diamagnetism at Magnus's laboratory.
3. *Philosophical Magazine for Sept. 1851*: J. Tyndall, '(Second Memoir) On diamagnetism and magnecrystallic action', *Phil. Mag.*, 2 (1851), pp. 165–88.
4. *I was again in Berlin*: Tyndall left London for the continent on 11 July 1853, spending time with Magnus in both Berlin and Paris and conducting some research while in Berlin.
5. *Exeter Meeting of the British Association*: the thirty-ninth meeting of the BAAS was held in Exeter from the 18–25 August 1869 (*Brit. Assoc. Rep. 1869*).

To Edward Youmans 13 April 1870 3175

April 13, 1870.

MY DEAR YOUMANS: I thank you more fully for the friendly interest you have taken in my affairs than for the money which has resulted to me[1] through the exercise of your kindness.

I have had many letters of the most gratifying description from the United States, and this is why I mentioned lectures in my last note to you.[2] I am not, however, certain whether it would not be better to pay you a visit without any thought of lecturing. I love freedom, and a scamper through the States, without the incubus of lectures, would be as instructive to me as it would be pleasant.

I saw Huxley last night. To him you have been acting as you have to me. The philosophers of England have much to thank you for. I was sorry to hear from Huxley that his little book[3] is not so successful in America as it might be. This surprises me, for it is an excellent piece of work. I wish I had time to do something similar in physics.

When I last saw Spencer he was flourishing. He told me he had written to you regarding an amanuensis. He endeavors to persuade me to lighten my labors in this way. But with me an amanuensis would not be so successful as with him. I have to rasp and rasp at my work myself before it pleases me.

With regard to the future I have to say that I am pinned this year by the meeting of the British Association at Liverpool;[4] next year I am pinned by my lectures and researches. If I go to the States without lecturing I could probably fly off in 1872.[5] But should I lecture, the needful preparations would throw the visit back to 1874. This is a long time to look forward to.

But whether I go or tarry, or whether I go as a lecturer or as a friendly visitor, it will make no difference in the feelings with which I reciprocate the kindness shown to me by your countrymen and yourself.

With best regards to Mrs. Youmans,[6] | Believe me, yours ever faithfully, | JOHN TYNDALL.

E. A. Youmans, 'Tyndall and his American Visit', *Popular Science Monthly*, 44 (February 1894), pp. 502–13, on p. 507

---

1. *the money which has resulted to me*: Youmans promoted Tyndall's work in the United States and to American publishers, from whom he received royalties.
2. *my last note to you*: letter missing.
3. *his little book*: possibly T. H. Huxley, *On the Physical Basis of Life* (New Haven: Yale College, 1869), but more likely T. H. Huxley & W. M. J. Youmans, *The Elements of Physiology and Hygiene; a Text-Book for Education Institutions* (New York: A. Appleton and Co.,

1868), which he co-authored with Edward Youmans' brother. Youmans often entered into agreements with Appleton to publish scientific works.

4. *meeting of the British Association at Liverpool*: the fortieth annual meeting of the BAAS was held in Liverpool from 14–21 September 1870 (*Brit. Assoc. Rep. 1870*).
5. *I could probably fly off in 1872*: Tyndall indeed visited the United States in 1872 for a multicity lecture tour, arriving in New York on 9 October and departing from there to England on 5 February 1873 (see *Ascent of John Tyndall*, pp. 303–13; A. S. Eve and C. H. Creasey, *Life and Work of John Tyndall* (London: MacMillan & Co. Ltd., 1945), chapter 14; K. Sopka, 'An Apostle of Science Visits America: John Tyndall's Journey of 1872–1873', *The Physics Teacher*, 10 (1972), pp. 369–75; and K. Sopka, 'John Tyndall: International Populariser of Science', in W. H. Brock, N. D. McMillan, and R. C. Mollan (eds), *John Tyndall: Essays on a Natural Philosopher* (Dublin: Royal Dublin Society, 1981), pp. 193–203; and E. A. Youmans, 'Tyndall and his American Visit', *Popular Science Monthly*, 44 (1894), pp. 502–13. Letters during the America lecture tour will be published in the forthcoming thirteenth volume of Tyndall's correspondence).
6. *Mrs. Youmans*: Catherine E. Youmans (née Newton Lee, *c.* 1819–94), the widow of lawyer and jurist, William Little Lee. She married Edward Livingston Youmans in 1861 and was his amanuensis and assistant in editorial work of her husband, who suffered from poor vision. There were no children from the marriage (E. A. Youmans, 'Sketch of Edward L. Youmans', *Popular Science Monthly*, 30 (1887), pp. 688–97).

To Louisa Baring [16 April 1870][1] 3176

White Hart Salisbury.[2] | Saturday ev[eni]ng. | off in a quarter of an hour.

I wish you a bright sunshiny heart this bright sunshiny morning. I am trying to get a little sunshine into my heart and a little strength and cleanness into my head. Yesterday I spent on Salisbury plain:[3] lying on the barrows, listening to the sough of the wind in the adjacent pine-wood, and wandering afterwards among the stones. The wind rusting through the pass has a more ghostly sound than elsewhere—curious thoughts and feelings fill the mud in the place. I had a good 20 mile walk yesterday.

I am now on my way to Swanage[4] where I wait till Thursday, and then return to town. I shall cross the Forest[5] not far from you, and shall be drawn towards you in spirit if not in body. Nor shall I forget Mazy,[6] indeed of often think of the child. I must some day chase away that timidity which I thought I noticed in her when I was last in Melchet.[7] I must not be myself when a child feels timid towards me—

Give my love to her, and accept so much of it as is right and proper yourself.

I am walking with an old fellow,[8] a German student: rough & plain outside, but deep & pure–and true at the core.

Yours ever | J. Tyndall

National Library of Scotland Ashburton Papers Acc.11388 N0.101

1. *[16 April 1870]*: the letter has been dated to Saturday, 16 April 1870. Tyndall writes that he is on his way to Swanage and writes letter 3177 on 19 April from Swanage.
2. *White Hart Salisbury*: White Hart was a hotel located in Salisbury.
3. *Salisbury plain*: Salisbury plain is a chalk plateau in central southern England located mainly in the county of Wiltshire. It is well known for its archaeological remains including Stonehenge.
4. *Swanage*: see letter 3045, n. 6.
5. *Forest*: the route from Salisbury to Swanage passes near Melchet Court, Hampshire, one of Louisa Baring's homes. Its construction was completed in 1868. The region to the south and south-west was extensively forested.
6. *Mazy*: Louisa Baring's daughter Mary Florence.
7. *Melchet*: see n. 5, and letter 2971, n. 5.
8. *an old fellow*: Henrich Debus. Tyndall went walking with Debus in Dorset over Easter (see letter 3186).

To the Editor of the *Times* 19 April 1870 3177

THE GERM THEORY OF DISEASE.[1] |
TO THE EDITOR OF THE TIMES.

Sir,—Amid the morass of wordy discussion and contradictory argument in which the germ theory of disease and the question of spontaneous generation are entangled, the fact and inference brought forward by Professor Lister[2] regarding the putrefaction of wounds appeared to me to present a spot of perfectly firm ground. They by no means involved the demonstration of the germ theory of disease in all its generality, and I thought that even those who are most hostile to the theory of its wider claims and bearings would be willing to concede that the origin of putrefaction, at least, was placed beyond dispute. Dr. Bastian's[3] letter in *The Times*[4] of last Wednesday causes me to fear that this admission on the part of those whom he represents is still distant, and that it was a delusion on my part to hope that it was near.

A similar delusion may cleave to my judgment when I say that in a letter which fills nearly a whole column of *The Times* Dr. Bastian has not, so far as I can see, brought forward anything beyond an opinion affecting, in one way or the other, the position of Professor Lister. He deprecates "indirect evidence," and speaks of the ambiguous experiments of Pasteur;[5] he refers to the virtues of turnip infusions and to the killing of germs by heat; he expresses amazement, and protests against laxity; he exhorts, he warns, he promises results "which will go very far towards the settlement of the much-vexed question as to the possibility* of 'spontaneous generation.'" But he successfully avoids offering the shadow of an argument calculated to influence either affirmatively or negatively the distinct issue raised in my letter of Thursday week.[6] Will Dr. Bastian forgive me for expressing the opinion that irrelevant writing has done, and is likely to do, more damage to the cause of exact medicine than the distinct statement of even erroneous views? In the one case we have a definite error on which to place the iron heel of logic or of fact; in the other we are met by a phantom which eludes both. The question raised in my letter lies in a nutshell. Either Dr. Bastian denies the inability of lung-filtered air to produce putrefaction or he admits it; if the former he can state his denial in three words; if the latter, I ask him to explain the fact.

Philosophical physicians have before now rendered immortal service to the cause of physical science; and, should the opportunity arise of making some small return for this service, it is not the philosophical physician who will scorn our offered aid. The germ theory of disease, says Dr. Bastian, "appertains most to the biologist and the physician." Granted. But it is, nevertheless, true that the very best modern work connected with this subject has been executed, not by a biologist, not by a physician, but by a chymist.[7] This is not the opinion of chymists only. Some of the most eminent biologists of the present time speak with no uncertain voice of the demonstrative force of the experiments of Pasteur.

Dr. Bastian refers to these experiments as "ambiguous." As I perused his statement the words of an honest Catholic writer,[8] familiar to me when a boy, came to my mind. "There is," said he, "more sound theology in one page of Luther than in all the writings of Thomas Aquinas." In like manner I would say that there is more solid science in one paper of Pasteur than in all the volumes and essays that have been written against him. The simple desire to render honour to whom honour is due induces me to write thus. But I would add that the suspense of judgment invoked by Dr. Bastian is already in exercise. If the "fragment of demonstrated truth" referred to in my letter be accepted, I am unpledged to more than a clear statement of the germ theory; and, should Dr. Bastian succeed in upsetting that theory and in establishing on its ruins the doctrine of spontaneous generation, no cheer that greets his success will be heartier than mine. I must, however, remark that his letter of Wednesday last contains nothing which would lead me to conclude that he is likely either to destroy the one doctrine or to establish the other.

I quite agree with Dr. Bastian that there is no proof that the organic matter of London air is all germinal matter. I do not believe it to be such. In fact, I have twice or thrice written down what the researches of others make the floating matter of London air to be, but tenderness for the public has withheld the description from the press. In the lecture, a report of which you did me the honour to publish, it is stated that "there is no respite from this contact with dirt, and the wonder is, not that we should from time to time suffer from its presence, but that *so small a portion of it* would appear to be deadly to man."[9] It is with reference to this "small portion" that I described the germ theory of disease. The really important point here is that the germs which are mixed up with the floating matter disappear when it disappears.

Permit me to place in a connected form before the readers of *The Times* the "indirect evidence" on which, to the surprise of Dr. Bastian, I rely. Schroeder[10] and Pasteur have demonstrated that air filtered through cotton wool is deprived wholly or in part of its power to produce animalcular life.[11] Why? Experiment with a beam of light answers the question; for, while it proves our ordinary air to be charged with floating matter, the beam pronounces air which has been carefully filtered through cotton wool to be visibly pure. There are no germs afloat in it; hence its impotence as a generator of life.

Again, Pasteur prepared 21 flasks, each containing a decoction of yeast, which he boiled, in order to destroy whatever germs it might contain. While the space above the liquid was filled with pure steam he sealed the necks of his flasks with a blowpipe. He opened ten of them in the damp still caves of the Paris Observatory, and 11 of them in the court-yard of the same establishment. Of the former only one showed signs of life subsequently. In nine out of the ten flasks no organisms of any kind were developed. In all the others organisms speedily appeared. Pasteur ascribed this unexpected result to the subsidence of the germs in the motionless air of the caves. Is his surmise correct?

The beam of light enables us to answer this question also. I have had a chamber constructed, the lower half of which is of wood, and the upper half of glass. On the 6th of February this chamber was closed, and every crevice that could admit dust or cause a disturbance of the air was carefully stopped. The electric beam, when sent through the glass, showed the air at the outset to be loaded with floating matter. The chamber was examined almost daily, and a gradual diminution of the floating matter was observed. At the end of a week the chamber was optically empty. The floating matters, germs included, had wholly subsided, and the air held nothing in suspension. Here again the ocular demonstration furnished by the luminous beam goes hand in hand with the experimental result of Pasteur.

Next comes Professor Lister's observation regarding the filtering power of the lungs. Turning once more to the luminous beam, we find that it declares air so filtered to be free from floating germs. Hence its inability to produce putrefaction, and the accompanying manifestation of animalcular life. Finally, as if to clinch the purely scientific experiments by a practical result, we find Professor Lister actually employing cotton wool to ward off the atmospheric germs, and, by its agency, keeping a large class of wounds sweet and pure. It is this intimate welding together of two independent series of observations that Dr. Bastian calls indirect evidence; it is my reliance on such evidence that fills him with "amazement." In matters of science I always sacrifice the emotions, otherwise I might permit myself a little amazement too.

For a long time to come I shall be unable to devote my attention to this subject, and this has caused me to write at what I fear you will consider inexcusable length. Were the question of less practical interest to humanity, I should not have troubled you either with this or my former communication.

I am, Sir, your obedient servant, | JOHN TYNDALLl. | Swanage, Dorset, April 19.

*The question relates to *facts*, not to possibilities.

Times, 21 April 1870, p. 8
RI MS JT/6/5/1/40–4

1. *The Germ Theory of Disease*: see letter 3170.
2. *Professor Lister*: Joseph Lister.
3. *Dr. Bastian*: Henry Charlton Bastian.
4. *Dr. Bastian's letter in The Times*: H. C. Bastian, 'The Germ Theory of Disease', *Times*, 13 April 1870, p. 4.
5. *Pasteur*: Louis Pasteur.
6. *my letter of Thursday week*: letter 3170.
7. *chymist*: chemist (see letter 3122, n. 4).
8. *honest Catholic writer*: Desiderius Erasmus Roterodamus, or Erasmus of Rotterdam (1466–1536), a Dutch philosopher and Christian scholar. Tyndall misquotes him: the words 'if I would but write against Luther; but he is a man of too great abilities easily to encounter and to say the truth, I learn more from one page of his, than from all the volumes of Thomas Aquinas' are attributed to Erasmus (*Catholic Encyclopedia*, vol. 5, online at https://www.newadvent.org/cathen/05510b.htm [accessed 4 August 2021]); T. Curtis, *The London Encyclopaedia; of Universal Dictionary of Science, Art, Literature and Practical Mechanics* (London: Thomas Tegg, 1829), p. 556).
9. *'there is no respite . . . would appear to be deadly to man'*: see 'Professor Tyndall on dust and disease', *Times*, 22 January 1870, p. 6. See also letter 3122, n. 1.
10. *Schroder*: Georg Friedrich Heinrich Schröder (1810–85), a German physicist, chemist, and mathematian. Schröder was professor of physics at various technical schools in Bavaria and Switzerland. In 1853 heand Theodor von Dush (1824–90) developed a process of air sterilisation using cotton as a filter. It could also be used for food conservation. Pasteur turned to this discovery for support (*CDSB*; *NDB*).
11. *animalcular life*: see letter 3170, n. 2.

To Thomas Archer Hirst 23 April 1870 3178

Royal Institution | Saturday | April 23/70.

My dear Tom.

I have just come in from Swanage[1] and am right grieved to learn on my arrival that you are not well. I would advise you to remain away were it not better under the circumstances that you should be here, where you can have good medical advice to heal your outward man, & true affection for your inner one.

From yours ever | John.

RI MS JT/1/T/987
RI MS JT/1/HTYP/563

1. *Swanage*: see letters 3045, n. 6, and 3176.

From Edward Sabine 25 April [1870][1] 3179

The Royal Society, | Burlington House, London. W. | Ap[ril] 25th

Dear Tyndall:

I have written to Wheatstone,[2] suggesting that he should propose Plateau for For[eign]: Mem[bership]:[3] in a letter to me, which I may read to the officers first, and then with their approval, to the Council. I have also written to Dr Robinson[4] to say that you have undertaken to propose Ångström,[5] and that I shall be glad if he will endeavour to find and send me the Memo which he wrote some months ago at my request regarding Ångström. We have 2 vacancies, and Plateau & Ångström would be most deserving names to fill them.

Sincerely yours | Edward Sabine

RI MS JT/1/S/36
RI MS JT/1/TYP/4/1336

1. *[1870]*: in 1870 Tyndall campaigned for Plateau and Ångström to be admitted to the RS as ForMemRS. See letter 3181.
2. *Wheatstone*: Charles Wheatstone.
3. *For[eign]: Mem[embership]*: of the RS. Tyndall had made extensive efforts to have Plateau elected (see letters 2975, n. 4).

4. *Dr Robinson*: (John) Thomas Romney Robinson (1793–1882), an Irish physicist and astronomer. Born in Dublin and raised in Belfast, Robinson studied at Trinity College, Dublin, where he met Ireland's Protestant scientific elites. He was appointed astronomer at the Armagh Observatory in 1823, where he was able to purchase through patronage instruments from some of the best optical instrument makers in Ireland for the observatory. Under Robinson the observatory focused on observational astronomy and star catalogues. He was President of the Royal Irish Academy from 1851 to 1856. Robinson's daughter, Mary Susanna, married George Gabriel Stokes (*DIB*; *ODNB*).
5. *Ångström*: Anders Jonas Ångström (1814–74), astronomer and physicist. He was professor of physics at the University of Uppsala from 1858 until his death. His work focused on emission spectra and spectral wavelengths were measured in angstroms. Elected ForMemRS on 24 November 1870, Ångström received the Society's Rumford Medal in 1872 (*CDSB*).

From George Hector Tyndale 25 April 1870 3180

Philadelphia April 25. 1870

My dear John

At his request I have taken the liberty of giving a letter of introduction to you to my friend Pliny E. Chase Esq.[1] of this City. Mr: Chase is an excellent man of the highest tone and honor. He is connected with the best families here and knows all the best of this country and is loved and respected by all. Very sincere and modest I could not say in my letter of him what I say here. But, if you can dear John, make him meet and know the best men and institutions and so help to make his visit as pleasant as you did Julius Hilgard's last year,[2] of which I wrote to thank you some time ago, please do it and add another obligation to the number I owe you.

Mr: Chase is well known here as a Man of Science, a prominent member of the American Phil[osophical]: So[ciety]: and of other Scientific Societies. He is a good Mathematician, and has written much—treating mathematically, social and physical statistics. He has lately made Meteorology his specialty. But more than this I have already said of his worth and modesty as a gentleman whom I very truly regard. I believe you will like him. Please introduce him to Tom Hirst and any others you can.

The other day came another "Times" from you with a letter of yours on the air germs[3] which I handed to my friend Lesley,[4] who is now editing a Mining &c. journal[5] here. I loaned your pamphlets on experiments upon the decomposition of vapors and light,[6] which you sent me some time ago, to Rev. Doc. Furness[7] who is charmed with them as has been every one else who saw your papers. By the way Doctor Furness told me that he wrote you some time ago,[8] and I think I wrote to you about him[9] at that time. You are known here as well, I fancy, as you at home and would meet many friends were you to come over. I speak now of the Northern States where you are best known.

Give my love to Tom Hirst who, should you come, will I hope come with you, and to Debus. Mrs: Tyndale[10] is well and sends her respects.

Always affectionately | your Cousin | Hector Tyndale

John Tyndall Esq | London

RI MS JT/1/T/62
RI MS JT/1/TYP/5/1682

1. *Pliny E. Chase, Esq.*: Pliny Earle Chase (1820–86), an American physicist and educator. He graduated with an A.B. from Harvard in 1839. While entering into business, Chase engaged in research in his leisure time and published articles in a number of journals including the *Amer. Phil. Soc. Proc.* and the *Phil. Mag.* He was also appointed Professor of Philosophy and Logic at Haverford College, Pennsylvania in 1871, remaining there until his death (*ANB*; P. C. Garrett, 'Memoir of Pliny Earle Chase', *Amer. Phil. Soc. Proc.*, 24:126 (1887), pp. 287–95).
2. *Julius Hilgard's last year*: see letters 2987, 2996, and 2998.
3. *a letter of yours on the air germs*: see letters 3170 and 3177. For published versions of these letters see *Times*, 7 April 1870, p. 5; and *Times*, 21 April 1870, p. 8.
4. *Lesley*: Joseph Peter Lesley (see letter 2987, n. 13).
5. *editing a Mining &c. journal*: in May 1869 Lesley became editor of the *United States Railroad and Mining Register* (M. Lesley Ames, *Life and Letters of Peter and Susan Lesley*, 2 vols (New York/London: Knickerbocker Press, 1909), p. 71).
6. *pamphlets on experiments upon the decomposition of vapors and light*: possibly 'On a New Series of Chemical Reactions Produced by Light' (see letter 2981, n. 3), 'The Generation of Clouds by Actinic Action' (see letter 2961, n. 1), and 'On the Action of Rays of High Refrangibility' (see letter 2981, n. 3).
7. *Rev. Doc. Furness*: William Henry Furness (1802–96), an American Unitarian minister. Furness was born in Boston and graduated from Harvard in 1820 and subsequently Harvard Divinity School in 1823. He served as a minister of the First Unitarian Church of Philadelphia from then until retirement in 1875. He had a lifelong friendship with Ralph Waldo Emerson who he first met at Boston Latin School as a child. His most significant area of research was the life of Jesus, and he published several books on his life (*ANB*).
8. *he wrote you some time ago*: letter missing.
9. *I wrote to you about him*: letter missing.
10. *Mrs: Tyndale*: see letter 2987, n. 5.

From Edward Sabine 26 April [1870][1] 3181

13. Ashley Place | Ap[ril]. 26.

Dear Tyndall

I have been for many years past under the impression that a medal had been awarded to Mr Joule[2]—If indeed it is not so, his researches on the Mechanical Equivalent of Heat[3] (specially) were most meritorious. In any case if his name is to be now brought forward for that, or any kindred researches, you are specially the person[4] by whom the proposition should be made.

I should make the same remark in reference to Dr. Mayer[5] of Heilbron.[6] If neither of these gentlemen have received suitable distinction from the R[oyal]. S[ociety]., their researches have not been wanting in the due merit; but those who by their engagement in corresponding enquiries ought to be their advocates, have not hitherto fulfilled their duty.

I think it not unlikely that Wheatstone,[7] to whom I have written about Plateau,[8] may be pleased at the opportunity of advocating Plateau's claims himself, which will release you from being the principal person to bring his name forward, whilst it leaves you quite free to advocate his claims.[9] So also in the case of Ångstrom;[10] Dr Robinson[11] writes to me today to say that he is completing a statement of his claims to be ready before the Council is summoned to elect into the two Vacancies. In such case, and supposing you propose Dr Mayer, there will be 3 Candidates, & two Vacancies—the one who is not chosen will stand over to the next election with every prospect of succeeding then.

<Rest of letter missing>

RI MS JT/1/S/37
RI MS JT/1/TYP/4/1337

---

1. *[1870]*: in 1870 Tyndall campaigned for Plateau and Ångström to be admitted to the RS as ForMemRS. See letter 3179.
2. *a medal had been awarded to Mr Joule*: James Prescott Joule had, indeed, been awarded the Royal Medal of the RS for his 1850 paper on the mechanical equivalent of heat. See n. 3.
3. *Mechanical Equivalent of Heat*: J. P. Joule, 'On the Mechanical Equivalent of Heat', *Phil. Trans.*, 140 (1850), pp. 61–82.
4. *you are specially the person*: this letter continues a discussion begun in letter 3179. Presumably Tyndall responded to that letter, but his response is missing.
5. *Dr. Mayer*: Robert Mayer.
6. *Heilbron*: Heilbronn, a city in the southern German state of Baden, now Baden-Württemberg.

7. *Wheatstone*: Charles Wheatstone.
8. *Plateau*: Joseph Plateau.
9. *you quite free to advocate his claims*: Tyndall was campaigning for Plateau to be elected ForMemRS (see letters 2975, n. 4).
10. *Ångstrom*: Anders Jonas Ångström (see letter 3179, n. 5). Ångström was elected ForMemRS on 24 November 1870.
11. *Robinson*: Thomas Romney Robinson (see letter 3179, n. 4).

To Ellen Busk[1] 29 April 1870 3182

Royal Institution of Great Britain | 29th Ap[ril] '70

My dear Mrs. Busk

You will "confound" my stupidity. On Monday last I bound myself to the engagement referred to in the accompanying note;[2] and hence must forego what would have been a great pleasure to me by seeing just */you/* once more.

Ever yours | John Tyndall

Waller Manuscript Collection, Uppsala University, gb-01756

1. *Ellen Busk*: see letter 3109, n. 1.
2. *the accompanying note*: enclosure missing.

From William Frederick Pollock 30 April [1870][1] 3183

Exchequer Office | Lincolns Inn | Saturday 30th April

Dear Tyndall,

Many thanks for the sight of your contribution to the Academy[2]—& for your kind allusion to my humble performance in it.

In all that you have written about the unfortunate affair of the Royal Society election[3] there is not one word that the warmest friend & admirer of Faraday could object to. It would indeed have been strange if what is written by the very warmest friend & admirer—to wit John Tyndall—could have afforded matter of objection to any other friend & admirer.

Whether Davy's[4] friends & admirers will be equally pleased with what you have written I will not venture to conjecture.

All I know of the matter is derived from Bence Jones' biography[5]—neither do you bring forward any fresh facts.

Of course I may been wrong in ascribing a motive for conduct in itself unseemly at best—and requiring explanation—I endeavoured to apply the ordinary rules of judging human conduct—& may have been wrong in using the words "private jealousy"—& should be glad to feel certain that Davy—so great in himself, & almost always so good to Faraday—whom he backed very loyally in the affair with Wollaston[6]—was quite exempt from human weakness in the matter of Faraday's election. But after all you only reduce "jealousy" to pique—& a feeling of surprise & irritation—natural enough—that a man so recently in a very subordinate position—& still his own humbly paid assistant should be mounting the same platform, as a scientific discoverer, as himself.

The feeling was, as I say, natural—but was it worthy of such a man as Davy, to give vent to it as he did?

Would Faraday have done so? Would you have done so? I know not.

What a jolly night we had on the philosophic heights with you—that mulled claret was capital—& enabled one to breathe the attenuated atmosphere at those altitudes with greater ease.

Juliet[7] wished to read the slips—& would send them back to you.

Yours ever | W FP

RI MS JT/1/P/254
RI MS JT/1/TYP/6/1879

---

1. *1870*: this letter has been dated to 1870 through its references to Henry Bence Jones' *Life of Faraday* and the second edition of Tyndall's *Faraday as a Discoverer*, both of which were published in 1870.
2. *your contribution to the Academy*: see letter 3159, n. 2.
3. *unfortunate affair of the Royal Society election*: a reference to Humphry Davy's objection to Michael Faraday's nomination for membership of the RS (see *Life of Faraday*, pp. 306–14, 339–41).
4. *Davy*: Humphry Davy.
5. *Bence Jones biography*: see n. 1.
6. *in the affair with Wollaston*: in 1821 Faraday was accused of stealing his electro-magnetic ideas and experiments from British chemist and physicist William Hyde Wollaston (1766–1828). Although Wollaston seems not to have perceived Faraday's work in this way, the accusation followed Faraday and emerged again, when he was seeking nomination to the RS (*Life of Faraday*, pp. 336–45).
7. *Juliet*: Juliet Pollock.

To William Sharpey[1] [1 May 1870][2] 3184

<letter begins here> of solid angles by those films. He studied and expounded the influence of viscosity. His experiments & reasoning have naturally led him to enquire into the radius of sensible molecular attractions. He finds the limit within which such attraction is exerted to be 1/17000th of a millimeter. This question, together with that of the size of molecules themselves, is now in the hands of William Thompson.

He afterwards examined the question of the limit of stability of liquids; & attaching for example two circles of wire to his sphere of olive oil, withdrawn from the action of gravity, and withdrawing the circles from each other, the oil by its adhesion also draws the 2 spheres in forms & becomes elongated. Plateau shows that the equilibrium is stable as long as the ratio of the length of the figure to its diameter does not exceed the quantity π or the ratio of the circumference of a circle to its diameter. Various other curious, and interesting papers might be referred to, among them his researches on the flow of liquid veins, and on the problem of Mähomed's Coffin.[3]

In 20 years & more Plateau has worked at these subjects through the hands & eyes of his assistants. For twenty years & more he has been totally blind. His experiments notwithstanding are of the most exquiste delicacy; and the reasonings by which they form the material are as acute *[1 word illeg]* as the experiments themselves are beautiful.

He is a member of principle scientific societies of the continent. He is corresponding member of the Institute of France,[4] of the Berlin Academy,[5] of the Academy of Belgium,[6] and numerous other societies & academies. In his old age he feels, I hear, keenly the apparent neglect of the Royal Society.[7] I hope the council will elect him.[8]

Yours ever John Tyndall
Dr. Sharpey Ser. R.S.

RI MS JT/1/T/1547

---

1. *William Sharpey*: William Sharpey (1802–80), a Scottish physiologist who studied humanities, natural philosophy, and medicine at the University of Edinburgh. From 1836 to his retirement in 1874 he held the chair of anatomy and physiology at University College, London. He was elected FRS in 1839, served as Secretary of the RS from 1853 to 1872, and vice-president from 1872 to 1874 (*ODNB*).
2. *[1 May 1870]*: this letter is placed between 26 April 1870, when Sabine approved Tyndall's campaign in support of Plateau's nomination as ForMemRS (letter 3181) and 23 May, when Plateau responded to a (now missing) letter of Tyndall's informing him that the

nomination was successful (letter 3192). Given the time needed to submit this letter to the RS prior to the election and the time needed for Tyndall's happy news to reach Plateau, the editors have placed this letter in the earlier part of the possible time span.

3. *Mähomed's Coffin*: a popular experiment in which a small metal object is held suspended in the air by magentism. The name references a legend that the coffin of the Prophet Mohammed hung suspended between heaven and earth.
4. *Institute of France*: see letter 3138, n. 4.
5. *Berlin Academy*: Prussian Academy of Science. See letter 3168, n. 26.
6. *Academy of Belgium*: Royal Academy of Science, Letters and Fine Arts of Belgium. See letter 3168, n. 25.
7. *the apparent neglect of the Royal Society*: Plateau expressed his dissatisfaction that he had not yet been elected ForMemRS in letters 3138 and 3168.
8. *will elect him*: for more on Tyndall's efforts to have Plateau elected ForMemRS, see letter 2975, n. 4.

From William Thomson 2 May 1870 3185

Largs | Ayrshire[1] | May 2, 1870

My Dear Tyndall

Your letter of the 27th[2] reached me after a detour by Edinburgh and therefore Largo[3] (however clear the s).

I shall be happy to second your proposal that Mayer[4] be elected a foreign member of the Royal Society.[5]

I am very glad to hear of the prospect of the Copley medal[6] being awarded to Joule—which I presume your suggestion to General Sabine[7] will bring to pass.[8]

Yours always truly | William Thomson

RI MS JT/1/T/16
RI MS JT/TYP/5/1554

---

1. *Largs | Ayrshire*: Largs, is a town on the Firth of Clyde, North Ayrshire, Scotland, located about 33 miles west of Glasgow. Thomson built a home there is 1875 and his later title was Baron Kelvin of Largs.
2. *your letter of the 27th*: letter missing.
3. *Largo*: Largo is a parish in Fife, Scotland, which contains two villages: Upper Largo and Lower Largo. Thomson is complaining that the postman sent his letter to Largo instead of Largs.
4. *Mayer*: Robert Mayer.
5. *elected a foreign member of the Royal Society*: see letters 3179 and 3181.

6. *Copley medal*: the RS's oldest and most prestigious medal for scientific achievement. First awarded in 1731 to recognise important discoveries made by experiment. It is thought to be the oldest scientific prize in existence.
7. *your suggestion to General Sabine*: see letter 3181.
8. *will bring to pass*: Tyndall later leveraged his championing of Joule's nomination in this year to lobby for the medal being awarded to Mayer the following year (*Ascent of John Tyndall*, p. 278).

To George Hector Tyndale 7 May 1870 3186

Royal Institution | 7th May 1870

My very dear Hector

Anything I can do to render the stay of your friend[1] in London agreeable to him shall be done with pleasure. Your letter[2] has just come, but he has not yet arrived.

"He hides in pure transparency".[3] This line was in my mind repeatedly during the preparation of that lecture. By the way a few nights ago I saw a couple of volumes from Emerson[4] on Carlyle's table at Chelsea.[5] I opened one of them and there was glad to see Emerson's own hand writing addressing the book "with unchangeable affection" to the old warrior. I carried the volumes home with me I was so pleased. I like this loyalty to a grand old man.

Tom,[6] I am sorry to say, has been recently suffering from a severe skin attack, brought on by lowering himself through overwork. The worst is now past—He is bright & cheerful and on the straight road to recovery.

I spent a few pleasant days at Easter with Debus on the Dorset[7] coast. The poor fellow is also overworked at Clifton College.[8] Tom has lately got a permanent appointment in connexion with the university of London which will probably give him more time.

I heard Motley[9] deliver a capital speech[10] at the Royal Academy[11] dinner. I met his wife[12] sometime ago at dinner, and at first I did not know that we should get on together. But we did. She is blunt, but full of good feeling.

I have had many invitations to go to America, and I think I will run over within the coming two or three years. Whether I will lecture or not is a question. I think I should enjoy myself more if I were unfettered. The kindness of the American people is beautiful & touching to me. I believe you are right when you say that on intellectual grounds I have as many friends there as here.

To Hilgard give my best remembrance.[13] He was altogether a bright phenomenon. Cheerful intellectual & vigorous. I liked him much.

And kindest remembrances to Mrs. Tyndall.[14]

Believe me dear Hector | Yours affectionately | John.

RI MS JT/1/T/1450
RI MS JT/1/TYP/4/1683–4

1. *your friend*: Pliny E. Chase (see letter 3180, n. 1).
2. *Your letter*: letter 3180.
3. *"He hides in pure transparency"*: R. W. Emerson, 'Wood Notes II' (1841), in R. W. Emerson, *The Major Poetry*, ed. A. J. von Frank (Cambridge, MA: Harvard University Press, 2015), line 374.
4. *Emerson*: Ralph Waldo Emerson.
5. *Chelsea*: Thomas Carlyle and his wife lived in the south west London neighbourhood of Chelsea, close to the Thames River.
6. *Tom*: Thomas Hirst.
7. *Dorset*: a county in South West England on the English Channel coast.
8. *Clifton College*: see letter 2996, n. 8.
9. *Motley*: John Lothrop Motley (1814–77), an American historian and diplomat. Motley specialised in the early Dutch Republic, and his three-volume *Rise of the Dutch Republic* (1856) was well-received. He was a member of the Boston intellectual circle, the Saturday Club, and travelled widely. In 1869 he was appointed US ambassador to the United Kingdom, but a falling out with President Ulysses Grant led to his recall in 1870, although he continued to reside in Europe (*ANB*).
10. *a capital speech*: Motley gave a toast on 'The Prosperity of the United States' at the annual dinner of the Royal Academy of the Arts on 2 May 1870. Dinner speaker was Charles Dickens, in his last public appearance; he died on 9 June 1870. Dickens spoke on 'Literature', and Prime Minister William Gladstone also gave a speech, on 'Her Majesty's Ministers' (*The Works of Charles Dickens: Letters and Speeches*, 2 vols (London: Chapman and Hall, 1908), vol. 2, pp. 528–30).
11. *Royal Academy*: Royal Academy of the Arts.
12. *his wife*: Mary Benjamin.
13. *To Hilgard give my best remembrance*: Julius Erasmus Hilgard visited London in May 1869 and was entertained by Tyndall (see letters 2996 and 2998).
14. *Mrs. Tyndall*: Julia Tyndale (see letter 2987, n. 5).

From Heinrich Debus 8 May 1870 3187

Clifton Coll[ege] | May 8. 1870

My dear John,

I enclose with this note[1] Budds[2] Certificate for the Royal Society[3] He informs me that Paget,[4] Sir T. Watson,[5] Dr Farr[6] of Registrar General Office, Lankester,[7] Carpenter,[8] Parke's[9] & Dr Acland[10] are men likely to sign it. Will you kindly look at it and see whether it is in order, & if so, either return it to me or get some of your friends to sign it,[11] just as you think it best.

I am now without the comfort of your company, without the cheering influence of your conversation. Test tubes, bottles, boys & smells surround me, and what else comes near me does not do me much good. I have just commenced work at the college[12] and my head commences to know again.

Yours always | H. Debus.

RI MS JT/1/D/39

---

1. *I enclose with this note*: enclosure missing.
2. *Budds*: William Budd.
3. *Certificate for the Royal Society*: nomination certificate for election as FRS. Tyndall and Debus were campaigning for Budd's membership. See also letters 3189 and 3196.
4. *Paget*: James Paget (1814–99), First baronet, an English surgeon. He apprenticed in Great Yarmonth and in 1834 enrolled in St Bartholomew's. In May 1836, Paget qualified as a member of the Royal College of Surgeons. He practiced medicine in London and published extensively on medical topics (*ODNB*).
5. *T. Watson*: Thomas Watson (1792–1882), a British physician. Watson received his BA (1815) and MA (1818) from St John's College, Cambridge. He obtained a university licence from Cambridge in 1822 and an MD in 1825. In 1859 he was elected FRS. He became president of the Pathological Society in 1857 and the Clinical Society in 1868 (*ODNB*).
6. *Dr Farr*: William Farr (1807–83), a British statistician and epidemiologist. He became a licentiate of the Society of Apothecaries in 1832 and made a living from public lectures and medical journalism, before joining the General Register Office as a statistician in 1839. His work not only resulted in promotion to superintendent of the statistical department but also allowed him to develop methodologies which became standard for those working in vital statistics and epidemiology. He was a recurrent witness before parliamentary investigations such as the Royal Commissions on Mines (1864) and the Royal Sanitary Commission (1868–9). Concerned with the discovery of laws of disease, and the waste of human life due to preventable conditions, Farr used his reports to advocate for public health reform. He was elected FRS in 1855 (*ODNB*).
7. *Lankester*: presumably Edwin Lankester (1814–74), an English surgeon and natural philosopher. Lankester studied medicine at the Univesity of London, where he also excelled in botany. From 1842–56 he held a lectureship at Grosvenor Place medical school and obtained a position at the New College of London in natual science in 1850. Lankester wrote on medical reform for the *Lancet* and the *Athenaeum* and was active in medical societies and public health and medical reform initiatives (*ODNB*).
8. *Carpenter*: William B. Carpenter.
9. *Parkes*: Edmund Alexander Parkes (1819–76), an English physician and hygenist. Parkes studied medicine at University College, London, graduating with his MB degree in 1841. He became a member of the Royal College of Surgeons and entered the Army Medical

Service. Parkes researched and published on aetiology and pathology. He was an important figure in the reform of military medicine during and after the Crimea war and was most prominent for his role as a hygienist (*ODNB*).

10. *Dr Acland:* Henry Wentworth Dyke Acland (1815–1900), an English physician. Following undergraduate studies at Oxford, he studied medicine at St George's Hospital, London before undertaking clinical training in Edinburgh. He completed his medical examinitions in 1846, took a DM in 1848, and was elected a fellow of the Royal College of Physicians in 1850 (*ODNB*).
11. *get some of your friends to sign it*: Budd was elected FRS on 8 June 1871. His nomination certificate was signed by Edward Frankland, Henry Salter, Samuel James Augustus Salter, Robert Hunt, William James Erasmus Wilson, Thomas Watson, George Burrows, George Busk, Henry Holland, Richard Partridge, William B. Carpenter, Francis Sibson, John Simon, and Edmund Alexander Parkes (RS, Certificates of Election and Candidature, EC/1871/02). This document has been digitised and can be found in the archival catalogue of the RS at https://catalogues.royalsociety.org/CalmView [accessed 20 December 2020]). See also letter 3189.
12. *the college*: Clifton College (see letter 2996, n. 8).

From Louis Pasteur [9 May 1870][1] 3188

Villa Vicentina 9 mai 70. | près Cervignano, Illyrie, (Autriche) | Voie de Vienne.

Monsieur,

Vous devez être surpris de n'avoir pas encore reçu de réponse à l'aimable lettre que vous m'avez adressée. C'est que j'ai voulu attendre la publication d'un ouvrage que je viens de terminer, afin d'avoir le plaisir de vous faire hommage d'un des premiers exemplaires. Vous devrez le recevoir en même temps que cette lettre.

Il n'a fallu rien moins que cette circonstance pour que je tarde tant à vous remercier des paroles flatteuses que vous adressez à quelques unes de mes observations, et pour vous féliciter d'être entré dans la voie où vous êtes pour le moment. Si vous persistez à la suivre elle sera semée des plus heureux résultats. Je ne connais personne qui ait au même degré que vous le talent de l'expérimentation. D'autres ont peut-être votre rigueur, mais il n'en est aucun qui ait votre brillante imagination; aussi tout ce que vous avez touché jusqu'à présent s'est réduit en or. Je serai donc toujours extrêmement flatté d'apprendre que mes recherches ont votre approbation motivée.

Permettez-moi de terminer ces quelques lignes que je dois dicter, vaincu que je suis par la maladie, en vous faisant observer que vous rendriez service aux colonies de la Grande-Bretagne en répandant la connaissance de ce livre et des principes que j'établis touchant la maladie des vers à soie. Beaucoup de ces colonies pourraient cultiver le mûrier avec succès et en jetant les yeux sur mon ouvrage vous vous convaincrez aisément qu'il est facile aujourd'hui, non seulement d'éloigner la maladie régnante mais en outre, de donner aux récoltes de la soie une prospérité qu'elles n'ont jamais eue.

Veuillez agréez, Monsieur, l'assurance de mes sentiments les plus distinguées. | L. Pasteur

Villa Vicentina, 9 May 70. | near Cervignano, Illyria, (Austria) | Voie de Vienne.

Sir,

You must be surprised to not yet have received a reply to the kind letter that you sent me.[2] It is that I wanted to wait until the publication of a book[3] that I have just finished, in order to have the pleasure of honouring you with one of its first copies. You should receive it at the same time as this letter.

It would have taken nothing other than such a circumstance for me to take so long to thank you for the flattering words that you sent me about some of my observations, and to congratulate you on embarking on the path where you are currently. If you continue to follow it, it will be full of the most desirable results. I do not know anyone who is as talented as you in experimentation. Others might have your rigorousness, but there is no one who has your brilliant imagination; also, everything that you touched so far has turned into gold. Thus, I will always be extremely flattered to learn that my research has your enthusiastic approval.

Allow me to finish these few lines that I have had to dictate, as overcome as I am by sickness, by showing you that you would render great service to the colonies of Great Britain by spreading the knowledge of this book and the principles that I established regarding the disease affecting silkworms. Many of these colonies could grow mulberry successfully, and by perusing my book you will be easily convinced that today it is easy not only to drive off the reigning disease, but also to give silk crops a prosperity that they have never had before.

Please accept, Sir, the assurance of my highest regards | L. Pasteur

RI MS JT/2/10/478–9

---

1. *[9 May 1870]*: date added by a different hand.
2. *the kind letter that you sent me*: letter missing.
3. *a book*: L. Pasteur, *Etudes sur la maladie des vers à soie, moyen pratique assuré de la combattre et d'en prévenir le retour* (Paris: Gauthier-Villars, 1870).

To Heinrich Debus [10 May 1870][1] 3189

Monday

My dear Heins,

I have not a full sheet of paper on which to spread my thanks for your last most odoriferous communication[2]—for it brought to the bricks and mortar, the dull air, and the dirty windows of London bloom from the country, a clear atmosphere, and those shiny windows which are never seen here, but which are pleasant to look upon in their own native home. I do not think you can be very badly off when you are able to write in such melodious strains. Might I ask when you will be free, and if when free you will do me the inexpressible pleasure of accompanying me to Switzerland?

We must make a strong effort to get in our friend[3] Budd.[4] If he does not get in on a first attempt it will be through the apathy or opposition of people who are not half as fit for the R[oyal] S[ociety]. as he is—But I think we shall carry him.

Yours ever | John Tyndall

RI MS JT/1/T/325
RI MS JT/1/TYP/7/2372

---

1. *[10 May 1870]*: this letter is a response to letter 3187 of 8 May 1870. 10 May 1870 was the next Monday.
2. *your last most odoriferous communication*: see letter 3187.
3. *get in our friend*: Tyndall and Debus were canvassing for Budd to be nominated and confirmed as FRS. See letters 3187 and 3196.
4. *Budd*: William Budd.

From Maria Ellery MacKaye[1] 17 May 1870 3190

Everett Street | Cambridge U.S.A. | May 17th. 1870
Prof. Tyndall

Dear Sir,

You may have entirely forgotten an American mother and child to whom you were exceedingly kind one beautiful day at Zermatt, July 29th. 1867.[2]

Talking to Mr Emerson a little while ago of our wonderful experiences that day and the way in which you spoke of him as we walked over the "Riffelhorn"[3] together, he was very much pleased and gave me his photograph just taken to send to you, adding his autograph on my suggestion.

I also enclose a freshly gathered blossom of the "Rhodora"[4] celebrated by him you remember.

My "edelweiss" is carefully treasured and is historical as well as symbolical to me.

You spoke too very kindly of my young cousin Clarence King.[5] It may interest you to know that he will soon send to press the Report of the joint Survey[6] on which he has been engaged for the last three years.

With cordial remembrances from my son Harry[7] and myself and an earnest hope that we may some day before long welcome you on this side of the Atlantic.

Ever yours | Maria Ellery McKaye.

Prof. Tyndall | F.R.S.

RI MS JT/1/TYP/2/787
LT Typescript Only

---

1. *Maria Ellery MacKaye*: third wife of James Morrison MacKaye (or McKay) (1804–88), an American lawyer and abolitionist who founded Wells Fargo and Western Union (W. Ellis, *Norwich University, 1819–1911: History, Graduates, Roll of Honor*, 3 vols (Montpelier, VT: Capital City Press, 1911), vol. 3, p. 731).
2. *You may . . . 1867*: Tyndall was quite taken with the child and his father (James Morrison MacKaye), but Maria Ellery MacKaye does not seem to have made a lasting impression on him. See letter 2627, *Tyndall Correspondence*, vol. 10.
3. *Riffelhorn*: a mountain in the Pennine Alps, south of Zermatt in the Swiss canton of Valais. See letters 3030, n. 15, and 3027, n. 9.
4. *"Rhodora"*: an 1834 poem by Ralph Waldo Emerson (R. W. Emerson, 'The Rhodora,' in *Poems* (Boston: James Munroe and Company, 1847), p. 59).
5. *Clarence King:* Clarence King (1842–1901), an American geologist, mountaineer, and first director of the United States Geological Survey. A student of geologist Josiah Whitney (1819–96), King worked on the California Geological Survey in the early 1860s and, as he was influenced by Tyndall's descriptions of the Alps, he named Mount Tyndall in the Sierra Nevada in 1864 in his honor. He was central to the establishment of the Geological Exploration of the Fortieth Parallel (n. 6). King retired in 1881 but continued to focus on his geological research, as well as being active in the mining industry as an engineer and consultant. His book *Mountaineering in the Sierra Nevada* (Boston: James Osgood & Co., 1872) was well received; he recounts the ascent and descent of Mt. Tyndall in chapters 3 and 4, pp. 49–93 (*ANB*; *CDSB*).
6. *Report of the joint Survey*: C. King, *Report of the Geological Exploration of the Fortieth Parallel*, vol. 1 (Washington: Government Printing Office, 1870). This was the first study of the natural resources of the mountainous areas near the Union and Central Pacific railways and the Cordilleran Mountain System. Six more volumes were published through

1880 (E. C. Carter II (ed.), *Surveying the Record: North American Scientific Exploration to 1930* (Philadelphia: American Philosophical Society, 1999), pp. 65–6).

7. *Harry*: Harry Godwin MacKaye (1856–1913).

To Margaret Ginty 22 May 1870 3191

Royal Institution | 22nd. May 1870.

My dear Mrs Ginty,

It was a cause of regret to me that you left London without my seeing you, and I have often thought of writing to you since. But this London, with its demands upon one's time, so harries one's thoughts from one subject to another that many a good intention is forgotten or has to be postponed. It grieved me when you were here to learn the serious state of your daughter's[1] health. Indeed I felt more for you far than I expressed. I trust since your return to Rio,[2] that you find matters agreeable to you. If matters can be arranged peaceably it is best that they should be so arranged. Of course it is your duty to look after your own interest and the interest of your other children; but if this can be done without bitterness or quarrelling it is best that it should be so done.

Your dividend for the last half year has been already in my hands for some weeks; but I have as yet received no instructions[3] what to do with it. Mr Worthington[4] is also unable to aid me, and I believe Mr James[5] is not now in London. Mr Worthington promised to write to you for instructions, and the probability is that he will know your wishes before this note reaches you. If he should not have written to you, then pray let me know what you wish to be done with the money.

I hope your son Willie[6] is going on to his mother's satisfaction—He bore a very good character at Queenwood,[7] and to my own observation of him he appeared to be a modest, gentle, kindly young fellow. I feared to give you any advice that would tend to draw him into a scientific life. For though genius in science is almost sure to be successful, still genius is so rare, and mere cleverness is so common among young men devoted to science, that the risk in making one's life and fortune dependent on science is very great.

Again I hope you are well and that you find your home pleasant to you. The letter of your son-in-law,[8] written when your husband died, gave me a very high opinion of him. It was with great concern that I heard that matters were not quite smooth between you. I hope that he will show towards you the good feeling and consideration due to your relation to him.

With this and other good wishes believe me | very faithfully yours | John Tyndall.

RI MS JT/1/TYP/11/3661
LT Typescript Only

1. *your daughter*: Kathleen Ginty, daughter of Margaret and William Ginty (*c.* 1820–66).
2. *since your return to Rio*: Tyndall's friend William Ginty and his family moved to Rio de Janiero, Brazil, in 1854. William died there in 1866.
3. *I have as yet received no instructions*: Tyndall assisted Margaret Ginty in legal matters after the death of her husband. See letters 3062 and 3113.
4. *Mr Worthington*: Edward Worthington (see letter 3062, n. 1).
5. *Mr James*: Bartlett James (see letter 3113, n. 2).
6. *Willie*: William Robert Ginty, son of Margaret and William Ginty.
7. *Queenwood*: Queenwood College, Hampshire, was established by George Edmondson (Society of Friends) and was one of the first colleges in England to teach science in the laboratory. Tyndall began teaching there in 1847 and taught surveying, mechanics, algebra, and geometry until September, 1848, when he left for Marburg to pursue a Ph.D. He returned in 1851 and remained until 1853. Thomas Hirst replaced Tyndall as mathematics teacher there in 1853 (W. H. Brock, 'Queenwood College Revisited,' article XVII in *Science for All: Studies in the History of Victorian Science and Education* (Aldershot: Ashgate Variorum, 1996); D. Thompson, 'Queenwood College, Hampshire', *Annals of Science*, 11 (1855), 246–54).
8. *The letter of your son-in-law*: letter missing; though presumably from a Mr. Holman who later married Kathleen. Tyndall refers to this letter in letter 2470, in M. D. Barton, I. R. Morus, and J. Ungureanu (eds), *The Correspondence of John Tyndall, Volume 9: The Correspondence, February 1865–December 1866* (Pittsburgh: University of Pittsburgh Press, 2022). Future letters between Tyndall and Margaret Ginty show that Holman was rather unsympathetic to the family, and on the passing of Bartlett James, who had taken on the role of executor to her late father's will, Kathleen's sister Fanny believed that her husband James Unwin should take over the role rather than Holman (see RI MS JT/1/TYP/11/3666, RI MS JT/1/TYP/11/3671, RI MS JT/1/TYP/11/3672, and RI MS JT/1/TYP/11/3682).

From Joseph Plateau 23 May 1870 3192

Gand, 23 Mai 1870

Mon cher Monsieur Tyndall,

Recevez mes chaleureux remerciements pour la bonne nouvelle que vous avez bien voulu me communiquer, et pour toutes les peines que vous vous êtes données au sujet de ma nomination. Je n'ai pas besoin de vous dire que cette nouvelle m'a causé une vraie satisfaction, augmentée encore par le fait, inattendu pour moi, de l'unanimité des suffrages. Dès que l'annonce officielle de mon élection me sera parvenue, je m'empresserai d'écrire au Secrétaire perpétuel pour le prier de transmettre mes remerciements à la Société.

Agréez, mon cher Monsieur Tyndall, l'assurance de tous mes sentiments d'affection et de haute considération.

J. Plateau.

Ghent, 23 May 1870

My dear Mr Tyndall,

Please accept my heartfelt gratitude for the good news that you have kindly delivered[1] to me, and for all the trouble that you have gone through for the matter of my nomination.[2] I do not need to tell you that this news gave me true satisfaction, amplified even more by the fact of the unanimous vote, which was unexpected for me.[3] As soon as the official announcement of my election will arrive, I will hasten to write to the Permanent Secretary[4] to ask him to pass on my gratitude to the Society.

Please accept, my dear Mr Tyndall, the assurance of my warmest affection and regards.

J. Plateau.

RI MS JT/1/P/108

---

1. *the good news that you have kindly delivered*: Tyndall supported Plateau's nomination as ForMemRS (see letter 2975, n. 4).
2. *matter of my nomination*: see n. 1.
3. *unexpected for me*: Plateau's previous nominations had been unsuccessful. See letters 2995 and 3138. For more on Tyndall's efforts in support of Plateau, see letter 2975, n. 4.
4. *Permanent Secretary*: the RS did not have a position of 'Permanent Secretary'. Plateau may be referring to Walter White (1811–93), who became RS library attendant in 1844 and was promoted to Assistant Secretary and Librarian in 1861.

From Henry Bence Jones [26 May 1870][1] 3193

84 Brook S^t^ | W

Dear Tyndall

Sir H Holland tells me that you say I refused to shake your hand & that therefore you have not spoken to me about the lecture of Capt Noble[2] & your own.[3]

I have no recollection that I did anything of the kind

After you seconded the vote of thanks to me at the annual meeting[4] I ought to have offered you my hand & I am, and was even at the time, sorry that I did not do so but my soreness at your words in the library and far greater soreness at those in the theater when I spoke to you on Institution[5] business made me unwilling even to thank you for your kind words at the meeting.

Pray accept this note as an apology for what you consider as my refusal of your friendship.

& Believe me | Yours very truly | H Bence Jones

Give Mr Vincent[6] whatever directions you please about your Friday evening.[7]

RI MS JT/1/J/113
RI MS JT/1/TYP/3/694

---

1. [*26 May 1870*]: several allusions allow this letter to be precisely dated. The RI annual managers meeting of 1870 took place on 2 May (*Roy. Inst. Proc.*, 6 (1870–72), p. 142), and the next one not until 1 May 1871 (*Roy. Inst. Proc.*, 6 (1870–72), p. 325). Captain Noble's lecture was given between the two, on 3 March 1871. Letters 3194 of 27 May 1870 and 3195 of 28 May 1870 both refer to a minor affront between Tyndall and Bence Jones that was further resolved.
2. *lecture of Capt Noble*: William Noble (1828–1904), a British astronomer. Elected fellow of the Royal Astronomical Society in 1855, Noble served on its council (1866–80) and also led a group of younger members wanting reform of the society, expressed through published critiques in the *English Mechanic*. He retired from the army with the rank of Captain in 1858 and settled down as a country gentleman in Sussex, devoting his time to astronomy and local government. His astronomical interests were wide-ranging, and he was a significant figure in the promotion of amateur astronomy and popular understanding of the science (*ODNB*). Noble delivered a Friday Evening Discourse at the RI, 'On the Pressure of Fired Gunpowder', on 3 March 1871 (*Roy. Inst. Proc.*, 6 (1870–72), pp. 273–83).
3. *your own*: Tyndall delivered a Friday Evening Discourse at the RI, 'On the Colour of Water, and on the Scattering of Light in Water and in Air', on 20 January 1871 (*Roy. Inst. Proc.*, 6 (1870–72), pp. 189–98).
4. *annual meeting*: annual managers meeting of the RI on 2 May 1870 (*Roy. Inst. Proc.*, 6 (1870–72), p. 142).
5. *Institution*: the RI.
6. *Mr Vincent*: Benjamin Vincent (1818–99), Assistant Secretary (1848–89) and Librarian (1849–89) of the RI. Vincent worked as a translator and editor prior to joining the RI. While there he was responsible for the publication of a new catalogue of the library and new editions of Haydn's *Dictionary of Dates* and *Dictionary of Biography*. Like Michael Faraday he was a Sandemanian and served as a Church Elder from 1849 to 1864 (*Faraday Correspondence*, vol. 5, p. lvii).
7. *your Friday evening*: see n. 3.

To Henry Bence Jones 27 May 1870 3194

Royal Institution | 27th May 1870.

Dear Friend

I liked your note[1] and thank you for it—it was written without a trace of irritation.

I think moreover that you had reason to feel sore with me. My reply to you in the theatre was I fear short if not harsh.

But the fact is that I also was fearfully sore at the moment—for I thought that you simply meant to ride me down in the library, and no power that I possess could keep my blood down under the circumstances.

I wish earnestly we could get on without these brushes: but attrition will sometimes bring sparks out of hot natures. Whatever may have been your feelings towards me I never nursed anger towards you—never anything the reverse of affection.

Yours ever | John Tyndall

Dr Bence Jones.

I have been very seedy all day or I should have written to you sooner. My time has been employed reading Pasteur.[2]

RI MS JT/1/TYP/2/695
LT Typescript Only

---

1. *your note*: letters 3193 and 3195.
2. *Pasteur*: Louis Pasteur.

From Henry Bence Jones 28 May 1870 3195

84 Brook Street W.

My dear Tyndall

I had not read your note[1] yesterday when we met. I looked only to see what was to be my conduct. Had I read it all I should have thanked you most warmly for it & should have also expressed my most earnest wish that our tempers might be more even. However I will say only many thanks for your letter

Ever yours truly | H Bence Jones | May 28. 1870

RI MS JT/1/J/63

---

1. *your note*: letter 3194.

From Heinrich Debus 29 May 1870 3196

Clifton | May 29. 1870

My dear John,

Two or three weeks ago I sent Dr Budd's certificate[1] to you, with the request to do, what you considered to be best. You have fulfilled this request of mine in excellent manner, far better than I could have done myself, & for this my thanks lie humbly at your feet. Your lectures, no doubt, were finished some weeks ago. Will you send me the notes?[2] Please do so. Here in the west of England, surrounded by my daily work, I am like one of the lesser members of the human body, situated far away from the heart, the seat of life. So, if I were to invoke all the Muses, to embellish my thoughts, I am afraid my children would cut a poor figure, for from where are the roots to draw the copious saps? Or what is the good to carry water to the well, or shine light into the sun? So you will be satisfied with a dull letter, with the performance of a country man. I am trotting along as best I can. I have something like two hundred boys to teach in my different classes, & about 16 volunteers of what they call V & VI form. You will think that I ought to know Chemistry by this time. The stars go in circles, human life has its periods, why should a schoolmaster complain when he has to move like a water wheel? People on the whole are very kind to me, and for this one ought to be thankful at least. How is your head? Here everything is in blooming condition

Yours Always | H Debus.

RI MS JT/1/D/40

1. *Dr Budd's certificate*: nomination certificate for William Budd's election as FRS. He was elected on 8 June 1871. See letters 3187, n. 3 and n. 11, and letter 3189.
2. *the notes*: for Tyndall's lecture series at the RI on 'Electrical Phenomena and Theories', see letter 3126, n. 3. These lectures were published as J. Tyndall, *Notes on a Course of Seven Lectures on Electrical Phenomena and Theories, delivered at the Royal Institution of Great Britain, April 28–June 9, 1870* (London: Longmans, Green, and Co., 1870).

To Heinrich Debus [30 May 1870][1] 3197

<letter begins here>

I am still lecturing—I have two more to give—and have let myself in for another Friday evening.[2] Tell D^{r}. Budd that I have received a most important volume from Pasteur.[3] I shall give a resume of it on the 24th of June.

Pasteur has also written me a very sweet nice letter.[4]

I send you the "Notes"[5] as far as they are ready. Longmans is just publishing those on Light[6]—price one shilling—I think that will do. Now I want your advice. When you have looked over these on Electricity just say whether you think it would be of any use to publish them. Would they be of any service to teachers or learners? There are a good many new experiments—But of course I could not make the notes a systematic resumé of electric science; as my limits of 7 lectures, and the fact of the subject being so frequently treated here determined my mode of treatment.

<letter ends here>

RI MS JT/1/T/361
RI MS JT/1/TYP/7/2372

1. *[30 May 1870]*: this letter has been dated through a series of references: Tyndall's final three lectures of the spring programme took place on 27 May, 2 June, and 9 June (RI Guard Book III, p. 89); 'Notes' on light were published in early June; and in his letter of 29 May (letter 3196), Debus requests the notes for Tyndall's lectures on electricity.
2. *another Friday evening*: see letter 3193, n. 3.
3. *a most important volume from Pasteur*: see letter 3188, n. 3.
4. *a very sweet letter*: letter 3188.
5. *"Notes"*: see letter 3196, n. 2.
6. *those on Light*: see letter 3097, n. 7.

From Lucien de la Rive[1] [late May 1870][2] 3198

1870

Cher Monsieur,

Je vous remercie vivement de m'avoir envoyé votre remarquable article sur Faraday; j'en ai lu le commencement avec beaucoup d'intérêt et j'espère en lire bientôt la suite. Il me semble que vous avez la raison de votre coté, quand vous dites que Davy ne pouvait pas être jaloux du futur grand homme, mais rien n'est plus difficile que de faire abstraction de ce que nous savons, pour juger une époque où on ne le savait pas.

Vous recevrez prochainement deux exemplaires d'une traduction d'Enoch Arden dont je suis l'auteur. Je vous prie de bien vouloir accepter l'un des deux et de me rendre le très grand service de faire parvenir l'autre à Mr Tennyson.

Veuillez, cher Monsieur, agréer l'expression de mes sentiments très dévoués. | L. de la Rive.

P.S. Je vous demande de vouloir bien faire parvenir à Mr Tennyson la lettre ci-incluse.

Dear Sir,

I thank you very much for sending me your remarkable article[3] on Faraday,[4] I have read the beginning with much interest and hope to soon read the rest of it. It seems that you have Reason on your side, when you say that Davy[5] could not be jealous of the future great man, but nothing is more difficult than ignoring what we know, to judge a time when it was not known.

You will soon receive two copies of a translation[6] of Enoch Arden,[7] of which I am the author. I would ask that you accept one of the two and be so kind as to send the other to Mr Tennyson.[8]

Please, dear Sir, accept the expression of my highest regards. | L. de la Rive.

P.S. I ask that you please forward Mr Tennyson the enclosed letter.[9]

RI MS JT/1/D/111
RI MS JT/1/TYP/1/383

---

1. *Lucien de la Rive*: Lucien de la Rive (1834–1924), a Swiss physicist and writer, son of Auguste de la Rive. He published on mathematical physics and the theory of relativity and translated the poems of Alfred Tennyson (*HLS*).
2. *[late May 1870]*: the reference to Tyndall's article on Faraday (n. 3), places this letter at the end of May. A. de la Rive also thanks Tyndall for sending these papers in his letter of 1 June 1870 (letter 3199).
3. *your remarkable article*: Tyndall's review of Henry Bence Jones's *Life of Faraday*. A. de la Rive mentions this publication in his letter of 1 June 1870 (letter 3199). See also letter 3159, n. 2.
4. *Faraday*: Michael Faraday.
5. *Davy*: Humphry Davy.
6. *a translation*: see L. de la Rive and H. Wadsorth (eds), A. Tennyson, *Essais de traduction poétique* (Paris: Libraire de Ch. Meyrueis, 1870).
7. *Enoch Arden*: 'Enoch Arden' is an 1864 poem by Alfred Tennyson.
8. *Mr Tennyson*: Alfred Tennyson.
9. *the enclosed letter*: letter missing.

From Auguste de la Rive 1 June 1870 3199

Presinge, le 1 juin, 1870.

Mon cher Tyndall,

Permettez-moi d'introduire auprès de vous l'un de mes jeunes compatriotes, Monsieur Raoul Pictet, qui s'occupe avec succès de physique. Il est engagé dans ce moment dans des affaires scientifiques-industrielles avec l'Egypte &

il va en Angleterre en vue de s'instruire dans la partie des sciences qui se rapporte surtout aux applications de la physique. Il a fait déjà de jolis travaux, soit théoriques soit pratiques et nos savants de Genève en particulier <*1 word missing*> & Soret font grand cas de lui. C'est un jeune homme ardent, plein de foi dans la science et qui a plutôt besoin d'être retenu, ce qui n'est pas un grand défaut. Je le recommande donc à votre bonne amitié pour nous en vous priant de vouloir bien l'aider de vos conseils et de vos bons offices.

Nous avons reçu mon fils Lucien et moi les exemplaires que vous nous avez envoyés du Journal l'Academy qui renferme votre article sur l'ouvrage de Bence-Jones. Vous avez fait usage d'une manière très naturelle des renseignements que je vous avais donnés et je vous remercie bien de votre note aimable et bienveillante sur vos amis de Genève qui ne vous oublient point, je vous assure. Merci aussi de l'envoi que vous m'avez fait de vos articles sur Dust and Disease et de la promptitude parfaitement aimable avec laquelle vous avez bien voulu traduire et insérer ma lettre dans le Phil. Mag.

Je vous enverrai très incessamment un mémoire que je viens de terminer sur la polarisation rotatoire magnétique; il renferme des résultats que je crois intéressants. Peut-être le Phil. Mag. voudra-t-il l'insérer, soit en totalité soit seulement en partie. Ce travail m'a occasionné beaucoup de peine et pris beaucoup de temps; ce sont des expériences difficiles et délicates qu'il faut répéter très souvent avant d'être certain du résultat. Vous pourrez du reste mieux que personne, en juger vous-même.

Ma femme et tous mes enfants se rappellent à votre bon souvenir.

Je vous prie, my dear friend, de recevoir l'assurance de mes sentiments les plus affectueux. | Aug. de la Rive.

Presinge, 1 June, 1870

My dear Tyndall,
Allow me to introduce to you one of my young compatriots, Mr Raoul Pictet,[1] who is successfully pursuing physics. He is engaged at the moment in industrial-scientific business with Egypt, & he is going to England in order to learn about the area of science that is especially concerned with applications of physics. He has already done some nice work, both theoretical and practical, and our experts in Geneva, especially <*1 word missing*> & Soret[2] think highly of him. He is a passionate young man with a lot of faith in science and who instead needs to be held back, which is not a big flaw. Thus, I recommend him to you, for your good friendship with us, and ask you to be kind enough to help him with your advice and good offices.

We received, my son Lucien[3] and me, the copies that you sent us of the Academy Journal, which contains your article on the work of Bence-Jones.[4] You have used the information that I gave you in a very natural way, and I thank you for your kind and caring words about your friends in Geneva, who do not forget about you, be assured of it. Thank you also for sending me your articles on Dust and Disease,[5] and for the incredibly kind promptness with which you were kind enough to translate and insert my letter into the Phil[osophical]. Mag[azine].[6]

I will send you, very shortly, a paper that I just finished[7] on magnetic rotatory polarization. It contains results that I think are interesting. Perhaps the Phil[osophical]. Mag[azine]. would like to include it either entirely or only in part.[8] This work caused me a significant amount of grief and took a lot of time; these are difficult and delicate experiments that need to be repeated often in order to be sure of the result. You, better than anyone, can judge it for yourself.

My wife[9] and all of my children[10] remember you fondly.

I ask you, my dear friend, to please accept the assurance of my most affectionate sentiments. | Aug. de la Rive.

RI MS JT/1/TYP/1/375
LT Typescript Only

---

1. *Mr Raoul Pictet*: Raoul Perre Pictet (1846–1929), a Swiss chemist and physicist. He served as a professor of industrial physics at the University of Geneva from 1879 until 1885 and professor of physics the University of Berlin from 1866. His major contribution was his research into the physics of low temperatures and the first liquefaction of an atmospheric gas, the latter achievement shared with Louis Paul Cailletet. At the time of this letter, he was still studying at the Universities of Geneva and Paris (*HLS*).
2. *Soret*: Louis Soret.
3. *my son Lucien*: Lucien de la Rive (see letter 3198, n. 1).
4. *your article on the work of Bence-Jones*: J. Tyndall, 'Life and Letters of Faraday' (see letter 3159, n. 2, and letter 3198).
5. *your articles on Dust and Disease*: 'On Haze and Dust' (see letter 3149, n. 3) and J. Tyndall, 'On Dust and Disease', *Fraser's Magazine* (March 1870), pp. 302–10.
6. *my letter into the Phil[osophical]. Mag[azine]*: see letter 3133.
7. *a paper that I just finished*: A. de la Rive, 'Recherches sur la polarisation rotatoire magnétique des liquids', *Archives des sciences physiques et naturelles*, 38 (1870), pp. 209–54.
8. *like to include it either entirely or only in part*: A. de la Rive, 'Researches on the Magnetic Rotatory Polarization of Liquids', *Phil. Mag.*, 40:4 (1870), pp. 393–420.
9. *My wife:* Louise-Victoire-Marie de la Rive (see letter 3096, n. 9).
10. *my children*: see letter 3096, n. 10.

To Maria Ellery MacKaye[1] 2 June 1870 3200

Royal Institution, Albemarle Street, 2d *June,* 1870

MY DEAR MADAM,—I have by no means forgotten our meeting[2] at the Riffel,[3] nor our reciting the poetry of your eminent countryman,[4] coming down the slope from the glacier to the hotel. I think we sounded Monadnoc.

> "Hither we bring
> Our insect miseries to the rocks;
> And the whole brood with pestering wing[5]
> Vanish, and end their murmuring,—
> Vanish beside those dedicated blocks."[6]

I quote from memory, for long ago I lodged these lines and many others of Emerson's in the book and volume of my brain.

I always thought those lines on Rhodora[7] exquisite. But what the rhodora was "I never knew." Some time ago I was staying with a friend in the country, and while under shelter of a pine wood a group of us talked of the rhodora, but none of us knew anything about it. I had quoted some of the lines regarding it in a little book of mine about the Alps,[8] written ten years ago: hence the conversation.

Many thanks to you for the flower; no other flower could be more acceptable to me.

Some time ago Mr. Emerson[9] gave me a pleasure of which he had necessarily no knowledge. I go down from time to time to Chelsea,[10] to see that grand old man Thomas Carlyle. When I was there last, two books of Emerson were on the table, addressed, "With unchangeable affection to Thomas Carlyle." It did my heart good to see this loyalty.

Poor Mrs. Carlyle handed him over to my safe-keeping when he went to Edinburgh to be installed as rector of the University. She died while he was in Scotland. I afterwards went with him to Mentone.[11] A few weeks ago I was with him in the country. It was a wild day, and we got into a clearing in the middle of a wood, where we sat in calm while the storm rolled around us. I plucked a cushion of ferns for the old man, placed it on the stump of a tree, helped him to light his pipe, and there we talked of death, and the privilege of being released from the fear of it.

I was so much pleased with Emerson's books and their superscription that I carried them away with me; they are here beside me.

I think my own single example would demonstrate the futility of all attempts to sever intellectual progress from moral influences, as Buckle[12] tried to do some years ago. For even my science owes a great debt to Emerson, Fichte,[13] and Carlyle,—three men who care little for science. But there were stirred the forces that were latent within me, and that these forces took the scientific direction was a mere accident.

I rarely write so long a letter. Good-bye. | Your most truly, | JOHN TYNDALL.

Give Mr. Emerson my thanks.[14] I might with truth offer more.

Atlantic Monthly, 73 (1894), p. 281

---

1. *Maria Ellery MacKaye*: see letter 3190, n. 1.
2. *our meeting*: see letter 3190, n. 2, and letter 2727, *Tyndall Correspondence*, vol. 10.
3. *Riffel*: see letter 3030, n. 15 and n. 16; and letter 3190, n. 3.
4. *your eminent countryman*: Ralph Waldo Emerson.
5. *And the whole brood with pestering wing*: Tyndall slightly misquoted the original line 'And the whole flight, with pestering wing'.
6. *Monadnoc*: R. W. Emerson, 'Monadnoc' in *Poems* (Boston: James Munroe and Company, 1847), pp. 94–114, lines 374–80.
7. *Rhodora*: poem by Ralph Waldo Emerson. See letter 3190, n. 4. This poem was very dear to Tyndall. See R. Jackson, N. Jackson, and D. Brown (eds), *The Poetry of John Tyndall* (London: UCL Press, 2020), pp. 62–3.
8. *quoted some of the lines regarding it in a little book of mine about the Alps*: *Glaciers of the Alps*, p. 213. See letter 2990, n. 4.
9. *Mr. Emerson*: Ralph Waldo Emerson.
10. *Chelsea*: see letter 3186, n. 5.
11. *Mentone*: in December 1866, Carlyle and Tyndall visited Louisa Baring's house in the south of France at Mentone, some twenty miles east of Nice, on the Côte d'Azur. See letter 2535 (from Tyndall to Ellen Busk), *Tyndall Correspondence*, vol. 10.
12. *Buckle*: Henry Thomas Buckle (1821–62), an English historian who was considerably influenced by German higher criticism and became a deist with a strong interest in rationalism and natural science. He is mostly remembered for his unfinished *History of Civilization in England*, 2 vols (London: John W. Parker and Son, 1857–61) in which he sought to outline the laws of human progress, following his larger goal of producing a science of human society. In 1858 Buckle delivered his only public lecture at the RI entitled 'The Influence of Women on the Progress of Knowledge', which was later printed in *Fraser's Magazine* for April 1858 (*ODNB*).
13. *Fichte*: Johann Gottlieb Fichte (1762–1814), a German philosopher who is considered one of the founding figures of post-Kantian German Idealism. His work was concerned with the conditions under which revealed religion is possible. Fichte termed his Idealism *Wissenschaftslehre*, which translates to 'Doctrine of Scientific Knowledge' (*NDB*). Tyndall was interested in his notions of individual identity and vocation.
14. *Give Mr. Emerson my thanks:* MacKaye did more and showed Emerson Tyndall's letter.

Emerson's response of 27 June 1870 was published in the same issue of *Atlantic Monthly*: 'I have been much interested by Professor Tyndall's letter, which you have so kindly allowed me to read. The good will he expresses toward myself is highly gratifying to me, as I know well his own eminent worth. I could heartily wish that, since his scientific researches make him so much of a traveler, they may, one of these days, bring him to America, where he has already, I doubt not, a larger public of readers than in England. He will, no doubt, like also to give new lessons to that young scholar of his at Riffel formerly, who, I am glad to hear, has come to value his letter. Mr. Tyndall's notices of Carlyle are especially interesting to me,—every word' (*Atlantic Monthly*, 73 (1894), p. 281).

From Mary Adair [4 June 1870][1] 3201

My dear Mr. Tyndall,

I feel sure that I might accompany my kind unthinking Friends to the R[oyal]. I[nstitution]. where I have spent so many happy hours.

Thank you for your note and promise.

Of "Diamagnetism"[2] I am terribly ignorant.

"Faraday as a Discoverer"[3] and the Bakerian Lecture for /55[4] comprising all I have read on the subject.

But even this slight preparation will help to the enjoyment of your treatise[5] which cannot be "dry" though doubtless, it will prove far beyond my depths.

You will write my name in it! and, believe me, in returning to the home duties and affections which are my Life. I shall carry with me no more valued Souvenir of my summer holiday than this pledge of continued friendship.

Ever very sincerely your's | H. M. Adair | Saturday night—

RI MS JT/1/A/21

1. *[4 June 1870]*: the date of this letter is estimated: Tyndall sent copies of his work *Diamagnetism* to friends and colleagues starting in early June 1870. Assuming that Adair is referring in her first sentence to the possibility of attending the 9 June lecture, this letter would have been written on Saturday, 4 June 1870.
2. *"Diamagnetism"*: *Diamagnetism* (see letter 3076, n. 8).
3. *"Faraday as a Discoverer"*: *Faraday as a Discoverer* (see letter 2951, n. 3).
4. *Bakerian Lecture for /55*: the Bakerian Medal is one of the RS's most prestige medals which requires the awardee to give a lecture, the Bakerian Lecture. Tyndall delivered the 1855 Bakerian Lecture on 25 January (J. Tyndall, 'The Bakerian Lecture.—On the nature of the Force by which Bodies are repelled from the Poles of a Magnet,' *Phil. Trans.*, 145 (1855), pp. 1–51). See letter 3235, n. 10.
5. *your treatise*: see n. 2.

To Heinrich Debus [7 June 1870][1] 3202

Tuesday | 1870

My dear Heins

The senior physician of Guys Hospital[2] has been asking my advice about their chair of Chemistry. Taylor[3] it appears has resigned.

It is worth in all about 500 or 600 a year. But they may divide it into two—I have written very sharply about you.[4] Odling also backs you, what a blessing it will be if we caught you here.

You would not object to this would you?

Yours ever | John Tyndall

There is no lecture on the 26th they would not let me lecture—they say I have done enough for the Institution & want rest. I will write an analysis of Pasteur for the Times[5]

RI MS JT/1/T/332

1. *[7 June 1870]*: the date of this letter has been estimated in reference to letters 3209 and 3211, both written on 9 June 1870, which respond to this letter and discuss the open position at Guy's Hospital. 7 June 1870 was a Tuesday.
2. *senior physician of Guys Hospital*: George Owen Rees. Guy's Hospital was a hospital and medical school in London.
3. *Taylor*: Alfred Swaine Taylor (1806–80), an English medical jurist, toxicologist, and FRS. Taylor built his public reputation as an expert witness in several high-profile murder trials. At Guy's Hospital, he was lecturer of medical jurisprudence (1831–77) and lecturer in chemistry (1832–70). He published on a wide range of subjects, including medical jurisprudence, chemistry, thermometry, and photography (*ODNB*).
4. *I have written very sharply about you*: Tyndall and Debus discuss this in letters 3209, 3211, 3214, and 3220. See also letter 3213.
5. *analysis of Pasteur for the Times*: Tyndall had already written to the *Times* about Pasteur (see letter 3177), but in this instance no record can be found.

To Thomas Archer Hirst 7 June 1870 3203

Royal Institution, Tuesday | June, 7th. 1870[1]

My dear Tom

Col. Strachey[2] has come from India and will be here for a few days only.

I have received the sweetest of notes[3] from Lady Colville:[4] it was brought to me by Strachey and Mrs Strachey[5] and it, backed by their persuasions, has proved irresistible they have entangled me for Sunday.

Now could we not spend our Sunday in some place nearer than Folkestone?[6] I would go with you there on Saturday but I am entangled in the city.

Yours affect[ionately]. | John

RI MS JT/1/HTYP/563
LT Typescript Only

---

1. *June 7th. 1870*: Hirst's journal confirms that he and Tyndall dined with the Stracheys at the Colville residence on Sunday, 12 June (RI MS JT/2/32d/1877).
2. *Col. Strachey*: Richard Strachey.
3. *sweetest of notes*: letter missing.
4. *Lady Colville*: Francis Elinor Colvile.
5. *Mrs Strachey*: Jane Maria Strachey (née Grant, 1840–1928), a British suffragist, author, and the daughter of John Grant (1807–93), a colonial administrator. In 1859 she married Richard Strachey, his second wife, and who was twenty-three years her senior. She was influenced by John Stuart Mill, and a friend of the suffragist Millicent Fawcett (1847–1929) and George Eliot (1819–80). She was a member of the National Union of Women's Suffrage Societies, and was host to the numerous political and scientific guests who frequented the Stracheys' home (*ODNB*).
6. *Folkestone*: see letter 3021, n. 20.

From Mary Egerton 7 June [1870][1] 3204

Mountfield Court, | Robertsbridge, | Hawkhurst. | June 7th

Dear Mr Tyndall,

It is much too good of you to send me your book![2] I feel exalted to all the dignity of a leading member of some "learned society", only unfortunately minus the learning! Really & truly though, I do not half like accepting such a valuable one! It is mere not for the fear of offending such a millionaire as you profess yourself to be! I should send at once to Longman's[3] the bill. (This is worse than paying for the monkey![4]) Well! Notwithstanding a sore conscious, I shall read the book with the greatest interest. Its earlier pages helped me through some hours of a very melancholy week last Christmas, and I shall be so glad to take the subject up again, and understand what I can—No. I am wrong it was "Light"[5] I had at Thoresby,[6] & this later.

We are coming up on Thursday, May & I, & hope to hear y[ou]r last lecture.[7] I wish you would make it two hours long—it would fill up a little more of the time till our return train!

By the by, some of the trains are changed this month, so if you sh[oul]d be "rushing" do not trust my list.

The Claret shall be forthcoming, as to the other item you required it had been proposed to raise a subscription or sort of "Voluntary rate", whereby through the joint contributions of the family a supply might be collected—but it is not said that the amiable tendency of human nature is to hate those to whom it is under obligations? if so any contributions must decidedly be exchanged for a very opposite one! Still however I think we might find sufficient to furnish a welcome for your visit & perhaps a little to take away with you—always provided it is considered as a loan—(i.e. to be returned.) Forgive the temptation of a weak friend! and the writing of an atrocious pen.

Believe me y[ou]rs truly & most gratefully | M F Egerton

RI MS JT/1/E/38

1. *[1870]*: the date of this letter has been determined from Egerton's mention that she planned to attend Tyndall's final lecture. The RI spring lecture series usually finished at the beginning of June, but on three occasions the final lecture took place after 7 June: on 8 June in 1871 and 1876, in which case Egerton would likely have written tomorrow', and 9 June in 1870, which was a Thursday. The black mourning border on the letter also lends weight to this dating, since Egerton's husband had passed away less than a year previously, in late summer 1869.
2. *your book!*: probably *Diamagnetism* (see letter 3076, no. 8).
3. *Longman's*: Tyndall's publisher, Longmans, Green, and Co.
4. *paying for the monkey!*: reference is not identified.
5. *"Light"*: *Notes of a Course of Nine Lectures on Light* (see letter 3097, n. 7).
6. *Thoresby*: Thoresby Park (see letter 3163, n. 1).
7. *y[ou]r last lecture*: the last of Tyndall's seven lectures on 'Electrical Phenomena and Theories' occurred on 9 June 1870, a Thursday (see letter 3126, n. 3).

From John Frederick William Herschel — 7 June 1870 — 3205

Collingwood | June 7/70

My dear Sir,

Pray accept my best thanks for your work on Diamagnetism and Magne-crystallic Action[1] which I rec[eived]d this morning. Of course I can only in these few hours time run my eye over the table of contents & now & then a page—but there is enough to shew me what a rich treasure of research is in store for me when I go through it seriatim which I shall do with all possible eagerness as it is a subject on which I have long desired to get some distinct knowledge. It is now some 40–45 years that I devised & began to prepare for a set of experiments on the possible magnetization of other metals besides Nickel & Cobalt thinking it likely that as Iron & those two differ only in degree it was exceedingly probable that more delicate means of enquiry would carry the analogy still farther down in the scale—(of course I never contemplated such a thing as a reversed action). My plan was suspension of bars at the two ends of an exceedingly light & delicately suspended torsion thread with very long arms to the suspended bar—& the approach of powerful magnets under proper precautions against extraneous disturbance &c &c.—I suppose I might have blundered out something—only as it happened the exp[erimen]ts were never made.

One exp[eriment] has caught my eye and given me a fuller impression than I have ever had of the immense resource placed in the hands of an experimental demonstrator by that same beam of electric light—I mean the exp[eriment] at the bottom of p. 247 of water rendered muddy by suspended magnetic oxide in fine division, and suddenly cleared (in one direction) by the polarity given to its particles by an electro magnet. How I envy those who can come and see all these wondrous things at the R[oyal] I[nstitution]—yourself waving the magic wand and educing the hidden powers and making them manifest themselves! I think the passage which forms the next ¶ after the description of that exp[erimen]t (in p. 248) must have been a very telling one.

Wishing you many years of health to employ the powerful means at your disposal with ever increasing efficiency.

believe me | My dear Sir | Yours very faithfully | J. F. W. Herschel

RI MS JT/1/H/117
RI MS JT/1/TYP/2/571
RS HS 24.317

---

1. *your work on Diamagnetism and Magne-crystallic Action*: *Diamagnetism* (see letter 3076, no. 8).

From George Biddell Airy 8 June 1870 3206

Flamsteed House, Greenwich Park, | London, S.E. | 1870 June 8

My dear Sir

I thank you much for the copy of your book on Diamagnetics &c.[1] I look to it with a gratitude allied to the old definition of political gratitude: the lively hope of future advantage.

I have just written the last sentence of a book on Magnetism,[2] but it will be one of a different class and utility from yours. It is intended to allude to almost every part of cosmical and experimental magnetism, but principally in the mathematical parts, which are best suited for my Cambridge friends.[3]

I am, my dear Sir, | Yours very truly | G B Airy

Professor Tyndall

RI MS JT/1/A/53
RI MS JT/1/TYP/1/44

1. *your book on Diamagnetics &c.*: *Diamagnetism* (see letter 3076 n. 8).
2. *a book on Magnetism*: G. B. Airy, *A Treatise on Magnetism* (London: MacMillan and Co., 1870).
3. *my Cambridge friends*: Airy had a long association with Cambridge. He was elected Lucasian Professor of Mathematics there in December 1826, and his textbook *Mathematical Tracts on Physical Astronomy*, published in the same year, was used as a textbook there.

From James Prescott Joule 8 June 1870 3207

Broughton Manchester June 8 1870

My dear sir

I beg you to accept my best thanks for the handsome present of your work on Diamagnetism[1] you have made me. The papers now collected I shall read again and with renewed interest

Believe me | Yours most truly | James P Joule

Prof: Tyndall &c. &c. &c.

RI MS JT/1/J/140
RI MS JT/1/TYP/2/664

1. *your work on Diamagnetism*: see letter 3076, n. 8.

From Felix Pouchet 8 June 1870 3208

Muséum d'Histoire Naturelle de Rouen
Le Dr Pouchet, directeur correspondant de l'Institut
(Académie des Sciences), Professeur à l'Ecole de Médecine
et à l'Ecole supérieure des Sciences, Officier de l'ordre
impérial de la Légion d'honneur de celui du Lion et du Soleil;
Chevalier de l'ordre des Saints Maurice et Lazare, &c.

à Monsieur Tyndall, Professeur à la Royale Institution
de Londres, Membre de l'Institut de France, &c.

Monsieur et très honoré confrère,

J'ai lu avec un vif intérêt, dans nos journaux français, la reproduction de vos dernières conférences sur les corpuscules de l'air. Je vous remercie de m'y avoir cité avec bienveillance et j'apprécie hautement l'importance des nouveaux moyens que vous proposez pour constater optiquement l'état de ces mêmes corpuscules.

Mais je ne suis pas tout-à-fait d'accord avec vous relativement au rôle du coton; si vous vouliez vous reporter aux expériences nombreuses qui ont été faites, soit en France, soit à l'étranger, vous reconnaîtrez que, quel que soit le moyen employé pour épurer l'air, celui-ci n'en est pas moins fécond. Je puis vous assurer que, sous peu de jours, je montrerai, dans mon amphithéâtre, des ballons optiquement vides dans lesquels ce fluide rentrera en traversant du coton, et que tous seront féconds. Cela dépend de la nature des corps que l'on emploie.

Dans quinze jours environ, devant un public nombreux, composé de toutes nos sommités scientifiques et de personnes d'élite, je ferai moi-même l'histoire des corpuscules de l'air, que de longues observations m'ont appris à connaitre; et comme ma leçon sera sténographiée, j'aurai l'honneur de vous l'adresser.

Mais vous me convertiriez tout-à-fait à vos doctrines, et je prononcerais bien sincèrement mon abjuration, si vous pouvez, avant ma conférence, me procurer la faveur de voir les spores que M. Dancer dit avoir découvert en si grand nombre dans l'atmosphère. Il serait très facile de m'en adresser, entre deux lames de verres, dans une lettre. Ma longue habitude me permettra, je l'espère, d'en débrouiller les espèces. Malgré les moyens d'une extrême précision que j'emploie, je n'ai jamais rien recueilli de semblable.

En voyant de telles choses, j'en suis certain, il n'est pas un hétérogéniste qui ne renonce à ses illusions. Quand MM. Joly et Musset et moi, nous avons sommé M. Pasteur de nous montrer ses œufs et ses semences atmosphériques—car je ne veux pas employer le mot germes qui a une autre acception biologique—il ne pût le faire. Quiconque le fera, je vous l'assure, nous convaincra tous, car nous n'avons qu'un but: la Vérité.

Recevez, Monsieur et très honoré confrère, l'assurance de ma vive admiration. | Pouchet

8 juin 1870

Museum of Natural History of Rouen.[1]

To Mr Tyndall, Professor at the Royal Institution of London, Member of the French Institute, &c.[2]

Sir and very honoured colleague,

I read with much interest, in our French newspapers, the reproduction of your last discourse[3] on air particles. I thank you for having cited me so kindly, and I highly appreciate the importance of the new means that you propose to optically account for the state of those same particles.

But I do not entirely agree with you regarding the role of cotton;[4] if you would like to look at the numerous experiments that have been done, either in France, or abroad, you will see that regardless of the means used to purify the air, it is no less clean. I can assure you that, in a few days, I will demonstrate, in my lecture hall, the passage of fluid through a cotton barrier into balloons which are optically empty and that each of these will be purified. This depends on the nature of the bodies we are using.

In about fifteen days, I will myself explain the matter of air particles that long observations taught me to understand, in front of a large audience composed of all our leading scientific experts and elite persons. Since my lecture will be taken down in shorthand, I will have the honour of sending it to you.

But you would convert me completely to your doctrine, and I would pronounce sincerely my abjuration, if you can, before my conference, do me the favour of seeing the spores that M. Dancer[5] says he has discovered[6] in such a great number in the atmosphere. It would be very easy to send me some, between two glass slides, in a letter. My familiarity will allow me, I hope, to sort out the species. Despite the means of extreme precision that I use, I have never collected anything similar.

In seeing such things, I am certain that there is not one heterogenist who would not *[renounce]* their illusions. When Messrs. Joly[7] and Mousset[8] and I asked Mr. Pasteur[9] to show us his atmospheric eggs and seeds—for I do not want to use the word <u>germs</u> that has another biological meaning—he could not do it. Whoever will do it, I assure you, will convince us all, because we only have one goal: the Truth

Please accept, Sir and very honoured colleague, the assurance of my profound admiration. | Pouchet

8 June 1870

RI MS JT/1/P/257

---

1. *Museum of Natural History of Rouen*: Museum of Natural History of Rouen. | Dr Pouchet, Correspondent director of the Institute (Academy of Science) Professor at the School of Medicine and at the Graduate School of Sciences, Officer of the Imperial Order of the Legion of Honour of the Lion and the Sun, Knight of the Order of Saints Maurice and Lazare, &c. Museum headed notepaper.
2. *French Institute*: the Institute de France is a French learned society founded in 1795 which is comprised of five academies including an Academy of Sciences which was formed in 1666.
3. *your last discourse*: possibly Tyndall's Friday Evening Discourse at the RI, 'On Dust and Disease', delivered on 21 January 1870 (see letter 3122, n. 1).
4. *role of cotton*: Tyndall experimented with using cotton to filter germs from air. See letters 3121, 3122, 3126, 3127, 3149, and 3188. He also later used cotton to produce a respiratory apparatus for firemen (J. Tyndall, 'On Some Recent Experiments with a Fireman's Respirator', *Roy. Soc. Proc.*, 22 (1874), pp. 359–61).
5. *M. Dancer*: John Benjamin Dancer (see letter 3121, n. 8).

6. *M. Dancer says he has discovered*: although Dancer's paper was not published until the following year, the studies were undertaken in 1867/68 and the results were already being discussed (J. B. Dancer, 'Microscopical Examination of the Solid Particles Collected by Dr. Angus Smith from the Air of Manchester', *Proceedings of the Literary and Philosophical Society of Manchester*, 10 (1871), pp. 270–4).
7. *Joly*: Nicolas Joly (1812–85), a French biologist. He received degrees in natural science and zoology in 1840 and medicine in 1851. From 1840 to 1878 he was Professor of zoology and the faculty of sciences at the University of Toulouse and Professor of physiology at the school of medicine and pharmacy ('Nécrologie Scientifique: Nicolas Joly', in L. Figuier (ed.), *L'Année Scientifique et Industrielle* (Paris: Librairie Hachette, 1886), pp. 532–3).
8. *Mousset*: Charles Musset (dates unknown), a student of Joly's whose thesis focused on spontaneous generation. In 1863 Pouchet, Joly, and Musset conducted an experiment in the Spanish Pyrenees to disprove Pasteur's hypothesis about the origins of fermentation and to show that spontaneous generation did, indeed, happen (see N. Roll-Hansen, 'Experimental Method and Spontaneous Generation: The Controversy between Pasteur and Pouchet, 1859–64', *Journal of the History of Medicine and Allied Sciences*, 34:3 (July 1979), pp. 273–92).
9. *Mr. Pasteur*: Louis Pasteur.

To Heinrich Debus 9 June [1870][1] 3209

9th June

My dear Heins,

They are thinking of dividing the Chair at Guys[2]—or at least dividing the duties & the pay.

Would you accept the principal post suppose it were possible to get it for you at a salary of £300 a year?[3]

This division of the work would leave you 6 months of the year absolutely to yourself, and the other six months would be by no means slavish.

Mind I do not know how things will turn up, but I want to know, supposing the chair is to be divided, whether you would accept its principal half.

All your laboratory expenses would be paid & also your assistant

Yours ever | John Tyndall

RI MS JT/1/T/314

1. *[1870]*: this letter has been dated through its place in a series of correspondence between Tyndall and Debus discussing Debus' application to the Chair of Chemistry at Guy's Hospital in 1870. It was written between letter 3202 (7 June 1870) and letter 3211 (9 June, 1870). Letter 3211 is a response to this letter. See also letters 3213, 3214, and 3220.
2. *Guy's*: Guy's Hospital Medical School in London.
3. *to get it for you at a salary of £300 a year?*: see letter 3211.

To John Frederick William Herschel 9 June 1870 3210

Royal Institution of Great Britain | 9th. June 1870.

My dear Sir John

Nothing gratifies me more than your sympathy and approbation.[1] I am glad to see your writing so firm and your spirit so fresh—long may they both continue so.

Believe me | Ever yours faithfully | John Tyndall

RS HS 409
RI MS JT/1/TYP/2/572

1. *your sympathy and approbation*: see letter 3205.

From Heinrich Debus 9 June 1870 3211

Thursday | June 9 1870

My dear John,

I wrote yesterday, and I suppose you have received my letter.[1] Your note of to day[2] has just been delivered to me.

Certainly, I would accept the principal post and 300 a year. My present place is not quite 450 and very much more laborious. I say this for your guidance. My opinion is: that it is not wise to divide the duties and the pay, because there will be two poor places. Yet the position at Guys[3] is better as regards work, it will afford an opportunity to do some original work, and will enable a man to pursue that self improvement without which any place is poor. above all it would bring me to London. There is also the chance of making some money by analysis. However, your judgment is in such matters better than my own and I am quite thankful to be guided by your wisdom

Always your | H Debus.

P.S. I want to save the post which is just going out.

RI MS JT/1/D/41

1. *my letter*: letter missing. Debus and Tyndall exchanged a series of letters regarding Debus's application for a position of Chair of Chemistry at Guy's Hospital. See letters 3202, 3209, 3213, 3214, and 3220.
2. *Your note of to day*: letter 3209.
3. *Guys*: Guy's Hospital Medical School in London.

From William Thomson 9 June 1870 3212

Largs | Ayrshire | June 9, 1870

My dear Tyndall

Many thanks for your beautiful volume of diamagnetism.[1] It will be a great pleasure to me to have your papers in this convenient form—I have had a similar enterprise relatively to my own electrical papers on hand for more than two years but have not been able for the last ¾ of that time to give it any attention until recently when I have put it into the hands of an able (newly found) assistant and so after standing for a year about p 300 there is a prospect that with little farther intervention on my part it may appear before Christmas. I thought of including our magnetic correspondence[2] and I presume you will have no objection that so much should be common to the two volumes.

Yours very truly | William Thomson

RI MS JT/1/T/17
RI MS JT/TYP/5/1555

---

1. *your beautiful volume of diamagnetism*: see letter 3076, n. 8.
2. *our magnetic correspondence*: 'Correspondence with Professor Tyndall', in W. Thomson, *Reprint of Papers on Electrostatics and Magnetism* (London: MacMillan & Co., 1872), pp. 540–47.

To Owen Rees 10 June [1870][1] 3213

10th June.

My dear Rees.

I wrote to Dr. Debus in the sense indicated by your last few words of conversation with me.

He has replied to me, and expressed himself willing to be guided by my advice under the circumstances.[2]

His income from Clifton[3] is about £450 a year. Taking the increased cost of living in London into account, what you are disposed to offer him is not more than half his present income. Let me frankly say that I do not think the position approaches one commensurate with Debus's standing in Science.

Debus also expresses to me his own doubts as to the wisdom of dividing the duties of a chair whose whole emolument is by no means inordinate.

Assuredly the wisest plan would be to offer him the whole chair.[4] His life would be thus rendered a purely scientific life; he would not have to seek after commercial work to eke out a necessary income. His whole power as a teacher

& investigator would be brought to bear in the lecture rooms and the laboratory of your Hospital.[5] Both he and you would assuredly be stronger in having the chemical work organised and executed by a single head. There would be more integrity—and efficiency in work thus organised than could possibly come out of a divided responsibility.

These are my convictions frankly stated. Still my desire to see Debus amongst us in London is so strong that should the division of the chair be a necessity I will still advise him to accept your offer. I know he will act as I advise; so I do not think a word more on my part is needed.

I am my dear Rees | truly yours | John Tyndall

Dr Owen Rees. | &c. &c. &c.

RI MS JT/1/T/1264
Transcript Only

---

1. *[1870]*: this letter has been dated through its place in the discussion between Tyndall and Debus on the position at Guy's Hospital. See letters 3209 and 3214.
2. *my advice under the circumstances*: see letters 3202, 3209, 3211, 3214, and 3220.
3. *Clifton*: see letter 2996, n. 7 and n. 8.
4. *offer him the whole chair*: the chair of chemistry was split between Debus and Thomas Stevenson, with Debus taking charge of practical chemistry (Anon, 'Changes in the Hospital and Medical Schools', *BMJ*, 2:265 (1870), p. 265).
5. *your Hospital*: Guy's Hospital Medical School in London.

To Heinrich Debus [10 June 1870][1] 3214

Dear Heins

I have sent the enclosed letter[2] to the senior physician[3] of Guy's[4]—I hope it meets your approval.[5]

Yours ever | John Tyndall

RI MS JT/1/T/360

---

1. *[10 June 1870]*: the enclosed letter Tyndall mentions was sent on 10 June 1870, and this one was likely sent on the same day or shortly after. See letter 3213.
2. *the enclosed letter*: letter 3213.
3. *the senior physician*: George Owen Rees.
4. *Guy's*: Guy's Hospital Medical School in London.
5. *your approval*: Tyndall and Debus discussed Debus's application for a position at Guy's in letters 3202, 3209, 3211, 3213, and 3220.

From William Robinson[1] 12 June 1870 3215

5 Glentilt Street, Belfast. | June 12th. 1870.

Dear Sir

I am heartily grateful for your more than kind note[2] & above all for the expression of sympathy that note contains. For that sympathy I thank you from my very soul.

I delayed replying to yours of the 6th. Inst.,[3] thinking you would be naturally inclined to hear what I intended to do & in what manner follow your ever good advice.

I wrote on Monday (8) to a Dublin gentlemen—Professor Haughton,[4] who replies to day from Euston Hotel, London, & most good-naturedly informs me that he has written to Mr Wilson, F.R.C.S.[5]—St. Mark's Ophthalmic Hospital, Dublin, about my case, & advises me to go up to Dublin at once to have the advantage of surgeon Wilson's skill whom Dr Haughton styles 'the best ophthalmic Surgeon in Dublin. This I intend to do as soon as circumstances permit, which may be in a few days.

Dear Sir, you will not be sorry to hear this news, as I feel assured your heart has too much true manhood in it not to feel sympathy for any one no matter how humble who is in distress.

I may inform you that my good eye has not been so sore these last few days—the pain when present being for the most part situated in the less sensitive eye. I am almost certain that the disease is 'Cataract.'

Hoping you will pardon the liberties taken in this note | Believe me to remain | now & always your obliged & grateful Servant | W[illia]m. Robinson

Prof. Tyndall F.R.S. | etc.

Dr. Haughton says: | 'I have forwarded your letter to Mr Wilson & have asked him to give you a bed in the Hospital & to take charge of your eyes'. . . .

'You will find him (Mr Wilson) very kind & skillful.'

RI MS JT/1/R/34
RI MS JT/1/TYP/3/1020

1. *William Robinson*: not identified.
2. *our more than kind note*: letter missing.
3. *replying to yours of the 6th. Inst.*: letter missing.
4. *Professor Haughton*: Samuel Haughton (1821–97), an Irish mathematician, geologist, anatomist, and physiologist who published a series of scientific and mathematical textbooks with his Dublin colleague, James Allen Galbraith. He was professor of Geology and Mineralogy at Trinity College Dublin (1851–81) where he and Joseph Galbraith taught the growing number of students who were taking public examinations to enter the Indian

Civil Service, the Royal Artillery, and Royal Engineers. In 1863 he was appointed registrar of medicine at Trinity College and introduced substantial reforms (*DIB*; *ODNB*). Tyndall supported his admission to the RS in 1858 and advised Haughton about presenting research to the RS in 1857.

5. *Mr Wilson, F.R.C.S.*: Henry Wilson (1838–77), an ophthalmic surgeon. Appointed assistant surgeon at St Mark's Hospital in 1861, Wilson was promoted to surgeon in 1868. He was a specialist in eye and ear surgery and was appointed the founding professor of ophthalmic and aural surgery at the Royal College of Surgery of Ireland in 1871 (*DIB*).

To Adelaide Sartoris[1] [13 June 1870][2] 3216

Royal Institution of Great Britain | Monday

My dear M[rs]. Sartoris

Some evil genius has thwarted my letter. I sent it by hand to 9 Park Place.[3]

As I stated in it I shall be on the downs of Folkestone on the 19th. My friend Bence Jones is driving me away as he says I must rest myself after my work.

Yours truly | John Tyndall

Wellcome Library MS. 7777/67

1. *Adelaide Sartoris*: Adelaide Sartoris (née Kemble, 1815–79), was briefly an opera singer, before she married in 1842 Edward John Sartoris (1814–88), a British landowner and later MP (*ODNB*).
2. *[13 June 1870]*: Tyndall visited Folkestone from 16–20 June 1870. The Monday before his departure was 13 June.
3. *9 Park Place*: 9 Park Place, St James's, London was the Sartoris home (Royal Institution of Great Britain, *List of Members, Officers and Professors* (London, 1863)).

From Thomas Henry Huxley 15 June 1870 3217

June 15th. 1870.

My dear Tyndall,

As I am going to take your name in vain in the enclosed[1] I wish you would just take the trouble to cast your eye over it, and tell me if you think there is anything that needs alteration, barring mere verbal corrections.

I am weary and not very confident in my own judgement just now.

Ever yours | T. H. Huxley.

RI MS JT/1/TYP/9/2932
IC HC 8.78
Typed Transcript Only

1. *in the enclosed*: see letter 3217encl.

From Thomas Henry Huxley 14 June 1870 3217encl

14.6.1870. | From R. Clay | Printing Office | Bread Street Hill | 1 A.P. G.H.[1]

My dear Tyndall,

I should have liked to provide this collection of "Lay Sermons, Addresses, and Reviews,"[2] with a Dedication and a Preface. In the former, I should have asked you to allow me to associate your name with it, chiefly on the ground that the oldest of the papers in the book is a good deal younger than our friendship. In the latter, I intended to comment upon certain criticisms with which some of these Essays have been met.

But on turning the matter over in my mind, I began to fear that a formal dedication at the beginning of this volume would look like a grand lodge in front of a set of cottages; while a defence of any of my old papers would simply amount to writing a new one—a labour for which I am, at present, by no means fit.

The book must go forth, therefore, without any better substitute for either Dedication or Preface than these few paragraphs, in concluding which it is only necessary for me to notify you or any other reader of two or three matters.

The first is, that the oldest Essay of the whole, that "On the Educational Value of the Natural History Science,"[3] contains a view of the nature of the differences between living and not living bodies out of which I have long since grown.

Secondly, in the same paper, there is a statement concerning the method of the mathematical sciences, which, repeated and expanded elsewhere, brought upon me at Exeter[4] the artillery of our eminent friend Professor Sylvester.[5]

No one knows better than you do, how readily I should defer to the opinion of so great a mathematician if the question at issue were really, as he seems to think it is, a mathematical one. But I submit, that the dictum of a mathematical athlete upon a difficult problem which mathematics offers to philosophy, has no more weight than the verdict of that great pedestrian Captain Barclay[6] would have had in settling a disputed point in the physiology of locomotion.

The genius which sighs for new worlds to conquer beyond that surprising region in which "geometry, algebra, and the theory of numbers melt into one another like sunset tints, or the colours of a dying dolphin,"[7] may be of comparatively little service in the cold domain (mostly lighted by the moon, some think) of philosophy. In fact, I confess, the more I think of it, the more clearly does our friend fall into the position of one of those "verständige Leute,"[8] about whom he makes so apt a quotation from Goethe.[9] For he evidently does not understand, or has not duly considered, two points. The first is, that I am in no way answerable for the origination of the doctrine he criticises: and the second is, that if we are to employ the terms observation, induction, and experiment in his sense, logic is as much an observational, inductive, and experimental science as mathematics; and that, I confess, appears to me to be a reductio ad absurdum of his argument.

Thirdly, the paper upon the Physical Basis of Life[10] was intended by me to contain a simple statement of one of the great tendencies of modern biological thought, accompanied by a protest from the philosophical side against what is commonly called materialism. The result of my well-meant efforts I find to be, that I am generally credited with having invented "protoplasm" in the interests of materialism. My unlucky "Lay Sermon"[11] has been attacked by microscopists, ignorant alike of biology and philosophy; by philosophers, not very learned in biology or microscopy; by the clergy of most denominations; and by some few writers who have taken the trouble to understand the subject. I trust that these last will believe that I leave the essay unaltered from no want of respectful attention to all they have said.

Fourthly, I wish to refer all who are interested in the topics discussed in my address on "Geological Reform,"[12] to the reply with which Sir William Thomson has honoured me.[13]

And lastly, let me say that I reprint the review of "The Origin of Species"[14] simply because it has been cited as mine by a late President of the Geological Society.[15] If you find its phraseology in some places more vigourous than seems needful, recollect that it was written in the heat of our first battles over the Novum Organon of Biology;[16] that we were all ten years younger in those days; and last, but not least, that it was not published until it had been submitted to the revision of a friend for whose judgment I had then, as I have now, the greatest respect.

Ever my dear Tyndall | Yours very faithfully | T. H. Huxley.

RI MS JT/1/TYP/9/2934–5
IC HP 8.79
Typed Transcript Only

---

1. *14.6.1870 . . . 1 A.P. G.H.*: added to upper left-hand corner of IC HP typescript and included in RI typescript.
2. *"Lay Sermons, Addresses, and Reviews"*: this is a draft version of T. H. Huxley, 'A Prefatory Letter', *Lay Sermons*, pp. v–viii.
3. *On the Educational Value of the Natural History Science*: T. H. Huxley, *On the Educational Value of the Natural History Science* (London: John Van Voorst, 1854).
4. *Exeter*: see letter 3008, n. 10.
5. *mathematical sciences . . . Professor Sylvester*: James Joseph Sylvester was president of the mathematics and physics section of the BAAS meeting in Exeter in 1869. His address built upon Huxley's views of mathematical science which had been published in *Macmillan's Magazine* and the *Fortnightly Review* (A. MacFarlane, *Lectures on Ten British Mathematicians* (England: Chapman and Hall, 1916), pp. 68–77; *Brit. Assoc. Rep. 1869*, pp. 1–9).
6. *Captain Barclay*: Robert Barclay Allardice (1779–1854), a famous Scottish 'Pedestrian', or endurance walker. He was known for his record of walking 1,000 miles in 1,000 hours in 1809 (*ODNB*).
7. *"geometry, algebra, and the theory of numbers melt into one another like sunset tints, or the colours of a dying dolphin"*: Huxley was quoting from J. J. Sylvester, 'A Plea for the Mathematician. II', *Nature*, 1 (1870), pp. 261–3, on p. 261.
8. *verständige Leute*: reasonable people (German). This is a reference to a line from Goethe's 'Zahme Xenien V': 'Verständige Leute kann du irren sehn, | Zu Sachen, nämlich, die sie nicht verstehn.' (You can find reasonable people erring in matters that they do not understand.) (M. Hecker (ed.), *Goethe's Sprüche in Reimen. Zahme Xenien und Invektiven* (Leipzig: Insel Verlag, 1908), p. 130).
9. *Goethe*: Johann Wolfgang Goethe.
10. *Physical Basis of Life*: T. H. Huxley, *On the Physical Basis of Life* (New Haven: The College Courant, 1869).
11. *"Lay Sermon"*: T. H. Huxley, 'On the Advisableness of Improving Natural Knowledge', *Lay Sermons*, pp. 3–22.
12. *my address on "Geological Reform"*: T. H. Huxley, 'Geological Reform', *Lay Sermons*, pp. 251–79.
13. *reply with which Sir William Thomson has honoured me*: W. Thomson, 'Of Geological Dynamics', *Transactions of the Geological Society of Glasgow*, 3 (1869), pp. 215–40.
14. *review of "The Origin of Species"*: T. H. Huxley, 'The Origin of Species' (*The Westminster Review*, April 1860), *Lay Sermons*, pp. 255–98.
15. *a late President of the Geological Society*: there are several possible candidates, depending on whether 'late' refers simply to past presidents or to deceased past presidents. For a list of Geological Society Presidents see www.geolsoc.org.uk/About/History/Past-Presidents [accessed 18 May 2021]. Most likely are Leonard Horner (1785–1864) and Andrew Crombie Ramsay (1814–91), whereby Horner was both past and deceased.
16. *Novum Organon of Biology*: Novum Organon is a reference to Francis Bacon's *Novum Organum, sive indicia vera de Interpretationes*, a treatise on scientific method published in 1620. With the 'Novum Organon of Biology', Huxley was referring to Charles Darwin's *On the Origin of Species* (1859).

To Thomas Henry Huxley 16 June 1870 3218

Langhorne Gardens, | Folkestone, 16th June | 1870.

Dear old Hal,

This "Preface"[1] certainly bears no marks of weariness that I can detect. It makes me think that you are either favoured with uncommonly lucid intervals of strength and freshness, or that the assumption of weariness is all a sham.

I call it an exceedingly good piece of writing—It was bound to be under the circumstances.

This book will enrich you if you beware of the publishers.

I have hardly any thing to say further. If you could omit the words "does not understand," [2] at the bottom of page VI the passage would be less frictional to old Sylvester. He is a good creature and does not require a hacked edge.

The touch of the moon is admirable—I discovered it <u>at once</u>!

The paragraph in which you give Beale[3] and Stirling[4] that backhanded stroke is exceedingly good and appropriate.

I hope some day to see you develop the last page, and more particularly the last paragraph of the "Physical Basis of Life." I probably am the only living man who sees the meaning shadowed forth in that page and paragraph!![5]

However let the book go as it stands—It will live. I will not say a word of the feeling which your allusion to our friendship aroused. God bless you my dear old fellow—that's more than you dare say in my favour.

Yours ever | John Tyndall.

RI MS JT/1/TYP/9/2933
IC HP 8.81
Typed Transcript Only

1. *"Preface"*: see letter 3217encl. and 3217.
2. *If you could omit . . . understand"*: these words were removed from the final printed version which reads 'Surely he has not considered two points'.
3. *Beale*: Lionel Smith Beale (1828–1906), a British physician and microscopist. He was chosen over Huxley among other candidates as professor of physiology at King's College (1858–69). He was known for his research on microscopy, histology, and chemical pathology, and identified the pyriform nerve ganglion cells, known as 'Beale's cells'. From 1868 to 1870 he presented and published his ideas on disease and germs, which were in opposition to Pasteur's germ theory (*ODNB*).
4. *Stirling*: James Hutchinson Stirling (1820–1909), a Scottish philosopher. The inheritance provided after his father's death in 1851, enabled Stirling to devote his time to literary

pursuits. He published the first comprehensive study of Georg Wilhelm Friedrich Hegel in English, *The Secret of Hegel* (1865), which was widely read. His subsequent publications, including *As Regards Protoplasm* (1869), which critiqued the work of Huxley, were not so well received (*ODNB*).

5. *I probably am the only living man who sees the meaning shadowed forth in that page and paragraph!!*: Tyndall was commenting on their shared view of the limits of science and materialism which Huxley acknowledges in this paragraph (*Ascent of John Tyndall*, p. 266).

To Thomas Archer Hirst 17 June 1870 3219

Folkestone | 17th. June, 1870

Dear Tom

Bring an empty Brandy flask with you as I find I must invest in a bottle—They won't sell less and I must divide it with you.

To morrow's post I suppose will bring me intelligence of your wishes—There is a hotel close at hand here—also lodgings I doubt not.[1]

Yours affectionately | John

RI MS JT/1/HTYP/564
LT Typescript Only

---

1. *To morrow's post . . . doubt not*: no letter from Hirst has been found, but according to Hirst's journal, he and his niece Lilly stayed at the West Cliff Hotel in Folkestone (RI MS JT/2/32d/1877).

To Heinrich Debus 19 June [1870][1] 3220

4 Langhorne Gardens | Folkestone | 19 June

Dear Heins

I have forwarded your work to Rees.[2] Matters look promising—How glad I shall be when you are near me.

I am down here for a breath of fresh air—London became perfectly unbearable. I do not know how the world of dinner parties can get on. I return next week.

Yours ever | John Tyndall

RI MS JT/1/T/315
RI MS JT/1/TYP/7/2372

1. *[1870]*: this letter is part of a series of letters discussing Debus's appointment as Chair of Chemistry at Guy's Hospital in 1870. See letters 3202, 3209, 3211, 3213, and 3214.
2. *Rees*: George Owen Rees.

To William Thomson 19 June [1870][1] 3221

Folkestone | 19th June

My dear Thomson.

Would you send me a short and strong letter that I might read to the Council[2] next Thursday regarding the claims of Joule.[3] The officers have taken me by surprise twice this year. I had no idea that the matter was to come in on Thursday, and now while staying here at Folkestone I am informed that a written statement of claims will be required. I hope you will speak out: I certainly will.

Yours ever | John Tyndall

Pray write to me to Albemarle Street:[4] I return there on Wednesday.

Add 7342/T625, Kelvin Correspondence, Cambridge University

1. *[1870]*: the year is given by reference to Tyndall's efforts to have James Joule awarded the Copley Medal (see n. 3).
2. *the Council:* Council of the RS.
3. *the claims of Joule:* this letter is part of Tyndall's efforts to have James Joule awarded the Copley Medal, the most prestigious award of the RS. See letters 3181 and 3185.
4. *Albemarle Street:* location of the RI.

To Juliet Pollock 20 June 1870 3222

Folkestone[1] | 20th June 1870

My dear Friend,

I share your sorrow and most deeply do I sympathize with your friend Sir Henry Taylor[2] in his great bereavement.[3] Such woe almost renders one reconciled to the negative side of life which if joyless is also painless. I have been stay[in]g here for a week and tomorrow I run across for a day to Lady Mary Egerton's. On Wednesday I return to London.

By quitting London I have saved myself from a breakdown—I know not how reasonable people with any intellectual work to perform live through the life of London at this season of the year.

Remember always that I am your friend, and most your friend when you are in sorrow.

Yours ever | John Tyndall

RI MS JT/1/T/1152
RI MS JT/1/TYP/6/2125

1. *Folkestone*: this letter was written on RI letterhead.
2. *Sir Henry Taylor*: Henry Taylor (1800–86), a British poet, playwright, public servant, and friend of the Pollocks. He went to sea as a midshipman in 1814 but was discharged before the end of the year. In 1824 he secured a clerkship with the Colonial Office and the following year was promoted to senior clerk, a position he held until his retirement in 1872. Taylor also published a considerable catalogue of verse dramas and poems. He received a DCL from the University of Oxford in 1862 and was made KCMG in 1869 (*ODNB*). Beginning in 1861, Taylor and his family spent summers at The Roost, his home in Bournemouth on the southeastern English coast (U. Taylor, *Guests and Memories: Annals of a Seaside Villa* (London: H. Milford, Oxford University Press, 1924), p. 14).
3. *his great bereavement*: Taylor's eldest son, Aubrey Ashworth Taylor (1846–70), who went to Balliol College, Oxford, passed away on 16 May 1870 (*ODNB*).

From John James Drysdale[1] 21 June 1870 3223

36a Rodney Street, | Liverpool | 21 June 1870.

Dear Sir

I feel quite flattered by the receipt of your courteous note[2] as I had no idea of troubling you to write a personal acknowledgement of my pamphlet[3] for it was merely sent to you in your official capacity.

I had at first the intention of getting a friend in London to ask if you would object to any allusion to that controversy in 1863[4] but afterwards thought it better not to intrude upon you as the matter is a public one & your position in it is one of which any one may well be proud.

It is very good of you take such a charitable view of the motives of both Sir W[illia]m Thomson & Prof[essor]r Tait, and I hope they are as you say: but every man must be judged by his acts and still I cannot but feel that a taint of meanness must cling to our idea of these characters. And also whatever Professor Tait's excuse may be for letting prejudice distort his intellectual vision in respect to Mayer, it is impossible for an impartial bystander like myself to help resenting his tone of downright impertinence towards yourself.

I am glad to see that Mr Joule is elected to the French Academy,[5] but it is a pity they lost the opportunity of doing as you say, a peaceful act in bracketing him with Mayer when there happened to be a double vacancy. The claim of Joule was surely better than that of Kirchhoff.[6]

I perceive with pleasure that your name is on the list of those to be proposed, & I trust that you will be the next to receive that high honour,[7] for there is no one who has a better claim.

Believe me | Yours very sinc[ere]ly | J Drysdale

RI MS JT/1/D/141

---

1. *John James Drysdale*: John James Drysdale (1815–92), a Scottish physician. He was president of the British Homeopathic Congress (1870) and consulting physician to the Liverpool Homeopathic Dispensary. Drysdale was the editor of the *British Journal of Homeopathy* and author of *Elements of General Pathology* (1842) and *The Protoplasmic Theory of Life* (1874), among other texts. (Jonathan Davidson, *A Century of Homeopaths: Their Influence on Medicine and Health* (New York: Springer Science & Business, 2014), p. 97).
2. *your courteous note*: letter missing.
3. *my pamphlet*: probably Drysdale's 1870 publication *Life and the Equivalence of Force* (London: Turner and Co., 1870).
4. *that controversy in 1863*: the Mayer-Joule priority dispute over the theory of the mechanical equivalent of heat (see 'Introduction' to *Tyndall Correspondence*, vol. 8, pp. xvii–xix). Drysdale discusses it in his chapter 'Nature and definition of force' in *Life and the Equivalence of Force* (London: Turner and Co., 1870), pp. 76–109.
5. *Mr Joule is elected to the French Academy*: James Joule was elected a corresponding member of the Parisian Académie des sciences on 30 May 1870 ('Correspondance', *Paris, Comptes Rendus*, 70 (1870), p. 1267).
6. *better than that of Kirchhoff*: Gustav Kirchoff was elected a corresponding member of the Parisian Académie des sciences on 27 January 1870 ('Nominations', *Paris, Comptes Rendus*, 70 (1870), p. 138).
7. *I trust . . . high honour:* Tyndall was among a number of candidates put forth that year for the section in physical sciences (his name is given in a list with Julius R. Mayer, Anders Jonas Ångström (see letter 3179, n. 5), French physicist Félix Billet (1808–82), German physicist Heinrich Dove (1803–79), William Grove (see letter 3299, n. 4), Joseph Henry, German physicist Moritz von Jacobi (1801–74), James P. Joule, Gustav Kirchhoff, German physicist Peter Theophil Riess (1804–83), George Gabriel Stokes [misspelled as 'Stockes'], William Thomson, and Italian physicist Paolo Volpicelli (1804–79), with Joule and Kirchhoff elected (see n. 5 and n. 6) ('Comité Secret', *Paris, Comptes Rendus*, 70 (1870), p. 54). This honor does not appear to have ever been bestowed upon Tyndall. Charles Darwin, though nominated in 1870, was not elected until 1878 (with numerous nominations in the years between); Thomas Huxley, also nominated in 1870, was later elected in 1885; while Joseph Hooker was never elected ('Comité Secret', *Paris, Comptes Rendus*, 70 (1870), p.

1417–8). The delay in Darwin's election had everything to do with anti-Darwinian views by the French, so it is possible the delay in Huxley's election, and lack of membership for Tyndall and Hooker, may have had to do with their being closely associated with Darwin (see F. Burkhardt, et al. (eds), *The Correspondence of Charles Darwin*, 28 vols (New York: Cambridge University Press, 2010), vol. 18 (1870), pp. xxi–xxii; P. Corsi, 'Recent Studies of French Reactions to Darwin', in D. Kohn (ed.), *The Darwinian Heritage* (Princeton, NJ: Princeton University Press, 1985), pp. 698–711, on p. 699; F. G. Henry, 'Anti-Darwinism in France: Science and the Myth of Nation', *Nineteenth-Century French Studies*, 27:3/4 (1999), pp. 290–304); and R. E. Stebbins, 'France', in T. F. Glick (ed.), *The Comparative Reception of Darwinism*, 2nd edn (Chicago: University of Chicago Press, 1988), pp. 117–63, on pp. 145–51).

From James Thomson[1] 21 June 1870 3224

Largs, Ayrshire, 21st June 1870.[2]

Dear Sir,

My brother[3] has received this morning your letter of the 19th inst[ant].[4] regarding the statement of claims in favour of Joule[5] to be submitted to the Royal Society.

In consequence of the death of Lady Thomson,[6] which took place on Friday last, he has asked me to write to you. He wishes me to say that in consequence of your previous letter to him and of one which he received from Archibald Smith,[7] he has been in communication with Mr Smith, and has written to him at considerable length regarding Joule's merits.

He had been informed that the discussion would not come on in the Royal Society till November, and had intended to write very fully to you and others on the subject in good time.

He now asks me to send you the enclosed list[8] of Papers by Joule, being a copy of one he had sent lately to Mr Smith. He wishes me to say that the large masses of highly important work done by Joule before 1842, give him a seniority over all others. Between Joule and Helmholtz[9] the discussion ought not, under any circumstance to be made a question of comparing merits. In regard to Helmholtz the highest honour[10] the Royal Society can give could not possibly be given to any person more deserving it. But Joule's early merits and the great arrears owing to him by the Royal Society on consequence of their rejection of the papers by which he laid the foundation of thermo-dynamics in electro-chemistry, electro-magnetism and pneumatics, render the claims of his proposers on the present occasion quite paramount.

My brother, you will understand, is quite unable to enter upon any business at present, but on receiving your letter this morning asking for a statement in time for Thursday, he felt very anxious that the consideration of

Joule's merits should not suffer in any degree from his incapacity to plead them, and he therefore wished me to write to you and express the hope that you will make an effort to bring forward some points which perhaps you as mover in some sense might have left for him to say as seconder of the proposal had he been able to do his part.

I am, Dear Sir, | Yours truly, | James Thomson.

Professor Tyndall LL.D.

P.S. My brother would be much obliged if you would let him have the enclosed list again at your leisure: but no hurry at all about this.

Add 7342/T626, Kelvin Correspondence, Cambridge University
RI MS JT/1/TYP/5/1556–7

1. *James Thomson*: James Thomson (1822–92), a British mechanical engineer and older brother of William Thomson (later Lord Kelvin). James Thomson worked as an engineer at firms in Glasgow, London, and Belfast before he became professor of engineering at Queen's University Belfast in 1854 and remained until 1875, when he took a position at the University of Glasgow. At Belfast, Thomson worked on improving the efficiency of centrifugal pumps, as well as working on water pumps for the Belfast water commissioners (*ODNB*).
2. *Largs, Ayrshire, 21st June 1870.*: handwritten above date in the same hand 'A copy of the sent letter. J.T.'.
3. *My brother*: William Thomson.
4. *your letter of the 19th inst[ant]*: letter 3221.
5. *Joule*: James Prescott Joule.
6. *Lady Thomson*: Margaret Thomson (née Crum, d. 1870), was William Thomson's second cousin, who he married in 1852, when she was 22. She was the daughter of cotton manufacturer and calico-printer Walter Crum (*ODNB*). She was ill for most of their marriage and they had no children.
7. *Archibald Smith*: Archibald Smith (1813–72), a Scottish mathematician who studied at Glasgow University and the University of Cambridge from where he graduated BA in 1836 and MA in 1839. He was senior wrangler in 1836 and was elected a fellow of Trinity College, Cambridge. Following studies at Lincoln's Inn he was called to the bar on 25 January 1841. His interest in science was already manifest while an undergraduate at Cambridge and in 1837 he co-founded the *Cambridge Mathematical Journal*. He put his mathematical skills to good use in several publications on navigation, particularly on correcting compass deviations, and was awarded a LLD from the University of Glasgow in 1864 for this work (*ODNB*).
8. *the enclosed list*: enclosure missing.
9. *Helmholtz*: Hermann Helmholtz.
10. *highest honour*: the Copley Medal is the RS's most prestigious award. Joule received it in 1870, Helmholtz in 1873.

From Auguste de la Rive 22 June 1870 3225

Genève, le 22 Juin, 1870.

Mon cher Tyndall,

Je vous remercie infiniment de vos deux lettres et de toutes les explications que vous voulez bien m'y donner. Je suis très touché de tout ce que vous me dites d'amical et de toutes les offres obligeantes que vous me faites. Cela ne m'étonne pas de votre part; j'ai toujours compté sur votre amitié comme vous pouvez compter sur la mienne à tout jamais.

Je n'accepte de toutes vos offres qu'une seule, celle d'avoir l'obligeance de faire insérer dans le Phil. Mag. la lettre ci-jointe que je vous envoie. Je l'ai faite aussi courte que possible. Ayez la bonté seulement de la faire traduire sous vos yeux afin qu'il n'y glisse pas d'erreurs, ce à quoi je tiens beaucoup. Je vous enverrai dès que je l'aurai terminé, le résumé des observations que j'ai faites depuis quelques années sur la transparence de l'eau. Peut-être aurez-vous l'obligeance de la faire insérer dans le Phil. Mag., en y joignant la description du photomètre nouveau, je ferai cette description un peu plus détaillée qu'elle ne l'est dans le C. R. de la Rev. des Sciences et dans les Annales de Ch. et de Phys., quoique vous puissiez déjà la trouver là aux endroits indiqués dans ma lettre assez complète. Si vous croyez qu'il vaille mieux l'insérer tout de suite; il vous sera facile de la trouver au Tome 61 p.1221 du C.R. et au T. 12, p.243 des Ann.—Faites comme vous le jugerez bon. Il va sans dire que vous pouvez faire suivre une lettre de toutes les réflexions et observations que vous jugerez convenables.

J'espère que vous avez fait mes remerciements à Mr Bence Jones pour l'envoi des Lettres de Faraday; rappelez moi aussi à son bon souvenir et croyez-moi votre tout dévoué & affectionné | A. de la Rive.

Geneva, 22 June , 1870

My dear Tyndall,

I am deeply grateful for your two letters[1] and for all of the explanations that you were kind enough to give me. I am very touched by all your words of friendship, and by all of your kind offers. This does not surprise me coming from you; I always counted on your friendship as you can count on mine forever.

I only accept one of your offers, that of having the kindness to insert into the Phil[osophical]. Mag[azine].[2] the enclosed letter[3] that I am sending you. I made it as short as possible. Kindly have it translated under your supervision so that no errors slip into it, which I much desire. I will send you, as soon as I have finished it, the summary of the observations that I have made for several years on the transparency of water. Perhaps you would be kind enough to insert it into the Phil[osophical]. Mag[azine]. and to add there a description of the new photometer. I will give a more detailed description than is provided in the C[omptes]. R[endus]. de la Rev[ue]. des Sciences and in the Annales de Ch[imie]. et de Phys[ique].,[4] although you can already find a reasonably comprehensive one at the places indicated in my letter. If you think that it is better to insert it right away; it will be easy for you to find it in volume 61, pg. 1221 of the C[omptes]. R[endus]., and in vol. 12, pg. 243 of the Ann[ales]. Do as you see fit. It goes without saying that you can follow up with a letter detailing the reflections and observations that you deem appropriate.

I hope that you gave my thanks to Mr Bence Jones for sending Faraday's Letters;[5] remind him of me, and know that I am, your devoted and affectionate | A. de la Rive.

RI MS JT/1/TYP/1/376
LT Typescript Only

---

1. *your two letters*: letters missing.
2. *insert into the Phil[osophical]. [Mag]azine.*: it does not appear that this was translated and published in the *Phil. Mag.*
3. *the enclosed letter*: enclosure missing.
4. *C[omptes] . . . Phys[ique]*: A. de la Rive, 'Note sur un photomètre destiné à mesurer la transparence de l'air', *Paris, Comptes Rendus,* 63 (1867), pp. 1221–5, and *Annal. Chim. et Phys.,* 12 (1867), pp. 243–9.
5. *Faraday's Letters*: *Life of Faraday.*

From François-Napoléon-Marie Moigno 22 June 1870 3226

Les Mondes | Revue Hebdomadaire | Des Sciences et de leurs applications aux arts et à l'industrie | par M. l'Abbé Moigno | Rédaction: | 2, Rue d'Erfurth, 2 | Paris, le 22 juin 1870

Mon cher Mr Tyndall

J'ai reçu votre volume sur le magnétisme et je vous en remercie cordialement. J'ai traduit les notes sur les leçons d'électricité à mesure qu'elles ont paru dans les Chemical News, et je vais *[*commencer*]* leur publication dans Les Mondes. J'en ferai une petite actualité.

Je voudrais bien *[*avoir*]* le petit volume des notes sur les *[*<u>douze</u>*]* leçons de la lumière, je les traduirai sur le champ.

Je travaille *[*sans*]* *[1 word illeg]* à mes Splendeurs de la Foi, accord de la science et de la religion, voilà pourquoi je néglige un peu les actualités. Mr Gauthier-Villars ne parle pas encore de publier la Chaleur, j'en suis désolé.

Votre humble et bien dévoué serviteur | l'abbé F. Moigno

Les Mondes[1] | Paris, 22 June 1870

My dear Mr Tyndall,

I received your volume on magnetism[2] and I am deeply grateful for it. I translated the notes on the lessons about electricity[3] as they appeared in the Chemical News,[4] and I will begin to publish them in Les Mondes.[5] I will make /a bit of news out of it/.

I would like to procure the small volume of notes on the /twelve/ lessons on light,[6] I will translate them immediately.

I work without *[1 word illeg]* on my Splendeurs de la Foi[7] a union of science and religion, and this is why I have been neglecting the news a bit. Mr Gauthier-Villars[8] no longer speaks of publishing Heat,[9] I am devastated by that.

Your humble and very devoted servant | Father F. Moigno

RI MS JT/1/M/119

---

1. *Les Mondes*: Les Mondes | Weekly Review | Of the sciences and their applications to the arts and the industry | by Father Moigno | Editorial offices: | 2 Erfurth street, 2. Headed notepaper.
2. *your volume on magnetism*: *Diamagnetism* (see letter 3076, n. 8).
3. *Notes on the lessons about electricity*: these were published first in Chemical News and Journal of Industrial Science, with the first appearing as 'Notes of a Course of Seven Lectures on Electrical Phenomena and Theories. By Professor Tyndall, LL/D. Lecture I)', *Chemical News and Journal of Industrial Science,* 21 (January 1870), pp. 210–12. They were then published as a book, J. Tyndall, *Notes on a Course of Seven Lectures on Electrical Phenomena and Theories, delivered at the Royal Institution of Great Britain, April 28–June 9, 1870* (London: Longmans, Green, and Co., 1870). The French translation was published as J. Tyndall, *Programme d'un cours en sept leçons sur les phénomènes et les théories électriques*, trans. M. l'abbé Raillard (Paris: Gauthier-Villars, 1871).
4. *Chemical News*: see letter 3106, n. 4.
5. *Les Mondes*: *Les Mondes* was a scientific journal founded by Moigno in 1862. He published Tyndall's lessons on electricity in this journal as: 'Sur les phénoménes et les theories électriques', *Les Mondes*, 23 (1870), pp. 361–6 (Lesson I), 428–34 (II), 564–71 (III), 622–31 (IV), 719–27 (V), 757–68 (VI–VII).
6. *small volume of notes on the [twelve] lessons on light*: probably J. Tyndall, *Notes on a Course of Nine Lectures on Light* (London: Longmans, 1870). The translation was published as J. Tyndall, *La lumière: notes d'un cours de neuf leçons. Sur le rôle scientifique de l'imagination*, trans. M. l'abbé Raillard (Paris: Gauthier-Villars, 1872).
7. *Splendeurs de la Foi*: F. Moigno: *Splendeurs de la Foi: accord parfait de la révélation et de la science de la foi et de la raison* (Paris: Gauthier-Villars, 1879). This title translates to 'Splendors of Faith: the perfect harmony between revelation and science of faith and reason'.
8. *Mr Gauthier-Villars*: Jean-Albert Gauthier-Villars (see letter 2949, n. 6).

9. *publishing Heat*: the second edition of Tyndall's *Heat Considered as a Mode of Motion*, translated by Moigno, was indeed published by Gauthier-Villars in 1874.

To Alfred Tennyson 24 June [1870][1] 3227

24th June.

My dear Tennyson.

I was very sorry to miss you when you called. But at that moment I was under the "thousand emeralds"[2] of a time reading either Shelly's[3] Skylark, his Cloud, Childe Harold,[4] Enone, the Sleeping Beauty, Sir Galahad or Saint Agnes[5]—one of these must have occupied me at the time. I had never brought you & Byron[6] so close together before—He used to be a grand favourite with me but I confess, without meaning to flatter you in the least; that he looked shady beside Enone & Galahad

I hope you are strong & fresh: were you not surfeited with Switzerland I would try to coax you there. I am aiming to be off on the 1st of July.

I had a most pleasant visit to Herschel[7] on Wednesday.

Yours ever | John Tyndall

TRC Letters, 5029

---

1. *[1870]*: this letter is dated in regard to Tennyson's reply of 27 June 1870 (letter 3230).
2. *"thousand emeralds"*: 'A million emeralds break from the ruby-budded', A. Tennyson, 'Maud', in *Maud and Other Poems* (London: Moxon, 1855), p. 14.
3. *Shelly*: Percy Bysshe Shelley (1792–1822), an English Romantic poet well known for interest in social justice. His time at Oxford was cut short in 1811, when he was expelled for suspected authorship of the tract *The Necessity of Atheism*. 1812 saw him in Ireland agitating for repeal of the Union and Roman Catholic Emancipation. Shelley's sense of social justice was increasingly borne out in his poetry. By 1815, Shelley secured recognition as a poet with several well-received publications in that and subsequent years (*ODNB*).
4. *Skylark, his Cloud, Childe Harold*: 'To a Skylark', 'The Cloud', and 'Child Harold's Pilgrimage' are from P. B. Shelley, *Prometheus Unbound. A Lyrical Drama in Four Acts* (London: C. and J. Ollier, 1820).
5. *Enone, the Sleeping Beauty, Sir Galahad or Saint Agnes*: 'Enone', 'The Sleeping Beauty', 'Sir Galahad' and 'Saint Agnes' are poems by Alfred Tennyson.
6. *Byron*: George Gordon Noel Byron (1788–1824), Sixth Baron Byron, an English Romantic poet. Byron was one of the most famous and influential poets of the nineteenth century, even though the suggestive nature of his poems often ran counter to Victorian high moralism (*ODNB*). Tyndall was deeply influenced by Byron's poetry and by the figure of the independent, defiant Byronic hero.
7. *Herschel*: John Herschel.

To Josiah Whitney[1] 25 June 1870 3228

Royal Institution of Great Britain | 25th. June <u>1870</u>.

My dear Sir,

Some days ago your truly beautiful "Guide Book of the Yosemite Valley"[2] reached my hand[3]— Accept my best thanks for it. It makes me long for a sight of those regions; but I find as one gets older duties thicken, and long continued pleasure excursions become more difficult. It is needless for me to express the pleasure with which I read Mr. King's[4] account of his ascent of Mount Tyndall.[5] Truly I thank both you & him.

Yours most faithfully | John Tyndall | To Prof. Whitney.

Yale MS 555, Box 15, 1–2

1. *Josiah Whitney*: see letter 3151, n. 1.
2. *Guide Book of the Yosemite Valley*: letter 3151, n. 5.
3. *reached my hand*: see letter 3151.
4. *Mr. King*: Clarence King (see letter 3190, n. 5).
5. *Mount Tyndall*: see letter 3151, n. 2.

From Louis Philippe Albert d'Orléans[1] 25 June 1870 3229

York House, | Twickenham, | Middlesex. June 25th 1870

Dear Sir

I thank you most cordially for the volume[2] which you have so kindly send to me and which will be a most valuable addition to my library, doubly valuable thanks to the inscription which it bears. Allow me to offer to you a book[3] which my brother[4] and myself we have just published, and which was written by our father[5] some thirty years ago. As a traveller you may be interested in finding in it an history of the first years of our Domination in Algeria.

If you are disengaged on Wednesday the 6th of July could you do us the pleasure of coming to dine here with us at 8 pm.?

I remain yours truly | L P A D'Orléans |C[om]te de Paris

RI MS JT/2/10/478–9
RI MS JT/2/13c/1380–1

1. *Louis Philippe Albert d'Orléans*: Louis Phillipe Albert D'Orleans (1838–94), the exiled Count of Paris, who lived in York House, Twickenham between 1848–71. He was the eldest son of Ferdinand Philippe Louis, Duke of Orléans. His grandfather, Louis Philippe

I, was the last King of France from 1830–48 until he was deposed by Napoleon III. In 1865, D'Orleans was the claimant to the throne during the second French empire under the rule of Napoleon III. From 1861–2, he served for the Union (the North) during the American Civil War (1861–5), and was known as 'Captain Perry'. After the end of the Franco-Prussian War in 1871 and the defeat of Napoleon III, D'Orleans returned to France, but was expelled again in 1886 (*ODNB*; M. Burlingame, *With Lincoln in the White House: Letters, Memoranda, and Other Writings of John G. Nicolay, 1860–1865* (Carbondale and Edwardsville, IL: Southern Illinois University Press, 2000), p. 211, n. 117).

2. *the volume*: the volume has not been identified, but Tyndall had been sending copies of his *Diamagnetism* to numerous individuals.
3. *a book*: Le Duc D'Orléans, *Campagnes de l'armée d'Afrique, 1835–1839* (Paris: Michel Lévy frères, 1870).
4. *my brother*: Robert Philippe Louis Eugéne Ferdinand d'Orléans (1840–1910), the second son of Ferdinand Philippe Louis, Duke of Orléans, and grandson of Louis Philippe I, King of France. Like his brother he served with the Union during the American Civil War (1861–5), and was known as 'Captain Chatters' (M. Burlingame, *With Lincoln in the White House*, p. 211, n. 117).
5. *our father:* Ferdinand Philippe Louis d'Orléans (1810–42), the eldest son of Ferdinand Philippe Louis, Duke of Orléans, styled king of the French under the July monarchy (1830–48). He was a soldier and served in Algeria from 1834 to 1836, returning again in 1839. Unfortunately this second campaign led to ill health, and he died at Neuilly in 1842 (*ODNB*).

From Alfred Tennyson 27 June 1870 3230

June 27th 1870

My dear Tyndall

I am sorry for my own sake that you were not in when I called, for you always have something scientifically interesting to show me, when I do.

Your call to the Alps[1] I cannot follow however much I may wish to do so: guests, engagements &c are in the way. I wonder whether you are going in the eclipse-ship:[2] if so I shall enjoy your company.[3]

ever yours | ATennyson

RI MS JT/1/T/4

1. *Your call to the Alps*: see letter 3227.
2. *eclipse-ship*: Tyndall traveled to Oran, Algeria from Portsmouth on the steam-ship *Urgent* to view the 22 December 1870 total solar eclipse, but the weather caused the expedition to abandon hope of witnessed the celestial event (see 'Voyage to Algeria to Observe the Eclipse', in *Hours of Exercise in the Alps*, pp. 429–73; he published this again in *Fragments of Science for Unscientific People: A Series of Detached Essays, Lectures, and Reviews*, 5th

edn (London: Longmans, Green, and Co., 1876), pp. 186–217). As other letters indicate, Tyndall at first planned to join an eclipse expedition to the Italian volcano Mt. Etna, but he changed his mind. For letters mentioning the eclipse expedition and Tyndall's travels to Algeria, see 3313, 3318, 3322, 3323, 3326, 3327, 3328, 3329, 3331, 3332, 3333, 3334, 3335, 3336, 3338, and 3347.

3. *I shall enjoy your company*: Tennyson decided against travelling with one of the eclipse ships after he discovered that he was expected to write a poem about it (H. Tennyson, *Alfred Lord Tennyson. A Memoir*, 2 vols (Cambridge: Cambridge University Press, 2012), p. 100).

From Mary Francis Egerton [29 June 1870][1] 3231

Dear M[r]. Tyndall,

We bring you a few Mountfield[2] Roses as you approved of them last time.

I dine out tonight, but no more this week in case you should look in. I am sure your party last night was not one to be left early unless you were dead tired, so I never expected you.

If we do not meet, let me hear just the "when & where" of your proceedings, & send me a line from the Engadine[3] if you sh[oul][d]. have some bad weather & nothing to do. Though you profess to take no interest in the world, there are those who take some slight degree of friendly interest in your existence in it!

Don't think I mean to entail a correspondence upon you. It is fate & not I, that brought about the necessity for sending notes & letters this last year.[4]

You will come to Mountfield when you come back?

Y[ou][rs] most sinc[ere][ly] | M F Egerton

Eaton Place | Wed[nesda][y]

RI MS JT/1/E/99e

---

1. *[29 June 1870]*: the date of this letter has been estimated through its reference to Tyndall's travels. Tyndall and Egerton met several times in June 1870. Egerton travelled from Mountfield to London for the day to hear Tyndall lecture on 9 June (see letter 3204). Tyndall visited Egerton at Mountfield on 21 June during a sojourn in Folkestone (*Hirst's Journal*, RI MS JT/2/32d/1878) and left London to begin his annual journey to Switzerland on 1 July. The only Wednesday between these two dates is 29 June. The weather in the Engadine was, unfortunately, unfavourable, so Tyndall changed his plans. See letter 3242.
2. *Mountfield*: see letter 3040, n. 17.
3. *Engadine*: see letter 3146, n. 19.
4. *this last year*: Egerton's husband Edward Christopher Egerton died suddenly in late August 1869 (see letter 3040, n. 2).

From Georg Hermann Quincke[1] 30 June 1870 3232

39 Brüderstrasse Berlin, den 30ten Juni 1870

Hochgeehrter Herr Professor!

Für die gütige Übersendung Ihres Buches über Diamagnetismus und die dazu gehörige Tafel sage ich Ihnen besten Dank. Das Studium desselben interessirt mich jetzt um so mehr, als ich selbst wieder mit Versuchen über die magnetische Drehung der Polarisationsebene beschäftigt gewesen bin, und die verschiedenen Arbeiten von Ihnen einem Deutschen, der nicht alle Journale selbst besitzen kann, schwer zugänglich sind, abgesehen davon, dass auch leicht ein wichtiger Aufsatz übersehen wird. Ich selbst habe auch schon daran gedacht meine eigenen Arbeiten im Zusammenhange herauszugeben, sobald ich dazu gekommen bin den Rest meiner optischen Untersuchungen fertig zu redigiren.

Leider kostet Berlin einem Gelehrten der allerlei Vorlesungen tagtäglich zu halten hat, womöglich noch mehr Zeit als London, und so bin ich in letzter Zeit viel weniger als ich wünschte zu wissenschaftlichen Arbeiten gekommen.

Zu meinem Bedauern habe ich Sie bei Ihrem letzten Berliner Aufenthalte nicht gesehen. Ich hätte Sie sonst gerne mündlich über einen Versuch gesprochen, den Sie in dem Werke „on Heat" pag. 16 beschreiben; Mir selbst ist es mit einem trocknen Blasebalge niemals gelungen, eine Erwärmung der Thermosäule nachzuweisen; ich erhielt stets eine Erkaltung, wie wenn Luft aus einem Gefäss, worin sie comprimirt worden, entweicht. Ich gestehe, dass ich auch erwarten würde a priori, dass kein Unterschied besteht, mag die Luft mit einer Druckpumpe oder zwischen den Backen eines Blasebalges comprimirt worden sein. Natürlich spricht das Resultat meiner Versuche nicht gegen die Grundsätze der mechanischen Wärmetheorie, und würde ich nur meinen, dass bei denselben die Temperaturerhöhung nicht beobachtet wird, welche der zerstörten Geschwindigkeit der Lufttheilchen aequivalent ist. Mit einem feuchten Blasebalg habe ich eine Erwärmung der Thermosäule erhalten, wohl durch die latente Wärme der auf der Thermosäule condensirten Wasserdampfer.

Da ich weiss, wie sehr bei dergleichen Versuchen kleine Abänderungen des Verfahrens ein anderes Resultat geben können, so hätte ich gern Ihre Meinung darüber gehört, um so mehr als sich dieselbe im persönlichen Verkehr viel schneller und besser, als schriftlich austauschen lässt.

Mit grösster Hochachtung | Ihr ergebenster | Georg Quincke

Professor Tyndall, London.

39 Brüderstrasse Berlin, 30th June 1870

Dear Professor!

I thank you kindly for sending me your book on diamagnetism[2] and the corresponding table. I am even more interested in studying them, now that I have myself been again occupied with experiments on the magnetic rotation of the polarisation plane, and it can be difficult for a German, who cannot possess all the journals himself, to gain access to your various works, not to mention that an important paper can easily be overlooked. I myself have thought about publishing my own works on the topic, as soon as I finish editing the rest of my optical investigations.

Unfortunately, Berlin costs a scholar who has to hold all kinds of lectures every day more time than London, and of late I come to scientific work much less than I had hoped.

To my regret, I did not see you during your last stay in Berlin.[3] Otherwise, I would have liked to speak to you about an experiment that you describe in your work "on heat",[4] pg. 16. I myself have never been able to detect the warming of a thermal column using a dry bellows; I always obtained a cooling, as when air escapes from a container in which it has been compressed. I admit that I would also a priori expect to find no difference whether the air has been compressed by a pump or within the bag of a bellows. The result of my experiments does not, of course, deny the principles of the mechanical theory of heat, and I would only suppose that in them one does not detect the increase in temperature that is equivalent to the disrupted velocity of the air particles. I obtained a warming of the thermal column with a moist bellows, possibly due to the latent heat of the steam that condensed on the thermal column.

As I know well that in such experiments, small variations in the procedure can lead to different results, I would like to have heard your opinion on this, particularly since one can exchange views much more quickly when talking in person than when writing.

With the highest respect | your humble | Georg Quincke

Professor Tyndall, London.

RI MS JT/1/Q/3

1. *Georg Hermann Quincke*: Georg Hermann Quincke (1834–1924), a German physicist. Quincke received his Ph.D. from the University of Berlin in 1858. In 1865 he became extra-ordinary professor of physics at the Gewerbeakademie (Trade Academy) and Bauakademie (Architecture Academy). In 1872, he was offered a chair in physics at the University of Würzburg and later Heidelberg, after Kirchhoff's retirement. Quincke

published numerous works on optics, electricity, and colloids. Quincke was a member of the Royal Prussian Academy of Sciences and was well-regarded by his peers, particularly in England, where his friends included John William Strutt (1842–1919), George Gabriel Stokes, and Tyndall (*NDB*).

2. *your book on diamagnetism*: *Diamagnetism* (see letter 3076, n. 8).
3. *your last stay in Berlin*: Tyndall visited Berlin with Thomas Hirst in July 1869 (see letters 3021 and 3023).
4. *"on heat"*: see letter 3013, n. 4.

From Rudolf Clausius 3 July 1870 3233

Bonn, 3/7/70.

Lieber Tyndall,

Ich habe Dir noch gar nicht gedankt für die Zusendung Deines schönen Buches über Diagmagnetismus und Magnetcrystallische Wirkung. Es ist sehr recht, dass Du Deine interessanten und wichtigten Untersuchungen über diese noch so jungen Zweige der Physik zusammen gefasst hast, und ich freue mich sehr sie als zusammenhängendes Werk zu besitzen, und sage Dir meinen besten Dank dafür.

Zugleich ist mir das Erscheinen dieses Buches, welches Dir doch gewiss ziemlich viel Arbeit gemacht hat, ein sehr erfreuliches Zeichen dafür gewesen, dass Deine Gesundheit nach dem vorjährigen Unfalle wieder recht gekräftigt sein muss. Wenn Du Dich durch das viele Arbeiten nur nicht zu sehr angestrengt hast.

Du wirst doch wahrscheinlich in diesem Sommer wieder reisen. Solltest Du es wieder so einrichten können, dass Du mit Hirst über Bonn kämest so würden wir uns natürlich sehr freuen. Wir wohnen jetzt in unserem neuen Hause in der Quantius-Strasse (No. 13). Hier haben wir zwei Stuben ganz frei und unbenutzt, und wir bitten Euch daher wenn Ihr nach Bonn kommt, sei es jetzt oder später, nicht nach einem Gasthofe zu fahren, sondern direct zu uns zu kommen und bei uns zu logiren.

Wir haben vor Kurzem wieder eine Photographie von Johnny und dem kleinen Alfred machen lassen, und ich lege Dir ein Exemplar davon bei. Wir finden es hübsch, hin und wieder Photographien der Kinder machen zu lassen, um uns später an der Reihe von Bildern ihre Entwicklung vergegenwärtigen zu können.

Mit freundlichen Grüssen von meiner Frau und mir an Dich und Hirst | Dein | Clausius.

Bonn, 3/7/70

Dear Tyndall,

I have not yet thanked you for sending me your fine book[1] on diamagnetism and magne-crystallic action. It is quite well that you have summarized your interesting and important investigations in this still very new branch of physics, and I am delighted to own them as one cohesive work, for which I thank you very much.

At the same time, the publication of this book, which certainly cost you much work, was a very welcome sign to me that your health has quite improved since your accident last year.[2] If only you have not exerted yourself too much with all this work.

You will probably travel again this summer. Should you be able to arrange to come to Bonn with Hirst[3] again, we would naturally be very glad. We now live in a new house on Quantius-Strasse (No. 13). We have two rooms that no one uses and we therefore beg of you, should you come to Bonn whether now or later, that you not to go to an inn but come directly to our house and to stay with us.

We recently had another photograph of Johnny and little Alfred[4] taken, and I am enclosing a copy[5] for you. We find it pleasant to have the children's pictures taken from time to time, so that we can later visualise their development by looking at the series of pictures.

With best wishes from my wife[6] and me to you and Hirst | Your | Clausius.

RI MS JT/1/TYP/7/231
LT Typescript Only

1. *your fine book*: *Diamagnetism* (see letter 3076, n. 8).
2. *your accident last year*: see letters 3030, 3031, 3032, 3037, 3041, 3048, 3049, and 3060.
3. *Hirst*: Thomas Hirst.
4. *Johnny and little Alfred*: Rudolf John and Alfred Clausius, sons of Rudolf and Adelheid Clausius. John was Tyndall's godchild (see letter 3146, n. 6). For their other children at this time, see letter 3146, n. 20.
5. *enclosing a copy*: enclosure missing.
6. *my wife*: Adelheid Clausius.

To Heinrich Rudolf Vieweg 10 July [1870][1] 3234

The Bel Alp | Brieg Canton de Valais | Switzerland | 10th July

Dear Sir,

The "notes on Light"[2] and on Electricity[3] were published purely for the use of Teachers & pupils in England. The Department of Science and Art wished them to be published & I had myself several applications of permission to publish them. Therefore as stated in the little preface I placed them in the hands of Messrs. Longman.[4]

I had begun a volume on Light, which I intended to be a boy's book[5] but I have no hope of being able to finish it for a long time to come.

Yours sincerely | John Tyndall

TU Braunschweig/Universitätsbibliothek/UABS V 1 T: 63

---

1. *10th July*: the year '1870' was written in pencil on the letter next to the month and day. Additional information confirms this date: Tyndall wrote to Stokes from Brieg on 12 July 1870 (letter 3235), and both books referred to were also published in 1870.
2. *"notes on Light"*: see letter 3097, n. 7.
3. *on Electricity*: see letter 3196, n. 2.
4. *Mess[rs]. Longman*: William Longman (1813–77) and Thomas Longman (1804–79) were two brothers who ran the publishing firm Longmans after the sudden death of their father, Thomas Norton Longman (1771–1849). While Thomas, the elder brother, served as chairman of the company, William was reputed to be more skilled in the business of publishing. Longmans was one of the largest publishing firms in nineteenth-century Britain, publishing books on history, theology, travel, sociology, novels, and science.
5. *a boy's book*: see letter 3097, n. 7.

To George Gabriel Stokes 12 July 1870 3235

Hotel Belle Alpe | Brieg. | Canton de Valais, Switzerland | 12th July 1870.

My dear Stokes.

I carried with me to this place the remarks of yourself and the other referees,[1] intending to make use of them. Since my arrival your letter has reached me.

I knew right well that some points w[oul][d] have to be looked to in proof. I was weak and weary from my wound[2] when I wrote the paper,[3] and was only too anxious to get it out of my hands and into the printer's.

The obscurity regarding the dark rays and platinum disappears in the introduction & repetition of the word white-hot before "platinum".

I have somewhat modified par. 3.

With regard to Draper[4] I simply avoid raising the question by striking out the passage. But is it not a fact that the leaves of trees when photographed produce not the slightest chemical effect on surfaces eminently sensitive to the blue rays? This would show that the blue rays are appropriated by the leaves. This of course is not a conclusive proof of the decomposing action of the rays; but when I wrote my paper at Clifton[5] I has a more or less distinct remembrance that somebody besides Draper had worked at the subject. Marcet[6] I think has worked at it. However for the present it is safer to leave the question in abeyance and this I have done.

I have added a note regarding phosgen.[7]

You surely do not imagine that I am a believer in the vesicular theory![8] I thought we had talked this matter over long ago—I no more believe in the theory than you do. But to avoid misapprehension I have struck the passage out.

Faraday once told me that he had seen the bubbles and confirmed the statement of de Saussure.[9] Can it be that he saw germs or spores out of focus; or that what he saw were bodies within his own eyes to the sight of which a fog is often favourable.

I have avoided referring to one point which has to some extent occupied my thoughts. Many vapours and gases which maintain their stability in the presence of a powerful flux of heat nevertheless absorb that heat. Is this absorption molecular or atomic? The experiments recorded in the Bakerian Lecture for 1855[10] show that as regards radiant heat a similar relation subsists between liquids & their vapours to that subsisting between the nitrite of amyl & its vapour as regards the chemical rays. Hence the reasoning applied in the case of the nitrite would also apply to other cases, and if this reasoning be valid the absorption of the radiant heat must also be atomic. Why then does it not effect decomposition?

I am getting strong here, but I wish Liverpool[11] was not in front of me. It will take a great deal of my holiday to get up my lecture

Yours faithfully | John Tyndall

CUL SC—add 7656 RS749

---

1. *remarks of yourself and the other referees*: most likely comments on his publication 'On the Action of Rays of High Refrangibility', *Roy. Soc. Proc.*, 160 (1870), pp. 333–65. See letters 3096, n. 8, 3269, and 3271.
2. *weak and weary from my wound*: Tyndall severely injured his leg while hiking in September 1869. See letters 3030, 3031, 3032, 3037, 3041, 3048, 3049, and 3060.
3. *wrote the paper*: see n. 1.
4. *Draper*: John William Draper (1811–82), an English-born American chemist known for contributions to early photography, photochemistry, and intellectual history. Born at St Helens, near Liverpool, England, Draper was the son of a scientifically inclined Methodist minister. He went to study chemistry at the University of London, which had recently been inaugurated in 1829. Draper immigrated to America in 1832 after his father's death, and thereafter studied medicine at the University of Pennsylvania. In 1836 he was appointed professor of chemistry, physiology, and natural philosophy at Hampden-Sydney College, followed by a call to the University of the City of New York in 1839 as

professor of chemistry and natural philosophy. There he founded the university's School of Medicine in 1840. In 1850 he became the school's president, which he held until 1873. In 1837 he began experimenting with photography and in the period 1839 to 1840 took the first known photographs of the moon. Draper had a strong interest in all types of radiant energy and published results in 1847 that all solid bodies became incandescent at the same temperature. Draper also spoke at the BAAS meeting at Oxford in 1860, which ignited the infamous exchange between Bishop Samuel Wilberforce and Thomas Huxley. In 1869 Draper also became founding president of the American Union Academy of Literature, Science and Art, and in 1876 founding president of the American Chemical Society. In 1875 he was awarded the Rumford Medal by the American Academy of Arts and Sciences for his research on radiant heat—an honor Tyndall himself held in 1864 for his work on the absorption and radiation of heat by gases and vapours. Draper is perhaps known for his massive *A History of the Intellectual Development of Europe* (1863), and probably best known for his *History of the Conflict between Religion and Science* (1874), which many historians of science claim is an exemplar of the "conflict thesis," the idea that science and religion are fundamentally and irrevocably in conflict (*ANB*; *ODNB*).

5. *Clifton*: see letter 2996, n. 7.
6. *Marcet*: François Marcet.
7. *phosgen*: also known as phosgene, as Stokes remarked. See letter 3269.
8. *vesicular theory*: the vesicular theory is based on the theoretical development of a protocell as the basis for spontaneous generation.
9. *de Saussure*: Horace Bénédict de Saussure (see letter 3133, n. 4).
10. *Bakerian Lecture for 1855*: J. Tyndall, 'The Bakerian Lecture.—On the Nature of the force by which Bodies are repelled from the Poles of a Magnet; to which is prefixed, an Account of some Experiments on Molecular Influences', *Phil. Trans.*, 145 (1855), pp. 1–51. Tyndall read this before the RS on 25 January 1855, and delivered the lecture as a Friday Evening Discourse at the RI on 26 January 1855 (*Roy. Inst. Proc.*, 2 (1854–58), pp. 13–6).
11. *Liverpool*: the fortieth annual meeting of the BAAS was held in Liverpool from 14–21 September 1870 (*Brit. Assoc. Rep. 1870*). Tyndall gave an evening lecture on 16 September, on 'The Scientific Use of the Imagination' (*Brit. Assoc. Rep. 1870*, p. xxxvii). This lecture was published along with his 1868 BAAS lecture in Norwich ('Scientific Limit of the Imagination') as: J. Tyndall, *Essays on the Use and Limit of the Imagination in Science* (London: Longmans, Green, and Co., 1870).

From Henry Bence Jones 14 July 1870 3236

July. 14. 1870

My dear Friend

This morning brought your note[1] and I am glad to hear that you are safe. I do not think there is an atom of news to tell you or rather nothing reaches me. Your paper on Pasteur's work[2] is out & reads well. The "stamping out" system for worms & cattle is a sad evidence of our ignorance. I wonder someone does not propose it for the small pox in Paris & elsewhere. Compared with a war the killing would be nothing. When on the cattle plague commission[3] I strove for the chance of some further knowledge, but in vain & our ignorance is too great at present to give us a chance of treating worms beasts & humans rationally.

Dr Gull[4] spent half his Harveian oration[5] on "Dust & disease" but I cannot get a thought out of it & I do not think you will get one if you ever read it. However he has "tact" & that is worth much in this world.

Huxley's sermons[6] have not appeared. I have had a long talk with Matthew Arnold[7] but cannot get him to lecture. I am getting tired of asking people. I have got Faraday's MSS lectures.[8] The 4 he sent to Davy & the ring with which he first got his current from the magnet. Cotterell[9] tells me you thought Sir James South[10] had the apparatus with which he first got the electric spark from the magnet, but I cannot find it among Sir J. South's things. Let me know if you are sure. I think the residue will be sold in a few days. I have found & got the first electric machine & I have a very interesting note about from the Duke of Somerset[11] saying how Faraday was touched when he saw it first at Sir James South's.

I have not been able to get South's papers yet, but I am in hopes of them. I expect sundry interesting things regarding Davy and Faraday. The life is to be reviewed in the Edinburgh Review[12] I see on Saturday. By the way I found your new volume at the Institution.[13] It makes a splendid volume. Sir H Holland[14] was praising it & offered to lend me his copy but I hope some day you will bestow one on me.[15] We began yesterday to be at the theatre again. I carried all you wished at the managers meeting. The old stair case is gone & large room! is now in its place. Cotterell is away as we settled most of the clearances very quickly. Instead of we I ought to say he & Odling. Mr Berkeley[16] came to the managers committee but nothing fresh was said or done regarding the theatre & so he & Odling will look after it until you come back. I had to give Brown[17] the porter my mind about the visitors report on the urinal & I told him if they reported on it again I should advise the managers to get another porter in his place.

Sometimes since you spoke about Odling coming to the managers meetings & as we wanted him on the theatre & clearing committee & he was in some respects to be in charge during your and my absence I thought it would add to his authority if it was on the minutes that he was to be invited to the managers meetings so I told them that you had proposed it some time since & it met with general approval.

I wish I had some better new to give you

Pasteur I see is much better. I wish I was with you in Switzerland.

I hope your staircase & bookcase will look to your liking.

I have finished all my talk which is worth nothing. The Thames Embankment[18] is a glorious work. The light up & down the river very good & St Pauls & Somerset House[19] first rate.

Yours most truly | H. Bence Jones

Pray take care of yourself.

I have some most interesting things in the Faraday MS books.

RI MS JT/1/J/64
RI MS JT/1/T/TYP/2/696–7

---

1. *your note*: letter missing.
2. *Your paper on Pasteur's work*: J. Tyndall, 'Pasteur's Researches on the Diseases of Silk-worms', *Nature*, 2 (21 July 1870), pp. 181–3.
3. *cattle plague commission*: Bence Jones was a member of the commission appointed to inquire into the origin and nature of the cattle plague (rinderpest) which broke out in Great Britain in the middle of the nineteenth century. The commissioners published their first report in 1866.
4. *Dr Gull*: William Withey Gull (1816–90), an English physician. Gull received his MB from London University in 1841 and began teaching materia medica and later natural philosophy at Guy's Hospital. From 1847 to 1849 he was Fullerian professor of physiology at the RI and became assistant physician in 1851 and full physician at Guy's in 1856. He garnered public attention in 1871 when he was the attending physician for the prince of Wales who was severely ill with typhoid fever (*ODNB*).
5. *Harveian oration*: an annual lecture held at the Royal College of Physicians of London, begun in 1656 by and named for the English physician William Harvey (1578–1657). William Gull's 1870 lecture focused on recent research on dust and its role in the propagation of contagious disease and engaged with the debate on the germ theory of disease (W. Gull, *The Harveian Oration. Delivered at the Royal College of Physicians, London, June 24, 1870* (London: John Churchill & Sons, 1870).
6. *Huxley's sermons*: *Lay Sermons* (see letter 3217encl).
7. *Mathew Arnold*: Matthew Arnold (1822–88), an English poet, writer, and inspector of schools. Arnold was a Francophile, and French culture influenced his literary and social criticism. In 1847 he became personal secretary to Land Lansdowne, a leading Whig politician, whose influence led to his appointment as inspector of schools in 1851. By the 1870s he had become a public figure due to his reputation as a critic in social and educational matters. He was awarded a DCL in 1870 by Oxford University and was granted a civil-list pension of £250 in 1883 (*ODNB*).
8. *Faraday's MSS lectures*: Bence Jones was researching Faraday's life for the second volume of his biography, *Life of Faraday*.
9. *Cottrell*: John Cottrell.
10. *Sir James South*: James South (1785–1867), a British astronomer. South was elected FRS in 1821 and awarded the RS's Copley Medal in 1826. He was a founding member of the

Royal Astronomical Society and was elected its president in 1829. South had died in 1867 prior to this letter, and after his death, an array of instruments from his observatory on Campden Hill, Kensington were sold (*CDSB*; *ODNB*).

11. *Duke of Somerset*: Edward Adolphus St Maur (Seymour) (1804–85) succeeded his father as twelfth duke of Somerset in 1855 (*ODNB*).
12. *life is to be reviewed in the Edinburgh Review*: this appeared as a review of three works on Michael Faraday, see 'Art. VII.–1. *The Life and Letters of Faraday*. By Dr. Bence Jones, Secretary of the Royal Institution. Second edition. 1870. 2. *Faraday as Discoverer*. By John Tyndall. New edition. 1870. 3. *Éloge historique de Michel Faraday*. Par M. Dumas, Secrétaire perpétuel de l'Académie des Sciences. Paris: 1868', *The Edinburgh Review*, 132 (July–October 1870), pp. 176–208.
13. *Institution*: the RI.
14. *Sir H Holland*: Henry Holland.
15. *I hope some day you will bestow one on me*: Tyndall had already sent a copy of his book (presumably *Diamagnetism*), but Bence Jones had forgotten he had received it. See letters 3076, n. 8, and 3268.
16. *Mr. Berkeley*: George Berkley.
17. *Brown the porter*: William Brown.
18. *Thames Embankment*: the Thames Embankment was built between 1861 and 1874. The embankment radically altered the riverscape of London by replacing the tidal flats of the city with a smooth wall that hide infrastructure such as sewers and pipes. The embankment not only improved navigation but also provided a river-edge thoroughfare. See D. H. Porter, *The Thames Embankment: Environment, Technology, and Society in Victorian London* (Akron, OH: University of Akron Press, 1998).
19. *Somerset House*: Somerset House is located near the Thames River, in central London, near Waterloo Bridge. Construction of a palace for Edward Seymour began there in 1547. The current building dates from 1775 when the original structure was demolished. The new building was completed in 1801. It had originally stood on the bank of the Thames until the completion of the Thames Embankment.

To Rudolf Clausius [15 July 1870][1] 3237

My dear Clausius

Here on the Belle Alpe[2] I have received the photograph of Johnny and Alfred.[3] It is charming. You are a man to be envied. Had I been wise I might also perhaps have had a little Johnny & Alfred to look at—and surely it would have made me a stronger and a better man. You must feel ten times the interest in life that I do. Give my love to the children all, and my kindest regards to their mother.[4]

As soon as I could get away from London I came over here. I have been here already nearly a week. The mountain air does me a world of good; and

I am really learning to take better care of my health than I used to do. I have been just correcting the proofs of a paper for the Royal Society[5] which ought to have been printed some months ago. I am very glad that the book on Diamagnetism[6] has reached you. It is a help to workers to have the labours of workers in a collected form. Thomson told me some time ago that he was collecting & revising his papers[7] with a view to their publication.

There was an amusing occurrence at Oxford some weeks ago. Every year they have what they call a "commemoration" at the university, and it is then usual among other things to confer the degree of D.C.L.[8] A number of such degrees were conferred this year; But the names of two or three heretics were proposed by the Chancellor, and one member of the Board before which the claims of the persons to be doctored are laid, objected strongly to the heretics—He would have nothing whatever to do with me, and so my offences against orthodoxy have been punished. The member referred to was I believe Dr Pusey,[9] the founder of the so named Puseyite Section of the English clergy. Out of respect for his religious feelings the matter was not pressed by the other members of the board.

The British Association meets in Liverpool this year[10]—Is it not possible to tempt you over? Huxley presides[11]—and I give one of the evening discourses[12] which will give me much trouble to prepare. They asked me to preside next year, but I have declined it.[13]

Goodbye dear Clausius | & believe me always | your friend | John Tyndall

RI MS JT/1/T/239

---

1. *[15 July 1870]*: in letter 3238 of 17 July 1870, Tyndall writes that he had been in Belalp for 10 days, and letter 3234 of 10 July was sent from Belalp. Hence he arrived in around 7–8 July, but before 9 July. He writes here that had been there for nearly a week, which would place this letter around 14/15 July.
2. *Belle Alpe*: Belalp (see letter 3030, n. 20).
3. *Johnny and Alfred:* Rudolf John and Alfred Clausius, two of Rudolf and Adelheid Clausius's children. For their other children at this time, see letter 3146, n. 20.
4. *their mother*: Adelheid Clausius.
5. *a paper for the Royal Society*: 'On the Action of Rays of High Refrangibility upon Gaseous Matter' (see letter 2980, n. 7).
6. *book on Diamagnetism*: *Diamagnetism* (see letter 3076, n. 8).
7. *he was collecting & revising his papers*: see letter 3212.
8. *D.C.L.*: Doctor of Civil law is a higher doctorate bestowed by the University of Oxford for significant contribution to scholarship.
9. *Dr Pusey:* Edward Bouverie Pusey (1800–82), an English clergyman and theologian. Pusey was appointed regius professor of Hebrew at Oxford University in 1828, and became a leading figure in the Oxford Movement, which sought to revive liturgical and devotional

customs that were either pre-reformation or from the contemporary Roman Catholic Church (*ODNB*).

10. *British Association meets in Liverpool this year*: see letter 3235, n. 11.
11. *Huxley presides*: Thomas Huxley was President of the BAAS for the Liverpool meeting. In his Presidential Address on 14 September, he took a materialist stance by discussing the controversial topic of spontaneous generation, describing recent experiments by Tyndall and Louis Pasteur, and coining the term *abiogenesis* to represent the origin of organic life from inorganic matter ('Address of Thomas Henry Huxley, LL.D., F.R.S., President', *Brit. Assoc. Rep. 1870*, pp. lxxiii–lxxxix). This was also published on its own as: *Address to the British Association for the Advancement of Science. Delivered by the President, Thomas H. Huxley, LL.D., F.R.S., &c., Professor of Natural History in the Royal School of Mines, at Liverpool, September 14, 1870* (London: Taylor and Francis, 1870); and later as 'Biogenesis and Abiogenesis', in T. H. Huxley, *Critiques and Addresses* (London: Macmillan, 1873), pp. 218–50. Huxley's address was reported in the *Times* the following day ('British Association for the Advancement of Science', *Times*, 15 September 1870, p. 4).
12. *I give one of the evening discourses*: see letter 3235, n. 11.
13. *I have declined it*: Tyndall eventually did preside over a BAAS meeting, at Belfast in 1874.

To Thomas Archer Hirst [17 July 1870][1] 3238

Sunday July 17/70

My dear Tom

Here is a rapidly running account of my life since I left London.

The passage to Boulogne[2] was uncomfortable, though it did not inflict the last humiliation. At Boulogne I set all right.

I halted at the old place in Paris. My notion was to go to Lausanne by Pontarlier,[3] so I waited till the Sunday evening, but on driving to the station could not disentangle the bewilderments of the time tables: the clerk moreover added his bewilderments to theirs. Stayed in Paris till Monday morning and then set off for Bale,[4] reached it, and next morning took train to Lausanne. There I stayed at the Hotel Richemont; it being quieter than the hotel Gibbon.

And on the following morning I was on my way to Sierre;[5] sweeping by Chillon[6] and Villeneuve[7] where you and I wandered many years ago. From Sierre by Diligence[8] to Brieg,[9] taking the conductor's banquette, and thus securing a little air. At Brieg I spent the night.

And next morning I was on my way to the Bel Alp;[10] taking an unfrequented path through the Blind Thal;[11] gray and mossy cliffs right and left, the valley being an apparent rent in the mountains. It was grand & lonely. My room at the Bel Alp was rapidly prepared, and here I have been now 10 days.

For sometime my London sleep, and that was bad, clung to me. But there is a clear lake within a quarter of an hour of the hotel. For many days its temperature was delicious. So I went there early: from a stone which emerges from the water an oblique header can be taken, and I took many such. My strength augmented. I took many walks, sometimes in company, sometimes alone. I am safer in company than alone. For alone I wander on and on and sometimes get into difficulty.

Yesterday I started with my axe and soup, which latter consisted of a bag of milk and a slice or two of Schwartzbrod. I left my vest behind; but I had not gone far when I found my coat too heavy; so I climbed to a difficult nook and left the coat there. Then I crossed the debris of glaciers, wandered among rocks strias and montaines. Got into a defile where the small juniper, rhododendron and heather were matted amicably together. It led upwards to a col,[12] and I wondered why people had not already found and used so pleasant a track. I reached the col and there the reason revealed itself: it fell into abominable steepness along a couloir: still it was plain from a faint streak here and there that at one period of the mountain's history a track had existed down this couloir. The weather had washed the traces away save at distant intervals. There seemed no other way of getting on; so with my accustomed caution in ugly places I managed to get to the bottom. There across another valley and up the opposite mountain side: I doubled round the breast of the hill and was at last stopped by a thundering torrent. I doubled back but still went upward and finally reached the end of the glacier whence the torrent issued: from the end not one but seven cascades fall, which afterwards united. The trunk torrent was far too much for me; so I surmounted its branches in detail. Then further along the mountains with the Aletsch glacier[13] in view on the right until I was stopped by a spur which it was too late to cross.

At about 3 P.M. I returned—keeping higher up in the hope of being able to get home by a new and easier way. I wanted to get upon a glacier which had to be crossed to reach the path to the Bel Alp. With great labour I got above this glacier, and a good way down toward it; but the precipices were so formidable that I was obliged to retreat. I then by an easier, but still by a very steep way got into a deep ravine—ascended the steep high slope at its opposite side. Got upon a moraine, which proved too steep to be descended. Some distance further down it became practicable. I had examined the glacier and thought I had discovered a quiet place where it might be crossed. I got upon the ice, crossed the glacier; but was cut off at its further side by formidable chasms. The day was getting on. I retraced my steps across the glacier; and plunged down for 2 or 3000 feet among the most detestable loose boulders. Came in front of my couloir of the morning; but it looked so ugly that I took to the Aletsch glacier instead. I am not sure that I could have disentangled myself from the crevasses, which when I scanned them afterwards seemed very formidable: but scanning the precipice which flanked the glacier I discovered a streak; the streak soon declared itself to be a path: "fortune favors the prudent" I exclaimed; and by this path I rescued myself.

It was a hard day: and today I am resting: there is however an additional reason. The fog is so thick that a distance of 20 or 30 feet obscures every thing.

Miss Wood,[14] whom we met at Folkestone[15] arrived yesterday. Her card came up to me immediately after my return last night.

And now my gossip is ended. I have been exceedingly disturbed by this news of war.[16] There is something so incredibly atrocious in the matter that it is still unreal. I cannot imagine it. Pon my soul I sh[oul][d]. not object to entering the Prussian service; if it were worth accepting I would give Prussia 1000 pounds tomorrow to carry on the war. I do not say this in consequence of any particular sympathy with Prussia; if the case were reversed, & if she behaved toward France as France is behaving towards her, I would in the same way join France. I fear Louis Napoleon is a sublime scoundrel. That human blood sh[ould][d]. be shed as it is likely to be through such contemptible jealousy, is abominable to the last degree.

The behaviour of the Council of University College is exceedingly gratifying to me—*[Gluck]*![17]—you cannot speak to the boys without saying something that it will be good for them to hear.[18]

I cannot send you the title of my lecture[19] just yet; as it is still unimagined—goodbye & excuse all this trumpery gossip.

Yours affectionately | John

BL Add MS 63092, ff. 67–9
RI MS JT/1/HTYP/565–6

---

1. *[17 July 1870]*: the date of this letter was written in a second hand, perhaps that of Hirst. 17 July 1870 was a Sunday, and the dating also fits well with Tyndall's description of being at Belalp for ten days.
2. *Boulogne*: see letter 3044, n. 10.
3. *Pontarlier*: a town located in the Jura Mountains in the Bourgogne-Fanche-Comté region of eastern France.
4. *Bale*: French name for the Swiss city of Basel, which is located along the river Rhine in northwestern Switzerland.
5. *Sierre*: see letter 3037, n. 5.
6. *Chillon*: an island castle located on Lake Geneva in the Swiss canton of Vaud.
7. *Villeneuve*: a town in the Swiss canton of Vaud.
8. *Diligence*: a public stagecoach (*OED*).
9. *Brieg*: see letter 3030, n. 17.
10. *Bel Alp*: Belalp (see letter 3030, n. 20).

11. *Blind Thal*: also Blindtälli, a shadowy valley beneath Belalp.
12. *col*: a depression in the summit-line of a mountain which normally provides a pass from one slope to another (*OED*).
13. *the Aletsch glacier*: see letter 3037, n. 1.
14. *Miss Wood*: not identified.
15. *Folkestone*: see letter 3021, n. 20.
16. *news of war*: the Franco-Prussian war began on 19 July 1870. The war saw the defeat of France by a coalition of German states. In turn, it ended French hegemony and led to the creation of a united German state. The war and its effects are a recurring theme in letters within this volume from mid-July 1870 through May of 1871 (3240, 3242, 3243, 3244, 3245, 3247, 3248, 3249, 3250, 3251, 3252, 3253, 3254, 3261, 3262, 3265, 3267, 3283, 3287, 3289, 3297, 3303, 3305, 3309, 3310, 3312, 3313, 3314, 3330, 3351, 3356, 3359, and 3364).
17. *[Gluck]!*: What luck! (German).
18. *behaviour . . . hear*: Hirst had been asked by the University College School Council to preside over the ceremony for the distribution of student prizes (RI MS JT/2/32d/1879).
19. *the title of my lecture*: Tyndall decided on the title 'On the Scientific Use of the Imagination'. See letters 3235, n. 11, and 3241.

To Thomas Archer Hirst [26 July 1870][1] 3239

The Bel Alp Tuesday 1870

My dear Tom,

I should like much to hear how, in your own estimation, you get on with the boys.[2] Still you are probably busy, and I should not like to add unnecessarily to your hours.

My life here is very uneventful. I get at all events plenty of fresh air, for I dawdle a good deal over the hills. I am reading Bain's Logic.[3] Here and there there is a flush of emotional strength in the book, but for the most part it is as dry as I am myself. It may however give my imagination that mechanical help which Emerson ascribes to the reading of a Dictionary, and thus aid me in the preparation of my Liverpool discourse.[4]

I have hardly an idea of what it is to be. Of the treatment I am quite in the dark, for a single page of it has not been yet written down. I thought I should at once find my thoughts clarified on coming here; but they are not clarified; they are rather softened and rendered nebulous by indolence—perhaps I am unconsciously laying up strength which will come to the surface by and by. Thus far however it has taken care to keep itself latent.

I have read all Bastian's articles in Nature.[5] They are in some parts very clever, but they carry an exclusiveness. The man is evidently in earnest and he has been working hard. both these circumstances give him an interest in my eyes. I thought of sending him the enclosed note[6]—cast your eye over it, and if nothing has occurred to render the sending of it undesirable have it posted for me.

This war[7] I fear will deflect public attention from Huxley's book.[8] It is a pity. I predicted for it an enormous sale. Have you read and compared the speeches of the two men, Napoleon and William of Prussia?[9] How characteristic they are! I was greatly moved by the speech of the latter:[10] it is so solid & dignified—full of calm strength. I read it on the open alp and felt refreshed and bettered by it.

I see my Hungarian friend[11] is once more exposing his sores in "Nature." What a fool I was to permit such a man to come near me but on the other hand who was to know that the human genus embraced such an individual. If useful I will double him up in a quick dignified way by and by.

I know as regards letter writing that you are almost as lazy as myself—or at least the spirit which in days gone by used to speak your thoughts over folios has now retreated, and left its bed like that of an Alpine torrent at this season,—dry. But where you can throw even an assemblage of gossip fragments together you would do right to lead them to me. If you were here I would take your advice as to the building of a small chalet,[12] containing four rooms, on the Bel Alp. I have had an estimate made of the cost, and found that the thing could be completed on a modest scale for one hundred pounds. I have sometimes thought of getting Carlyle over here; the chalet would enable him to sleep without any noise save the lullaby of an alpine stream.

Debus I trust will find me here, and then we shall go to Pontresina[13] together; but I do not like to quit this place without setting my thoughts straight in the matter of Liverpool.[14] So yet I am quite in the air; no purpose; no consolidation. Under the circumstances it is of course impossible for me to state the proper title. Perhaps when Debus[15] comes collision with him will produce fire.

I am very much obliged to you for your supervision of my letters: I suppose but few come; and this accounts for the drought of the post of the Bel Alp.[16] We have now a telegraph from this place! I have written very few letters, and those few short ones, since I came here. Common kindness and good feeling must however soon induce me to write to Lady Mary Egerton, Lady Belper,[17] and two or three other friends.

I send you many feelings and good wishes. If that lecture contemplated by Carpenter[18] be really a desirable thing you may engage me to give it but not unless the need is pressing.

Yours affectiona[te]ly John Tyndall

BL Add MS 63092, ff. 79–80
RI MS JT/1/HTYP/569

1. *[26 July 1870]*: this letter was written between the start of the Franco-Prussia War on 19 July and Tyndall's letter to Hirst of 1 August 1870 (letter 3241), by which time he had decided on a topic for his BAAS discourse. The only Tuesday remaining in the month was the 26 July.
2. *get on with the boys*: this is reference to the Council of University College. See letter 3238.
3. *Bain's Logic*: A. Bain, *Logic* (London: Longmans, Green, Reader, & Dyer, 1870).
4. *my Liverpool discourse*: Tyndall was preparing an evening lecture for the annual meeting of the BAAS at Liverpool (see letter 3235, n. 11).
5. *Bastian's articles in Nature*: H. C. Bastian, 'Sensation and Perception.—I.', *Nature*, 1 (1869), pp. 213–4; 'As regards Protoplasm, in relation to Professor Huxley's Essay on the Physical Basis of Life', *Nature*, 1 (1870), pp. 424–6; 'Facts and Reasonings Concerning the Heterogenous Evolution of Living Things*', *Nature*, 2 (1870), pp. 170–7; 'Facts and Reasonings Concerning the Heterogenous Evolution of Living Things', *Nature*, 2 (1870), pp. 219–28; 'Sensation and Perception: II', *Nature*, 2 (1870), pp. 309–11.
6. *enclosed note*: letter missing.
7. *this war*: the Franco-Prussian War (see letter 3238, n. 16).
8. *Huxley's book*: *Lay Sermons* (see letter 3217encl).
9. *William of Prussia*: Friedrich Wilhelm Louis (1831–88), Crown Prince of Prussia, son of Emperor Wilhelm I, who was lauded for his leadership and successes during the Franco-Prussian War. He later became the King of Prussia as Friedrich III for just 99 days in 1888 (*NDB*).
10. *the speech of the latter*: King William of Prussia gave a speech on 19 July 1870 in response to France's declaration of war.
11. *my Hungarian friend*: C. K. Akin, a Hungarian physicist. Tyndall and Akin had an increasingly acrimonious dispute in 1864 over priority in infrared experiments. Although the relations were originally cordial, they soured as Akin first privately and then publicly accused Tyndall of plagiarizing his research. Akin, who was less well-connected in the scientific community at this time, lost out to Tyndall, who was able to marshal his large professional network in his support. See *Tyndall Correspondence*, vol. 8, pp. xx–xxii; letters 1925, 2075, 2076, 2077, 2165, 2167, and 2170. Tyndall was likely referring to C. K. Akin, '"Kinetic" and "Transmutation"', *Nature*, 2 (14 July 1870), p. 211.
12. *the building of a small chalet*: Tyndall and his future wife Louisa Tyndall eventually had a chalet built on Belalp in 1877, which they named Alp Lusgen.
13. *Pontresina*: see letter 3104, n. 6.
14. *Liverpool*: see n. 4.
15. *Debus*: Heinrich Debus.
16. *Bel Alp*: Belalp (see letter 3030, n. 20).
17. *Lady Belper*: Amelia Harriet Strutt (née Otter, d. 1890), the youngest daughter of William Otter (1768–1840), bishop of Chichester. In 1837 she married politician, FRS, and first Baron Belper, Edward Strutt (1801–80). They had four sons and four daughters (G. E.

Cokayne, et al. (eds), *The Complete Peerage of England, Scotland, Ireland, Great Britain and the United Kingdom, Extant, Extinct or Dormant*, 2nd edn, 13 vols (in 14) (London: St Catherine Press Ltd, 1910–88), vol. 2 (1912), p. 113).

18. *Carpenter*: William Carpenter.

From Auguste de la Rive 27 July 1870 3240

Genève, le 27 juillet, 1870.

Mon cher Tyndall,

Vous recevrez incessamment un exemplaire du mémoire que je viens de publier sur la polarisation rotatoire des liquides. J'en envoie aussi un exemplaire à la Rédaction du Phil. Mag. dans l'espérance qu'ils voudront bien en insérer un extrait dans leur Journal. Vous verrez que ce n'est qu'un premier travail que je compte continuer l'hiver prochain quand je me serai procuré les substances dont j'ai besoin.

J'ai à vous remercier du beau volume que vous m'avez envoyé. Cette collection de vos travaux sur le Diamagnétisme est pour moi d'un grand intérêt. Je n'oublierai jamais que j'ai fait votre connaissance scientifique en lisant vos premières recherches faites avec Knoblauch. Ce fut pour moi comme un trait de lumière qui me parut immédiatement offrir la solution du problème de l'action magneto-cristalline étudiée par Plücker & Faraday. J'en rendis compte immédiatement dans les Archives; c'était, si je ne me trompe, en 1849 ou 1850. Dès lors que vous avez ajouté des choses à ce premier travail, entre autres la démonstration de la polarité diamagnétique dont j'eus le plaisir d'être témoin avec Faraday en 1855 dans la chambre de la Royal Institution. Que de précieux souvenirs j'ai là!

J'espère que vous continuez à être bien & complètement remis de votre accident de l'année dernière. Je voudrais bien que la guerre ne vous empêchât pas de venir nous faire une visite cet été. La Suisse sera cette année un asile pacifique pour tous les touristes & surtout pour tous les hommes de science qui ne recherchent pas la guerre. Venez donc passer quelque temps chez nous; nous nous distrairons ensemble par de bonnes conversations scientifiques des horreurs de la guerre. Où en sommes nous en 1870 qu'on puisse encore jouer à cet épouvantable jeu. Vantons nous après cela de notre civilisation & du progrès des lumières.

Agréez, très cher monsieur, l'assurance de mes sentiments les plus affectueux. | Votre tout dévoué Aug. de la Rive.

Avez-vous continué vos recherches sur les gaz & sur les vapeurs? Dans ce cas ne les publierez-vous pas bientôt?

Geneva, 27 July, 1870

My dear Tyndall,

You will shortly receive a copy of the report[1] that I just published on the rotatory polarization of liquids. I am also sending a copy to the editors of the Philosophical Magazine in hopes that they will insert an excerpt[2] in their Journal. You will see that this is only a first draft that I intend on continuing next winter when I get the substances that I need.

I am grateful for the beautiful volume[3] that you sent me. This collection of your works on Diamagnetism is of great interest to me. I will never forget that I first became acquainted with your scientific research by reading your first works done with Knoblauch.[4] For me, it was like a ray of light that appeared to immediately offer a solution to the problem of the magneto-crystalline action studied by Plücker[5] & Faraday.[6] I immediately *[reported it]*[7] in the Archives;[8] that was, if I am not mistaken, in 1849 or 1850. Since the time that you added more things to this first work, including the demonstration of diamagnetic polarity, which I had the pleasure of witnessing with Faraday in 1855 at the chamber of the Royal Institution. What precious memories I have of that!

I hope you continue to be well & fully recovered from your accident last year.[9] I wish the war would have not prevented you from coming to visit us this summer. This year, Switzerland will be a peaceful refuge for all the tourists & especially for all the men of science who do not seek war.[10] Come spend some time with us; we will distract ourselves from the horrors of war with good scientific conversations. Where are we in 1870, that we can still play this appalling game? Let us boast after this about our civilization & the progress of the Enlightenment.

Please accept, very dear Sir, the assurance of my most affectionate sentiments. | Your very devoted Aug de la Rive.

Have you continued your research on gases & on vapours? In this case, will you not be publishing it soon?

RI MS JT/1/TYP/1/377
LT Typescript Only

---

1. *the report*: 'Recherches sur le polarisation rotatoire magnétique des liquids' (see letter 3199, n. 7).
2. *an excerpt*: this paper was published as A. de la Rive, 'Researches on the Magnetic Rotatory Polarization of Liquids', *Phil. Mag.*, 40 (1870), pp. 393–420.
3. *beautiful volume*: *Diamagnetism* (see letter 3076, n. 8).
4. *your first works done with Knoblauch*: Tyndall was Knoblauch's student and collaborator on the diamagnetism work during his time at Marburg. They wrote two co-authored papers in 1850, Tyndall's first scientific publications: J. Tyndall and H. Knoblauch, 'On the Deportment of Crystalline Bodies between the Poles of a Magnet', *Phil. Mag.*, 36 (1850),

pp. 178–83; and J. Tyndall and H. Knoblauch, 'Second Memoir on the Magneto-Optic Properties of Crystals, and the Relation of Magnetism and Diamagnetism to Molecular Arrangement', *Phil. Mag.*, 37 (1850), pp. 1–32.

5. *Plücker*: Julius Plücker (1801–68), a German mathematician and experimental physicist and professor of physics at the University of Bonn (1847–68). After focusing on analytical geometry, he turned his attention to electricity and magnetism (*CDSB*; *NDB*). He and Tyndall clashed in the 1850s, when Tyndall challenged Plücker's ideas on diamagnetism and the behaviour of crystals in magnetic fields, but they later resolved their differences.
6. *Faraday*: Michael Faraday.
7. *[reported it]*: probably 'Polarisation cristallique du bismuth et des autres corps, et leurs rapports avew la forme magnétique de la force, par Mr. Faraday', *Archives Sci. Phys. Nat.*, 10 (1849), pp. 225–7.
8. *Archives*: de la Rive was editor of the journal *Archives Sci. Phys. Nat.*
9. *your accident last year*: see letters 3030, 3031, 3032, 3037, 3041, 3049, and 3060.
10. *war*: the Franco-Prussian War (see letter 3238, n. 16).

To Thomas Archer Hirst [1 August 1870][1] 3241

Dear Tom

The title of my lecture at Liverpool[2] will be:—

On the scientific use of the Imagination.

Yours ever | J. Tyndall

RI MS JT/1/T/1001
RI MS JT/1/TYP/568

1. *[1 August 1870]*: the date of this letter is based on LT's notes. In his letter to Hirst of [17 July] (letter 3238) Tyndall stated that he still had not decided on a title for the lecture.
2. *Liverpool*: Tyndall was preparing an evening lecture for the annual meeting of the BAAS at Liverpool (see letter 3235, n. 11).

To Auguste de la Rive[1] [early August 1870][2] 3242

Bel Alp Brieg | Canton de Valais

My dear Friend

Your most kind and welcome letter[3] reached me here in my temporary mountain home. I look forward with the greatest interest to the reading of your memoir;[4] and if I can do any thing to induce Dr Francis to give it an English dress it shall be done with pleasure. Probably Francis of his own spontaneous act will have the paper at once translated.[5]

You are quite right in saying that you were one of the first to recognise whatever merit these papers on diamagnetism may possess. I have not forgotten your reference to this work in the testimonial which you were good enough to write for me when Huxley and myself were fellow applicants for posts in Toronto.[6]

This war[7] is a terrible affair—and nobody is able to assign its limits. The causes of it seem so inadequate that it is difficult to invest the thought of it with reality. It is simply barbarous and nothing else. The energy and determination shown by Switzerland is the best guarantee of the safety of this country. Neither of the combatants is likely to meddle with a power which shows itself so resolute to defend its frontiers.

We had a splendid thunderstorm here a few days ago. The lightning struck the pineforest in two places—not upon the tops but upon the steep slopes of the hills. It set both forests on fire The greatest fire speedily burnt itself out, but the other extended and kept burning for several days. A night or two ago it was perfectly grand. The whole mountainside was burning; and now and then when the flames seized a large tree it blazed forth like a pyramid of fire. The heavy rain that has fallen during the last 48 hours has however extinguished the fire.

The mountains were wonderfully beautiful this morning. All above 9000 feet were covered with new snow. The Mont Cervin[8] is like a pyramid of snow. I have never seen it so white before.

Many thanks to you for all the kind expressions in your letter. I am expecting a friend[9] to join me here, and then I purpose going to the Engadin.[10] But in bad weather the Engadin is very cold, so that unless the weather becomes warmer and brighter than it is at present I must give up the excursion to Pontresina.[11]

I am much occupied with a discourse[12] to be given before the British Association in Liverpool. It takes up almost all my holiday time. I have nothing in the way of experiment sufficiently complete for a lecture, and the consequence is that I have to develop one out of the depths of my own consciousness.

Pray give my kindest regards to Madame De la Rive,[13] and to both your sons and daughters.

I have the proofs of a paper for the Phil[osophical] Trans[actions] with me here. It is entitled "On the Action of Rays of high refrangibility on Gaseous Matter."[14] It ought to have been printed long ago but the Royal Society is slow.

With best wishes | Ever Yours | John Tyndall

RI MS JT/1/TYP/1/378
LT Typescript Only

1. *Auguste de la Rive*: this letter is addressed 'My dear friend', but Tyndall requests that the recipient give his regards to Madame de la Rive. This and the connection to letter 3240 from de la Rive strongly indicate that he was the intended recipient.
2. *[early August 1870]*: Tyndall received a letter from de la Rive in Belalp dated 27 July 1870 and sent one to Heinrich Debus on 10 August 1870 mentioning that he had decided not to travel to the Engadin. See letters 3240 and 3243.
3. *your letter*: see letter 3240.
4. *your memoir*: 'Recherches sur le polarisation rotatoire magnétique des liquids' (see letter 3199, n. 7).
5. *will have the paper at once translated*: 'Researches on the Magnetic Rotatory Polarization of Liquids' (see letter 3240, n. 2).
6. *Huxley and myself were fellow applicants for posts in Toronto*: Huxley and Tyndall applied for a professorship at the University of Toronto in 1851. For Tyndall's application to Toronto and his correspondence with de la Rive about it, see 'Introduction' (pp. xxiv–xxv) and letter 522, in R. Barton, J. Rankin, and M. Reidy (eds), *The Correspondence of John Tyndall, Volume 3: The Correspondence, January 1850–December 1852* (Pittsburgh: University of Pittsburgh Press, 2017).
7. *this war*: the Franco-Prussian War (see letter 3238, n. 16).
8. *Mont Cervin*: see letter 2950, n. 3.
9. *a friend*: Heinrich Debus. Debus eventually decided not to join Tyndall (see letter 3243).
10. *Engadin*: see letter 3146, n. 19.
11. *Pontresina*: see letter 3104, n. 6.
12. *a discourse*: Tyndall was preparing an evening lecture for the annual meeting of the BAAS at Liverpool (see letter 3235, n. 11).
13. *Madame De la Rive*: Louise Marie de la Rive.
14. *"On the Action . . . Matter"*: 'On the Action of Rays of High Refrangibility' (see letter 2980, n. 7).

To Heinrich Debus 10 August 1870 3243

10th. Aug[ust], 1870.

My dear Heins

I wish you much joy and much permanent advantage from your Scotch Journey. It will be a great comfort to me to have you near me in London.[1] *[This]* is a consummation for which I have often devoutly wished.

I will remain here and give up the Engadin[2] for the present year. In fact the preparation of my Liverpool lecture[3] will take up all my spare time.

What surprising intelligence this is that we receive from the seat of war.[4] Germans had the reputation of being slow coaches but they are proving themselves to be of very different metal—It is truly wonderful and beyond all expectation.

Your letter which was written on the 29th[5] was 8 days in coming here. Letters come with extreme slowness through Belgium. Via France is the quickest. But with your present sympathies unless you toned them down your letters might not escape the hands of justice.

I am ever Yours | John Tyndall.

RI MS JT/1/TYP/7/2372–3
LT Typescript Only

1. *have you near me in London*: consulting Tyndall, Debus applied for the position of Chair of Chemistry at Guy's Hospital Medical School in London and was successful. He began there in the autumn of 1870 (for the winter session that commenced in October).
2. *Engadin*: see letter 3146, n. 19.
3. *Liverpool lecture*: Tyndall was preparing an evening lecture for the annual meeting of the BAAS at Liverpool (see letter 3235, n. 11).
4. *the seat of war*: the Franco-Prussian War (see letter 3238, n. 16).
5. *your letter which was written on the 29th*: letter missing.

To Thomas Archer Hirst [13 August 1870][1] 3244

Saturday

My dear Tom

Your Capital letter[2] has just come to me & I will not delay my reply a minute—you are as usual acting the part of the Good Samaritan—All I have to say is that whenever you think an outside helping hand would be desirable you can calculate on mine. Since I wrote my last letter[3] to you I have been at work on my lecture.[4] I find it will not do to wait till I get into a state of superhuman strength; for the matter is not to be cast off by a sudden effort of strength. It is a small affair but it has cost me much trouble to get my thoughts into it. And I conquer the terrain, not as the Prussians are doing[5] but bit by bit, and by slow degrees. However I have already got into into[6] matter which will keep me close upon an hour, if not an entire hour. I will add another half hour's material to it and then close my books; give my head rest and my limbs more action than I have hitherto had. I have been already more than five weeks here and have done no climbing: in fact that old defect of mine came powerfully into play; namely, the defect of being unable to subdivide my time so as to devote different portions of it to totally different things.

I am exceedingly glad to hear that you get on so well with Carpenter.[7] I did not doubt that you would do so. But your giving way gave me some concern. I knew it must come; for the time of restoration which you allowed yourself was quite insufficient to pin the molecules in their places; they were loose and naturally, jiggled themselves asunder again. Debus[8] gave up the idea of coming to Switzerland: he was right and I am not sorry. For I must finish this lecture before I allow myself to do anything else.

One thing I hardly ever omit though our weather here for the last fortnight or more has been atrocious. I either jump into a lake, get under a cascade; or immerse myself in a tub with a sandy bottom every morning. I am in very fair health. But this damned lecture is a terrible fetter upon me. Huxley[9] will have much to answer for to posterity and the heavenly powers for committing me to it. But then I compare myself with Lubbock;[10] or with Huxley himself. Surely if I find matters difficult they ought to find them still more so. This mollifies me: but my task is harder than theirs: Huxley will sit down among his books and give a digest of a controversy which is all before him. I have not a single book with me, so that every mortal word of my lecture has to be dragged from the depth of consciousness. Lubbock's difficulty must be great.

But never mind, I shall get over it. A M^r^. Huggins[11] has written to me about Konig[12] but I was unable to give him any information. It might be as well if you fixed me for the Friday.[13] I suppose Rankine[14] would then come on on Monday.[15] I should then get the matter early out of my hands and head & be free for the enjoyment of the sections[16]—free to run away also if I found the sections not enjoyable.

I wish you every prosperity dear Tom as regards your health. I feel fine however that you do not allow your body time to repair itself. Nor do you always take the best means to repair itself. You are now in a good way; persist in it, remembering that the life is more than being and the body than raiment.

I have been thinking two or three times of trying to get a passport to one of the Armies and to do something like what you did at Solferino.[17] We have Galignani,[18] the journal to Genève[19] and some German papers here daily; but the Times[20] is also acceptable. I almost pity the French Emperor;[21] though I was extremely exasperated with him at first. But the Prussians[22] will not pity him or anything else: they will remember only being dragged from their homes they will bear in mind that the aim of France was and is to break the back of their country, and they will act accordingly. It is perfectly wonderful. But do you not remember the saying of old Clausius[23] as we climbed the Drachenfels,[24] "Depend upon it when the day of trial comes with France Prussia will be found as well prepared as she was in the case of Austria." The ole chap was right. And now my dear boy goodbye—Make yourself strong and let me find you in first rate condition when I return to England.

Yours affectionately | <u>John.</u>

BL Add MS 63092, ff. 72–3
RI MS JT/1/HTYP/570–1

---

1. *[13 August 1870]*: the date of this letter has been estimated from Tyndall's mention that he had been at Belalp for more than five weeks. Tyndall arrived at Belalp before July 10 (letter 3234), and probably around 7–8 July (letter 3238). 13 August 1870 was a Saturday.
2. *Your Capital letter*: letter missing.
3. *my last letter*: probably letter 3241.
4. *my lecture*: Tyndall was preparing an evening lecture for the annual meeting of the BAAS at Liverpool (see letter 3235, n. 11).
5. *as the Prussians are doing*: the Prussians had a number of victories in the ongoing Franco-Prussian War (see letter 3238, n. 16).
6. *into*: Tyndall repeated the word 'into'.
7. *Carpenter*: William Carpenter.
8. *Debus*: Heinrich Debus, who had just decided not to join Tyndall (see letter 3243).
9. *Huxley*: Thomas Huxley, who was President of the BAAS for this year and convinced Tyndall to give the discourse. On Huxley as President, see letter 3237, n. 11.
10. *Lubbock*: John Lubbock, who was scheduled to give the popular lecture to the 'Operative Classes' on 17 September, on the subject of 'Savages' (*Brit. Assoc. Rep. 1870*, pp. xxxvii, lxxii).
11. *A Mr. Huggins*: not identified; letter missing. Possibly William Huggins (1824–1910), an English astronomer (*ODNB*), who was a member of the Council of the BAAS for this year.
12. *Konig*: possibly Rudolph König (1832–1901), a German physicist. König is primarily remembered for his work in acoustics. His work was heavily influenced by Hermann Helmholtz, and he improved upon and developed many scientific instruments (*CDSB*).
13. *fixed me for the Friday*: Tyndall's BAAS lecture indeed fell on a Friday (see n. 4).
14. *Rankine*: William John Macquorn Rankine (see letter 2938, n. 6).
15. *come on on Monday*: Rankine delivered his BAAS discourse on 'Stream-lines and Wave in Connexion with Naval Architecture' at 8:30 pm, on Monday, 19 September (*Brit. Assoc. Rep. 1870*, p. lxxii).
16. *the sections*: BAAS meetings were divided in sections. The sections in 1870 were: Section A. Mathematics and Physics, Section B. Chemistry and Mineralogy, Section C. Geology, Section D. Biology, Section E. Geography, Section F. Economic Science and Statistics, Section G. Mechanical Science.
17. *Solferino*: a small town in the province of Mantua, Lombardy, Italy. It was the site of the Battle of Solferino during the second Italian War of Independence on 24 June 1859, fought between the combined forces of the French army under Napoleon III and the Piedmont-Sardinian army under Viktor Emmanuel II against the army of the Austrian Emperor Franz Joseph I. According to his journal, Hirst was in Italy in June 1859 and visited Solferino immediately after the battle (RI MS JT/2/32d/1504–9).
18. *Galignani*: colloquial name for the *Galignani's messenger*, an English-language newspaper published daily in Paris from 1814 to 1884 (continuing as the *Daily Messenger* until 1904).
19. *journal to Genève*: *Journal de Genève* was printed in Geneva, Switzerland from 1826 until 1991. It was a French language newspaper which became a daily in 1912.

20. *The Times*: the London-based *Times* was first published in 1785 as a daily newspaper. It is considered a paper of record.
21. *the French Emperor*: Charles-Louis Napoléon Bonaparte.
22. *the Prussians*: the Franco-Prussian War broke out while Tyndall was in Belalp (see letter 3238, n. 16).
23. *Clausius*: Rudolf Clausius.
24. *Drachenfels*: see letter 3020, n. 37.

From Henry Bence Jones 17 August [1870][1] 3245

5 Albion Villas | Folkestone | Aug 17

My dear friend

I have just got your letter[2] and as you talk of staying ten days more I will write at once. We shall be most pleased to see you & hope you will stay here as long as you can. I want to hear all your lecture, all the molecular arrangements & all the varieties of eternal motion that your imagination can put together. The hereditary fears of the tadpoles seem to me no more & no less wonderful than the motion of the monads & vibrios which only the highest powers of the microscope can see or the twinkling of the stars which I cannot see without my telescope.

Who would think from your peaceful letter that somewhere between you & me that Sunday murder by the thousand[3] was going on. You seem on your mountain to have been transformed & by prayer & fasting[4] to have forgotten this world & the horrors of it. I was expecting another battle all night & each currant of a fruit pudding I eat for dinner became in my dreams a dying soldier. However the Times again puts off the battle & the Prussians[5] seem likely to have some champagne gratis to keep up their fighting powers. The collapse of personal 'rule' seems as great as it could be. Prussian organisation has come out wonderfully & their educational organisation will some day come out equally good also.

I have heard nothing about the Institution[6] for some time but I expect Odling[7] down to day for a breath of sea air and I shall doubtless hear all that has been done from him.

I have not got on with my lecture arrangements as I have failed with Kingsley[8] & Ruskin[9] & Mathew Arnold[10] but however there is plenty of time. I have asked[11] Froude[12] & Sterndale Bennett[13] (the musician) & I shall probably try Sharff.[14] Richmond[15] has been here alone & has come in the last few evenings but I have not been able to get any artistic suggestions from him. I find that Scott's weather lectures[16] were thought dull—Seeleys[17] duller.[18]

I have been here about three weeks & have been hard at work on Count Rumford's life[19] & with the help of the new life of him[20] coming out in

America & the American papers in the Royal Institution[21] and other things I think I shall get a wonderful picture of the man. I fully purposed to make a totally different arrangement from the Life of Faraday[22] but alas I have no power of word painting. Good use of scissors & searching are all I can do, so I must be content with accuracy & give up all hope of higher praise.

I shall make a short sketch of Davy[23] up to the 5th of April 1813 when he resigned the chair of Chemistry. I shall make him draw himself also. In some things he is very like you. The syllabus of Young's lectures[24] in 1802 was not known to Peacock.[25]

The best thing I have found in it is a Scholium on Space. "Scholium-Space is a mode of existence, incapable of definition and supposed to be understood by tradition," one of his friends says. I remember also his taking one with him to the Royal Institution to hear him lecture to a number of silly women and dilettante philosophers. But nothing would show less judgement than the method he adopted; for he presumed like many other lecturers and preachers on the knowledge & not on the ignorance of his hearers.

The Life of Garnett[26] will have very little interest in it. A glimpse or two of Coleridge[27] & his lectures will come out as new altho' they will be old but the new place will be a new frame to an old picture.

I have absolutely nothing new to tell you so I have spun all my yarn out of my clippings. It will take a long time before they are crystallized into a book. But there is no hurry in this as there was in the Life of Faraday. A year or two hence will do just as well as next season & I am very lazy & feel unfit for any work & as neither eating drinking nor sleeping come pleasantly to me I am a useless mortal & had better be helping the worms & the grass to grow faster than they otherwise would do for the benefit of some body else.

Write to me soon again & tell me the day I am to go to the boat to meet you. Night or day we shall have a bed ready for you.

My wife is proud to be in the same boat as Lady Emily Peel. We shall all be glad to see you but don't hurt yourself in the bath[28] before you come this year.

Goodbye dear friend

I hope everything will go well with you | Ever yours | H. Bence Jones

RI MS JT/1/J/82
RI MS JT/1/TYP/2/698–700

1. *[1870]*: this letter was dated through its reference to the Franco-Prussian War (see letter 3238, n. 16).
2. *your letter*: letter missing.
3. *that Sunday murder by the thousand*: the battle of Mars-la-Tour (Metz) took place from 15 to 16 August 1870.

4. *on your mountain to have been transformed & by prayer & fasting*: while this is clearly a biblical reference, it is less clear whether Bence Jones was referring to Moses or Jesus or even thinking about a specific biblical mountain.
5. *the battle & the Prussians*: this letter was written during the Franco-Prussian War (see letter 3238, n. 16).
6. *the Institution*: the RI.
7. *Odling*: William Odling.
8. *Kingsley*: Charles Kingsley.
9. *Ruskin*: John Ruskin.
10. *Mathew Arnold*: see letter 3236, n. 7.
11. *I have asked*: the editors have been unable to confirm whether Bence Jones was eventually successful with his requests in the longer term, although discussions with Froude had been ongoing since September 1869 (letters 3038 and 3039).
12. *Froude*: James Froude. He did not lecture at the RI in 1871.
13. *Sterndale Bennett*: William Sterndale Bennett (1816–75), an English composer. Bennett was admitted as a chorister in King's College chapel at the age of seven in 1824 and at the age of nine entered the Royal Academy of Music. From 1848 he regularly taught harmony at Queen's College, London, and in 1856 he was elected professor of music at Cambridge University. Later that year he received his MusD and was instrumental in reforming the award of music degrees at Cambridge. He was appointed principal of the Royal Academy of Music (1866), received honorary degrees from Cambridge (1867) and Oxford (1870), and knighted at Windsor (1871). Bennett is regarded as the leading English composer of the early Victorian period (*ODNB*). He did not lecture at the RI in 1871.
14. *Sharff*: possibly George Scharf (1820–95), an English artist, critic, and gallery director. He was made secretary of the National Portrait Gallery in 1857 and director in 1882 (*ODNB*). A recognized authority on historical portraiture, he gave several lectures on the subject at the RI: a Friday Evening Discourse on 2 March 1866, 'On Portraiture: its Fallacies and Curiosities as connected with English History (*Roy. Inst. Proc.*, 4 (1862–66), pp. 543–9); three lectures, 'On National Portraits', on Saturdays from 14–28 April 1866 (*Athenaeum*, 16 December 1865, p. 825; *Roy. Inst. Proc.*, 4 (1862–66), p. 485); and six lectures 'On Historical Portraiture of Various Times and Countries', on Tuesdays and Thursdays from 3–19 March 1868 (*Roy. Inst. Proc.*, 5 (1866–69), p. 197. Scharff did not lecture at the RI in 1871.
15. *Richmond*: George Richmond (1809–96), a British portrait painter. His portrait of Thomas Dyke Acland, later eleventh baronet (1840) increased his popularity and commissions. He was elected a member of the Athenaeum Club (1856) and the Royal Academy (1866). A portrait of Tyndall, for display at the RI, was created by Richmond in 1864 (G. M. Prescott, 'Forging Identity: The Royal Institution's Visual Collections,' in F. A. J. L. James (ed.), *Science and Society at the Royal Institution of Great Britain* (London/New York: Routledge, 2017), p. 87). Richmond did not lecture at the RI in 1871.
16. *Scott's weather lectures*: Robert Henry Scott (see letter 3038, n. 25). One such lecture

delivered at the RI by Scott is R. H. Scott, 'On the Work of the Meteorological Office, Past and Present', delivered on Friday, 30 April 1869 (*Roy. Inst. Proc.*, 5 (1866–69), pp. 535–47). Scott did not lecture at the RI in 1871.

17. *Seeleys*: John Robert Seeley (1834–95), an English historian and essayist. He published on a range of topics ranging from Jesus Christ (1866) to British imperialism (1883) to Goethe (1894). His anonymous work, *Ecce homo: a Survey of the Life and Work of Jesus Christ* (1865) was controversial, appreciated by William Gladstone, but few others. Seeley succeeded Charles Kingsley to the regius chair of modern history at Cambridge University in 1869 with Gladstone's support (*ODNB*).
18. *duller*: Seeley delivered a course of three lectures at the RI, 'On Some Principal Characteristics of the Present Period of English History', on 24 and 31 May, and 7 June 1870, ('Literary and Scientific Meetings for the Ensuing Week', *Morning Post*, 23 May, p. 7, 28 May, p. 7, and 4 June 1870, p. 5; *Notes and Queries*, 21 May 1870, p. 479; *Roy. Inst. Proc.*, 5 (1866–69), p. 611).
19. *Count Rumfords life*: Bence Jones discussed the life of Benjamin Thompson, Count Rumford, considered the founder of the RI, in the first two chapters of his book, *The Royal Institution: its Founder and its First Professors* (London: Longmans, Green and Co., 1871).
20. *the new life of him*: G. E. Ellis, *Memoir of Sir Benjamin Thompson, Count Rumford, with Notices of his Daughter* (Boston: American Academy of Arts and Sciences, 1871).
21. *the American papers in the Royal Institution*: presumably the papers of Benjamin Thompson, Count Rumford held by the RI (Royal Institution, GB 0166 Benjamin Thompson, Count von Rumford Collection).
22. *Life of Faraday*: Bence Jones's *Life of Faraday*.
23. *a short sketch of Davy*: Bence Jones discussed Humphry Davy in the fourth chapter of his *The Royal Institution* (see n. 19).
24. *Young's lectures*: T. Young, *A Syllabus of a Course of Lectures on Natural and Experimental Philosophy* (London: The Press of the Royal Institution, 1802). Thomas Young (1773–1829), physician and natural philosopher, studied medicine from 1794 to 1799 and established his practice in London. He was appointed professor of natural philosophy at the RI in 1801, delivering a course of fifty lectures in 1802 and 1803. He is primarily remembered for his work in optics, his role in the RS, and his interest in Egyptology (*CDSB*; *ODNB*).
25. *Peacock*: George Peacock (1791–1858), an English mathematician and university reformer. He studied at Trinity College, Cambridge, and taught there from 1815, becoming Lowndean professor of astronomy and geometry in 1837. He was elected FRS in 1818 and was a founding figure in the Cambridge Philosophical Society (*ODNB*). In this instance Bence Jones was referring to G. Peacock, *Miscellaneous Works of the late Thomas Young* (London: John Murray, 1855).
26. *The Life of Garnett*: Bence Jones discussed Thomas Garnett in the third chapter of his *The Royal Institution* (see n. 19). Thomas Garnett (1766–1802), an English chemist and physician who was professor of natural philosophy and chemistry at the RI. He published several works on the nature and medicinal properties of the spa waters at Yorkshire. When

he died penniless, the RI published *Popular Lectures on Zoonomia* (1804), the proceeds of which were to benefit his orphaned daughters (*ODNB*).

27. *Coleridge*: Samuel Coleridge (1772–1834), an English poet, critic, and philosopher. An 'inquiring spirit,' he wrote on several topics, including religion, politics, and literary theory, but is best remembered for his poems, 'The Rime of the Ancient Mariner' (1798), 'Christabel' (1797, 1800), and 'Kubla Khan' (1816), among others. His influence can be felt in the work of Wordsworth, Byron, Keats, Emerson, and Poe. After a life-long addiction to laudanum (opium) which increased in later years, Coleridge died from heart failure in 1834 (*ODNB*).
28. *don't hurt yourself in the bath*: in 1869 Tyndall slipped while bathing in a mountain stream in the Alps and suffered a serious injury requiring prolonged recovery. See letters 3030, 3031, 3032, 3037, 3041, and 3048.

To Thomas Archer Hirst [19 August 1870][1] 3246

19th Aug[ust].[2]

My dear Tom.

Mr. Laird,[3] the member for Birkenhead[4] invited me[5] long ago to stay with him during the meeting of the B.A.[6] But I said I was not quite free.

Is there any just cause or impediment to my accepting his invitation: or is it expected that I should hold on to our party?

Mr. Laird has written to me a second time,[7] and I must now decide.[8]

Do you know a man of gentlemanly tone who would be able & willing to accept a chair of Physics and Mechanics (elementary) in an American University.[9] His salary w[oul]d. be 2000 dollars in gold

Yours aff[ectionatel]y. | John

Reply to me soon, as I shall be quitting this place[10] in a few days.

RI MS JT/1/T/919
RI MS JT/5/12/173a

---

1. *[19 August 1870]*: the year of this letter was determined through references to Laird, who was MP for Birkenhead from 1861 to 1874, and the BAAS meeting, which took place in Liverpool in 1870.
2. *19th Aug[ust].*: added in a second hand, possibly Hirst's.
3. *Mr. Laird*: John Laird (1805–74), a British shipbuilder and politician. Laird was educated at the Royal Institution, Liverpool, and subsequently trained as a solicitor. He entered his father's boiler works in 1828 and became a pioneer in the construction of iron vessels, including a gunboat, the *Nemesis*, that became famous during the opium wars with China. He retired from the business in 1861 and became the first MP for Birkenhead, as a conservative, that year. He held this post until his death in 1874 (*ODNB*).

4. *Birkenhead*: a town located on the west bank of the River Mersey, opposite the city of Liverpool. Historically it was located in Cheshire. William Jackson (1805–76), the great-great-great-grandfather of one of the general editors of *The Correspondence of John Tyndall*, was created the first baronet of the Manor House, Birkenhead in 1869. Unlike Laird, he was a Liberal MP (*Debrett's Illustrated Baronetage, with the Knightage, of the United Kingdom of Great Britain and Ireland* (London: Dean & Son, 1871), p. 261; 'Sir William Jackson' in C. G. Mott, *Reminiscences of Birkenhead* (Liverpool: Henry Young and Sons, 1900), pp. 45–6).
5. *invited me*: letter missing.
6. *B.A.*: the BAAS meeting in Liverpool (see letter 3235, n. 11).
7. *written to me a second time*: letter missing.
8. *must now decide*: see letter 3254.
9. *an American University*: not identified.
10. *this place*: Tyndall was at Belalp (see letter 3242).

From Jules Jamin 23 August 1870 3247

Paris 23 août 1870

Mon cher ami

Vous êtes un digne Anglais. Je vous remercie et j'accepte avec honneur votre contribution qui sera versée à la caisse générale des blessés.

Permettez-moi de ne pas cacher votre nom. Vous êtes déjà bien connu en France. Vous y serez maintenant aimé par tous; comme vous [l'êtes] par moi!

À vous sincèrement | Jamin

24 rue Soufflot

Paris 23 August 1870

My dear friend,

You are a worthy Englishman. I thank you, and I accept with honour your contribution,[1] which will be paid to the general fund of the wounded.

Allow that I reveal your name. You are already well known in France. You will now be loved there by all; just as you are by me!

Sincerely yours, | Jamin

24 Soufflot Street

RI MS JT/1/J/13

1. *your contribution*: Tyndall had sent Jamin a donation to aid with the wounded from the Franco-Prussian war. See letters 3238, n. 16, and 3261.

From Robert Bunsen 25 August 1870 3248

Heidelberg den *[25]*[1] Aug 1870.

Mein lieber theuerster Tyndall!

Wenn etwas die herzliche Freundschaft, welche mich mit Ihnen verbindet, noch erhöhen könnte, so ist es die hochherzige Weise, mit der Sie meiner in diesen schweren Zeiten gedacht haben. Die wahrhaft fürstliche Gabe, mit der Sie den wahrhaft herzzerreißenden Jammer mitlindern helfen, hat mich innig bewegt und mit Dank erfüllt. Seit Wochen passiren hier fast täglich Züge mit Hunderten von Verwundeten durch unsere Stadt, in der an zwanzig Nothspithälen errichtet sind, in denen sich gegen 800 Verwundete befinden. Wo die Verluste wie bei Metz allein auf unserer Seite 30 bis 50 tausend Mann betrugen, da reichen menschliche Kräfte nur langsam noch aus, wenn es gilt, Hülfe zu schaffen. Man sieht hier Verwundete mit zerschmetterten Gliedern und zersplitterten Knochen, die erst hier ihres dürftigen Nothverbandes entledigt und in ärztliche Behandlung genommen werden konnten; da ist es dann nur zu oft zu spät zur Rettung.

Unsere jungen Mädchen und Frauen sind in den Spitälen thätig und mit Herstellung des mangelnden Verbandzeugs beschäftigt und es ist rührend zu sehen, wie die kleinsten Dörfer dazu ihre Beiträge spenden. Von *[Erlassene]* die eben vom Schlachtfelde zurückgekehrt, höre ich, daß das Elend dort grenzenlos sein soll.

Ich glaube in Ihrem Sinne zu handeln, wenn ich Ihre große Gabe dem Hülfscommitee übergebe, welche für die Pflege der Verwundeten und die Ausrüstungen zur ärztlichen Hülfe wirken. Und so, mein lieber theurer Tyndall nehmen Sie noch einmal meinen und aller Freunde herzlichen Dank für die hochherzige Bethätigung Ihrer Theilnahme für uns.

Bahn und Telegraphen sind wenigstens nur oft mit großen Unterbrechungen für Privatzwecke offen. Ich weiß daher nicht ob und wo Sie meine Nachrichten in der Schweiz *[fassen]* werden, und richte daher diese Zeilen nach London, von wo sie, wie ich glaube, eher und sicherer in Ihre Hände gelangen werden als von hier.

Von ganzem Herzen und in treuer Freundschaft | Ihr | RW Bunsen

PS. Hätten wir doch die Freude Sie auf der Rückreise hier zu sehen!

Heidelberg, *[25]* Aug[ust] 1870.

My dearest Tyndall!

If there is anything that could strengthen the cordial friendship that unites us, it is the generous way in which you have thought about me in these difficult times. The truly princely donation with which you are helping to ease truly heartbreaking misery has moved me thoroughly and filled me with gratitude. For weeks now, trains carrying hundreds of wounded have passed almost daily through our city, where about twenty emergency hospitals have been built to house around 800 wounded.[2] Where the losses, as at Metz,[3] add up to 30 to 50 thousand men on our side alone, human force can hardly suffice to provide help. Here, one sees wounded with shattered limbs and splintered bones, who only here are relieved of their pitiful emergency bandages and received into medical care; only then it is too often too late to save them.

Our young girls and women work in the hospitals and are occupied with the production of the scarce bandaging materials, and it is touching to see how even the smallest of villages make their contributions to this end. I hear from *[those released]* who have just returned from the battlefield that the wretchedness there is without bounds.

I believe that I am acting in your spirit if I give your donation to the relief committee that is responsible for the care of the wounded and the provision of medical care equipment. And so, my dear Tyndall, please accept once again my and all of our friends' sincerest thanks for the magnanimous act of solidarity with us.

At least, the railroads and telegraphs are open for private use, although all too often only with great disruptions. I therefore do not know if and where you will *[receive]* my news in Switzerland, and so send these lines to London from where, I believe, they will reach you faster and more certainly than from here.

From the bottom of my heart and in true friendship | Your | R W Bunsen

PS. If only we had the pleasure of seeing you here on your return journey!

RI MS JT/2/10/458–9/f

1. *[25]*: the date appears to be '25' written over a different date.
2. *800 wounded*: Tyndall made donations to help the wounded on both sides of the Franco-Prussian War. See letters 3238, n. 16, 3247, and 3356.
3. *Metz*: a city in northeast France located near the Germany and Luxembourg borders. It was the site of a key battle in the Franco-Prussian War. See letter 3252, n. 4

From Henry Russell[1] 27 August 1870 3249

Pau 14 rue Marca | Saturday Aug[ust]. 27th

Dear Mr. Tyndall

Will you allow me to trespass on your always valuable *[moments]* to remind myself to you for a few minutes, and let me ask you to thank in my name the Editor of the Times, whom I have not the honour of knowing even by sight, for his readiness and kindness in publishing my letter to him in July last,[2] describing my encounter with those very unchivalrous and ugly brigands, of whom I hear one has been caught. What a triumph for Spanish justice!

How am I to speak of this excecrable war?[3] It makes me feel so many different things. But I deeply, bitterly regret the attitude the English press has lately taken towards France. You can't imagine the harm it has done to the Entente Cordial[4] which I and so many considered the finest and happiest thing our century had realized. It is true that many foolish and fanatical Paris papers have talked very great nonsense in their irritation. But indeed it cannot be too much deplored that newspaper correspondents were excluded from the French camps, for the Prussian telegrams, pretty fair in the beginning, have become, during and since their imaginary "Metz" Victories,[5] a tissue of mystification. I would not say this lightly. We are deprived here of official news! but by the mass of private letters which daily come to us from the seat of war, and which came regularly from Metz in 24 hours at the very time when all the foreign papers affirmed that Metz and Bazaine[6] were surrounded, cut off etc etc., from those heaps of evidence which never discontinued, it is very hard not to conclude that with the purest intentions the English press has been deluded.

Let me add also, as one whose double origin makes it easy for him to speak on the subject without passion, that the power and incalculable resources of this country are singularly underrated in England, both as to money and as to men. Look at the loan! The city of Paris alone subscribed more than all Germany put together, and if another 30 millions sterling were asked, they would be found tomorrow evening.

As to men (I mean regular soldiers, not recruits) not one half of them have yet left their homes. They are swarming on all sides, even here. The truth is that for once France was late, slow and unprepared; but indeed, only eyes are wanted, to see that those who speak of her exhaustion and of peace know nothing of her vitality and powers.

I don't speak of the Future. For God alone holds the gates of it. Will I never have the pleasure of welcoming you to our Pyrenean glaciers? They are cold, but handsome: only English young ladies combine everything. If this excecrable war ceases and the two lions shake paws . . . , I hope to spend a few weeks in Dear England in October, and it will be very great pleasure to me, my dear Mr Tyndall, to renew your acquaintance.

In the meantime, once more let me ask you to give my best thanks to the Editor of the Times, to remind me very kindly to Sir Rod[erick] Murchison and my valuable friends of the Athenaeum[7] and to believe me always,

Very sincerely your's | C[om]t^e^ Henry Russell.

RI MS JT/1/R/82

1. *Henry Russell*: Henry Patrick Marie, Count Russell-Killough (1834–1909), an Irish writer and mountaineer. He was a pioneer of Pyrenean exploration, and climbed every Pyrenean summit of note, including the Vignemale which he summited over thirty times, and made the first winter ascent in 1869. He later created a series of caves carved into the side of the Vignemale where he hosted fellow mountaineers in style, replete with Persian carpets, silk cushion, and dinners of Bayonne ham and vintage claret (R. Bailey, *The Man Who Married a Mountain: a Journey through the French Pyrenees* (London: Transworld Publishers Limited, 2005); K. Reynolds, *The Pyrenees: The High Pyrenees from the Cirque de Lescun to the Carlit Massif* (Cumbria: Cicerone, 2004), pp. 219–20).
2. *my letter to him in July last*: H. Russell, 'Brigands in Aragon', *Times*, 21 July 1870, p. 12.
3. *excecrable war*: the Franco-Prussian War (see letter 3238, n. 16).
4. *Entente Cordial*: the belief that there was a 'cordial understanding' between the United Kingdom and France from the mid-nineteenth century. Despite these claims, there was much, predominately imperial, tension between them for the rest of the century. This should not be confused with the formal Entente Cordiale agreement of 1904.
5. *"Metz" Victories:* the siege of Metz and the surrender of Achille Bazaine and the French army that he commanded. See letter 3252, n. 4.
6. *Metz and Bazaine:* Marshal of France, Achille Bazaine, was appointed commander-in-chief of the French army during the Franco-Prussian War. Following the siege of Metz he surrendered his army of 140,000 men to the Prussians. For these actions he was sentenced to death on 10 December 1873. The sentence was later commuted to twenty years' imprisonment.
7. *Athenaeum:* a weekly periodical published in London from 1828 until 1923. It covered a broad range of topics including literature, fine arts, music, theatre, politics, and science.

From Johann Caspar Bluntschli[1] 29 August 1870 3250

Hochverehrter Herr.

Durch die gütige Vermittlung des Hr Gh Rath Herr Prof Bunsen haben wir von Ihnen zum /Beßten/ der Pflege der verwundeten Krieger die schöne Gabe von Liv. St. 75 (fünf & siebenzig) erhalten & sagen Ihnen dafür unsern herzlichen Dank.

Mit ausgezeichneter Hochachtung | Aufsichts-Commission für die Reserve-Lazarethe | Bluntschli

Heidelberg 29 Aug. | 1870.

Dear Sir.

We received, through the kind mediation of Mr Private Council Prof Bunsen,[2] your friendly donation[3] of 75 (seventy-five) P[oun]d. St[erlin]g. for the benefit of nursing injured soldiers and send you our deep thanks.

Respectfully | Supervisory commission for the military hospitals of the reserve | Bluntschli

Heidelberg 29 Aug[ust] | 1870.

RI MS JT/1/B/101

1. *Johann Caspar Bluntschli:* Johann Caspar Bluntschli (1808–81), Swiss-born jurist appointed to a professorship in constitutional law in Heidelberg in 1861, and later a politician in his adoptive Baden. Bluntschli's posthumous autobiography confirms that he was indeed in charge of the 'Aufsichts-Commission' responsible for supervising the care of wounded soldiers in reserve military hospitals in Heidelberg in 1870 during the Franco-Prussian War (J. C. Bluntschli, *Denkwürdiges aus meinem Leben. Teil III* (C. H. Beck: Nördlingen, 1884), p. 262; *HLS*).
2. *Prof. Bunsen*: Robert Bunsen.
3. *friendly donation*: see letter 3248.

To Henry Russell[1] 1 September 1870 3251

Royal Institution | 1st S[e]pt[mbe]r 1870

My dear Count Henry.

I have just reached England from the Alps, and will not delay a moment in acknowledging the receipt of your kind & hearty letter.[2]

I thought your wishes would be best met by sending it to the Editor of the Times.[3]

This war[4] is as you say truly execrable; and I deplore as much as you do many of the utterances of the English press—but the French press is really exasperating. France indeed from the first put herself clearly in the wrong. From my heart I deplore the sufferings of the brave men on both sides. I had a letter this morning from Bunsen[5] of Heidelberg sufficient to make the blood run cold.

Some time or other I trust to see your Pyrenean glaciers and whenever you come to England you may at all events calculate on one to whom your presence will give pleasure. No doubt it will to hundreds besides—but of one I am sure.

Yours very faithfully | John Tyndall

The Count Henry Russell | &c &c &c

RI MS JT/1/T/1267

1. *Henry Russell*: see letter 3249, n. 1.
2. *your kind & hearty letter*: letter 3249.
3. *sending it to the Editor of the Times*: not found. See letter 3249.
4. *This war*: the Franco-Prussian War (see letter 3238, n. 16).
5. *a letter this morning from Bunsen*: letter 3248. Tyndall had likely just read Bunsen's letter, which was sent to the RI.

From Rudolf Clausius 1 September 1870 3252

Bonn, 1/9/70.

Lieber Tyndall,

Du hast Dich gewiss gewundert, dass ich Deinen lieben Brief aus der Schweiz und die freundliche Depeche in welcher Du Dich erbietest, die Meinigen im Falle der Noth in London aufzunehmen, so spät beantworte. Der Grund liegt aber darin, dass ich etwa 4 Wochen von Bonn abwesend war. Ich bin nämlich mit einem vorzugsweise aus Bonner Studenten und einigen Professoren bestehenden Corps von freiwilligen Krankenpflegern und Nothhelfern auf dem Schlachtfelde mit in den Krieg gezogen. Wir gehörten zum III preussischen Armeecorps und haben mit diesem die Schlacht vom 16ten August bei Metz (bei den Dörfern Mars la Tour, Gorze und Vionville) mitgemacht. Auch bei der Schlacht vom 18ten August bei Gravelotte war ich zugegen, aber nicht thätig, weil bei dieser unser Armeecorps nicht betheiligt war. Unser (nämlich das III) Armeecorps hat am 16ten von morgens 9½ Uhr bis nachmittags 3 Uhr fast allein gegen die ganze französische Macht gestanden, und erst um 3 Uhr kam ihr das X Armeecorps zu Hülfe. Das III Armeecorps hat daher ungeheure Verluste erlitten, und wir hatten während der Schlacht und mehrere Tage nach der Schlacht unendlich viel zu thun, den Verwundeten die erste Hülfe zu leisten. Ich blieb nach der Schlacht noch 6 Tage in Vionville bis die meisten Verwundeten von dort fortgeschafft und in grösseren Lazarethen untergebracht waren. Dann bin ich mit dem grössten Theile der Studenten nach Bonn zurück gekehrt, weil unser Armeecorps wahrscheinlich keine Schlacht mehr mitmachen wird, sondern wegen der Verluste die es erlitten hat, in Reserve bleiben soll.

Die Erfolge sind für Deutschland so ausserordentlich günstig, dass durchaus keine Wahrscheinlichkeit ist, dass meine Familie in Bonn in Gefahr kommen könnte. Ich danke Dir aber in meinem und meiner Frau Namen herzlich für Dein freundliches Anerbieten.

Zur Versammlung der British Association zu kommen war mir unter den

Umständen natürlich unmöglich, und ich habe sogar versäumt, für die von Huxley und Hirst unterschriebene Einladung zu danken. Sei so gut, mich bei ihnen wegen der unvorhergesehenen Abhaltung zu entschuldigen. Ich hatte gehofft, Du würdest im nächsten Jahre präsidieren, und bedauerte sehr, aus Deinem Briefe zu ersehen dass Du abgelehnt hast.

Mit herzlichen Grüssen von meiner Frau und mir | Dein | Clausius.

Bonn, 1/9/70.

Dear Tyndall,

You must have wondered why I took so long to respond to your kind letter[1] from Switzerland and your friendly telegram,[2] in which you offered to accommodate my family in London in case of emergency. The reason is that I was absent from Bonn for about four weeks. Namely, I went to the battlefields[3] and was drawn into the war as part of a corps of volunteer nurses and emergency medical staff consisting primarily of Bonn University students and several professors. We were part of the 3rd Prussian army corps and saw action in the battle at Metz[4] on August 16th (near the villages of Mars-la-Tour,[5] Gorze[6] and Vionville).[7] I was also present at the battle at Gravelotte[8] on August 18th but not active, because our corps was not involved. Our army corps (namely the 3rd) stood practically alone against the whole French force from 9.30 a.m. until 3 p.m. on the 16th, and only at 3 p.m. did the 10th battalion come to our aid. The 3rd battalion therefore suffered terrible losses, and we were endlessly busy during the battle and many days thereafter, giving first aid to the injured. After the battle, I remained in Vionville for six days until most of the injured had been taken away and brought to larger military hospitals. I then returned to Bonn with most of the students, as our army corps will probably not participate in any more battles, but rather remain in reserve due to the casualties it has suffered.

The events are so extraordinarily favourable for Germany that it is unlikely that my family will be endangered in Bonn. But on behalf of my wife and myself, I thank you for your kind offer.

It was obviously impossible for me to come to the meeting of the British Association[9] under these circumstances; I even failed to express my thanks for the invitation signed by Huxley[10] and Hirst.[11] Please be so kind as to apologise to them on my behalf and explain that I had unforeseen obstacles. I had hoped you would preside in the next year and am very sorry to learn from your letter[12] that you had declined.

With best wishes from my wife[13] and me | Your | Clausius.

RI MS JT/1/TYP/7/2316–7
LT Typescript Only

1. *your kind letter*: probably letter 3237.
2. *your friendly telegram*: letter missing.
3. *I went to the battlefields*: during the Franco-Prussian War (see letter 3238, n. 16).
4. *battle at Metz*: also known as the battle of Mars-la-Tour (16 August 1870) this was a major battle between the French and German armies during the Franco-Prussian War. The French Force of 140,000 men was commanded by Marshal Achille Bazaine, while the Prussian commander was General Helmuth von Moltke. As a consequence of the battles of Mars-la-Tour and Gravelotte (16–8 August 1870), Bazaine's forces fell back to the French fortress of Metz.
5. *Mars-la-Tours*: the town of Mars-la-Tour is located in the commune of Mars-la-Tour in north-eastern France.
6. *Gorze*: a village located in north-eastern France.
7. *Vionville*: a village in north-eastern France.
8. *Battle at Gravelotte*: one of the largest battles of the Franco-Prussian War, on 18 August 1870.
9. *meeting of the British Association*: the BAAS meeting in Liverpool (see letter 3235, n. 11).
10. *Huxley*: Thomas Huxley.
11. *Hirst*: Thomas Hirst.
12. *your letter*: letter 3237.
13. *my wife*: Adelheid Clausius.

From [Juliet Pollock][1] 2 September [1870][2] 3253

S^t^. Julians | Sevenoaks | Sept[embe]r 2.

fear = rude = tender = true | ever Johny = old Boreas

Your letter was a relief and a blessing in the midst of complicated sadness. You will know before this reaches you that my husband[3] and I have lost both the guardians of our childhoods within a few days of each other[4] and you will easily conceive that we have both had much to go through. Happily our affections were interweaved between those two lives, as the two deaths were so strangely coincidental—I certainly loved Fred[eric]^k's^ Father[5] not less than he did; and Fred[eric]^k^[6] was also very much attached to my aunt,[7]—nor did any difficulty or misunderstandings ever arise at any time between us four, so that now we can talk together of the past, freely and quietly.

My husband's health has suffered from all this wear and tear more than mine has, but I hope to see him much better before long: it is unfortunate that the great public outcry of the day[8] is one which casts a gloom over the thoughts and that the diversion from personal sadness is the history of carnage and of the misery of a great nation undergoing dying agonies. I hear that our friend Gueneau de Mussy[9] is in the deepest depression, he is sure to join that pretty little kitten his wife at Brussels.—

We hope to have our boys here next week, and they will bring happiness with them—

We intend to stay here till the 15th and then we think of staying for a while at Dover, if we can find good lodging there, for change of air & scene, which we rather require.

good bye dear faithful admirable friend | I am ever your true | Eolian.

RI MS JT/1/E/46

---

1. *[Juliet Pollock]*: although the sender does not provide a proper name, the letter is from 'Eolian' to 'Boreas', the nicknames Tyndall and Juliet Pollock used for one another.
2. *[1870]*: although the letter is without year, it refers to the death of William Frederick Pollock's father, which occurred in August of 1870.
3. *my husband*: William Frederick Pollock.
4. *lost both the guardians of our childhoods within a few days of each other*: Juliet Pollock's father Henry Creed, vicar of Corse, Gloucestershire, died as this letter indicates in August 1870, while William Frederick Pollock's father Jonathan Frederick Pollock died on 23 August 1870.
5. *Fred[eric]k's Father*: Jonathan Frederick Pollock (see letter 3015, n. 1).
6. *Fred[eric]k*: William Frederick Pollock.
7. *my aunt*: probably Isabella Maria Herries (*c.* 1789–1870), daughter of Charles Herries and Marry Anne Johnson. She was a sister of Juliet Pollock's mother, Catherine Creed (née Herries). Isabella raised Juliet Pollock and her cousin Maria Julia Herries (*c.* 1820–1857) after the deaths of their mothers.
8. *great public outcry of the day*: the Franco-Prussian War (see letter 3238, n. 16).
9. *Gueneau de Mussy*: Noel Gueneau de Mussy (1813–85), a French physician with ties to the United Kingdom through marriage (*BMJ*, 1:127 (1885), p. 1064).

To Thomas Archer Hirst 3 September 1870 3254

Royal Institution | Sep[tember] 3rd. 1870

My dear Tom.

I am here. Got through Paris without molestation:[1] read all the French papers to my immense moral deterioration—they are so full of delusion; so full of frivolity, so full of lies. Hardly one sentence from beginning to end worthy of the dignity of an average human soul.

I have written to Mr. Laird[2] declining[3] his invitation;[4] so I am now committed to the X.[5] I hope Hamilton[6] has not committed the mistake of supposing me to be like him, competent to sleep through any thing. And "entre nous,"[7] I hope he will keep himself at night far away from me. He is one of the most uproarious snorers I ever met.

I intercepted the postman on the slopes of the Bel Alp[8] as I descended towards Brieg,[9] and extracted your letter from his bag—While reading it I went astray. In fact I think Lilly[10] was to blame for this; for I pictured her sitting as quiet as a mouse and then sending me that kindly message. Give her my love in return.

Yours ever affec[tionate]^{ly}. | John <u>Tyndall</u>

I am going to run down to Folkestone.[11] I shall remain there a week: if you write to me the post office will answer for an address.

<u>Saturday</u>

Miss Robertson[12] sends you her kind regards.

RI MS JT/1/T/680
RI MS JT/1/HTYP/572

1. *got through Paris without molestation*: due to the Franco-Prussian War (see letter 3238, n. 6), travel on the continent was difficult.
2. *Mr. Laird*: John Laird (see letter 3246, n. 3).
3. *declining*: the typescript version from LT has 'accepting' here. This was Laird's second invitation to Tyndall to stay with him during the BAAS meeting in Liverpool. Tyndall declined the first, as he told Hirst in letter 3246.
4. *his invitation*: letter missing.
5. *the X*: the X Club (see letter 3045, n. 5).
6. *Hamilton*: not identified. Possibly Sir Charles Hamilton (1810–92), a manager at the RI, and a relation of Mary Adair. In 1833, he married Katherine Emily Wynne, the second daughter of William Wynne of Dublin (Edward Walford, *The County Families of the United Kingdom,* 5th edn (London: Robert Hardwicke, 1869), p. 450).
7. *Entre nou*: between us (French).
8. *Bel Alp*: Belalp (see letter 3030, n. 20).
9. *Brieg*: see letter 3030, n. 17.
10. *Lilly*: Hirst's niece.
11. *Folkestone*: see letter 3021, n. 20.
12. *Miss Robertson*: not identified.

To Charles Darwin 7 September 1870 3255

Folkestone | 7th. Sept[ember]. 1870.

My dear Darwin,

I have asked Spottiswoode & Co.[1] to send you the proof of a lecture[2] which I purpose giving in Liverpool on the 16th.

I have had no books beside me in writing it; and I would therefore thank you much if you would glance at that portion of the proof where your name is introduced, and kindly inform me whether it is rightly introduced. Prior to page 31 I do not think your name occurs: so that you will not have much to read.

Would you be good enough to send the proof to me with your remarks, if any, to the Royal Institution some time this week? I am very anxious to get it wholly of my hands.

I hope you are well. When I saw you last I was far from well—indeed I was very ill—this accounts for my unsociability. Thank the gods I am now in a fair way.

Yours ever | John Tyndall.

I have been staying with Bence Jones for the last few days.

That remark of Elie de Beaumont[3] where he requires you to recant prior to being admitted into the Academy[4] is simply intolerable—This spirit has much to do with the present condition of France.

RI MS JT/1/TYP/9/2808
LT Typescript Only

---

1. *Spottiswoode & Co.*: Spottiswoode and Co. was a publisher located in London.
2. *a lecture*: Tyndall lecture for the BAAS meeting in Liverpool, 'On the Scientific Use of the Imagination' (see letter 3235, n. 11).
3. *Elie de Beaumont*: Jean Baptiste Armand Louis Léonce Elie de Beaumont (1798–1874), A French geologist. Beaumont studied mineralogy at the Ecole de Mines from 1819 and in 1822 was chosen, alongside Sufrénoy, to tour England in order to study mines and British geological surveying. From 1827 Beamont served as professor of geology at the Ecole des Mines and from 1823 he occupied the same chair at the Collége de France. He became a member of the Académie des Sciences in 1835 (*CDSB*).
4. *the Academy:* the French Académie des Sciences. In the nineteenth century the academy was a prestigious organisation with much authority and patronage. Darwin was eventually elected a corresponding member in 1878 after being nominated several times. While de Beaumont felt that Darwin's theory of evolution had much value, he also thought that 'he had spoiled [this] by dangerous and unfounded speculations' which he felt should be renounced prior to membership being granted ('Sketch of Charles R. Darwin, LL.D.', *Popular Science Monthly*, 21 (June 1882), p. 263).

From Charles Darwin 8 September 1870 3256

Down. | Beckenham | Kent. S.E.

Sep[tember] 8th. 1870

My dear Tyndall

Your whole discourse[1] strikes me as grand & most interesting. What you say about Pangenesis[2] is quite correct, & your expression, "not only is the organism as a whole wrapt up" &c is most happy. What you say about me, coming as it does from you, has pleased me extremely, so much that there must be clouds of vanity in my mind to bring out such pleasant sensations as your remarks passed through them. You are a rash man to say a word for Pangenesis, for it has hardly a friend amongst naturalists, yet after long pondering (how true your remarks are on pondering) I feel a deep conviction that Pangenesis will some day be generally accepted. I have been particularly delighted by your "as if" argument. These words have lately been fired at me, & if I had been forced to answer, I should not have known what to say, now I c[oul]d answer by a round of your artillery. I have ventured to mark (p. 12. & 40) two sentences which seem to me not so perfectly lucid as all the rest. I may also mention that in the discussion on "the particles of our sky" I read 2 or 3 pages thinking that you referred to the atoms of the air: w[oul]d it not be well for blunderers to shew in the early part, that you refer to foreign particles?

With hearty admiration & thanks, yours very truly | Ch[arles] Darwin

P.S. I return the proofs by this post.

When in Brazil I was much struck with the frequent blue haze & I quote (p. 32 of my travels)[3] Humboldt's[4] remark on "the thin vapour which, without changing the transparency of the air, renders its tints more harmonious <hereafter LT Typescript Only> & softens its effects."

It seemed to me that this phenomenon was much more strongly marked in the tropics than in our temperate regions—

On one day when the beautiful pale blue haze had come on very strong, the difference between the dew point & temperature by Daniel's hygrometer[5] had increased from 70.5. to 170—

RI MS JT/2/10/458–9
RI MS JT/2/13c/1366–7

---

1. *your whole discourse*: Tyndall asked Darwin to read a draft of his BAAS discourse 'On the Scientific Use of the Imagination' (see letters 3255 and 3235, n. 11).
2. *Pangenesis*: Darwin's mechanism for heredity. He proposed that gemmules expelled by the body's cells were tiny particles of inheritance that accumulated in the reproductive organs. The host organism environment could modify these gemmules and thus these changes

were passed on to offspring. The theory was first proposed in 1868 but lost popularity by the 1890s with the demise of the theory. See J. Hodge, 'The Darwin of Pangenesis', *Comptes Rendus Biologies*, 333:2 (2010), pp. 129–133; and K. Holterhoff, 'The History and Reception of Charles Darwin's Hypothesis of Pangenesis', *Journal of the History of Biology*, 47:4 (2014), pp. 661–95.

3. *p. 32 of my travels*: this quote appears in C. Darwin, *Journal of Researches into the Natural History and Geology of the countries visited during the voyage of H.M.S. Beagle round the world, under the Command of Capt. Fitz Roy, R.N.*, 2nd edn (London: John Murray, 1845), p. 32 (in the first edition (1839), as well as Darwin's volume of the three-volume *The Narrative of the Voyages of His Majesty's Ships Adventure and Beagle* (1839), edited by Robert Fitzroy, the quote appeared on p. 36).
4. *Humboldt's*: Alexander von Humboldt (1789–1859), the world-renowned Prussian naturalist and explorer, and by the 1850s one of the most notable scientific figures in Europe. From 1799 to 1804 Humboldt journeyed through South America with French botanist Aimé Bonpland (1773–1858), publishing thirty volumes upon his return. His meticulous attention to climate, temperature, and the ecology of his discoveries were made possible by the many scientific instruments that he took with him and he defined a new era of accuracy in science. His wide-ranging *Kosmos* appeared in five volumes between 1845 and 1862. Financial constraints stemming from the publication of his works forced Humboldt to spend the last thirty years of his life in Berlin working for the Prussian government (*CDSB*; *NDB*).
5. *Daniel's hygrometer*: an instrument for determining the dew point, invented by John Frederic Daniell (1790–1845).

To Thomas Archer Hirst 9 September 1870 3257

Folkestone Friday | Sep[tember] 9th / 1870

My dear Tom

I return to London today—and have just sent Spottiswoode & Co[1] a telegram[2] to send you a copy of my proposed address at Liverpool.[3] I wish you would look at it & send it up to me before you quit Yorkshire with any remarks you have to make upon it.

I do not think I will go down before Friday. The journey is long & the lecture afterwards will be rather a heavy draft: but it is best I think to secure a good night's rest previously. That secured all I trust will go well—At all events I am in no way anxious about the lecture.

I was out with Mr. Penn[4] yesterday in his yacht—we steamed to Dover[5] & Deal[6] & the Downs[7] & saw the Iron Clads.[8]

I have had a very civil invitation from Lord Derby:[9] but have told him that I form one of a band of heretics[10] which refuse to be dissolved.

Yours affectionately | John Tyndall

RI MS JT/1/T/681
RI MS JT/1/HTYP/572

1. *Spottiswoode & Co.*: see letter 3255, n. 1.
2. *a telegram*: letter missing.
3. *my proposed address at Liverpool*: 'On the Scientific Use of the Imagination' (see letter 3241, n. 2). He also asked Charles Darwin for commentary (see letters 3255 and 3256).
4. *Mr. Penn*: see letter 3043, n. 4.
5. *Dover*: see letter 3071, n. 4.
6. *Deal*: a town in Kent, England, located on the coastline near the North Sea and the English Channel.
7. *the Downs*: located off the coast of the Kent Downs, it stretches from the Surrey border to the cliffs of Dover, England.
8. *the Iron Clads*: iron-clad steamships.
9. *Lord Derby*: Edward Stanley (1826–93), a British politician. Fifteenth Earl of Derby, he was elected to parliament as a Tory for King's Lynn in December 1848, retaining the seat until he took his seat in the Lords in 1869. His lack of faith alienated him from High-Church Anglicans but was still an important figure within the Tory party being offered and declining leadership of the Lords in February 1870 (*ODNB*).
10. *a band of heretics*: a reference to Tyndall's membership in the X club (see letter 3045, n. 5).

From Mary Francis Egerton [9 September 1870][1] 3258

Mountfield Court, | Robertsbridge, | Hawkhurst. | Friday night

Dear Mr. Tyndall,

Many thanks for the books[2] which arrived quite safe. Could I have guessed you would find time to look them out so speedily, I w[ould]d have asked you to send them to Charley[3] to bring down & so saved your stamps, all I have had time to see in them yet, is that old William Tyndall's[4] portrait bears the most striking family likeness to you; it is neatly curious!

Wednesday, was a day of luxury, intellectual and physical, such as it would not do to indulge in too often. Certainly poetry from living lips has a peculiar power of stirring the heart's core! And then that fragment of dear Mrs Herschel's[5] ideal Paradise afterwards! You remember what she said about hearing clever men talk? But such days sh[oul]d not close at 6 o'clock!

Let me say one word with reference to a subject we were talking about. I do hope you will be careful how you go into Theology at Liverpool.[6] The World does not know, as your friends know, that you have an altar in your heart though "to the unknown God,"[7] and then they use that hard word that makes me shudder.

Will you not some day recognize that the "Vernunft" of Kant[8]—the

"Faith" of Fichte, the "indwelling Word" of Maurice,[9] all point to one truth, namely, that "Breath of Life" that God breathed into man, whereby he became a living soul—a spiritual existence—in harmony with the Spiritual World, as truly as by his understanding and his senses he is in harmony with the intellectual & the physical world.

I am afraid you will think me pertinacious for saying this, but I never feel so strongly convinced of it as after conversing with you.

Y[ou]$^{\underline{rs}}$ most truly | MF Egerton

The words I have written with a mutual friend but it was hard work.

I hope the R[oyal] I[nstitution]. meeting went off satisfactorily. I ought to have acknowledged Friday night the books earlier I forgot you could not get this till Monday.

RI MS JT/1/E/81

---

1. *[9 September 1870]*: this letter was probably written between Tyndall's return to England on 3 September and his address at the BAAS meeting in Liverpool on 16 September 1870. Tyndall travelled to Folkestone on 7 September and had, in the past, combined a visit to Folkestone with a visit to Egerton. 9 September is the only Friday between his arrival and the Liverpool meeting. In letter 3257 of 9 September to Hirst, he indicates that he intended to return to London that day, but his plans may have changed.
2. *the books*: not identified.
3. *Charley*: Charles Augustus Egerton (1846–1912), son of Mary Egerton. See thepeerage.com/p7808.htm [accessed 4 April 2020].
4. *William Tyndall*: also William Tyndale.
5. *M$^{\underline{rs}}$ Herschel*: Margaret Brodie Stewart (1810–84), wife of John Herschel.
6. *Liverpool*: the BAAS annual meeting took place in Liverpool in 1870 (see letter 3241, n. 2).
7. *"to the unknown God"*: this quote is from Paul before the Areopagus, Acts XVII.23.22. Also a reference to Herbert Spencer's 'unknowable', his theistic substitution for conventional religion based on the inability of man to truly know fully what underlies reality. See B. Lightman, 'Herbert Spencer and the Worship of the Unknowable', in *The Origins of Agnosticism: Victorian Unbelief and the Limits of Knowledge* (Baltimore: Johns Hopkins University Press, 2019 (1987), pp. 68–90.
8. *Kant*: Immanuel Kant (1724–1804), a German philosopher. Kant's work and, in particular, his philosophy of knowledge, were central to philosophical scholarship throughout the nineteenth and twentieth century synthesizing rationalism and empiricism (*NDB*).
9. *"indwelling Word" of Maurice*: F. D. Maurice, *The Kingdom of Christ; of, Hints to a Quaker* (London: Rivington, 1842). John Frederick Denison Maurice (1805–72), an English theologian, was Chaplain of Guy's Hospital and Professor of English literature and history at King's College, London (*ODNB*).

To Thomas Archer Hirst [14 September 1870][1] 3259

Well done old Tom! It[2] had already passed muster with Darwin,[3] Spencer & Debus;[4] but your applause[5] is best of all.

J.T.

I shall keep away till Friday.

RI MS JT/1/T/975

1. *[14 September 1870]*: on September 9, Tyndall wrote to Hirst, informing him that a copy of his upcoming discourse had been sent for his feedback. Considering the time needed for this letter and for Hirst's response to arrive, this letter has been given an estimated date of 14 September 1870. See letter 3257.
2. *It*: Tyndall sent copies of his proposed discourse for the BAAS meeting in Liverpool 'On the Scientific Use of the Imagination' (see letter 3235, n. 11).
3. *already passed muster with Darwin*: see letters 3255 and 3256.
4. *Spencer & Debus*: the responses from Spencer and Debus are missing.
5. *your applause*: letter missing.

To Thomas Henry Huxley 15 September 1870 3260

Royal Institution. | 15th. Sept[ember]. 1870.

My dear Huxley,

I have seen the Times Report of your Address.[1] In my opinion no Address delivered in my memory before the British Association[2] can match it in importance.

I am glad you singled out Sanderson[3]—He merits it: In fact the whole thing will be an immense encouragement to good men and a heavy blow to the evil doers.

Yours ever | John Tyndall.

Sought you at Jermyn St.[4] but could not find you. Tomorrow I shall be with you. It so happens that I fill in a portion of the outline which you have sketched.

IC HP 8.82

1. *your address*: see letter 3237, n. 11.
2. *British Association*: the BAAS.
3. *Sanderson*: John Burdon Sanderson (1828–1905), an English pathologist and physiologist,

graduated MD from the University of Edinburgh in 1851. During his work for the Cattle Plague Commission investigating the rinderpest outbreak in Britain he proposed that the disease might be caused by a transmissible germ. Between 1865 and 1870 Sanderson expanded his investigations into the causes of infectious disease concluding that some form of particulate transmitted the disease (*ODNB*).

4. *Jermyn St.*: location of the Royal School of Mines, where Huxley was professor of natural history from 1854–71.

From Jules Jamin 18 September 1870 3261

Paris 18 7bre 1870

Mon cher ami

Je ne sais si cette lettre vous parviendra, car nous sommes aujourd'hui assiégés. [Néanmoins] je veux vous dire combien j'ai été touché de l'offre que vous me faites de prendre soin de ma femme et de mes enfants s'ils étaient en Angleterre.

Je les ai envoyés en Bretagne, à Pornic car je les crois en sureté. Cependant je leur ai écrit que si le danger arrivait jusque [eux], ils *[1 word illeg]* à [passer] en Angleterre et à se mettre dans votre [possession], persuadé que le meilleur pour eux comme pour moi est d'avoir un ami sûr.

J'ai [versé] comme vous me l'avez recommandé les [4] livres que vous m'avez envoyé à la caisse des blessés sous votre nom. J'en ai *[1 word illeg]* reçu que je vous remettrai si nous en échappons.

Quand la [tourmente] sera passée, nous causerons de toutes les misères. Dieu nous [châtie] il faut *[expirer]* nos fautes, *[1 word illeg]* nous et tâcher de nous venir en [aide].

A vous de cœur | J. Jamin

Paris 18 September 1870

My dear friend,

I do not know if this letter will reach you, since today we are besieged.[1] [Nevertheless,] I wish to tell you how touched I was by the offer that you made me about taking care of my wife and children if they were in England.

I sent them to Pornic in Brittany,[2] because I believe them to be safe there. However, I wrote to them that if the danger reaches them, they *[1 word illeg]* to go to England and to give themselves into your care,[3] as I am convinced that the best for them, as it is for me, is to have a faithful friend.

As you requested,[4] I have [deposited] the [4] pounds that you sent me, into the funds for the wounded, under your name. I received *[1 word illeg]*, which I will give to you if we survive.

When the turmoil is over, we will talk about all the miseries. God is *[punishing]* us, we have to *[1 word illeg]* our sins, *[1 word illeg]* us, and to try to help us.

Yours truly | J Jamin

RI MS JT/1/J/14

1. *today we are besieged*: the siege of Paris (19 September 1870–28 January 1871) by the Prussian Army took place during the Franco-Prussian War (see letter 3238, n. 16).
2. *Pornic in Brittany*: Pornic is a town in south-east Brittany.
3. *to give themselves into your care:* for Eudoxie Jamin's response to this offer, see letter 3309.
4. *As you requested:* Tyndall sent Jamin a donation to aid with the wounded from the War. See letter 3247.

To Juliet Pollock 26 September 1870 3262

Shelbourne Hotel Dublin | Monday m[ornin]g | 26th Sept[ember] 1870

My dear Friend.

I am waiting for the train that is to take me to Kingstown,[1] where the steamer of the Board of Irish Lights[2] is waiting for me. I steam to one of the lighthouses on the coast; return by a second lighthouse, and then tell the Board of Trade my impressions of what I have seen.

Half an hour hence I start; and I fill this half hour by scribbling to you. I ought to have done so earlier. I ought to have thanked you for your last good note to me:[3] good & true and kind, as all your notes have been to me. I did not go to Liverpool till the Friday I preached my sermon[4] on Friday evening to a great audience, and I ran away from the racket on Saturday. I spent Sunday in a delightful ramble from Windermere[5] to Ambleside;[6] to Rydal,[7] and thence along a mountain path to Grassmere.[8] At Rydal we (Spencer & myself) met its present proprietor: He is a newspaper proprietor & editor in London: and he was civil to the last degree; he showed us every nook of the house and every square yard of the ground and gardens. He evidently took a pride in doing so, and thus all ends were served. We lunched at Grassmere—took a boat over the lake; crossed to Loughrigg[9] climbed the hill and lost our way amid the rocks and bracken at the top: reaching Ambleside about dusk, whence we drove home. On Tuesday I returned to Liverpool.

Thence I went to Manchester & stayed to see the first stone laid of Owen's College.[10] Thence to Dublin, which I reached on Saturday evening, and which I must contrive to leave as soon as I possibly can.

The Liverpool meeting was singularly successful—and assuredly Huxley[11] must have had his appetite for hero-worship satisfied.[12]

I shall soon return to London. In fact I long to end this straggling dissipated life, and to settle down closely & sternly to my work.

This horrible war[13] is it seems to go on. Poor unfortunate, corrected, grand, versatile, deluded, criminal France. What a dearth of greatness in the country! Hardly a sentence uttered worthy of a strong man. I hope she will come from the ordeal purer—assuredly she needed purification.

Yours ever | John Tyndall

RI MS JT/1/T/1153
RI MS JT/1/TYP/6/2130

---

1. *Kingstown*: see letter 3124, n. 6.
2. *Board of Irish Lights*: the Commissioners of Irish Lights (see letter 3019, n. 7). Tyndall traveled to Ireland at the request of the Board of Trade for a meeting after the BAAS meeting in Liverpool. For Tyndall's work on lighthouses more generally, see letter 2937, n. 6.
3. *last good note to me*: letter missing.
4. *Liverpool . . . preached my sermon*: see letter 3235, n. 11.
5. *Windermere*: the largest lake in England, located in the Lake District in the county of Cumbria.
6. *Ambleside*: a town in the Lake District in the county of Cumbria.
7. *Rydal*: the home of William Wordsworth (see letter 3263, n. 5) from 1813 until his death in 1850.
8. *Grassmere*: a lake and village in the Lake District in the county of Cumbria.
9. *Loughrigg*: Loughrigg Fell is a hill located above Grasmere, in the Lake District in the county of Cumbria.
10. *Owen's College*: Owens College in central Manchester was founded in 1851 from a bequeathment left by the industrialist John Owens (1790–1846) (*ODNB*). From the 1860s it followed the German model of a research-based university. It later became the Victoria University of Manchester and is now the University of Manchester.
11. *Huxley*: Thomas Huxley.
12. *appetite for hero-worship satisfied*: Thomas Huxley was president of the BAAS that year and gave an address on 14 September 1870 (see letter 3237, n. 11).
13. *This horrible war*: the Franco-Prussian war (see letter 3238, n. 16).

To Mary Egerton 26 September 1870 3263

Shelbourne Hotel | Dublin. | Monday morn[in]g. | 26th. Sep[tember]. 1870.

My dear Lady Mary

Ten minutes remain to me this morning which I fill by scribbling to you. I held back from L[iver]'pool until the Friday; then fired my shot,[1] and quitted the consequent racket on Saturday. I forsook the Lecturers for Windermere.[2] On Sunday my friend Herbert Spencer and myself had a delightful walk to Ambleside,[3] thence to Rydal Mount,[4] whose present proprietor showed us every nook and corner of Wordsworth's[5] nests. Thence by a beautiful mountain path to Grassmere[6] where we had luncheon. Thence by boat to the base of Loughrigg.[7] We climbed the hill and lost ourselves amid the rocks and bracken on its top. But my known skill in difficulty came into play and we managed to reach Ambleside before dusk.

Spencer quitted me on Monday, and I spent the day in a scamper to Coniston[8] and back: on Tuesday I returned to L[iver]'pool

On Thursday I went to Manchester: stayed with Whitworth,[9] who was kindness itself—and was induced to wait until the ceremony of laying the first stone of Owen's College[10] was completed

I then came over here on the business of the Board of Trade.[11] In half an hour I start for Kingstown[12] where a steam Yacht is waiting for me to take me to some light-houses in the Irish coast—to night I make experiments.

I shall soon be in London where I long to settle down closely and strictly to my work. I have had already too much straggling and dissipation.

Goodbye. & Kindest regards to all. | Yours ever | John Tyndall

RI MS JT/1/TYP/1/390
LT Typescript Only

---

1. *fired my shot*: Tyndall gave a discourse 'On the Scientific Use of the Imagination' at the Liverpool BAAS meeting on 16 September 1870. (see letter 3235, n. 11).
2. *Windermere*: see letter 3262, n. 5.
3. *Ambleside*: see letter 3262, n. 6.
4. *Rydal mount*: see letter 3262, n. 7.
5. *Wordsworth*: William Wordsworth (1770–1850), an English Romantic poet. By the time of his death, he was well recognised as a canonical figure in English poetry (*ODNB*).
6. *Grassmere*: see letter 3262, n. 8.
7. *Loughrigg*: see letter 3262, n. 9.
8. *Coniston*: a lake and village in the Lake District in the county of Cumbria.
9. *Whitworth*: Joseph Whitworth.
10. *Owen's College*: see letter 3262, n. 10.

11. *Board of Trade*: see letter 2937, n. 6. Tyndall traveled to Ireland at the request of the Board of Trade for a meeting after the BAAS meeting in Liverpool.
12. *Kingstown*: see letter 3124, n. 6.

To Henry Bence Jones 27 September [1870][1] 3264

Shelbourne Hotel, Dublin. | Tuesday, 27th. Sep[tember].

My dear Friend

Hearing that I was at Liverpool[2] the Board of Irish Lights[3] pounced down upon me. They wrote to the Board of Trade, and the Board of Trade commanded me to put myself in connexion with them and report on what they had to show me. Hence my presence in this land of Fenians.[4]

I stayed in London till the Friday. Then went down to Liverpool, fired my shot on Friday night, and in order to avoid the consequent racket, escaped on Saturday to Windermere.[5] Herbert Spencer accompanied me. On the Sunday we had a beautiful walk by Ambleside and Rydal to Grassmere.[6] Returned over Longhrigg,[7] where we went astray among the roots and bracken of the mountains. On Monday Spencer left me. He suffers far more than I do from sleeplessness; and though he seems to know every thing about his own condition and its requirements, I doubt whether he manages himself wisely. I went on the Monday to Coniston water.[8] The country round the head of the lake is indescribably lovely. In fact I had forgotten that the lake district was so beautiful as it really is.

I am here inspecting the application of gas to revolving lights. We had a lovely day yesterday, steamed first from Kingstown[9] to Howth Baily,[10] thence to an isolated mass of rock. "Rockabill"[11] on which stands one of the noblest lighthouses that I have seen. We had about 5 hours actual steaming in a yacht which was able to perform 15½ miles an hour. In the evening we were observing Howth Baily from Kingstown. I got home by 11 o'C[lock].

I am aiming to get away tomorrow, but I do not know whether this will be possible. I went to Manchester and stayed with Sir Joseph Whitworth. They laid the first stone of Owens College[12] and had a great luncheon afterwards. I hope the praises lavished upon Huxley[13] may not render his fleshly pride too high. But he deserved success for he managed matters admirably.[14]

I am longing to get to my work for I hate row and dissipation.

I hope you are well, and that all around you are well; and that you may return to London tough enough to defy the coming season.

Yours ever | John Tyndall

RI MS JT/1/TYP/3/796
LT Typescript Only

1. *[1870]*: Tyndall travelled to Ireland at the request of the Board of Trade for a meeting after the BAAS meeting in Liverpool. Letter 3263 for 26 September has him staying at the Shelbourne Hotel, Dublin. For Tyndall's work on lighthouses more generally, see letter 2937, n. 6.
2. *Liverpool*: the annual BAAS meeting took place in Liverpool (see letter 3235, n. 11).
3. *Board of Irish Lights*: the Commissioners of Irish Lights (see letter 3019, n. 7).
4. *land of Fenians*: Fenians was the generic name for separatist republican movements based in Ireland and the United States. The Irish branch, the Irish Republican Brotherhood, was known publicly by the name of its U.S. counterpart the Fenians. It had launched an unsuccessful rebellion in 1867.
5. *Windermere*: see letter 3262 n. 5.
6. *Ambleside and Rydal to Grassmere*: see letter 3262, n. 6 and n. 8.
7. *Longbrigg*: see letter 3262, n. 9.
8. *Coniston water*: Coniston Water is the third largest lake in the English Lake District.
9. *Kingstown*: see letter 3124, n. 6.
10. *Howth Baily*: see letter 3019, n. 3.
11. "*Rockabill*": a group off two islands located in the Irish sea of the coast of Skerries, County Skerries (see letter 3124, n. 7).
12. *Owens College*: Owens College, Manchester, was founded in 1851. In 1870 due to the expansion of the college, work started on new premises on Oxford Road. See letter 3262, n. 10.
13. *Huxley*: Thomas Huxley.
14. *managed matters admirably*: Huxley was president of the BAAS that year and gave an address at the annual meeting in Liverpool on 14 September 1870 (see letter 3237 n. 11).

From Louisa Baring 27 September 1870 3265

Portree—I[sle]. of Skye | Sept[embe]r 27th 70.

How very dear and kind it was of you to remember me—that address[1] which I long to read, arrived here today—& please accept my grateful thanks—As I got no word from Switzerland[2]—I thought you had forgotten all about me—so this was doubly precious—

Mary[3] and I are on our way to Lewis[4]—Wind & weather permitting, & then after a few weeks at Loch Liuchart,[5] I hope we shall be in the South—How many awful things are happening all over the world! & it seems scarcely credible that the great drama being enacted before Paris,[6] can be a reality—I hope you have been kept safe & well all this time—& are returned full of work. I hope for the winter—Mary never forgets you—nor do I—need I say so?—

Ever Yours | L[ad]y. Ashburton

Elton Hall, Ashburton Collection 13
RI MS JT/1/TYP/1/133

1. *that address*: presumably Tyndall's address to the BAAS at Liverpool (see letter 3235, n. 11).
2. *from Switzerland*: Tyndall spent the summer of 1870 in Switzerland.
3. *Mary*: Mary Baring.
4. *Lewis*: the Isle of Lewis, or Lewis, is the main island of the Outer Hebrides, Scotland.
5. *Loch Liuchart*: Loch Luichart is a loch (lake) located in the north of Scotland. There is a village, Lochluichart, located on its shore.
6. *the great drama being enacted before Paris*: Siege of Paris (19 September 1879–29 January 1870) during the Franco-Prussian War (see letter 3238, n. 16).

To Elizabeth Dawson Steuart [28 September 1870][1] 3266

[Dublin][2] | Wednesday.

My dear Mrs Steuart,

I had fully hoped to get away from Dublin this morning; but my investigation[3] has turned out more complicated than I expected. To-day I must remain here, and tonight spend some hours upon the water observing the Howth Baily Light.[4] I shall run down to Gorey[5] tomorrow; and shall make an effort to reach Leighlin;[6] but this may not be in my power as I have to meet the Light Commissioners[7] here on Friday and I must be in London on Saturday.

Lest I should not be able to get as far as Leighlin and should any letters have been addressed to you for me would you kindly redirect them to me on Thursday at the

Imperial Hotel | Sackville Street | Dublin

that I may have them when I return here on Friday?

Believe me dear Mrs Steuart | Yours ever faithfully | John Tyndall

RI MS JT/1/TYP/10/3398

1. *[28 September 1870]*: Tyndall travelled to Dublin on Monday, 26 September 1870 (letter 3263), wrote to Henry Bence Jones on Tuesday, 27 September (letter 3264), and wrote this letter a day later, on Wednesday, 28 September.
2. *Dublin*: added to the typescript in pencil.
3. *my investigation*: Tyndall was in Ireland to carry out investigations for the Board of Irish Lights. See letter 3264.

4. *Howth Baily Light*: see letter 3019, n. 3.
5. *Gorey*: see letters 2987, n. 7 and 2996, n. 29.
6. *Leighlin*: Leighlinbridge is a village on the River Barrow in County Carlow, Ireland. Tyndall was born and spent his childhood there.
7. *Light Commissioners*: members of the Board of Irish Lights.

From Thomas Carlyle 28 September [1870][1] 3267

5 Cheyne Row | Chelsea | 28 Sep[ember]. 18*[70]*

Dear Tyndall,

Your letter,[2] the solitary snipe, not without its bill (Der einzige Schnepf hat doch einen Schnabel)[3] was very welcome to me, especially when bringing, withal, the prospect of soon seeing you in person again. I heard repeatedly of your new discourse[4] at the Scientific Association;[5] but did not anywhere see it, nor a readable report of it: furthermore the copy of it did not accompany your letter[6] yesterday, nor has come today; so that I am still left in the dark, though not without my curiosities upon the matter.[7] If you have by you another copy, please dispatch it at once

I warmly agree with all you say about this War.[8] It is such an instance of Michael versus Satan[9] as I have never read in all History before. Satan getting himself smashed to impalpable pulp[10] in one month's time. Bismarck,[11] I suppose, will insist upon his Alsace and Lorraine;[12] nor can all the world hinder him if that is his determination. To poor France it is the frightfullest lesson she ever had; and unless she can learn by it, which is doubtful, may be the beginning of incalculable downbreak quod deus avertat.[13]

Return winged; and let us see you light at Chelsea at 9 p.m. the first night you have to yourself

Believe me | Yours always | T. Carlyle

RI MS JT/1/C/15

---

1. *[1870]*: although the year on this letter is illegible, Carlyle referred to Tyndall's recent BAAS discourse.
2. *your letter*: letter missing.
3. *Der einzige Schnepf hat doch einen Schnabel*: slight misquote of 'Der einzige Schnepf hat noch einen Schnabel' (German), which translates to: 'The lone snipe still has a beak . . .'. During the Reformation, a Nuremberg statesman described a contemporary with these words. The sentence continues '. . . to sing in a Christian and constant way'.
4. *your new discourse*: see letter 3235, n. 11.
5. *Scientific Association*: the BAAS.

6. *your letter*: letter missing.
7. *though not without my curiosities upon the matter*: Tyndall did send Carlyle a copy and could report later in letter 3287, 'The discourse has passed the censorship of "the General"'.
8. *this War*: the Franco-Prussian War (see letter 3238, n. 16).
9. *Michael versus Satan*: the book of revelation gives an account of the defeat of Satan by the Archangel Michael (Revelation, 12:1–9).
10. *smashed to impalpable pulp*: the war was going badly for France.
11. *Bismark*: Otto von Bismarck (1815–98), a German statesman. He was Minister President of Prussia and Chancellor of the North German Confederation. Bismarck used the Franco-Prussian War to align the south German states with the Prussian dominated confederation and was a crucial figure in the unification of the independent German sovereign states into a single national state (*NDB*).
12. *Alsace and Lorraine*: these were regions of eastern France until they were annexed by the newly created German Empire in 1871 following the Franco-Prussian War.
13. *quod deus avertat*: God forbid (which God may avert) (Latin).

From Henry Bence Jones 28 September 1870 3268

Folkestone | Sep[tember]. 28. 1870

My dear friend

I can only get hold of a scrappy paper & I have been in such a bustle getting my things ended to leave London that I could not write to you before. I hope I have done all your commissions to your satisfaction and that the papers of all kinds will reach you—not that there is anything in them. Both French and Prussians mean that we should know nothing but what they please so that lots of lying will be added to their thievish rascalities. It would firmly establish the truth of the Christian religion if god would work one more miracle and hang Louis Napoleon and Bismark.[1] But as this is not likely it is to be hoped that both peoples will improve each the other off the face of the earth & that all the fighting will be taken out of them for a generation or two: How false all the ideas of "advance of civilization" are shewn to be. Your letter[2] recalled to me about your book.[3] I brought it home and put it away most carefully & then forgot that I had ever had it.[4] I immediately found it and I did thank you for it at the time and thank you again now & yet again for saying you would give me another copy. I ought not to have forgotten that you gave me the first. It looks as if I had not valued it at all now I do value it very much. I left the Institution[5] yesterday in a very quiescent state. There is nothing yet done in the theatre. But the store rooms are finished & I have made your room very much nicer with the book case which will hold all your papers &c and keep your room much better in order. If you dont like it you can turn it out. I gave Brown[6] a rowing about the urinals and I told him if there were any more complaints from the visitors that the managers must get another servant in his place. And so they must. He is not a careful man. I wish I had something else to tell you. I have got a promise of a Friday evening from Dean Stanley.[7] I should like one of the infallibility of the Pope from the Archbishop of Westminster. Ruskin I saw today just landed after a passage in a gale too rough for the yacht to go out. Some of the sailors in the Battle will wish the soldiers on shore I expect. It blew all night most gustily & we shall hear of some harm done. The Emperor[8] left Paris at 10 this morning Thursday so by Sunday they ought to come to blows poor devils. I hope tomorrow we shall go to Bolougne.[9]

The lay sermons[10] were out I think when I last wrote[11] and the letter making you his audience must be very pleasant to you. I think I ordered the athenaeum[12] with a review of the sermons[13] to go to you. But they do not allude to the introduction.[14]

Sir H. Holland shewed me a letter from Herschell[15]—a review of a paper you have seen of Sir H's. It is to my weak mind a marvel to hear him (Herschell) and Wheatstone say that facts are best accounted for by assuming the Ether to be a solid. Herschell one day told this to Lyell[16] and L. was struck dumb in his geological mind. How little we know of the present or past even. You would think that the history of our Institution 70 years ago would be very well known; but alas the more I look the darker it gets as to the causes that were shaping it and the tempers & stomachs of the dead can not easily be got out of minutes of meetings or even private letters.

I have found Guizot's[17] life of Madame de Rumford.[18] His account of their separation is that she would be called Lavoisier Rumford & he would call her Madame de Rumford only!!! That is Rumford, he thought Rumford[19] was greater than Lavoisier.[20] He seems to have hated all Frenchmen except Cuvier[21] & showed his hatred also. Cuvier's eloge[22] strikes me as cleverer each time I read it. I should like simply to translate it but then it would not shew how Rumford's mind became what it did & so led him to act as he did regarding the Institution[23] & many other things. I do not like the new American life of Rumford[24] at all but there are many new things in it altho not of the great interest from the Institution point of view which is what I chiefly look at. I have found the three first prospectuses (volumes like your last[25]) of Garnetts,[26] Youngs[27] & Davys lectures Garnett 212 pages 1801[28] Young 190 pages 1802[29] and Davy 92 pages 1802.[30] The two last from the Press of the Royal Institution. Davys last four lectures as reported by Faraday will also make a first rate ending to my volume.[31]

Goodbye dear friend I hope you will flourish & look at us on your return home

Believe me yours | very sincerely | H. Bence Jones

RI MS JT/1/J/65
RI MS JT/1/T/TYP/2/701–3

1. *Bismark*: Otto von Bismarck (see letter 3267, n. 11).
2. *Your letter*: letter missing.
3. *your book*: most likely *Diamagnetism* (see letter 3076, n. 8).
4. *then forgot that I had ever had it*: see letter 3236, n. 15.
5. *the Institution*: the RI.
6. *Brown*: William Brown, a porter at the RI. Brown had already been warned about the poor condition of the urinals and his possible replacement in July 1870. See letter 3236.
7. *Friday evening from Dean Stanley*: Arthur Penrhyn Stanley (1815–81), an Anglican clergyman, was Dean of Westminster from 1864–81 and a supporter of religious toleration. Having served as a tutor at the University of Oxford, in July 1851 he was offered a canonry at Canterbury. In December 1856 he was appointed regius professor of ecclesiastical history at Oxford and in March 1858 installed as a canon of Christ Church. He wrote many books on the church. Following his marriage to (Lady) Augusta Bruce (1822–76) in 1863 he was a confidant of Queen Victoria on matters of church and state. Stanley was installed as Dean of Westminster on 9 January 1864 (*ODNB*). Stanley did not give a Friday Evening Discourse at the RI during its 1871 season, but he he did deliver one on 26 January 1872, 'On the Daemon of Socrates' (*Roy. Inst. Proc.*, 6 (1871–72), pp. 404–16).
8. *The Emperor*: Louis Napoléon.
9. *Bolougne*: Boulogne-sur-Mer, often referred to as Boulogne, is a coastal city in northern France.
10. *lay sermons*: from Thomas Huxley's *Lay Sermons* (see letter 3217encl).
11. *when I last wrote*: Bence Jones wrote to Tyndall on 17 August (letter 3245), but it is unknown whether there might be additional, missing letters.
12. *the athenaeum*: see letter 3249, n. 7.
13. *review of the sermons*: this was a review of Huxley's *Lay Sermons* (*The Athenaeum*, 2230 (23 July 1870), pp. 107–9).
14. *the introduction*: see letter 3217encl.
15. *Herschell*: John Herschel.
16. *Lyell*: Charles Lyell.
17. *Guizot's*: François Guizot (1787–1874), a French historian and leader of the constitutional monarchists during the July Monarchy.
18. *Madame de Rumford*: F. Guizot, *Madame de Rumford, 1758–1836* (Paris: De l'imprimerie de Crapelet, 1841). After the execution of her husband, Antoine Lavoisier, Marie Ann Paulze married the physicist and purported RI founder, Benjamin Thompson, Count Rumford (1753–1814) in 1805, but they separated within a few years.
19. *Rumford*: Benjamin Thompson (1753–1814), Count Rumford, was an American-born British physicist and inventor whose challenges to established physical theory became part of the nineteenth-century development of thermodynamics. He was one of the founders of the RI, and the RS's Rumford Medal was named after him (*CDSB*; *ODNB*).

20. *Lavoisier*: Antoine Lavoisier (1743–1794), a French chemist, was a significant figure in the chemical revolution of the eighteenth century. He helped to develop the modern system of naming chemical substances (*CDSB*).
21. *Cuvier*: Georges Cuvier (1769–1832), a French zoologist. Cuvier established the sciences of comparative anatomy and palaeontology. Despite believing that the earth had existed for a relatively short time and rejecting evolution, his work demonstrated a remarkable array of extinct species strengthening the theory of catastrophism (that the earth had experience a series of catastrophes which had destroyed entire species) (*CDSB*).
22. *Cuvier's eloge*: G. Cuvier, 'Éloges historiques de A.A. parmentier et du Comte De Rumford, 9 Janvier 1815', in *Recueil des Éloges Historiques Lus Dans Les Séances Publiques De L'institut Royal de France* (Paris, 1819), pp. 159–231.
23. *act as he did regarding the Institution*: Rumford was a founding member of the RI in 1799 and lived at the Institution's premises in Albemarle Street for two years in a supervisory capacity. However, he fell out with some of the other managers and returned to the United States.
24. *American life of Rumford*: see letter 3245, n. 20.
25. *volumes like your last*: probably either Tyndall's *Notes of a Course of Nine Lectures on Light* (see letter 3197, n. 6) or *Notes on a Course of Seven Lectures on Electrical Phenomena* (see letter 3196, n. 2).
26. *Garnetts*: Thomas Garnett (see letter 3245, n. 26).
27. *Youngs*: Thomas Young (see letter 3245, n. 24).
28. *Garnett 212 pages 1801*: T. Garnett, *Outline of a Course of Lectures on Chemistry* (London: T. Cadell, Jun. and W. Davies, Stand, 1801).
29. *Young 190 pages 1802*: see letter 3245, n. 24.
30. *Davy 92 pages 1802*: H. Davy, *A Syllabus of a Course of Lectures on Chemistry* (London: The Press of the Royal Institution, 1802).
31. *my volume*: see letter 3245, n. 19.

From George Gabriel Stokes 30 September 1870 3269

Observatory, | Armagh. | 30th Sep[tember]. 1870.

My dear Tyndall,

In looking over your paper[1] for press I have a few remarks.[2]

1. p. 335, 1st foot note. My eye seems to recollect phosgen[3] analogous to hydrogen oxygen nitrogen, but in Rees's encyclopaedia[4] it is spelt phosgene whichever be more correct this is a matter of very small moment.

2. p. 338. In the column headed

"Quantity of air" and beginning 15 inches, the reader might not unnaturally suppose that 15 cubic inches was what was meant. It might be well to insert in the text, "the quantity of air being measured by the height of the numerical column which measures its pressure" or something to that effect;

or else in this, the first, table, for "inches" might be put "in. of mercury"[5] for which there would be room without deranging the type. In the subsequent tables the reade[r] would at once understand "inches" to mean inches of mercury.

3. p. 341. As you have written phenomen<u>a</u>, and accordingly in the next line put <u>their</u> I have altered has to have.[6]

p. 343 the pencil mark was merely put down because a question occurred to me as to the use of terms. The sense in which the terms molecular and atomic are used is however virtually explained in the paper itself, so that the meaning is clear.[7]

4. p. 349. I take for granted that the tube is supposed to be viewed normally, or what may be deemed normally—that the difference of character B or W depends upon a difference in the state of the cloud, and not upon a difference in the obliquity of viewing. If so, st

5. p353. The expression "Colour of the Crystal" led me at first to ask myself was it the relative absorption exercised by the crystal that was dealt with? I saw afterwards what was meant, and I believe the expression to be scientifically correct, though the popular use of the expression of the colour of an object might possibly cause misapprehension.

q might it be well to put "The field of the Crystal was yellow green?"[8]

6. p.353. The use of the term reagent strikes me as altogether erroneous here. The incident beam of light which produced the effect is the reagent. The eye is the instrument of detection, or we may say the detective; but surely it is an utter abuse of terms to call it a reagent, inasmuch as it does not act on the vapours at all. The fact that chemical reagents are constantly employed by chemists, not for the sake of producing certain products with which to work further, but merely with a view of detecting a substance there already, does not render the word "reagent" equivalent to "means of detection." Would a chemist speak of spectral analysis as a reagent?[9]

Not impossible, for chemists of all men seem prone to make mighty free with the Queen's English, but it strikes me as a most improper use of the term if they do. In the present case surely in all scientific precision it <u>is</u> the actinic rays and <u>is</u> <u>not</u> the eye that is the reagent; it is the rays that, being acted upon (absorbed), in their turn react upon (decompose) the vapour.

7. p. 355. As a matter of English I should much prefer "Between the two"[10]

8. p. 358. I should feel inclined to <u>prefer was very curious</u>.[11] One regards it as the sudden expansion and contraction (the line stands for an algebraic vinculum); one thinks of the phenomenon of sudden expansion and contraction, not of the individual expansions and contractions.

9. p. 348. 3. I think this <u>undoubtedly</u> ought to be altered to the plane of polarization of the h.h.light is parallel to[12] . . . A plain matter of fact ought not to be described in terms implying the adoption of a particular theory, in terms

indeed which are unintelligible if it be not known to which of the belligerent theories the writer attaches himself. An experimentalist simply describing a matter of fact ought to exhibit the calm impartiality of a neutral nation.

I return you the paper by book post; please send it on to Taylor's.[13]

Yours very truly | G. G. Stokes | J. Tyndall Esq.

RI MS JT/1/TYP/4/1411–12
LT Typescript Only

---

1. *looking over your paper*: 'On the Action of Rays of High Refrangibility' (see letter 3095, n. 8). Tyndall adopted most of the suggestions in this letter in the published version of this paper.
2. *remarks*: see letter 3271. For more correspondence on this article see letters 3280, 3284, and 3286.
3. *recollect phosgen*: while retaining the spelling phosgen, Tyndall did note that Stokes had reminded him that the name is derived from the formation of the gas under the influence of light.
4. *Rees's encyclopaedia*: *The New Cyclopaedia, or, Universal Dictionary of the Arts and Sciences,* commonly known as Rees's Cyclopedia, was commissioned by Thomas Longman and produced by Abraham Rees (1743–1825). The first half-volume was published in 1802 and the last of the 39 volumes was printed in 1820.
5. *in. of mercury*: Tyndall adopted this revision in the published version.
6. *altered has to have*: neither of these words remain in the published version.
7. *molecular and . . . so that the meaning is clear*: Tyndall added an extensive footnote at this point to explain what he means by molecular and atomic.
8. *"The field of the Crystal was yellow green?"*: this is the exact wording used in the published version.
9. *The use of the term . . . reagent?*: this term is omitted from the published version.
10. *"Between the two"*: this expression appears in the published version.
11. *was very curious*: this expression appears in the published version.
12. *the plane of polarization of the h.h.light is parallel to*: this appears as 'The plane of polarization of the perfectly polarized light is parallel to . . . ' in the published version.
13. *Taylor's*: Taylor and Francis, publisher of the *Phil. Trans.*

To Heinrich Debus 2 October 1870 3270

Be it known to thee O Heins that

In not coming to see me today you have omitted your plainest duty, and proved yourself unworthy of the friendship of

John Tyndall

seen and approved | Thomas A. Hirst | Sunday 2nd Oct[ober]. 1870

RI MS JT/1/T/266

To George Stokes 2 October 1870 3271

Royal Institution of Great Britain | 2nd Oct[ober]. 1870.

My dear Stokes,

If to every paper of the Philosophical Transactions you devote the same trouble that you have bestowed upon mine,[1] then by Jove your position as secretary is no sinecure.[2]

I have acted upon all, or nearly all,[3] your suggestions.

Had you not come down so heavily upon my poor little term reagent I should not have said a word in its defence; but common chivalry compels me to say a word for a term which you have hit so mercilessly. I contend then that the term is a legitimate one under the circumstances.

I am speaking of the eye as a body apart from consciousness. That body undergoes change through the impact of the ether waves. Supposing the eye to be in the hands of an angel with vision of keen enough to see the changes of the retina, and who employs the eye, in the ether, as you might employ a piece of ozone paper in air, such an eye would undoubtedly be a reagent. But for the angel the consciousness of man, and for the molecular changes of the retina their equivalent subjection effects, the eye is just as much a reagent as before. I say point blank therefore that it is not an abuse of terms to use the word reagent—that it is a perfectly good term, defensible philosophically. Still lest others sh[oul]d find it a stumbling block I have behaved like a Christian and omitted it.

Yours truly | John Tyndall

CUL SC—add 7656 RS761

1. *bestowed upon mine*: see letter 3269.
2. *sinecure*: Any office or position which has no work or duties attached to it, esp. one which yields some stipend or emolument (*OED*).
3. *nearly all*: Stokes had made a number of editorial suggestions on Tyndall's article 'On the Action of Rays of High Refrangibility'. See letters 3096, n. 8, 3269, 3280, 3284, and 3286.

From Thomas Henry Huxley 2 October 1870 3272

October 2nd. 1870.

My dear Tyndall,

I got back yesterday and found a note awaiting me from Dr. Inman[1] of Liverpool to say that poor Miller[2] had died the day before. Inman assured me that in all his ravings[3] there was no reference to things or persons scientific which is a comfort.

I write a hasty line to suggest that you being on the Council[4] must look sharp after the appointment of his successor.[5] No-one it seems to me would be a better, a fitter man than Lubbock[6] for the Treasurership.[7] And if you think so and will take care to bring his name forward at once no-one can possibly oppose him. I should much like to talk to you about this, but I am off to Leicester and shall not be back till Tuesday afternoon.

I have not yet thanked you properly for your great contribution to the success of our meeting.[8] I was nervous over the passage about the clergy, but those confounded parsons seem to me to let you say anything while they bully me for a word or a phrase. It's the old story "One man may steal a horse while the other may not look over the wall."[9]

We have been staying with Whitworth[10] in Derbyshire where you ought to have been. It was very quiet and refreshing and we have taken a great liking to the old man. He presents a great contrast to Armstrong[11] but much as I like the latter, Whitworth seems to me to be a man of more natural power and breadth of genius.

Ever yours | T. H. Huxley.

IC HP 8.83
Typed Transcript Only

---

1. *Inman*: Thomas Inman (1820–76), an English physician. He graduated King's College, London, with an MB (1842) and MD (1844). In 1844 Inman was appointed house surgeon at the Liverpool Royal Infirmary and was subsequently appointed physician and part-time lecturer. He published extensively on medicine, antiquities, and mythology (*ODNB*).
2. *Miller*: William Allen Miller (1817–70), an English Chemist. Following an apprenticeship as a surgeon he worked in the medical department of King's College, London. In 1840 he spent some months in Professor Liebig's laboratory in the University of Giessen. Upon his return to Britain he was appointed a demonstrator of chemistry at King's College, London, and subsequently held the chair of chemistry at that institution. As well as academic publications his textbook *Organic Chemistry* (1857) was extremely popular. Elected FRS 1845, he served as treasurer of the society from 1861 until his death in 1870. He died on 30 September 1870 (*ODNB*).
3. *all his ravings*: Miller suffered from delusions in the weeks before his death (*Ascent of John Tyndall*, p. 385).
4. *the Council*: the Council of the RS.
5. *appointment of his successor*: see letter 3274.

6. *Lubbock*: John Lubbock.
7. *Treasurership*: the Treasurership of the RS, which became vacant on the passing of William Allen Miller on 30 September 1870. John Lubbock and William Spottiswoode were both considered, and Tyndall leaned toward Spottiswoode for the post. And while John Peter Gassiot vied for the position, too, in the end Spottiswoode was elected on 27 October 1870. See letters 3274, 3276, 3292, 3293, 3294, 3299, 3300, 3301, 3306, and 3327. On the political plays regarding the treasurership of the RS, see *X Club*, pp. 269–71.
8. *our meeting*: the annual meeting of the BAAS, where Tyndall presented a discourse 'On the Scientific Use of the Imagination' (see letter 3235, n. 11).
9. *One man . . . wall*: Proverb (first recorded 1546).
10. *Whitworth*: Joseph Whitworth.
11. *Armstrong*: William Armstrong (1810–1900), a British armaments manufacturer and industrialist, who also worked on hydraulics and electrostatics. Armstrong performed extensive service for the BAAS as well as a number of engineering institutions. He was knighted in 1859 for giving the patent for his design for the Armstrong gun to the crown. Later, in 1887, he was elevated to the peerage as Baron Armstrong of Cragside (*ODNB*).

To Joseph Henry 3 October 1870 3273

3rd Oct[ober]. 1870.

My dear Professor Henry,

Mr. Sharswood[1] sent me the accompanying note[2] which reached me in the Alps and to which I replied. My note to him[3] was intended to be a private friendly answer to a private friendly question. To my surprise I found on my return to London that he has published the note[4] in the American newspapers.

Now this may be a cause of trouble, disappointment & complaint and I should very much regret giving my American friends any cause of complaint against me. Had Mr. Sharswood communicated with me privately I should have tried to define the time when I might be able to go to America but all I can say now is that I have no immediate intention of going there, and that when I do go it will be probably in the character of a private visitor and not in that of a lecturer. My dislike of public racket and excitement is so great that I shall in all probability avoid them.[5]

If you could in any way make known that I am not in a position to go to America at present, and that I shall not be in such a position for some years to come, you would do me a favour, and perhaps spare some disappointment to those who are kind enough to wish to see me amongst them.

I am dear Prof. Henry | yours most faithfully | John Tyndall.

John Tyndall letter to John Henry, 1870, October 3, RH 3963, William Rhees papers, The Huntington Library, San Marino, California

---

1. *M^r. Sharswood*: William Sharswood (1836–1905), an American chemist, naturalist, and playwright who graduated from the University of Pennsylvania in 1856 and received a Ph.D. in Chemistry from the University of Saxony in 1859, publishing the thesis as *Studia Physica* or *Researches on Physical Science* (Vienna, 1859).
2. *the accompanying note*: enclosure missing.
3. *My note to him*: letter missing.
4. *the note*: this note is mentioned in letter 3281 and *The Chicago Tribute,* 10 September 1870. The writer of letter 3281 describes reading of a note from Tyndall to Sharswood in a newspaper called *The Independent*, but neither note nor newspaper have been identified.
5. *when I do go . . . avoid them*: Tyndall went on an extensive tour of the United States from October 1872 through January 1873 (he departed England on 28 September 1872 and returned on 19 February 1873). At the height of his fame as a scientist and public intellectual, he visited and lectured in seven cities; in summary, the opposite of what he sets out in this letter. See letter 3175, n. 5.

To Thomas Henry Huxley 3 October 1870 3274

Royal Institution. | 3rd. Oct[ober]. 1870.

My dear Huxley,

Lubbock or Spottiswoode.[1] Hirst goes in decidedly for Spottiswoode. Lubbock is already our banker; he has affairs of state to attend to and Hirst doubts his ability to attend the Council Meetings, which is a matter of great importance.

The parsons[2] know very well that I mean kindness; if I correct them I do it in love and not in wrath.

I glanced last night over that second part of Bastian's[3] paper:[4] His impudence is certainly astounding.

Yours ever | John Tyndall

Professor Huxley.

RI MS JT/1/TYP/9/2938
IC HC 8.84
Typed Transcript Only

---

1. *matter of great importance*: the treasurership of the RS (see letter 3272, n. 7).
2. *The parsons*: see letter 3272.

3. *Bastian*: Henry Charlton Bastian.
4. *paper:* H. C. Bastian, 'Reply to Professor Huxley's Inaugural Address at Liverpool on the Question of the Origin of Life', *Nature*, 2 (1870), pp. 431–4. The first part of this reply is on pp. 410–3 of the same volume.

To the Editor of the *Record* 4 October [1870][1] 3275

To the Editor of the RECORD.

SIR—It is a grave error on your part to represent me as calling Mr. Darwin 'one of the weaker brethren.'[2] Were I asked to name the highest representative of the stronger ones, I should probably name him. But in your article[3] you link my name with that of a writer[4] whom I do rank among the weaker brethren; weak through a defect common to him and his antagonists—the incompetence, namely, to look round a great question and see its bearings on all sides.

I am, Sir, your obedient servant, | JOHN TYNDALL.
ATHENAEUM CLUB,[5] October 4.

J. Tyndall, *Essays on the Uses and Limits of the Imagination in Science* (London: Longmans, Green & Co., 1870), p. 8

1. *[1870]*: this letter has been dated to 1870, since it responds to an editorial printed in the *Record* on 23 September 1870.
2. *one of the weaker brethren*: quoted from the editorial in the *Record,* 23 September 1870, which is reprinted in J. Tyndall, *Essays on the Uses and Limits of the Imagination in Science* (London: Longmans, Green & Co., 1870), pp. 5–6.
3. *your article*: see n. 2.
4. *a writer*: Ludwig Büchner (1824–99), a German physician, philosopher, and science historian. Considered the most prominent nineteenth-century German scientific materialist, Büchner promoted the works of Darwin, which aligned with his empiricist and materialist views. His first and most famous work, *Kraft und Stoff* (1855) presented his view of the world based on the interaction of matter and force and without divine intervention. It sold out in the first few weeks of publication and was published in multiple editions (*NDB*). See F. Gregory, *Scientific Materialism in Nineteenth Century Germany* (Boston: D. Reidel Publishing Company, 1977), pp. 100–21; and A. Kelly, *Descent of Darwin: The Popularization of Darwinism in Germany, 1860–1914* (Chapel Hill, NC: University of North Carolina Press, 1981), pp. 10–35.
5. *ATHENAEUM CLUB*: see letter 2944, n. 1.

From Thomas Henry Huxley — 4 October 1870 — 3276

25 Abbey Place, | October 4th. 1870.

My dear Tyndall,

I am just back from Leicester[1] where I was "saying a few words" last night.

No doubt we could not go wrong with either Spottiswoode or Lubbock in the Treasurership[2] but I am greatly disposed to think that under many contingencies Lubbock would be the more useful man and carry greater weight. Again you must remember that his father[3] was Treasurer and that he is likely to be ready to make some sacrifices to take the office if it comes in his way.

I wrote to him to ask him to turn the matter over in his mind, so that we might consider it at the 'x'[4] on Thursday, and to-night my wife has had a note from Lady Lubbock[5] which implies that her husband would greatly like the post.

However in writing to Lubbock I spoke entirely on my own hook, so as to commit no-one but myself, and we shall be none the worse for coming prepared to discuss our course of action.

My wife[6] sends her love (<u>second-best</u>, she adds carefully in parenthesis)

Ever yours faithfully, | T. H. Huxley.

IC HP 8.85
Typed Transcript Only

---

1. *Leicester*: a city in Leicestershire, in the English East Midlands. Huxley lectured for the Leicester Literary and Philosophical Society (founded in 1835) on 'What is to be Learned from a Piece of Coal' on 3 October 1870 (L. Huxley (ed), *Life and Letters of Thomas Henry Huxley*, 2 vols (New York: Appleton, 1901), vol. 1, p. 362).
2. *Treasurership*: the treasurership of the RS (see letter 3272, n. 7).
3. *his father*: John William Lubbock (1803–65), an English astronomer and banker. In 1833 he married Harriet Hotham (1810–73), and they had eleven children, including the eldest son, John. The elder Lubbock served as treasurer and vice-president of the RS from 1830 to 1835, and from 1838 to 1834 (*ODNB*).
4. *the 'x'*: the X Club (see letter 3044, n. 5). Tyndall attended the meeting.
5. *Lady Lubbock*: Ellen Francis (née Hordern, 1834–79), the daughter of an Anglican minister. She married John Lubbock in 1856 and they had three sons and three daughters (R. Birkbeck, *Notes on the History and Genealogy of the Family of Lubbock*, pp. 35–6).
6. *My wife*: Henrietta Huxley.

To William Crookes 5 October 1870 3277

THE ATHENAEUM | 5th Oct[ober]. 1870

My dear Mr Crookes,

Somebody has written in a very kind manner about my book on Diamagnetism[1] in the last number of the Quarterly Journal of Science.[2] Pray express my acknowledgements to the writer. It is a comfort that the effort to make things clear, which in my case is a very laborious one, is at all events appreciated by some good men.

Yours faithfully | John Tyndall

Wellcome Library MS. 7777/13

1. *my book on Diamagnetism*: *Diamagnetism* (see letter 3076, n. 8).
2. *Quarterly Journal of Science*: 'Notices of Scientific Works', *The Quarterly Journal of Science*, 7 (1870), pp. 501–4.

To Henry Bence Jones 5 October 1870 3278

Royal Institution | 5th, Oct[ober]. 1870.

My dear Friend

I have been all over the works in the theatre.[1] Mr Parkinson[2] assures me that he will have all ready by the 7th, of November. I will push him still further forward to day. The arrangements for the ventilation of the gallery promise to be very successful. No wonder that it was stuffy, seeing that there was absolutely no escape from that stagnant receptacle under the roof.

I had a long talk with Mr Berkeley[3] yesterday regarding ornamentation. He and I are going to visit St Bartholomew's[4] lecture room together, and he is to try and get Owen Jones,[5] who has been already very kind, to accompany us. The gas burners above have been changed, so that we shall have as good a light with a far less expenditure of gas.

With regards to discovery, that belongs not "to him that willeth nor to him that runneth but to God that giveth mercy."[6] You know your bible of course. Your questions <u>never</u> irritate me. Supposing you were to say to me in that hearty good humoured voice which you can employ sometimes "Tyndall dear friend I want a thousand pounds, can you let me have them? My reply would be "By all means my dear friend; with the greatest possible pleasure." But suppose you were to call upon me in your authoritative way "Stand and deliver;" I fear I should give you a rather irritating answer. But never mind, as long as there is no under dirt at the bottom of a man's character, the surface water though troubled for a time soon becomes clear again.

Reed,[7] the exconstructor of the Navy will give you a Friday evening.[8] I have written to Douglass of the Trinity House[9] who has lately superintended the erection of the finest revolving light in the world on the Wolf Rock[10] near the Lands End. I think it would be an admirable subject.[11]

The very best way to make me humble is to praise me: The Times article[12] (I should not for a moment ascribe it to Williamson)[13] will be useful in this way.

Come home strong and gladden | Yours ever | John Tyndall

I see you are near that rosebloom maiden.[14] | Kind regards to Dr Henry.[15]

RI MS JT/1/TYP/3/801
LT Typescript Only

---

1. *the theatre*: lecture theatre of the RI. See letter 3236.
2. *Mr. Parkinson*: not identified.
3. *Mr Berkeley*: George Berkley.
4. *St Bartholomew's*: St Bartholomew's Hospital, known as Barts, is located in the Smithfield area of London, England.
5. *Owen Jones*: Owen Jones (1809–74), an English-born Welsh architect. One of the most influential designers of the nineteenth century, Jones had a significant impact not only on design theory and education but also on consumer goods. He was a superintendent of works for the Great Exhibition and was appointed direct of decorations once the building was moved to Sydenham, Kent, in 1854. Jones also acted as a decorator throughout his career and this was a prominent role near the end of his life (*ODNB*).
6. *To him . . . mercy*: Romans 9:16.
7. *Reed*: Edward James Reed (1830–1906), a British naval architect. In 1860 he was appointed secretary of the newly formed Institution of Naval Architects and editor of its *Transactions*. Appointed chief constructor of the Navy in 1863, he helped develop thicker armored ships. In 1870 he resigned over disagreements about ship design (*ODNB*).
8. *a Friday evening*: Reed delivered a Friday Evening Discourse at the RI on 10 February 1871, 'On Some Fallacies connected with Ships and Guns,' *Roy. Inst. Proc.*, 6 (1872), pp. 211–4.
9. *Trinity House*: see letter 2937, n. 6.
10. *Wolf Rock*: Wolf Rock lighthouse. See letter 2969, n. 1.
11. *I think it would be an admirable subject*: Douglass delivered a Friday Evening Discourse at the RI on 17 February 1871, 'On the Wolf Rock Lighthouse,' *Roy. Inst. Proc.*, 6 (1872), pp. 214–27.
12. *The Times article*: 'The British Association and Professor Tyndall', *Times*, 3 October 1870, p. 6.
13. *Williamson*: Alexander William Williamson (see letter 3008, n. 7).

14. *rosebloom maiden*: an allusion to T. Carlyle, *Sartor Resartus* (London: Saunders and Otley, 1838), p. 38. The person Tyndall is referring to here has not been identified.
15. *Dr Henry*: probably Joseph Henry, who was on a four-and-a-half-month European tour on behalf of the Smithsonian Institution in the United States. He sailed from New York on 1 June 1870, and during his tour he visited England, Ireland, Scotland, Belgium, as well as parts of Germany and France (*Annual Report of the Board of Regents of the Smithsonian Institution, showing the Operations, Expenditures, and Condition of the Institution for the Year 1870* (Washington: Government Printing Office, 1871), pp. 45, 85, 141; *A Memorial of Joseph Henry* (Washington: Government Printing Office, 1880), pp. 307–8).

From Juliet Pollock 7 October [1870][1] 3279

Sunnyside | Freshwater | Oct[ober.] 7th

My dear rude kind tender cruel health giving Boreas Eolia[2] greets you well and thanks you for the last breeze stirring, the scientific imagination.[3] It is lent for the present to Mrs. Cameron[4] & Henry Taylor.[5] Freshwater[6] is charming as ever, itself, and its surroundings are all pleasant, our cottage fitted up, built & created by Mrs. Cameron, indicating her richly coloured genius at every step, in every nook, in every cranny. Henry Taylor has outwardly recovered from his sorrow;[7] inwardly there is the loss of a fond hope.

W. Clifford[8] arrived here yesterday & sends you word that you were expected both at the Spottiswoode's and at the Huths[9] with all of whom he has been staying & that to neither did you come; in fact that he has been all over the world trying to meet you in vain.

In half an hour we are all going to air ourselves on the East Down wh[ich]. you know is very pretty & which I once marched over with you in silence because I felt I should cry if I tried to speak; so battered and tormented I was by the wind.

That was one of the blasts of Boreas

Good bye Love from all | your ever friendly | Eolian

RI MS JT/1/E/54

1. *[1870]*: this letter was dated from the allusion to Tyndall's 1870 BAAS lecture on science and the imagination (see n. 3) and the death of Henry Taylor's son earlier in 1870 (see n. 7).
2. *Boreas Eolia*: Tyndall and Juliet Pollock used these nicknames for one another. Tyndall was Boreas; Pollock was Eolia.
3. *last breeze stirring, the scientific imagination*: Tyndall's address at the BAAS meeting in Liverpool, 'The Scientific Use of the Imagination" (see letter 3235, n. 11). By 'last breeze', Pollock was referring to Tyndall's lecture on 'The Scientific Limit of the Imagination',

delivered on 20 Aug 1868 at the BAAS meeting in Norwich. Both were published in: J. Tyndall, *Use and Limit of the Imagination in Science* (London: Longmans, Green, and Co., 1870).

4. *M^rs^. Cameron*: Julia Margaret Cameron (1815–79), a British portrait and genre photographer, was born in Calcutta, where she married Charles Cameron. She moved to England in 1845 and she became involved in artistic circles in London. In 1860 she and her family moved to Freshwater, Isle of Wight, and in 1863 she took up photography. She built a reputation for her portraits of famous male sitters, including a number of scientific and literary figures including Charles Darwin and Alfred Tennyson (*ODNB*).
5. *Henry Taylor*: see letter 3222, n. 2.
6. *Freshwater*: Freshwater Bay is a village on the Isle of Wight, where the Tennysons had a home, Farringford House, since 1853. From 1869, the Tennyson's main residence was Aldworth House near Haslemere. Sunnyside was a cottage owned by Julia Margaret Cameron adjacent to the Tennyson estate and as this letter indicates, used to host guests. She purchased it in October 1860 (C. Ford, *Julia Margaret Cameron. A Critical Biography* (Los Angeles: John Paul Getty Museum, 2003), p. 30).
7. *his sorrow*: Taylor's eldest child died on 16 May 1870, his son Aubrey Ashworth Taylor (1846–70). See also letter 3222, n. 3.
8. *W. Clifford*: William Kingdon Clifford.
9. *the Huths*: Henry Huth (1815–78), a noted English book collector (*ODNB*), And his wife Augusta Huth (née von Westenholz). She was elected a member of the RI on 2 March 1868 (*Roy. Inst. Proc.*, 5 (1866–9), p. 308).

From George Gabriel Stokes 8 October 1870 3280

Observatory Armagh | 8th Oct[obe]^r^ 1870

My dear Tyndall,

I just write a line to say that the question between us is merely one as to the use of words, and could only be decided by reference to a dictionary which should have for modern scientific terms the authority which Johnson's dictionary[1] for example possesses for ordinary English. My objection[2] to the use of the word "reagent" in the sense employed has nothing whatever to do (as from your argument you seem to have supposed) with the question of objective or subjective, but with the question of active or passive merely. I understand the word in accordance with its derivation, as implying action as well as being acted on. The ordinary chemical use of the term is in full accordance with this idea. I should consider it correct to speak of a mixture of starch and iodide of potassium as a reagent for ozone because the iodide of potassium acts on the ozone as well as the ozone on the KI. I should consider it incorrect to speak of the eye, or of a photographic camera if you will, as a reagent for dust, because though the dust acts (through the interaction of light) on the eye or sensitized plate, the eye or sensitized plate does not act on the dust.

A chemical reagent is constantly employed merely as a means of detection. This practice makes it probable that the term "reagent" may have assumed, or may hereafter assume, in violation of its derivation, a wider signification. Such a change constantly occurs in the meaning of words. The change I mentioned has not in my mind passed over the word "reagent", but I make no pretence to be the Johnson of science.

Yours very truly | G. G. Stokes

An "agent" which does not act rather grates on my ideas.

RI MS JT/1/S/251
RI MS JT/1/TYP/4/1413

---

1. *Johnson's dictionary*: Samuel Johnson's *Dictionary of the English Language* was first published in 1755 and was the most complex dictionary ever produced.
2. *My objection*: Stokes and Tyndall are discussing the draft of Tyndall's 'On the Action of Rays of High Refrangibility'. See letters 3096, n. 8; 3269, 3271, 3284, and 3286.

From William Wright[1] 9 October 1870 3281

Springfield Clack, Co. Ohio. | Oct[ober]. 9th 1870.

Professor Tyndall | London, England.

Dear Sir,

In the columns of the Independent[2] I noticed a reply of yours[3] to Mr Sherwood[4] of Philadelphia that there was a probability of your visiting America this fall. I hope your duties will permit you to accomplish it at no distant day, if for no other reason, than that of cementing the friendly feeling which should exist between two nations so closely allied in blood and tradition. The very fact of your being an Irishman would do good for unfortunately there are so many of our countrymen that only attempt to sow discord between the countries, and are the enemies of both. Many Americans, as well as Irishmen, would hail with delight the dawn of a better feeling, and particularly the coming winter for the Alabama question[5] will most likely be an exciting one at the next Session of Congress. Our new Minister to England—Gov. Morton of Indiana,[6] will probably receive special instructions in relation to it that will renew the agitation on both sides of the water. It behoves the friends of progress and civilization to be on the alert and exert their influence to prevent any misunderstanding that would be likely to arise in the discussion of the question. From the many flattering notices I have seen for years in our papers of Dr. Tyndall and his discoveries, I have no doubt but what his advent here would be productive of much good and none would hail it with more delight than he who has for many, many years felt an humble pride in being once the schoolfellow of one who had attained so proud an eminence on both sides of the Atlantic and held in high esteem by the scientific ones of each. The reception of this will probably surprise you at this time and will attempt an excuse for its intrusion. Now, as long as I resided in Ohio, I would have an opportunity of seeing you, should you cross the Atlantic, but now that I am contemplating making my future residence in the far West it is not so likely. Preparing to move next week, I have been overhauling my old papers &c. and amongst them find some of your letters written in 43, 45, 50 and *[55]*[7] had laid them aside for the purpose of addressing you once more before I left Ohio feeling if I did so that it would perhaps recall some of the pleasant recollections of our early years and not be considered a trespass or the more serious thoughts of the present and the multifarious studies that you must usually perform and if it should elicit a reply from you I would take it as a favour that would lighten the scene of my border home.

You are perhaps aware that for the past 25 years I have been connected with a Rail Road and also some outside trading, the latter has not been remunerative for some years, and the former not sufficient to give much guarantee for the future of my boys. We have all concluded to move out to a new country and take a fresh start. This day two weeks I expect to be on the Osage[8] Indian lands on the southern border of Kansas where the Walnut[9] empties into the Arkansas river.[10] The Indians have just ceded those lands[11] in Trust to the U.S. Government and they, the Indians, are to be removed further West this fall and winter when their land will be surveyed, and brought into market next spring at one dollar and a quarter per acre to actual settlers. Parties going there now are permitted to preempt 160 acres each, can make their selection of the best quality of land for either grazing or farming purposes, make improvements that will secure the location until the government are ready to make them a title. I was twice out there and feel satisfied that is the finest portion of our country for stock raising and cheap feeding, also at a convenient distance 40 to 60 miles of the Chisholm trail[12] over which immense herds of cattle are yearly driven from Texas and the Cherokee Indian country to Abilene[13] on the Kansas Pacific Railroad and from thence ship'd to St. Louis or Chicago (those fit for beef) those that require more feeding find a ready market from farmers in Missouri, Illinois, Indiana and Ohio. My two eldest boys have absented themselves from home since the close of the war.[14] Richard was last heard from in California about 4 years ago. Samuel 1 year ago on his way to Texas. No communication from either since. I have three more: Willie, Tommie, and George, that are getting about that age when our American youths are difficult to control and all of them having expressed a desire to go out West I let Willie go about 13 months ago and he has settled on those lands. Tommie went out this Spring and George I take with me now. I purpose going into the cattle trade and giving the two eldest boys an interest in it. My eldest daughter Kate is married and living in the eastern part of the State at Emporia on a farm. *[Emmie]* is also married to a farmer living within two miles of this city. Mrs Wright[15] and the three youngest children will remain here until next Spring when all will move out. Should you find time to write to me a letter directed to Arkansas city, Cowley County, Kansas, will be likely to reach me there until the close of the year. Should I learn that you will come out to America this fall or winter it would be likely to hasten my departure from there (as I have to return before Spring for the family)

Yours with respects | William Wright.

RI MS JT/1/TYP/5/1837–9
LT Typescript Only

---

1. *William Wright*: William Wright (1818/9–92), from Leighlinbridge, a close childhood friend of Tyndall's, with whom he subsequently maintained a long-term correspondence. In early 1841 he was working in Roscrea, County Tipperary, and courting Sarah Maria Neale, whom he married on 7 February 1841 at St Nicholas Parish Within, Dublin (Irish Church Records). Possibly due to opposition from her family, the newlywed couple rapidly emigrated to America, sailing from Liverpool in early March 1841. They settled in Cincinnati, Ohio, where William worked in the dry goods and grocery trades. He frequently urged Tyndall to join him in America.
2. *the Independent*: not identified.
3. *a reply of yours*: see letter 3273.
4. *Mr Sherwood*: William Sharswood. See letter 3273, n. 1.
5. *the Alabama question*: the British-built Confederate cruiser *Alabama* captured, sunk, or burned sixty-eight commercial ships belonging to the Union during the American Civil War and was a focal point for US grievances against the United Kingdom. A tribunal deemed the British held responsibility for the maritime losses the United States suffered due to the *Alabama*, and awarded the United States damages of $15,500,000 ('Alabama Claims', *New American Supplement to the latest edition of the Encyclopaedia Britannica* (New York: The Werner Company, 1902), vol. 25, pp. 110–11; H. Jones, *Union in Peril: The Crisis over British Intervention in the Civil War* (Chapel Hill, NC: University of North Carolina Press, 1992), pp. 146–7). The dispute over the damages was connected to wider tensions concerning Canadian fisheries and the north-western boundary of the United States, which were eventually resolved by the Treaty of Washington, 8 May 1871.
6. *Gov. Morton of Indiana*: Oliver Perry Morton (1800–77), governor of Indiana from 1861 to 1867 and U.S. Senator from 1867 to 1877. He accepted the post of ambassador to Great Britain on 23 September but then declined it around 25 October (*ANB*; B. Vincent, *Haydn's Dictionary of Dates,* 14th edn (New York: Putnam & Sons, 1876), p. 756).

7. *some of your letters written in 43, 45, 50 and [55]*: while the letters from Tyndall to Wright are missing, several from Wright to Tyndall for this period exist and offer insight into their discussions. See letters 0193 and 0216, G. Cantor and G. Dawson (eds), *The Correspondence of John Tyndall, Volume 1: The Correspondence, May 1840–August 1843* (Pittsburgh: University of Pittsburgh Press, 2016); and letters 0237, 0269, 0304, 0313, and 0324, M. Baldwin and J. Browne (eds), *The Correspondence of John Tyndall, Volume 2: The Correspondence, September 1843–December 1849* (Pittsburgh: University of Pittsburgh Press, 2016).
8. *the Osage*: the Osage, or Ni-u-Kon-ska, are a Native American tribe of the Ohio and Mississippi river valleys in North America.
9. *the Walnut*: the Walnut River, a tributary of the Arkansas River, is located in Kansas.
10. *Arkansas river*: the Arkansas River is a tributary of the Mississippi River and flows through the states of Colorado, Kansas, Oklahoma, and Arkansas.
11. *The Indians have just ceded those lands*: this was the second time that the Osage were forced to relocate by the US Government. Following the American Civil War their lands in the Neosho River valley, Kanas, were opened for emigrants and the tribe was moved to Oklahoma.
12. *Chisholm trail*: nineteenth-century trail for cattle drivers in the western United States.
13. *Abilene*: a city in the county of Dickinson, Kansas.
14. *the war*: the American Civil War between the northern United States (the Union) and the southern United States (the Confederacy) was fought from 1861 to 1865.
15. *Mrs Wright*: Sarah Maria Neale married William Wright on 7 February. See n. 1.

To Hermann Helmholz 10 October 1870 3282

10th. Oct[ober]. 1870.

My dear Helmholtz,

Accept my best thanks for the copy of the "Ton-Empfindungen"[1] which reached my hands a few minutes ago.

And with it came two copies of the translation of my tract on Faraday.[2] Let me offer you my best thanks for these also. I have looked over it, and the original appears to be improved by the translation. It is clear and forcible. In fact I cannot congratulate myself too much on the fact of coming before the German public under such auspices.

Wishing you years of health and strength to pursue your glorious scientific career.

Believe me always | most faithfully yours | John Tyndall.

RI MS JT/1/T/487
LT Typescript Only

1. *"Ton-Empfindungen"*: presumably, Helmholtz sent Tyndall the third edition of his 1863 book on acoustics, which was published in 1870. H. Helmholtz, *Die Lehre von den Tonempfindungen als physiologische Grundlage für die Theorie der Musik*, 3rd edn (Braunschweig: Vieweg & Sohn, 1870).
2. *my tract on Faraday*: J. Tyndall, *Faraday und seine Entdeckungen* (Braunschweig: Vieweg & Sohn, 1870). Although the publication lists Herman Helmholtz as translator, this book was actually translated by his wife, Anna Helmholtz, who translated several of Tyndall's works. See letter 2951.

To Rudolf Clausius 12 October 1870 3283

12th Oct[ober]. 1870.[1]

My dear Clausius.

Your last letter[2] was read by me with exceeding interest—what experiences you must have passed through during those terrible days at Metz![3] The whole thing hangs before one's mind as a horrible dream: it is hardly possible to realize it.

I was in Switzerland when you were at Metz, and reading in the newspapers that a great number of wounded soldiers were at Heidelberg,[4] and that the people of the town were doing all in their power to lessen the sufferings of the wounded men, I sent to my old dear Friend Bunsen a subscription[5] to aid in this good work. He was kind enough to take charge of it & send it to the proper quarter.

Last night while dining with my friend Captain Galton,[6] who has been recently in Germany to visit the hospitals, I learned that in relation to its population Bonn also has a very large number of wounded in its care, the information immediately caused me to resolve to do that which I had been thinking of doing previously; namely, to send you a subscription also, and to ask you to be good enough either to distribute it yourself, or to place it in the hands which you think most likely to turn it to good accounts.

With this letter therefore you will receive a cheque for fifty pounds (etwa[7] 333 Thaler).

I wish matters could have been arranged so as to make peace after Sedan.[8] The collapse of France was so complete that one would have thought a war indemnity and the demolition of the fortresses would have been a good ending. But it is hard for outsiders to form a judgement on the matter. Upon Paris rests the principal guilt of the war, and perhaps the notion was that Paris ought to be made to feel the consequences. I wish to Heaven the bloodshedding could be put an end to. And as I listen to the wind the thought of the numbers which must die of disease and exposure is very terrible.

And what a fearful lesson for France. What is to be the end of it no one knows. For there appears to be no organic force in the country. The Paris press fed the land with lies and delusions as long as possible, and now when the truth is known there is no man to be found strong enough or competent enough to take the helm of affairs and preserve the Country from Shipwreck. What is to become of France I know not. When the Germans go they will tear each other to pieces.

Old Carlyle[9] wrote to me thus some time ago, regarding the war. "It is such an instance of Michael versus Satan as I have never read in all history before. Satan getting himself smashed to impalpable pulp in one month's time."[10] The pith and force of the grand old man are wonderful.

I have begun to work a little. The first thing to be done is to clean up and complete my experiments on the action of small particles upon light. This is exceedingly curious. The colour and polarization of the sky can as you know be imitated in the most perfect manner. With regard to the neutral parts I think it at present probable that if the atmosphere at all places preserved the purity of its higher portions you would have no neutral points. It is I think the gross haze which wraps the earth close to its surface which gives rise to polarization by refraction—But how? I can imitate perfectly the atmospheric neutral points by varying the density of Dunst.[11] Some effects are very curious. Brown paper, Resin, turpentine and a host of other substances, may be caused to smoke without ignition. Let such smoke be illuminated by a beam of light, and let the illuminated beam be regarded through a Nicol[12] and a concave plate of dolomite. You see the circular bands. Now ignite the substance and examine the smoke arising from the flame. The plane of polarization is instantly reversed. This is invariable. The particles rising from the flame are finer than those which rise from the unignited substance and the consequence is a change of the polarization. I have been also making some experiments regarding the use of cotton wool and charcoal combined for preventing suffocation by smoke. Yesterday I sat for seven minutes in a glass chamber filled with smoke of extraordinary density; a single inhalation, without the cotton wool and charcoal, was sufficient almost to choke me. I hope the thing will be useful for our firemen.[13]

I see Carbolic acid has been used very much in the case of the wounded. Ask your friend Max Schultze whether he thinks it does not act by destroying the germs which cause putrefaction.

Give my love to the children,[14] best regards to your wife[15] and believe me ever

dear Clausius | Yours faithfully | John Tyndall

RI MS JT/1/T/169

1. *1870*: '1870' has been written over '1880'. Since this letter is a response to letter 3252, 1870 is clearly the correct date.
2. *Your last letter*: letter 3252.
3. *Metz*: battle and siege of Metz during the Franco-Prussian War (see letters 3238, n. 16, and 3252, n. 4).
4. *Heidelberg*: see letter 3024, n. 3.
5. *a subscription*: see letter 3248.
6. *Captain Galton*: Douglas Strutt Galton (1822–99), was a British engineer. A graduate of Royal Military Academy, Woolwich, Galton was commissioned second lieutenant in the Royal Engineers in 1840. He joined the British Ordnance Survey in 1846 and the following year was appointed secretary of the railway commission. In 1862 he was appointed Assistant Permanent Under-secretary for War and in 1869 director of public works and buildings. He was elected FRS in 1859, joined the BAAS in 1860 and was elected its president in 1895. Galton was active in the efforts of the Society for the Aid to Sick and Wounded in War (now the Red Cross) to assist the wounded of the Franco-Prussian War. The biostatistician and eugenicist Francis Galton was his cousin (*ODNB*).
7. *etwa*: approximately (German).
8. *Sedan*: the battle of Sedan took place on the 1 September 1870 and resulted in the defeat of the French and the surrender of Napoleon III during the Franco-Prussian War.
9. *Old Carlyle*: Thomas Carlyle.
10. *It is such an instance of . . . one month's time*: see letter 3267. Tyndall quoted Carlyle to Louisa Baring also (see letter 3287).
11. *Dunst*: haze (German).
12. *Nicol*: Nicol prism (see letter 2966, n. 4).
13. *I hope the thing will be useful for our firemen*: Tyndall continued working on this filter, eventually presenting a paper to the RS (J. Tyndall, 'On some recent Experiments with a Fireman's Respirator', *Roy. Soc. Proc.*, 22 (1874), pp. 359–61).
14. *the children*: for the children of Rudolf and Adelheid Clausius at this time, see letter 3146, n. 20.
15. *your wife*: Adelheid Clausius.

To George Gabriel Stokes 12 October 1870 3284

Royal Institution, | 12th Oct[ober]. 1870

My dear Stokes.

In the system of nature where spontaneity is shut out the terms antecedent and consequent are purely relative. If we fix on any link of the series of phenomena that link is a consequent in relation to all the proceeding links, but it is all antecedent in relation to all succeeding ones.

The same remark, I take it, applies to agents[1] and objects in a system which obeys the law of conservation. Starting, for example, from the sun we have here the agent which stirs the ether. The ether here is the object acted upon. But the ether in its turns stirs the optic nerve, and the optic nerve stirs the brain and thus brings the phenomena of light home to consciousness.

I do not wonder at "an agent which does not act"[2] grating on your ideas—But my agent does act, and thus eludes your baton.

Yours faithfully | John Tyndall

Prof Stokes

RI MS JT/1/T/1402
RI MS JT/1/TYP/4/1484

---

1. *agents*: Stokes and Tyndall are discussing Tyndall's draft of 'On the Action of Rays of High Refrangibility'. See letters 3096, n. 8, 3269, 3271, 3280, and 3286.
2. *"an agent which does not act"*: see letter 3280.

From Thomas Achilles Clifford[1] 13 October 1870 3285

New Orleans | U. S. A. | 13th. October, 1870. | Prof Tyndall

Sir,

The American Press have now and then informed us of your intention to visit America in company with Prof. Huxley[2] on a pleasure and profit expedition. These feelers to find out if you would be acceptable to the American public were very scientifically deployed and judging from the remarks elicited by these means, neither of you would receive any more attention than, any other arrogant, pretentious, workmen, whose brass and impertinence, combined with scientific slang and primeval gases stolen from dead asses and all such ignorant fellows, from a workshop has lifted them into a Professor's Chair—Some how or other, the stink, and drunken blaze of the workshop never leave you fellows, try how you will to be gentlemen of refinement and behaviour, when you get to a professors chair. Your work on Sound[3] shows what an ignorant ass you are to the learned and intellectual but the dammed fools you lecture to, being equally ignorant as yourself they like so many monkeys applaud when you please their eye with philosophic experiments: and then they sit, mouth and eyes patulous, wondering whether or not the Pons Asinorum[4] you are conducting them over, will not break down, or lead them to some physical demonstration which might be fatal to their senses—if not their lives—I cant imagine how the people of England can tolerate such Tom Noodles[5] as you and Huxley are.

I heard one of the New York savants[6] remark at a party of ladies and gentlemen at Brooklyn a few months ago—We dont want such blustering, gasey ignorant jackanapes[7] as Tyndall and Huxley here. They have nothing original about them. What they put forth as original and as discovered by themselves, we find on inquiry to be nothing more than a false philosophy buried long ago—or the rakings from some men's fields—which like thieves they apply to their own purposes without any acknowledgment whatever.

The idea said they, that Huxley should attempt an explanation of the origin of life, of spontaneous generation, of the laws which govern organic nature, when every work he writes or utters proves him to be as totally ignorant of the subject, as a monkey is of mathematics.

And as for Tyndall his experiments mislead the minds of his hearers and he is a dangerous man to entrust the minds of youth to. Look at the assertions and wild deductions he makes regarding the nature of sound. And the ignorant fool makes all these assertions and statements, with the most unblushing effrontery, but the word "Rascal" seems to glow all over his face when you speak to him, or when he gets in a quandary as he often does while lecturing

For since the Editorial of the London Lancet[8] appeared in September many American papers have noticed it at length, and they say, they are glad to find so able a critic as the Hon[ourable] Mr Wakley[9] to take you in hand, and prove to the English literate as well as foreign savants what an infernal fool you are. That you are an ass covered and wearing the skin of a lion, and ought to be kicked out of the company of any gentlemen; and so you, and Huxley will find if you show your impudent faces on the American Continent.

Yours &c. in tenebris.[10] | Thomas Achilles Clifford | F. R. S. *[ie.]* | Fried, Roasted & Stewed oysters.[11]

RI MS JT/1/TYP/9/2941
LT Typescript Only

---

1. *Thomas Achilles Clifford*: not identified, possibly a pseudonym.
2. *Prof. Huxley*: Thomas Huxley.
3. *Your work on Sound*: J. Tyndall, *Sound: A Course of Eight Lectures Delivered at the Royal Institution of Great Britain* (London: Longmans, Green, and Co., 1867).
4. *Pons Asinorum*: an obstacle or problem which will defeat an unskilled or foolish person (*OED*).
5. *Tom Noodles*: a fool; or, as given in E. C. Brewer, *A Dictionary of Phrase and Fable* (London: Cassell, Petter, and Galpin, 1870), a 'mere nimcompoop' (p. 899).
6. *New York savants*: not identified.
7. *jackanapes*: an ape or monkey, typically one kept as a curiosity or for entertainment (*OED*).

8. *Editorial of the London Lancet*: see *The Lancet*, 17 September 1870, pp. 407–10.
9. *Mr Wakley*: James Wakley (1825–86), a British journal editor and proprietor, joined *the Lancet* in 1852 to assist his father Thomas Wakley. Wakley had qualified as a medical doctor but never practiced. He focused on extending the public appeal of the journal, inviting many influential writers to contribute, and left the more technical aspects of editing scientific and medical content to his staff. His sub-editor Ernest Hart assisted in this until his appointment as editor of the *British Medical Journal* ('The Late Dr. Wakley', *Illustrated London News*, 11 September 1886, p. 282; *ODNB*).
10. *in tenebris*: in darkness, in a state of ignorance or doubt (Latin; *OED*).
11. *Fried, Roasted & Stewed oysters*: from 'Scraps from an Epicure's Note-book. Oysters, Lobsters, and Crabs', *Titan: A Monthly Magazine*, 28 (February 1859), pp. 214–33, on p. 223, in which an FRS from London meets another person with FRS attached to their name, only to discover that they are not a Fellow of the RS, but that they like oysters and FRS stands for 'Fried, Roasted, and Stewed'.

From George Gabriel Stokes 14 October 1870 3286

Observatory Armagh | 14th Oct[obe]r 1870

My dear Tyndall,

My notion of "reagent"[1] involves the prefix re as well as the termination ent When I wrote of an agent which does not act I assumed that the object of the action (if any) was to be that which produced the other of the two under contemplation. If A acts on B and B on C that does not according to my notion constitute B a reagent. For that it is necessary that A acting on B, B should also act on A. When I wrote of an agent which did not act I assumed that the only action (if any) of B which would justify the application of reagent to B must be on A The question before my mind was—is there or is there not an action on A.

I don't assume that I am right in this restricting the term; I do assert as a fact that such is my own idea connected with the term, and having that idea I did not find the word suitable in the sentence referred to

Yours very truly | G. G. Stokes

RI MS JT/1/S/252
RI MS JT/1/TYP/4/1414

1. *My notion of "reagent"*: Stokes and Tyndall are discussing the draft of Tyndall's 'On the Action of Rays of High Refrangibility'. See letters 3096, n. 8, 3269, 3271, 3280, and 3284.

To Louisa Baring 16 October 1870 3287

16th. Oct[ober]. 1870.

Though I did not write to you from Switzerland[1] I was far from forgetting either you or May.[2] Outside my own thoughts I had little to write about, and these thoughts were not worth ink and paper. Some of them I have already sent you. The discourse[3] has passed the censorship of "the General"[4] successfully and that was the only point I felt anxious about. I should not like to find myself opposed to him even on a speculative point, but he has justified me, and I am very glad he has done so.

The only thing in Switzerland out of the way was the burning of two forests which were struck by the same lightning storm. One of them blazed suddenly up but soon went out again. The other continued to burn for several days and nights. It was exceedingly grand. The flames used to creep stealthily across the brush wood & then leap upon the noble pines converting them for the time being into spires of flame. Then the flame would die out, and the red hot skeleton of the tree would be seen.

The beginning of this war[5] was an atrociously criminal act on the part of France: I was so exasperated with France and Frenchmen that had I been privy I should probably have gone as a volunteer to Germany. But France has fallen so low that I now pity her. I wish the Germans could have halted at Sedan.[6] They had it in their power to show an example for all time—a nobler example than history furnishes—by offering peace at Sedan. Demanding of course an indemnity & the demolition of threatening fortresses, but leaving France otherwise intact. It would have been a glorious thing could they have done this. But bystanders can poorly enter into the feeling of Germany which this war has converted into a land of mourning. And besides the blustering and the delusions of the Parisians were very hard to be borne. Had the Germans offered peace at Sedan it w[oul][d]. undoubtedly have been proclaimed at Paris that they were afraid to come there. Still I am sorry they could not lift themselves above all this. Now I feel very little interest in their success, save in so far as it may shorten the period of carnage and establish permanent peace. My sympathy would almost turn towards France if any thing noble could be discerned there. But nothing noble or strong has hitherto demonstrated itself. The general[7] writes "It is such an instance of Michael versus Satan as I have never read in all history before. Satan getting himself smashed to unpalpable pulp in one month's time."[8] Still you can see that he has some pity for France, and can pray God to give her peace to learn the lesson forced upon her.

Goodbye: my love to May & believe me always | Yours faithfully | John Tyndall

I wish you patience to read this scrawl—and peace & tenderness to keep you from denouncing me for giving you so much trouble.

National Library of Scotland Ashburton Papers Acc. 11388 N0.101

---

1. *Though I did not write to you from Switzerland*: Baring had complained about this in letter 3265.
2. *May*: Mary Baring.
3. *The discourse*: Tyndall's lecture at the BAAS meeting in Liverpool, 'On the Scientific Use of the Imagination' (see letter 3235, n. 11).
4. *the General*: Tyndall often referred to Thomas Carlyle as 'the General'.
5. *this war*: the Franco-Prussian War (see letter 3238, n. 16).
6. *Sedan*: see letter 3283, n. 8.
7. *The general*: see n. 4.
8. *It is such an . . . one month's time*: see letter 3267. Tyndall quoted Carlyle to Rudolf Clausius too (see letter 3283).

To Thomas Archer Hirst [19 October 1870][1] 3288

R. I. | Wednesday | Oct[ober] 19th | 70

My dear Tom

Strachey[2] dines with me <u>here</u> tomorrow at 7½ punctually—I wish you would join us![3]

Ask Lilly[4] to grant this favour to me![5]

RI MS JT/1/T/967

---

1. *[19 October 1870]*: the date on this letter was written in another hand, perhaps Hirst's.
2. *Strachey*: Richard Strachey.
3. *join us*: Hirst accepted Tyndall's invitation, as he wrote in his journal for 20 October 1870: 'Dined with Col. Strachey, Debus and Busk at Tyndall's' (W. H. Brock and R. M. MacLeod (eds), *Natural Knowledge in Social Context: The Journals of Thomas Archer Hirst FRS* (London: Mansell microfiche, 1980), p. 1886).
4. *Lilly*: Lilly Hirst.
5. *to me*: the lower right-hand section of this letter has been excised, and the signature is missing.

From Hermann Knoblauch 19 October 1870 3289

Halle 19.X.70.

Mein theuer Freund!

In frohen Tagen habe ich lange nicht an Sie geschrieben. An wen man in der Trauer schreibt, den hält man für einen treuen Freund.

So melde ich Ihnen und unserm Freund Hirst heute den schweren Kummer, welcher meine arme Frau u. mich in dem Verlust unseres lieben einzigen (3 Jahr alten) Töchterchens betroffen. Unsere kleine Henrietta war ein freundliches, dankbares, lebensfrohes Kind, das—neben den schon selbstständigeren Knaben—den Mittelpunct häuslicher Sorge, Aufmerksamkeit und Freude bildete.

Am Sonntag den 9ten October noch frisch und gesund, erkrankte unser Töchterchen an einem sehr heftigen Scharlach-fieber ganz unerwartet; und ehe wir noch zu vollen Bewusstsein der Gefahr gekommen waren, war es uns schon durch den Tod entrissen.

Gott gebe dass wir von weiterer Gefahr geschützt bleiben!

Es ist überhaupt ein schweres Jahr! erhebend für die Grösse des Vaterlandes, aber voll Jammer in den Familien, und noch immer kein Absehen des blutigen Kampfes, keine Aussicht auf Frieden und auf eine dauernde Zukunft!

Ich war kürzlich in Berlin und sah die dortigen Physiker. Magnus ist nun auch nicht mehr unter Ihnen. Er hat wenig Freunde hinterlassen, weil er keine offene und edle Natur war.

Gern denke ich an unser letztes Wiedersehen im vorigen Jahr: es war mit Ihnen und Hirst die alte treue schöne Marburger Freundschafts-Zeit, die vor mir stand, und die—wie Sie aus diesen Zeilen ersehen—mir unvergesslich bleibt mein Leben lang.

In alter Treue | Ihr H. Knoblauch.

Wir gehen einem stillen Semester entgegen. Wie anders gestaltet sich dieses 3te Rectorats-Jahr gegenüber den beiden verflossenen welche so freundlich für mich gewesen!

Halle 19.X.70.

My dear friend!

In happy times, I went long periods without writing to you. Those to whom one turns in mourning are those one considers true friends.

So I report to you and our friend Hirst[1] today of the heavy sorrow that has afflicted my poor wife[2] and me through the loss of our dear and only (3-year-old) little daughter—Our Henriette was a friendly, grateful, cheerful child who, together with our already more independent boys, was the center of our preoccupations, attention, and joy.

On Sunday, 9th October, our little daughter, who until then was fresh and healthy, quite unexpectedly became ill with very severe scarlet fever, and ere we became fully conscious of the danger, was wrested from us by death.

May God protect us from further danger!

This has in overall been a very difficult year! Uplifting for the fatherland,[3] but full of sorrow in the families, and there is still no end of the bloody battles in sight, no hope of peace and a lasting future![4]

I was recently in Berlin and saw the physicists there. Magnus[5] is now also no longer among them. He left few friends behind because he did not have an open and noble character.

I have pleasant memories of the last time we met last year:[6] with you and Hirst, it was the good old times of our friendship in Marburg[7] that stood before me and that, as you see from these lines, will remain unforgettable for the rest of my life.

Your faithful | H. Knoblauch.

We are about to begin a quiet semester. This 3rd year as rector will be very different from the previous two, which brought me so much joy.

RI MS JT/2/10/460–1
RI MS JT/2/13c/1369

1. *Hirst*: Thomas Hirst.
2. *my poor wife*: Bertha Marie Emilie Knoblauch (née Winter, 1835–87) (Jan Mende, Stadtmuseum Berlin, personal communication).
3. *Uplifting for the fatherland*: German troops emerged victorious from several major battles of the Franco-Prussian War (see letter 3238, n. 16).
4. *no hope of peace and a lasting future!*: the Franco-Prussian War was ongoing.
5. *Magnus*: Gustav Magnus had died on 4 April 1870. See letter 3174.
6. *last time we met last year*: see the beginning of letter 3024.
7. *our friendship in Marburg*: Tyndall, Hirst, and Knoblauch met at the University of Marburg in the late 1840s and early 1850s, when Tyndall (1848–50; 1851) and Hirst (1851–2) studied and Knoblauch (1849–53) taught there.

From William Hugh Spottiswoode[1] 19 October 1870 3290

Anstie Grange Dorking | Oct[ober] 19. 1870

My dear jolly old Tyndall

Do come down here & see us next Saturday to stay over Sunday & bring Mr Hirst[2] with you—

Father[3] & Mother[4] want to see you very much & so do Hughie & Cyril.[5]

If you dont come soon you won't have much chance of seeing dear old Anstie[6] again as we are going away in a fortnight.

It would be such fine fun for us to go to get some chesnuts—you & me together & perhaps go up on the top of Leith Hill[7] & get some more gingerbeer.

Goodbye jolly old Tyndall Mind you come |

Y[ou][r] most affectionate friend | Hughie Spottiswoode | T.O.

We endorse the whole of the preceeding. It arose every word of it out of the boy's own mind Do come if you can we shall be quite alone.

Y[rs] very truly— | L Spottiswoode.[8]

RI MS JT/1/S/193

1. *William Hugh Spottiswoode*: William Hugh ('Hughie') Spottiswoode.
2. *M[r] Hirst*: Thomas Hirst.
3. *Father*: William Spottiswoode.
4. *Mother*: Eliza ('Lise') Spottiswoode.
5. *Cyril*: Cyril Andrew Spottiswoode (1867–1915), son of William and Eliza Spottiswoode.
6. *dear old Anstie*: the Spottiswoode home, Anstie Grange, in Dorking, Surrey.
7. *Leith Hill*: a hill to the north of Broome Hall, the residence of Andrew Spottiswoode (1787–1886), the father of William Spottiswoode.
8. *L Spottiswoode*: see n. 4.

To Emily Peel 21 October 1870 3291

Royal Institution. | Friday, 21st Oct[ober]. /70.

Dear Lady Emily,

Thanks to your kindness, and to that of the Baroness Rothschild,[1] I have been able to get my own mind into clearness regarding the Mediterranean water.[2]

I send you a short article[3] which I wrote on the subject a few days ago.

Would you present my thanks and compliments to the Baroness, and have forwarded to her on some fitting opportunity the copy of "Nature" which bears her name?

I had both your names written in my first draught of the article, but I feared you might consider the use of them a liberty and therefore did not incur the risk.

Yours most truly, | John Tyndall.

RI MS JT/1/TYP/3/960
LT Typescript Only

1. *Baroness Rothschild*: Charlotte von Rothschild.
2. *Mediterranean water*: see letters 3156 and 3158.
3. *a short article*: 'On the Colour of the Lake Geneva and the Mediterranean Sea' (see letter 3156, n. 2).

To John Gassiot 22 October 1870 3292

Royal Institution of Great Britain | 22nd October 1870 | draft[1]

My dear Gassiot.

It is my desire that everything should be clear and straightforward between you & me, and this desire causes me to write you this letter.

Shortly after the death of poor Miller[2] some distinguished Fellows of the Royal Society spoke to me about a successor to the treasurership.[3] Spottiswoode was mentioned as the most fitting man. Lubbock[4] was respectfully mentioned; but as far as I could judge the expression of opinion was strongest in favour of Spottiswoode. And knowing his extreme fitness for the post I resolved to support him and expressed my intention of doing so.

After this your name was mentioned; but when first mentioned some of your intimate friends assured me that you would not accept the post. I frankly confess that I was glad of this. There is no Fellow of the Royal Society who stands personally higher in the consideration of the Fellows generally than you do; there is no man I believe for whom a more profound and general respect is entertained. But I think in going in for this office you would not be acting in accordance with the wishes of your staunchest friends. As I heard one of them say tonight, it can add no honour to your present position; it cannot augment your weight by an iota. You may be, but it will not be a unanimous election even by the Council, and many of the most rigorous outsiders are decidedly of opinion that this is the proper time to engraft young blood and energies upon the staff of the Society.

I could not take my part in the proceedings of next Thursday without letting you know in all honesty the views entertained by the best men of my acquaintance. It would grieve me mightily to find myself in any way opposed to you. It is a consideration which I never expected. But the views expressed by those which constitute the backbone of the society I must respect; and as they harmonize with my own they will determine my action on Thursday.

Pardon this plain-speaking

I know that this is a liberty,—but it is a liberty consistent with the deepest regard for you; and which will find its justification in the future.

Yours ever affectionately | J. Tyndall

RI MS JT/2/10/405

1. *draft*: this draft was sent to Hirst, who was also involved in discussions about the treasurership of the RS. See letter 3293. The version sent to Gassiot is missing.
2. *death of poor Miller*: William Miller. See letter 3272, n. 2.
3. *successor to the treasurership*: the treasurership of the RS (see letter 3272, n. 7).
4. *Lubbock*: John Lubbock.

To Thomas Archer Hirst 22 October 1870 3293

22nd Oct[ober]. /70

Dear Tom.

Here is another heavy blow to poor Knoblauch. Send the letter[1] back: it is very fine & beautiful though so full of woe.

I sent this morn[in]g the enclosed note to Gassiot.[2] I wished him to know beforehand the line I sh[oul]d. take. Send the draft back to me.

I run down to Lady Belper[3] to day, but shall be back on Monday.

Yours aff[ectionate]ly | John Tyndall

RI MS JT/1/T/968
RI MS JT/1/HTYP/573

1. *the letter*: Hermann Knoblauch's young daughter had just died from scarlet fever. See letter 3289.
2. *enclosed note to Gassiot*: letter 3292. On the treasurership of the RS, see letter 3272, n. 7.
3. *Lady Belper*: see letter 3239, n. 17.

To the Royal Society Council[1] [22 October 1870][2] 3294

I am persuaded Sir that every word of recognition and appreciation—every expression of esteem for his character and respect for his work—which you have uttered in reference to the gentleman you have brought forward will find a full response in the minds of the members of this council. For my own part the term esteem or respect would not adequately express the sentiments of which for nearly twenty years I have entertained towards him.[3] It would be therefore to me a matter of great difficulty if not of great pain to take a step or utter a word which might seem to oppose his accession to an honourable position.[4] But I do not think that his being brought forward on the present occasion is meant to be conferring of an honour. And were this a question of honorary distinctions I would join with you in placing that honour on the head of our Gassiot.[5] He is brought forward because he is considered to be the fittest man to do the work of treasurer.

Putting aside then all extraneous personal considerations, the question which each member of the Council has to ask himself is, I believe, this:—what, under the circumstances, is best, not only for the present health & strength, but also for the future health, strength, and development of the Royal Society? One condition of health in our case I imagine to be that growth and change should be gradual; that large or sudden alterations in our official staff ought, if possible, to be avoided; that the younger men with energies intact should be incorporated with our officers which those whose energies have been already expended in the service of the society are still there to instruct him. I think this condition, which is one of great importance, would be admirably fulfilled by the election of Mr. W[illia]m. Spottiswoode for the post of treasurer. His name in connexion with one or two other names raise spontaneously in many minds when the treasurership became vacant. He is a man of high scientific reputation; of large general culture; of perfect courtesy of manner; of thorough business qualifications. I know from experience how expertly he would discharge the duties of treasurer, and I am authorised to think that if elected he would accept the post with the desire and intention to promote to the upmost the interests of the Royal Society. Our veteran president[6] who accepted for so long a time the office now vacant, and who is therefore better fitted than any living man to give instruction as to its duties, would find in Mr. Spottiswoode a pupil both willing & competent to make the best of that instruction. As a man therefore in whom are confirmed the youth, the gentleness, the firmness, the scientific character, and the large general intelligence which all of us would wish to see near the core of the Royal Society I propose, with your permission, Mr. William Spottiswoode as our future treasurer.

John Tyndall

RI MS JT/2/10/p470–1/d

---

1. *To the Royal Society Council*: this letter is a draft.
2. *[22 October 1870]*: the date of this draft letter has been estimated through its reference to the election of a new Treasurer to the RS. The election took place on 27 October. Tyndall discusses his support for William Spottiswoode in letters 3292 and 3293 of 22 October 1870. See also letters 3272, 3274, 3276, 3299, 3300, 3301, 3306, and 3327.
3. *towards him*: John Peter Gassiot.
4. *an honourable position*: the treasurership of the RS (see letter 3272, n. 7).
5. *Gassiot*: John Peter Gassiot.
6. *our veteran president*: Edward Sabine.

From Charles Kingsley[1] 22 October 1870 3295

I have never seen the Lake of Geneva,[2] but I thought of the brilliant dazzling dark blue of the mid-Atlantic under the sunlight, and its black-blue under cloud, both so solid that one might leap off the sponson on to it without fear; this was to me the most wonderful thing which I saw on my voyages to and from the West Indies.

J. Tyndall, 'On the Colour of Water, and on the Scattering of Light in Water and in Air', *Roy. Inst. Proc.*, 6 (1870 –72), pp. 189–99, on p. 192.[3]

1. This letter fragment was taken from the published source cited above. The original letter is missing.
2. *Lake of Geneva*: for Tyndall's interest in Lake Geneva see letters 3156, 3158, and 3291.
3. This excerpt from a Kingsley letter was also published in: J. Tyndall, 'From Portsmouth to Oran to See the Eclipse', *Fortnightly Review*, 9 (new series) (1 March 1871), pp. 330–51, on p. 349; and J. Tyndall, 'Voyage to Algeria to Observe the Eclipse', in *Hours of Exercise in the Alps*, 1st edn (London: Longmans, Green, and Co., 1871), pp. 429–73, on pp. 467–8.

To Thomas Archer Hirst 23 October 1870 3296

Tuesday | Oct[ober] 23/70

Dear Tom.

I said Tuesday,[1] but I am engaged for that day—we must arrange another. I shall be in town tomorrow ev[enin]g,

Yours aff[ectionatel]y | John

RI MS JT/1/T/959
RI MS JT/1/HTYP/573/4

1. *I said Tuesday*: in letter 3293 Tyndall mentions returning to London on Monday. There may be a missing note between these two letters.

From Rudolf Clausius 23 October 1870 3297

Bonn, 23/10 70.

Lieber Tyndall,

Ich würde sofort nach Empfang Deines lieben Briefes geantwortet haben, um Dir den Empfang Deiner großmüthigen Gabe für die Verwundeten anzuzeigen, wenn ich nicht erst Max Schultz die mir aufgetragene Frage hätte vorlegen wollen. Ich bin zweimal bei ihm gewesen, habe ihn aber beide Male nicht zu Hause getroffen, und so will ich nun meinen Brief nicht länger verzögern, und werde Dir die Antwort von Schultz später schreiben. Zufällig finde ich unter den neuesten Doctordissertationen unserer Universität eine, welche sich auf die klinische Verwerthung der Carbolsäure bezieht, und obwohl ich nicht weiß, ob sie irgend etwas für Dich interessantes enthält, will ich sie Dir doch unter Kreuzband zuschicken, da ich sie gar nicht gebrauche, und sie bei mir nur unnütz herum liegen würde.

Deine Versuche über die Einwirkung kleiner Körperchen auf das Licht, und über die Filtrirung der Luft durch Baumwolle und Kohle sind sehr interessant, und die letzteren von der größten practischen Bedeutung. Nachdem einmal der wahre Grund der heilsamen Wirkung dieser filtrirenden Stoffe erkannt ist, wird ihre Anwendung sich gewiß sehr vermehren und wird in rationeller Weise geschehen.

Sehr gefreut habe ich mich über die Ansichten, welche Du über die großartigen politischen Vorgänge unserer Zeit aussprichst und über die Art wie Carlyle sich darüber äußert. Du kannst Dir denken daß die Deutschen der ganzen Entwicklung mit der größten Aufregung folgen, besonders diejenigen, welche links vom Rhein wohnen und selbst das Object des Kampfes waren. Wenn wir die Zustände, die hier stattfinden würden, wenn die Franzosen gesiegt hätten, mit den jetzigen Zuständen vergleichen, so müssen wir über den Ausgang des Kampfes sehr glücklich sein. Aber wir freuen uns noch ganz besonders, wenn so urtheilsfähige unpartheiische Männer unsere Sache gegenüber der der Franzosen als die gute Sache bezeichnen.

Deine so überaus reichliche Sendung für die Verwundeten hat mich in Erstaunen gesetzt, und ist mir ein neuer Beweis für die großartige Opferwilligkeit, die Du überall zeigst, wo etwas wirklich Gutes zu thun ist. Ich werde das Geld nach bestem Wissen unter solche Vereine, welche für die Leiden der Krieger wirken, vertheilen und werde Dir, sobald es geschehen ist über die Art der Vertheilung Rechenschaft ablegen.

Mit herzlichen Grüssen von meiner Frau und mir | Dein Clausius.

Soeben habe ich auch Deinen interessanten Artikel in der Nature über die Farbe des Genfer Sees und des Mittelmeers erhalten, und mit vielem Interesse gelesen. Ich danke Dir bestens für die Zusendung.

<u>Bonn, 23/10 70.</u>

Dear Tyndall,

I would have replied immediately after receiving your kind letter[1] to inform you that I had received your generous donation for the wounded, had I not intended to first ask Max Schultz the question that you requested me to pass along to him. I went to see him twice, but both times did not find him at home. I do not want to postpone my letter any longer and will send you Schultz's answer[2] later. By coincidence, I found among our university's most recent doctoral dissertations one that has to do with the clinical uses of carbolic acid and although I do not know if it is of any interest to you, I will send it to you[3] separately as printed matter, as I do not need it and it would only lie around here without purpose.

Your experiments[4] on the influence of small particles on light and on the filtering of air using cotton and coal are very interesting, and the latter are of great practical significance. Once the true reason for the healing action of these filtering substances has been found, their utilisation will certainly greatly proliferate and will happen in a rational way.

I was delighted about the views you expressed about the monumental political events[5] of our times and about the way in which Carlyle has discussed them.[6] You can imagine that the Germans are following the whole development with great agitation, particularly those who live left of the Rhine[7] and were themselves the object of the dispute. If we compare the current conditions with those that would have come to be, had the French been victorious, we must be very grateful for the outcome of the battle. But we are even more pleased when men who are impartial and capable of discernment pronounce our cause, over that of the French, to be the good one.

Your exceedingly liberal contribution for the wounded has astonished me, and I see it as further proof of that admirable willingness to make sacrifices that you show where good is to be done. I will distribute the money according to my best judgement among those societies that work for the sufferings of the soldiers, and will give you an account of the manner of distribution when done.

With heartfelt greetings from my wife and me, | Your Clausius.

I have also just received and read with great interest your fascinating article in Nature[8] on the colour of Lake Geneva and the Mediterranean. I thank you very much for sending it.

RI MS JT/2/10/460–1

---

1. *your kind letter*: letter 3283.
2. *Schultz's answer*: letter 3305.
3. *I will send it to you*: letter missing.
4. *your experiments*: see letter 3283.
5. *monumental political events*: the Franco-Prussian War (see letter 3238, n. 16).
6. *Carlyle has discussed them*: see letter 3283.
7. *Left of the Rhine*: the regions of Alsace and Lorraine, on the western side of the River Rhine in modern France, were controlled by France prior to the Franco-Prussian War (1870–71). The Peace of Westphalia seceded control of the region of Alsace to France.

However, the population spoke a German dialect known as Alsatian. The regions were annexed by the German Empire at the end of the War and merged to form the German territory of Alsace-Lorraine.

8. *article in Nature*: 'On the Colour of the Lake of Geneva and the Mediterranean Sea' (see letter 3156, n. 2).

From John Herschel 23 October 1870 3298

Collingwood | Oct[ober]. 23. 1870.

My dear Sir,

Many thanks for your lecture,[1] but more especially for your capital paper on the Action of Rays of high refrangibility on gases and vapors[2]—This paper opens out quite a new world of enquiry. The phenomena are most wonderful and the experiments must have (been)[3] delicate, difficult, and laborious in the extreme.

I cannot get myself quite into the clear about the reflection of an undulation on a particle so minute as to have a diameter incomparably smaller than the length of a wave. According to the Huygenian theory,[4] when a wave is reflected from a material obstacle every point in the surface receiving the wave becomes a centre of undulations spreading back from it spherically and the total reflected wave is the result of a combination of all these spherically reflected waves into a common tangent surface, the lateral emanations destroying each other by interference (producing however inflection fringes). Supposing the particle of infinitesimal dimensions we should have a spherical wave spreading all around and supposing it only very small as compared with the wave-length, the reflexion would not begin to degrade by interference till after it had spread over a certain angle. Still, so far as I can see, there is nothing here to alter the length of the period or interval of time between the reflexion of two successive crest of the waves—That is to say if the reflecting particle be absolutely fixed in space. But here comes in a consideration which complicates the affair. Suppose the particle movable by the pressure, or impact of the Ether, and (merely to simplify the conception) suppose the luminiferous undulations to be direct instead of transverse. Now—as in a wave so propagated there is a bodily advance and recess of the ethereal matter, so also will there arise a corresponding to and fro movement of the reflecting particle itself. Were the magnitude of that particle only infinitesimal its advance and recess would be precisely equal to and simultaneous with those of the ethereal molecules surrounding it. It would simply be carried to and fro with the ether and produce no reflexion whatever its density. (this is a consequence of the doctrine of resisting media). If its magnitude though minute be not infinitesimal let us see what will happen. During the advance of the undulating molecules it will be carried forwards by their pressure, its motion commencing simultaneously, but it will not be carried quite so fast as the molecules first in contact with it. In its first excursion it will lag behind them. But moreover when they begin to recede, it, (having inertia and having required momentum by their pressure during their joint advance) will still tend forwards, and will require to have its momentum destroyed by the receding ether before it can begin its own recess. Regarded then as a vibrating particle, and as an origin of a reflected (x) undulation, it will tend to propagate a wave having a longer period of undulation than that of the original ether wave.

This however is only a prima facie view and on reading it over again, to say the truth, I am not quite satisfied with it. It is incomplete since I have not followed out the receding movement or traced it to and fro over several alternations, and I think I perceive that there will enter into them, considerations of gradually increasing discordance of times and movements. I shall get clearer notions by thinking a little more about it. After all the analysis of the case as a dynamical problem does not seem very difficult; but I do not want longer to delay thanking you for your kindness in sending me these interesting papers.

Believe me, my dear Sir, | Yours very sincerely | J. H. Herschel.

(x) P.S. | Looking again over it I see that the wave so generated is after all not properly the true reflected wave, the real echo. This (the Echo) is a result of the abstraction of the motion from the ether in the act of pressing or impinging on the particle—the reaction of that particle. That here considered is a sort of secondary vibration. The theory of the true echo wave is by no means so clear when the motion of the echoing particle is considered.

RI MS JT/1/TYP/2/573–4
LT Typescript Only

---

1. *your lecture*: Herschel may be referring to Tyndall's BAAS discourse 'On the Scientific Use of the Imagination' (see letter 3235, n. 11).
2. *your capital paper... gases and vapours*: 'On the Action of Rays of High Refrangibility' (see letter 3096, n. 8).
3. *(been)*: this word is in parenthesis in the typescript.
4. *Huygenian theory*: from Christiaan Huygens (1629–95), a Dutch physicist, astronomer, and mathematician. Huygens published his *Treatise on Light* in 1690, in which he proposed a mathematically-based wave theory of light. Although it was sidelined by the Newtonian corpuscular theory in the seventeenth century, it experienced a revival in the eighteenth, in part because it could better explain the phenomena described by Herschel in this letter (*CDSB*).

From John Gassiot 24 October 1870 3299

CLAPHAM COMMON, | SURREY. | 24^{t} October 1870

My Dear Tyndall

I rec[eiv]^{ed} your note[1] late on Saturday Evening you and I have always been so clear and distinct with each other that for the moment I really cannot comprehend what is at issue between us

You say that "shortly after the death of poor Miller[2] certain distinguished F.R.S's spoke to you about a successor to the Treasurership" that "first our mutual friend Mr Spottiswoode and then Sir John Lubbock were named and that your opinion was the strongest in favor of M^{r} S—in this I fully reciprocate with you—I have the highest respect for Sir John and I believe I have the honor of his personal friendship—yet I cannot but think his multifarious engagements incapacitate him for what I consider the onerous duties of Treasurer to the Royal Society involving as they do far more serious considerations and responsibilities than mere details of finance.

You refer to my name having been subsequently mentioned to you, and that some of my most intimate friends, assured you that I would not accept the post—you frankly confess that you were glad to receive this assurance and you follow thus by some observations which I accept as a proof of personal kindness to myself although in many respects undeserved.

I fully agree with you that the offer of Treasurer to the R[oyal] S[ociety] can add no honor to my present position, the recognitions I have already received, I unhesitatingly acknowledge are more than my contributions to science deserved—nevertheless those who awarded what I received unsolicited & at the time most unexpected on my part accepted the responsibility, but the question at issue is not whether the acceptance of the Treasurership of the R[oyal] S[ociety] will or will not add to my own position—the real question at issue is what is best to be done for the R[oyal] S[ociety] under the peculiar position it is now placed by the unfortunate sudden death of Mr Miller—it is this and this only the Fellows have to consider.

A few Days since I met Mr Busk[3] in the City, he told me my name had been mentioned for the Vacant office—I instantly replied that I refused it 2 or 3 years since when our late friend Miller talked to me as to his then desire to retire, I explained to Mr Busk that at my time of life (in my 73 year) I had determined not to accept further duties—he mentioned Mr Spottiswoode & Sir John Lubbock as names which had also been mentioned to him intimating his preference for Sir John—I replied as already mentioned in this note explaining my reasons for preferring our friend Spottiswoode and so we parted

Shortly afterwards I rec[eiv][ed] a written communication which induced me to write a private note to M[r] Busk, which he is now at full liberty to shew you should you desire it

If my acceptance of the vacant office under the painful circumstances which have arisen (for a single year) can be of any advantage to the interest of science—I should be indeed ungrateful to the Royal Society if I did not consent should it be offered as for entering into a contest I will neither court it or fear it—I have too vivid a recollection of former times. The R[oyal] S[ociety] never made a greater mistake than when it rejected the services of M[r] Grove[4] (this was before your time) I have not the pretension of placing my position on a par with his, but I yield to none in an ardent and sincere desire to uphold the interests of science through the Royal Society—I should indeed be a coward if I shrunk from a contest because the election was not unanimous in the Council. Whatever the Council decides I shall feel it my duty to accept. I give you full liberty to read my letter to any of our friends or to the Council so let you and I have no further correspondence on it

Believe me | My Dear Tyndall | Ever yours | J. P. Gassiot

D[r] Tyndall FRS

RI MS JT/2/10/466–7
RI MS JT/2/13c/1372–3

---

1. *your note*: see letter 3292. This is part of a series of correspondence regarding the treasurership of the RS (see letter 3272, n. 7).
2. *Miller*: William Miller. See letter 3272, n. 2.
3. *M[r] Busk*: George Busk (1807–86), a British naval surgeon who retired in 1855 and studied zoology and palaeontology. He was one of the founders of the Microscopical Society in 1839 and edited the *Quarterly Journal of Microscopical Science* (1853–68). He served as treasurer of the RI (1873–86) and secretary of the zoological section of the Linnean Society (1857–68). He was also a member of the X Club.
4. *M[r] Grove*: William Grove (1811–96), a Welsh judge and electrochemist. He was called to the bar in 1835 but also took an interest in science. He held a professorship of experimental philosophy at the London Institution (1841–47) and developed a gas battery (the first fuel cell), for which he was awarded the RS's Royal Medal in 1847. In the late 1840s financial constraints turned his attention back to practising law. He became a Queen's Counsel in 1853, was appointed a judge of the court of common pleas in 1871, and then the Queen's Bench in 1880. He remained a highly respected member of the scientific community and was elected president of the BAAS in 1866.

From Henry Bence Jones [27 October 1870][1] 3300

Dear friend

The last paragraph of Gassiots letter[2] is bitter.

It looks as if he had made up his mind & would look only his own way. I could not imagine that you could look another way.

It reads as though he did not fully understand that you intended to propose Spottiswoode & did not like to do so without telling him (Gassiot)

After this note I should not write again but that is my temper

Ever yours | H Bence Jones

RI MS JT/1/J/105
RI MS JT/1/TYP/2/720

---

1. *[27 October 1870]*: the date of this letter has been estimated through its reference to letter 3299 of 24 October 1870.
2. *Gassiots letter*: Tyndall seems to have forwarded letter 3299 to Bence Jones for advice and support. This is part of an ongoing discussion regarding the treasurership of the RS (see letter 3272, n. 7).

From Henry Bence Jones [28 October 1870][1] 3301

Dear friend

I am so glad you have won[2]

It is a great success & will make the future much easier. Many thanks for your note.

I am sure you will write[3] to Gassiot beautifully

Ever y[ou]rs | HBJ

Write also to the Bishop of Winchester[4] for me

RI MS JT/1/J/102
RI MS JT/1/TYP/2/723

---

1. *[28 October 1870]*: the date of this letter is estimated through its reference to the election of William Spottiswoode as Treasurer of the RS over John Gassiot. The election took place at a meeting of the Council of the RS on 27 October 1870.
2. *you have won*: Tyndall supported Spottiswoode over Gassiot for the RS the treasurership of the RS (see letter 3272, n. 7).
3. *you will write*: it is unknown whether Tyndall wrote to Gassiot on this occasion.

4. *Bishop of Winchester*: Samuel Wilberforce (1805–73), Bishop of Oxford and Winchester, fifth son of William Wilberforce (1759–1833) and Barbara Ann (1777–1847). Following studies at Oriel College, Oxford, he became a priest in 1828, Bishop of Oxford in 1845, and Bishop of Winchester in 1869 (*ODNB*). A noted social commentator, he is remembered for his clash with Thomas Huxley during an 1860 Oxford debate on Darwin's theory of evolution (see I. Hesketh, *Of Apes and Ancestors: Evolution, Christianity and The Oxford Debate* (Toronto: University of Toronto Press, 2009); and N. K. L. Kaalund, 'Oxford Serialized: Revisiting the Huxley–Wilberforce Debate through the Periodical Press', *History of Science*, 52:4 (2014), pp. 428–53).

To Sedley Taylor[1] 29 October 1870 3302

Royal Institution of Great Britain | 29th Oct[ober]. 1870

My dear Sir,

I have read your exceedingly lucid essay "on French and German as substitutes for Greek"[2] with the highest satisfaction

Yours faithfully | John Tyndall

Cambridge University Library MS Add.6260/93

---

1. *Sedley Taylor*: Sedley Taylor (1834–1920), a British music scholar and benefactor, graduated BA (sixteenth wrangler) from Cambridge University in 1859 and MA in 1862. He was elected a fellow of Trinity College in 1861 and despite having no formal role remained there for the rest of his life. In his early years there he also suffered a publicly declared loss of faith. Music was his primary academic interest and he carried out scientific research into sound. In addition, he was also fluent in French and German (*ODNB*).
2. *"on French and German as substitutes for Greek"*: S. Taylor, *On French and German as Substitutes for Greek in University Pass Examinations* (London: Macmillan and Co., 1870).

From Auguste de la Rive 29 October 1870 3303

Presinge, Genève, le 29 Oct. 1870.

My dear sir,

Je n'ai pas encore répondu à votre bonne lettre datée de Bel Alp & cependant je veux vous en remercier en même temps que de l'envoie du mémoire sur l'action des rayons réfrangibles sur les matières gazeuses. J'ai reçu ce mémoire aujourd'hui même & je suis impatient de le lire car ce sujet & tous ceux qui ont en général pour objet les rapports entre la matière pondérable & les ondulations de l'éther m'intéressent vivement; c'est là qu'est l'avenir de la science aussi bien de la chimie que de la physique, comme vous le faites remarquer; j'en suis bien convaincu aussi pour ma part.

J'ai lu dans les journaux avec un grand intérêt votre discours à Liverpool; il a du faire beaucoup d'effet. Vous avez en Angleterre un grand public qui apprécie & qui comprend ce genre de questions. C'est un grand appui moral pour la science. C'est ce qui fait malheureusement défaut en France. Ce pauvre pays paye maintenant bien cher sa frivolité & sa *[*légèreté*]*. On voit combien la corruption avait pénétré profondément & quelle démoralisation générale il en était ressorti. Jusqu'à quand cette horrible guerre continuera-t-elle? Elle ne paraît pas près de finir; & en attendant l'anarchie règne partout plus ou moins en France. Espérons que les Puissances neutres, l'Angleterre en particulier, ne tâchent pas à faire finir cette boucherie.

J'ai reçu dernièrement d'Airy un travail accompagné d'une lettre explicative sur les courants terrestres & leur liaison avec les variations du magnétisme terrestre. Il arrive à des combinaisons qui me paraissent très remarquables & très favorables à la théorie que j'ai donné de ces phénomènes et de celui des Aurores polaires. Il paraît tout-à-fait contraire à l'idée de M. Wolf de la liaison entre les retours périodiques des taches solaires & des perturbations magnétiques. Il dit qu'au moins rien n'autorise à admettre l'existence de ces périodes de 10 à 11 ans. Qu'y croire? Avez-vous quelque idée sur ces questions & seriez-vous disposé à admettre les conclusions de M. Airy? C'est bien à vous, à ce que je présume que je dois le *[*N°*]* du Journal Nature qui renferme un morceau très intéressant de vous sur la couleur du lac de Genève & de la Méditerranée. Je viens d'envoyer ce mémoire à Soret à qui il fera un grand plaisir. Je crois que vous avez parfaitement raison lui et vous. Le dit Soret que j'ai vu hier me charge de vous remercier infiniment de votre mémoire et de le rappeler à votre bon souvenir.

Toutes les fois que vous avez quelque chose de nouveau, ne manquez pas de nous l'envoyer, comme vous le faites; nous en serons toujours très reconnaissants Soret et moi. Ma femme me charge particulièrement de la rappeler à votre bon souvenir; tous vos amis de Genève vous envoient leurs meilleures amitiés.

Votre tout affectionné | Aug. de la Rive.

Presinge,[1] Geneva, 29 October 1870.

My dear sir,

I have not yet replied to your kind letter[2] from Bel Alp[3] & nonetheless I wish to thank you for it at the same time as for sending me the dissertation[4] on the action of refrangible rays on gaseous material. I received this dissertation today & I am looking forward to reading it since this subject & all that which generally relate to ponderable matter & waves of ether, interest me greatly. That is the future of science as well as chemistry and physics, as you point out, and of which I am convinced as well.

I read your Liverpool speech[5] in the newspapers with much interest; it must have made a grand impression. You have a large audience in England that appreciates & understands these sorts of questions. It is a significant moral support for science. This is what is sadly lacking in France. This poor country now pays dearly for its frivolity & its flippancy. We see how deeply corruption had penetrated & what general demoralization had emerged from it. How long will this horrible war[6] continue? It does not seem to be nearing the end, & in the meantime, anarchy reigns more or less everywhere in France. Hopefully the Neutral Powers, England in particular, will hasten to end this slaughter.

I recently received a work from Airy,[7] accompanied by an explanatory letter on land currents & their link to the variations of terrestrial magnetism. Combinations arise, which, to me, seem remarkable and very supportive of the theory that I gave on these phenomena and about polar auroras. It seems at odds with Mr Wolf's[8] idea of the link between the periodic return of sunspots & magnetic disturbances. He said that at least there is no reason to admit the existence of these periods of 10 to 11 years. What to believe? Do you have any ideas regarding these questions & would you be willing to accept Mr Airy's conclusions? It is to you, I presume, that I owe the *[N°]* of the journal Nature, which contains a very interesting piece from you[9] about the colour of Lake Geneva & the Mediterranean. I just sent this dissertation to Soret,[10] whom it will please greatly. I think that he and you are absolutely right. Soret, whom I saw yesterday, asked me to thank you very much for your dissertation and to remind you of him.

Whenever you have something new, do not forget to send it to us, as you have been; we will always be very grateful, Soret and I. My wife[11] asked me especially to remind you of her. All of your friends from Geneva send you their best regards.

Your affectionate | Aug. de la Rive.

RI MS JT/1/TYP/1/379
LT Typescript Only

---

1. *Presinge*: a village located in the Swiss canton of Geneva.
2. *your kind letter*: possibly letter 3242.
3. *Bel Alp*: Belalp (see letter 3030, n. 20).
4. *the dissertation*: 'On the Action of Rays of High Refrangibility' (see letter 2980, n. 7).
5. *your Liverpool speech*: 'On Scientific Use of the Imagination' (see letter 3235, n. 11).
6. *this horrible war*: the Franco-Prussian War (see letter 3238, n. 16).
7. *a work from Airy*: the work from George Biddell Airy is not identified.
8. *Mr Wolf's*: Rudolf Wolf (1816–93), a Swiss astronomer best known for his work on sunspot activity and cycles (*HLS*).

9. *a very interesting piece from you*: 'On the Colour of the Lake of Geneva' (see letter 3156 n. 2).
10. *Soret*: Louis Soret.
11. *my wife*: Louise Marie de la Rive (see letter 3096, n. 9).

From John Herschel[1] 30 October 1870 3304

Collingwood | Oct[ober] 30/70

My dear Sir

I forgot to mention that I tried Goethe's experiment on the fluorescence of Horse Chestnut water[2] mentioned in your lecture[3] with much care and without obtaining the smallest sign of fluorescence.—The same result also when the Spanish or edible chestnut was tried—the result (in this latter case) being a strong solution of Gallic acid.

Yours very tr[uly]^{y} | JFW.

PS. I had heard before of "Esculine"[4] a substance */made/* from chestnut husks as a fluorescent substance.

RI MS JT/2/10/478–9

1. This letter is on a postcard, addressed to: 'Prof[esso]r Tyndall | Royal Institution | Albemarle Street | London W.', with the postmark as 'HAWKHURST | C | OC30 | 70'.
2. *Goethe's experiment . . . horse-chestnut water*: Goethe described putting a piece of fresh horse chestnut bark in a glass of water. The water in the glass took on a blue or yellow colour, depending on whether the glass was lit from the front or the back (J. W. Goethe, *Goethe's Werke. Vol. 59. Enthüllung der Theorie Newtons* (Stuttgart/Tübingen: Cotta, 1842), p. 53).
3. *your lecture*: 'On the Scientific Use of the Imagination' (see letter 3235, n. 11).
4. *Esculin*: also aesculin, a crystalline compound present in the bark and other parts of some trees (notably the horse chestnut, *Aesculus hippocastanum*), and used in tests to identify certain bacteria by means of its fluorescence (*OED*).

From Rudolf Clausius 1 November 1870 3305

Bonn, 1/11/70

Lieber Tyndall,

Nachdem ich heute Max Schultze gesprochen habe, beeile ich mich Dir seine Meinung über die Wirkung der Carbolsäure mitzutheilen. Er stimmt vollkommen mit Dir überein, dass die heilsame Wirkung derselben bei

Wunden darauf beruht, dass sie Keime, welche Fäulniss verursachen, zerstört. Er sagte auch dass er bei Leichen, um sie länger von Fäulniss frei zu halten, das Verfahren anwende, sie mit einer Lösung von Carbolsäure einzuspritzen.

Zugleich will ich Dir über die bisherige Verwendung des von Dir so grossmüthig für die Verwundeten geschickten Geldes Rechenschaft ablegen. Ich hatte zuerst daran gedacht das Geld an einige Vereine welche sich hier zur Pflege der Verwundeten gebildet haben zu vertheilen. Als ich aber einen Arzt der an der Spitze eines der hiesigen Lazarethe steht, zu Rathe zog, sagte mir der, dass das nicht zweckmässig sei. Diese Vereine sorgten nämlich nur im Grossen für diejenigen Bedürfnisse, die allen Kranken gemeinsam seien, wie Betten, die gewöhnliche Nahrung, Wäsche, etc. Es kämen aber häufig auch specielle Bedürfnisse für einzelne Kranke vor, dass z.B. ein Reconvalescent, der sehr schwach geworden wäre, besonders nahrhafte, und doch leicht verdauliche Speisen *[haben]* müsste, oder dass einer zur Stärkung seines Magens etwas guten alten Wein haben sollte, u.s.w. Für solche Fälle fehle es dann oft den Aerzten und Wärtern an den nöthigen Mitteln. Es sind nun hier 14 Lazarethe für verwundete und kranke Soldaten eingerichtet. 10 darunter sind Privatlazarethe, welche gut versorgt werden, die 4 grössten aber, welche zusammen 384 Kranke haben, sind Staatslazarethe. In diesen wird für jeden Kranken täglich ein gewisser Betrag bezahlt, der nur gerade für die gewöhnliche Verpflegung ausreicht, aber für besondere Bedürfnisse geschieht gar nichts. In diesen 4 Lazarethen habe ich nun den Personen, welche die Verpflegung vorzugsweise leiten, und zuverlässige Leute sind, für jedes der Lazarethe 50 Thaler für aussergewöhnliche Bedürfnisse gegeben. Ferner besteht hier ein sehr nützlicher Verein, welcher für die verwundeten und kranken Soldaten auf der Eisenbahn sorgt, sowohl für solche, die hierher gebracht werden, als auch für solche, die nur durchtransportiert werden, und unterwegs Erquickung bedürfen. Diesem Verein habe ich auch 50 Thaler gegeben. Es bleiben nun noch 90 Thaler zu vertheilen. Diese habe ich noch zurückbehalten, weil ich es für möglich halte dass jetzt von Metz au seine grössere Anzahl Verwundeter und Kranker hergebracht wird, um dann vielleicht die Hülfe noch ganz bensonders nöthig sein wird. Ueber die Verwendung dieser 90 Thaler werde ich Dir später schreiben. Die Personen (theils Herren theils Damen) welchen ich die vorherbesprochenen 250 Thaler übergeben habe, lassen Dir im Namen der ihrer Pflege untergebenen Kranken herzlich danken.

Der mittlerweile eingetretene Fall von Metz ist ein wahrhaft grossartiges Ereigniss, und ich freue mich jetzt doppelt, bei den Kämpfen welche dieses Resultat vorbereitet haben, zugegen gewesen zu sein.

Hoffentlich wird nun auch bald der Friede folgen.

Mit herzlichen Grüssen von meiner Frau und mir, | Dein | Clausius.

Bonn, 1/11/70

Dear Tyndall,

After having spoken with Max Schultze[1] today, I hasten to communicate to you his opinion on the action of carbolic acid. He completely agrees with you that its medicinal effect on wounds is based on it destroying germs that cause putrefaction.[2] He also said he uses a process of injecting corpses with a solution of carbolic acid in order to delay decomposition for a longer period of time.

I would also like to give you an account of how the money you so generously sent[3] for the needs of the injured has been used. Initially, I considered giving the money to various associations that were formed to provide care for the wounded. But when I consulted with a physician heading one of the local military hospitals, he told me that this would not serve the purpose. These associations only provide for those basic needs that are common to all patients, such as beds, ordinary food, linens, etc. However, oftentimes there arise special needs for individual patients; for example, that a convalescent patient who has become very weak needs especially nutritious but easily digestible food, or that a patient requires some good, aged wine to strengthen his stomach, etc. In such cases, doctors and caretakers often lack the necessary resources. Fourteen military hospitals for wounded and sick soldiers have now been established here. Ten among them are private and well supplied, but the 4 largest, which together have 384 patients, are public military hospitals. In these, a certain per diem is paid for each patient, which is just enough for ordinary provisions, but nothing is being done for special requirements. For these 4 hospitals, I have now given those persons who are chiefly in charge of provisioning, and who are reliable people, 50 thalers each for special needs. There is also a very useful organisation here that cares for injured and sick soldiers on the railways, both for those who are brought here as well as for those who are travelling through and are in need of refreshment. I gave 50 thalers to this association as well. There are still 90 thalers left to distribute. I am holding on to these, since I consider it possible that a larger number of wounded and sick soldiers will now be brought here from Metz,[4] and that any aid will then be especially necessary. I will inform you later about how these 90 thalers have been distributed. Those who received the above-mentioned 250 thalers (some gentlemen and some ladies) send you their heartfelt thanks on behalf of the patients under their care.

The fall of Metz, which has occurred in the meantime, is a truly grandiose event, and I am now doubly glad that I was present at those battles[5] that have laid the groundwork for this result.

Hopefully peace will be soon to follow.

With warmest wishes from my wife and myself, | Your | Clausius.

RI MS JT/1/TYP/7/2318–9
LT Typescript Only

1. *After having spoken with Max Schultze*: see letters 3283 and 3297.
2. *putrefaction*: see letter 3283.
3. *the money you so generously sent*: Tyndall entrusted Clausius with a donation to aid the wounded during the Franco-Prussian War. See letters 3283 and 3297.
4. *Metz*: see letter 3252, n. 4.
5. *I was present at those battles*: Clausius was part of a medical corps at the Battle at Metz, also known as the battle of Mars-la-Tour, and that at Gravelotte (16–8 August 1870). See letter 3252.

To Eliza Spottiswoode 8 November [1870][1] 3306

Tuesday. 8th Nov[ember].

My dear Mrs. Spottiswoode.

The fight[2] was in some respects painful, but the duty—was too clear to be shirked or compromised. It is I think the most important step that the Royal Society has taken for many years;—the first I trust in a long course of amelioration & improvement.

I will dine with you some day next week, as soon in fact as you can guarantee me a capital glass of claret—Shall I say Tuesday the 15th?

Give my best love to Hughie[3]—If I were less entangled I would rush down expressly to see the little man.

Yours ever | John Tyndall

RI MS JT/1/T/1296
RI MS JT/TYP/2/1271

1. *[1870]*: the year of this letter has been identified through its reference to Tyndall's efforts to secure the treasurership of the RS for William Spottiswoode (see letter 3272, n. 7).
2. *The fight*: see n. 1.
3. *Hughie*: William Hugh Spottiswoode.

From [Joseph Henry][1] 9 November 1870 3307

Smithsonian Institution | Washington, Nov[ember]. 9, 1870

My dear Sir:

Your letter of October 3d[2] was duly received but an accumulation of business during my sojourn abroad has prevented my giving you an earlier reply. I, however, immediately inserted in the Washington and New York papers the accompanying announcement,[3] which I believe has been extensively copied; I hope it will meet your approbation.

Bidding you farewell at Dublin,[4] we proceeded to Queenstown[5] and embarked on the following Sunday (October 2^{d}.) on the "Russia"—the vessel which took us to Europe—and after a somewhat boisterous passage of twelve days, arrived safely in New York. Fortunately, neither my daughter nor myself were sea-sick; and having a mercators chart and a sensitive aneroid barometer, which had been tested on the Kew Observatory,[6] I amused myself in plotting our course from day to day and studying the changes of the atmosphere. We encountered a storm which occupied six days in our passage through it; it began two days after sailing; the pressure of the air continued to diminish gradually for three days, when, suddenly, the wind changed from a South-Easterly to a North-Westerly direction and the pressure began to increase—continuing to do so until the end of the storm; this at the rate we were sailing would give an atmospheric disturbance of at least 1500 miles in diameter.

My friend the Honorable Hugh McCulloch[7] sails today for London, whither he goes to establish an 'American'-Banking-house connected with the negotiations of United States bonds. Under the administration of Mr. Lincoln,[8] and, subsequently, under that of Mr. Johnson,[9] he was Secretary of the United States Treasury, during which time he managed the financial affairs of the country in such a manner as to command the approbation of the intelligent portion of our people. I have given him a letter to you, with the intimation, however, that too many demands are made upon your time for him to expect any attention from you. Still, if you can, conveniently, invite him, on some occasion, to one of your lectures I would be gratified. He is a gentleman of wealth and influence and of excellent head and heart on whom attention will not be fruitlessly bestowed.

I remain | very cordially & truly, | your friend

Professor John Tyndall, | London—E.

Smithsonian Archives, Coll 7001, Box 12, Folder 4

---

1. *[Joseph Henry]*: Henry has been identified by reference to letter 3273.
2. *Your letter of October 3^{d}*: letter 3273.
3. *accompanying announcement*: enclosure missing.
4. *farewell at Dublin*: see letter 3278, n. 15.
5. *Queenstown*: Queenstown (from 1849–1920 otherwise known as Cobh) is a town located in Cork Harbour, Ireland. It was used as a point of departure for transatlantic shipping.
6. *Kew Observatory*: Kew Observatory is located near the River Thames, Richmond, Surrey. In the nineteenth century it was a significant centre of astronomical research. See L.

MacDonald, *Kew Observatory and the Evolution of Victorian Science, 1840–1910* (Pittsburgh: University of Pittsburgh Press, 2018).

7. *Honourable Hugh M^c^Colloch*: Hugh McCulloch (1808–95), an American financier, worked in banking prior to becoming comptroller of the currency in 1863. He served as secretary of the Treasury (1865–69) under US Presidents Abraham Lincoln and Andrew Johnson. Following the American Civil War (1861–5) he attempted to return the country to the gold standard but was unsuccessful. In 1870 he moved to England to manage the London operations of Jay Cooke, McCulloch & Co (*ANB*).
8. *Mr. Lincoln*: Abraham Lincoln (1809–65), sixteenth President of the United States (1861–65). Lincoln led the United States through the American Civil War (1861–65), abolishing slavery and preserving the Union. While watching a comedy at Ford's Theatre in Washington D.C., he was assassinated by Confederacy supporter John Wilkes Booth (*ANB*).
9. *Mr. Johnson*: Andrew Johnson, seventeenth President of the United States (see letter 2987, n. 20).

From James Fraser[1] 10 November 1870 3308

Manchester. Nov[ember]. 10. 70

My dear Professor Tyndall,

I really have not the courage, with my scant stores of information, & under the great & continuous pressure of local duties which would allow me no time to put into any presentable form the little that I do know, to appear before such an audience[2] as is wont to congregate at the Royal Institution.

Indeed, I hardly expect to spend any time in London next year, certainly not till late in the season.

And as for the Education subject—really it is worn threadbare, & the world (so far as it may care to know them) knows all my ideas upon the question already.

I hope that I may be allowed to call upon you some day when I am in London: or we may profitably meet at the Athenaeum.[3]

I devoutly wish that my work here were not so exacting; for I honestly feel as tho' I never should be able to look into a book again.

Yours very truly | J. Manchester

RI MS HBJ/A/31

---

1. *James Fraser*: James Fraser (1818–85), Bishop of Manchester, graduated BA from the University of Oxford in 1840 and was elected a fellow of Oriel College. He was a tutor from 1842 to 1847, graduated MA in 1842 and became a subdean and librarian in 1844. He entered the diaconate (1846) and the priesthood (1847) accepting the college living of

Cholderton, Wiltshire, which included a fellowship. Fraser became a leading figure in educational matters securing positions on several government commissions into education matters. On 25 March 1870 he was appointed Bishop of Manchester due to his expertise in educational matters (*ODNB*).

2. *appear before such an audience*: Fraser seems to be responding to an invitation to give a Friday Evening Discourse at the RI. No invitation letter has been found.
3. *Athenaeum*: the Athenaeum Club (see letter 2944, n. 1).

From Eudoxie Jamin[1] 11 November [1870][2] 3309

Pornic 11 Novembre

Monsieur,

J'ai été on ne peut plus touchée de votre bonne lettre et je ne saurais vous dire combien je vous suis reconnaissante de l'amitié que vous nous témoignez.

Mon pauvre mari m'avait fait part de votre gracieuse invitation avant que nos communications ne fussent *[coupées]*, en me recommandant de me confier à vous si je n'étais plus en *sécurité ici*. Jusqu'à présent je *[vis]* fort *[tranquille]* dans la retraite que je me suis *[choisie]* et je suis décidée à y rester, car je ne puis m'empêcher d'espérer que les Prussiens ne viendront pas jusqu'ici.

Il y a pour moi une raison grave qui me ferait quitter Pornic avec beaucoup de regrets, c'est que ne pouvant faire passer aucune lettre à mon mari depuis le 18 7bre, il ne saurait plus où je suis et je serais alors privée totalement de ses nouvelles que je ne reçois que fort rarement par ballons *[montés]*. Je lui ai envoyé des lettres par divers *[1 word illeg]* recommandés à cet effet, même par des pigeons-voyageurs; il n'a rien reçu, car il se plaint dans toutes ses lettres des souffrances et des inquiétudes que lui occasionne cette absence totale de rapports avec les siens.

J'ai reçu de lui hier un mot daté du 6; c'est la première fois que j'en reçois un de date aussi récente. Il me dit qu'il se porte toujours très bien et n'a pas encore eu à subir trop de privations; mais il souffre cruellement de l'isolement, de la pensée de nos désastres, et des inquiétudes pour l'avenir.

Permettez-moi, Monsieur, de vous exprimer de nouveau tous mes remercîments pour votre bonté, je ne puis en *[1 word illeg]* pour le moment, mais soyez assuré que si je me voyais dans la nécessité de quitter la France j'aurais recours avec vous avec la plus grande confiance.

Votre bien reconnaissante | E. Jamin

Pornic[3] 11 November

Sir,

I was greatly touched by your kind letter[4] and I cannot tell you how grateful I am for the friendship you have shown us.

My poor husband[5] told me about your gracious invitation[6] before our communications were cut, and advised me to confide in you if I was no longer safe here. So far I live very quietly in the retreat that I have chosen and I will stay here, because I cannot help but hope that the Prussians will not come this far.[7]

There is for me a serious reason that would make me leave Pornic, with great regret; because I have not been able to send my husband any letters since September 18th, he would not know where I am, and I would thus be entirely deprived of the news that I receive extremely rarely by air balloon mail. I sent him letters by various *[1 word illeg]* recommended for this purpose, even by carrier pigeons. He has received nothing since in all of his letters he complains about his suffering and the worries that this total absence of contact causes him.

Yesterday I received a word from him dated on the 6th; it is the first time that I receive one from such a recent date. He says he is still doing well and has not yet been deprived of much, but he suffers cruelly from the isolation, the thoughts about our disasters and the worries about the future.

Allow me, Sir, to reiterate my thanks for all your kindness, I cannot *[1 word illeg]* for the moment, but be assured that if I needed to leave France, I would take your counsel with the utmost confidence.

Your very grateful | E. Jamin

RI MS JT/1/J/6

1. *Eudoxie Jamin*: Thérèse Joséphine Eudoxie Jamin (née Lebrun, 1832–80), the wife of Jules Jamin.
2. *[1870]*: 1870 has been inserted by a second hand. The letter refers to the events of the Franco-Prussian War in 1870.
3. *Pornic*: a region in western France on the Atlantic Ocean.
4. *your kind letter*: any letter between Tyndall and T. J. E. Jamin is missing but this may refer to letters sent by Tyndall to her husband. See n. 6.
5. *My poor husband*: Jules Jamin.
6. *your gracious invitation*: see letters 3261 and 3364.
7. *Prussians will not come this far*: the Prussian army made considerable advances into France during the Franco-Prussian War but did not reach the west coast.

To Elizabeth Dawson Steuart — 14 November 1870 — 3310

14th Nov[ember]. 1870.

My dear Mrs Steuart,

It is a very long time since I had a line from you—very much too long indeed.

I made my usual journey to the Alps this year; but I had a good deal of indoor work to do, and I did not therefore derive as much benefit from the mountain air as I might have done.

I had thought of visiting those terrible battlefields[1] on my return—many of my friends were there giving aid to the wounded. But on reflection I felt that I could do no good, and that it was not a case for mere curiosity: so I came home through Paris.

The only delay encountered was one of five hours due to the furniture of the poor people surrounding Paris which quite encumbered the Railway Station.

Now of course the place is clasped by a band of steel. The whole matter is terrible to the last degree, and incongruous to the last degree in the year 1870.

I have had two notes lately on which I should like a word of your advice. Caleb Tyndall[2] and his son John Tyndall[3] have written to me the letters which I enclose.[4] Now I am willing to lend them a helping hand if it only aids in really bettering their condition and rendering them better and more useful members of society. Dr. Trayer[5] is referred to in one of the notes. Would you ask his opinion whether the money, if sent, would be properly laid out. If so I would send you or him 20 or 30 pounds; though the drafts upon my purse this year have been exceptionally heavy.

When I was at Liverpool[6] in September I received the instructions of the Board of Trade to go to Dublin and report on certain Light House experiments.[7] Could I have squeezed a day out of my time I should have run down to see you. As it was I ran down to Gorey[8] for two or three hours on the Sunday before my work began, and I found Emma[9] in excellent condition. Her health is completely restored, and I do not think I ever saw her so well and so happy.

Yours ever faithfully | John Tyndall.

RI MS JT/1/TYP/10/3388
LT Typescript Only

1. *terrible battlefields*: the battlefields of the Franco-Prussian War (see letter 3238, n. 16).
2. *Caleb Tyndall*: (*c.* 1798–1879), was an Irish farmer and John Tyndall's uncle. In 1841 he achieved notoriety by shooting Mary McAssey while facing a pro-Repeal mob. At his trial he was cleared of malicious wounding. He subsequently served as High Constable for the Barony of Idrone West (County Carlow) and Overseer of the Poor.
3. *John Tyndall*: Caleb's son.
4. *the letters which I enclose*: letters missing.
5. *Dr. Trayer*: James John Trayer.
6. *at Liverpool*: the BAAS annual meeting in September 1870 (see letter 3235, n. 11).

7. *Light House experiments*: see letter 2937, n. 6.
8. *Gorey*: see letter 2996, n. 29.
9. *Emma*: Emily Tyndall.

To Walter Tiffin[1] 14 November [1870][2] 3311

14th Nov[ember].

My dear Sir,

During the last seventeen years I have never once seen a picture resembling that to which your letter refers. We have a fine painting of Sir H. Davy[3] in the anteroom.

Yours faithfully | John Tyndall

W. Tiffin Esq[ui][r][e]

Staatsbibliothek zu Berlin, Slg Darmstaedter 1855 Tyndall Be. 1–58, 3–6

1. *Walter Tiffin*: Walter Francis Tiffin (1879–90), a miniature and landscape painter, print collector, and author. He lived and worked in Salisbury from 1844. Tiffin wrote two books on portraits (W. F. Tiffin, *Gossip about Portraits* (London: H.G. Bohn, 1866); and W. F. Tiffin, *Catalogue of a Collection of English Portraits in Mezzotint* (Salisbury: Bennett Brothers, 1883), the latter in a limited run for private circulation). ('Walter Francis Tiffin (1817–1890)', Royal Academy of Arts at www.royalacademy.org.uk/art-artists/name/walter-francis-tiffin [accessed 4 Jan. 2021]).
2. *[1870]*: this letter has been dated through Tyndall's reference to 'seventeen years' at the RI. He was elected Professor of Natural Philosophy in 1853.
3. *fine painting of Sir H. Davy*: probably the portrait of Humphry Davy painted by Henry William Pickersgill (1782–1875) for the RI in 1831.

From Emily Peel 16 November 1870 3312

VILLA LAMMERMOOR | SÉCHERON |
GENÈVE | Nov[ember]: 16/70

Dear Professor Tyndall,

Your letter & Papers[1] have been forwarded to me here & I gave the Baroness Rothschild[2] the copy destined for her.—She begs me to thank you extremely for having thought of her & only wishes that you were here to explain some of the terms w[hic][h] you make use of & w[hic][h] rather puzzle our uninitiated minds.—I shall look forward to that pleasure when I get back to London, or rather when the Prussians are kind enough to let me return through France[3]—for at present the journey home by Germany with so many children & servants is rather alarming at this season of the year—we are therefore delaying for a while to see how matters turn up—I hope you did not come by Geneva this year,[4] else I should not forgive you not having paid me a visit.

I remain yours | sincerely | Emily Peel

RI MS JT/1/P/53

---

1. *Your letter & Papers*: see letter 3291 and 3291, n. 3. Tyndall sent copies of his paper 'On the Colour of Lake Geneva'. Peel and Rothschild had helped him procure water from the lake. See letters 3156 and 3158.
2. *Baroness Rothschild*: Charlotte von Rothschild.
3. *when the Prussians . . . through France*: the Franco-Prussian War made European travel difficult (see letter 3238, n. 16). Tyndall was caught up in the wartime disruption during his continental travels as well. See letter 3314.
4. *come by Geneva this year*: Tyndall did not visit Geneva in 1870. See letter 3314.

From Mary Egerton 18 November [1870][1] 3313

Thoresby Park, Ollerton. | Nov[embe]r. 18th.

Dear Mr Tyndall,

I promised to let you know when we were returning home. We hope to be back at Mountfield on Wednesday next, and should you be disposed for a few days rest in the country, we shall be delighted to see you at any time; but don't come for civility because you have been asked! If you do come, I hope it may be before the 5th of Dec[embe]r, as May is going away then for a visit, & would not at all approve of missing you!

I suppose though by that time you will be starting for the Eclipse Expedition?[2] What a mess there seems to have been about it! but I hope it will be, "all's well that ends well"!

Are you really going to Etna?[3] Provided we are not at war with everybody I hope so—such a sight must be very grand from the top of a mountain! To watch the darkness spreading below & gradually creeping up, till it gathers you into its solemn gloom! I fancy you climbing the highest crag to enjoy it alone! It would require great self possession to seize the moment for Polarization Experiments—to study the forms of the Corona &c. How anxiously one shall look for the weather report! Fancy a snow storm at the critical moment! I hope there is a Telegraph along Italy, for one shall get no news across poor distracted France[4] at present. Alas! this "Warfare of Michael against Satan",[5] is

I fear introducing a somewhat Satanic Element into the character of the Arch Angel & changing him into a Destroying Angel! With my love for the Germans, I cannot help grieving for their misdoings, as much as for the sufferings of the poor, *[1 word illeg]* the politics of Europe hold out! But no more

Y[ou]rs Ever sinc[ere]ly. | M F Egerton

RI MS JT/1/E/61

---

1. *[1870]*: the year of this letter has been determined through its references to the Franco-Prussian War and Tyndall's planned travel to view the solar eclipse.
2. *Eclipse Expedition*: see letter 3230, n. 2.
3. *Etna*: Mount Etna is a volcano located on the east coast of Sicily, Italy. Although Tyndall considered participating in the British eclipse expedition to Etna, and it was reported in the newspapers that he would, he eventually travelled with another group to Oran, in Algeria. See letter 3318.
4. *poor distracted France*: this letter was written during the Franco-Prussian War (see letter 3238, n. 16).
5. *"Warfare of Michael against Satan"*: see letter 3267, in which Thomas Carlyle calls the Franco-Prussian War 'an instance of Michael versus Satan'. Tyndall was quite impressed with the phrase and repeated it to several of his correspondents. See letters 3283 and 3287.

To Emily Peel 20 November 1870 3314

Royal Institution. | Sunday, 20th Nov[ember]. 1870

Dear Lady Emily,

Some injustice is done me by the mere supposition that I could have passed through Geneva without calling to see you.[1]

At the Bel Alp[2] I had heard that the Crown Prince[3] was marching upon Paris.[4] I immediately set out to compete with his Royal Highness, and I am happy to say competed successfully. But I did not touch Geneva.

When you return to London I will make all that water question clear to you.[5] I have been working a little more at the subject, and shall write a little more about it in a few days.

At Canterbury a few days ago I saw some wonderfully blue water. It is first pumped from the body of the chalk, and is admirably transparent. But it is rendered hard by the chalk which it holds in solution. By the addition of pure lime water the carbonate of lime is precipitated, and allowed to subside. When it has subsided the water it leaves behind is a most marvelous blue. As blue in fact as the sky itself, and in part for the same reason.

I sent you a copy of a sermon[6] which I preached at Liverpool, in which such blueness is accounted for. Probably it has not reached you as it was sent to Whitehall.[7] But in a few days I hope to send both you and your friend the Baroness a second and somewhat expanded edition of the "sermon".[8]

Ever yours faithfully, | John Tyndall.

RI MS JT/1/TYP/3/961
LT Typescript Only

---

1. *without calling to see you*: see letter 3312.
2. *Bel Alp*: Belalp (see letter 3030, n. 20).
3. *Crown Prince*: Friedrich Wilhelm Louis, Crown Prince of Prussia (see letter 3239, n. 9), commanded the six army-corps (alongside the army of the Meuse under the Crown Prince of Saxony) which besieged Paris from September 1870 until January 1871 during the Franco Prussian War (see letter 3238, n. 16).
4. *marching on Paris*: see n. 3.
5. *water question clear to you*: see letter 3312, n. 1.
6. *a copy of a sermon*: 'On the Scientific Use of the Imagination' (see letter 3235, n. 11).
7. *Whitehall*: No. 4 Whitehall Gardens, the Peel home in London.
8. *second and somewhat expanded edition of the "sermon"*: J. Tyndall, *Essays on the Use and Limits of the Imagination in Science,* 2nd edn (London: Green, 1870).

From Elizabeth Dawson Steuart — 20 November 1870 — 3315

Nov[ember]. 20th. 1870.

My dear John,

I have thought it a very long time since I heard from you,[1] and when you were in Liverpool,[2] persuaded myself that you would run across, to see Emma[3] and me. I felt disappointed and sorry not to see you, but was sure that business alone prevented your coming. I have been very unwell or you should have heard before now, the result of my enquiries respecting the two letters you enclosed.[4] I went to the landlord Mr Newton,[5] and there happened to meet Dr. Trayer[6]—it is not true that Mr Newton is willing to accept of John Tyndall,[7] (nor any of his family) as tenants for the farm mentioned, he had never heard a word about it until I told him—there is a lease of half the farm only, and when it expires, Mr Newton will take possession and choose his own tenant, so you now can see the imprudence of throwing away money on such pretences. Under any circumstances, John Tyndall is very unsteady: went twice to America, (I believe at your expense)[8] but was too idle to remain. He applied to Mr Newton to get him a situation on the railroad, which he can do, as gate-keeper, and it is ready for him, if he choose to take it. Indeed you have been a great deal too liberal to that family, who, with very few exceptions, are unworthy of it. I need scarcely add that we all 3, Mr N[ewton]. Dr. Trayer and myself agreed in saying you will not be justified in complying with the request, and I am sure he, Mr N[ewton]. will act as I have already stated

Ever most sincerely yours | E. D. S.

RI MS JT/1/TYP/10/3389
LT Typescript Only

1. *I heard from you*: letter 3310.
2. *in Liverpool*: for the annual BAAS meeting in September 1870 (see letter 3235, n. 11).
3. *Emma*: Emily Tyndall.
4. *the letters you enclosed*: letters missing. See letter 3310.
5. *Mr Newton:* probably Philip Jocelyn Newton, former High Sheriff of Carlow who also served as Justice of the Peace and Deputy Lieutenant. He was the chairman of the Bagnalstown town commission and lived at Dunleckney, Bagnalstown (*Thom's Irish Almanac and Official Directory, 1870* (Dublin: Alexander Thom, 1870), pp. 1184–5, 1326).
6. *Dr. Trayer*: James John Trayer.
7. *John Tyndall*: see letter 3310, n. 2 and n. 3.
8. *(I believe at your expense)*: Tyndall wrote to Steuart on 23 March 1866: 'It was only the day before yesterday that I lost 12 pounds in sending a son of Caleb Tyndall's to America' (letter 2386, in M. D. Barton, I. R. Morus, and J. Ungureanu (eds), *The Correspondence of John Tyndall, Volume 9: The Correspondence, February 1865–December 1866* (Pittsburgh: University of Pittsburgh Press, 2022)). The financial requests did not end, and in June 1867, he wrote to Steuart, 'Caleb Tyndall's son has again written to me, but I have thrown his letter in the fire. He wants the means of carrying him home to Ireland. If his sister thinks it right to devote a portion of the twenty pounds to this purpose she may do so. But I can have no direct communication whatever with this very troublesome and very stupid person' (letter 2634, *Tyndall Correspondence*, vol. 10).

From Mary Adair [*c.* 23 November 1870][1] 3316

Heatherton Park | Wednesday

My dear Dr Tyndall,

When we returned home yesterday my table was burthened with a mountainous pile of letters, circulars &etc. accumulated during our three months absence—and beneath all lay Pandora's gift in the shape of your charming Discourse on the Pleasures of the Imagination.[2] I call it this advisedly, for it must be a real pleasure to exercise the imagination on the noble and fascinating Hypotheses of science. We read the Essay first of all among the beautiful Blair Athole Hills[3]—for it was very well reported in The Times[4]—and now I shall enjoy reading it again in its completed shape, and I thank you much for your kind gift.

We have been wandering all the autumn in the North of England and Scotland and greatly enjoyed our expedition. We heard of you on the Bel Alp[5] and again from Sir Charles Hamilton[6] as we passed through London a fortnight ago. I hope you enjoyed your mountain tour as much as of old.

With our very kind regards believe me always sincerely yours. | H. M. Adair.

RI MS JT/1/A/20

---

1. *[c. 23 November 1870]*: the date of this letter has been estimated: Adair notes she has been away 'all the autumn' and refers to Tyndall's discourse *On the Scientific Use of the Imagination,* which was published in late September 1870 (see letter 3235, n. 11). Since Adair thanks Tyndall for receipt of this publication again in letter 3324 of 30 November 1870, this letter was presumably sent shortly before that one. 23 November 1870 was a Wednesday.
2. *your charming Discourse on the Pleasures of the Imagination*: see n. 1.
3. *Blair Athole Hills*: Blair Atholl is a mountain in the Grampian Mountains of Scotland.
4. *reported in The Times*: most likely 'The British Association for the Advancement of Science', *Times*, 17 September 1870, p. 7.
5. *Bel Alp:* Belalp (see letter 3030, n. 20).
6. *Sir Charles Hamilton*: see letter 3254, n. 6.

From Thomas Henry Huxley 23 November 1870 3317

Athenaeum Club | Nov[ember]. 23rd. 1870.

My dear Tyndall,

About an hour ago, I received a note from Mr Gassiot,[1] informing me that he intends to-morrow to call attention to something or other (what I know not) which I have said, which stands recorded on the minutes of the Philosophical Club.[2] Mr Gassiot has not the ordinary courtesy to inform me what the matter complained of is[3]—, and you observe that, having had two or three weeks before him, he gives me just twenty-four hours notice of his attack.

The result of this unusual procedure is that I am unable to attend; as I made an engagement for Thursday evening, last week.

Considering that I have been on friendly terms with Mr Gassiot, for fifteen or sixteen years; and that I am not aware of having done anything to interfere with perfect openness between us, I cannot say that I greatly relish this proceeding. I have replied to Mr Gassiot, and I have sent an official letter to the secretary requesting him to bring the matter forward in my absence. I hope, if you are present, you will insist on this being done.

I certainly shall not take the trouble to defend myself from proceedings of this kind, but leave the whole matter to be judged by the good sense and the good feeling of the Club.

I am | Ever Yours | T. H. Huxley.

IC HC 8.87
Typed Transcript Only

---

1. *Mr Gassiot:* John Peter Gassiot.
2. *Philosophical Club*: the Philosophical Club of the RS was a dining club consisting of some members of the RS. It was founded in 1847 by William Robert Grove in the aftermath of the reform of the RS's governance. The RS's President was chairman of the Club. Tyndall and Huxley became members in 1855.
3. *what the matter complained of is*: Gassiot objected to a comment Huxley had made at the June meeting about limiting the term of the President of the RS (T. G. Bonney, *Annals of the Philosophical Club of the Royal Society, Written from Its Minute Books* (London: MacMillan and Co., 1919), pp. 63–4).

To Elizabeth Dawson Steuart [*c.* 24 November 1870][1] 3318

Royal Institution

My dear Mrs Steuart,

Many thanks to you for both your last letters.[2] I acted immediately in accordance with your last, by sending back John Tyndall's[3] letter[4] with the remark written on the face of it that I could have nothing whatever to do with people who do not speak the truth.

If while discouraging the untruthful any thing could be done towards aiding the very few good members of this family I should be very glad, and very willing to contribute practically to so desirable a result.

I should be much obliged to you if you would devote the five pounds to the necessities of the poor of Leighlin.[5]

With regards to my going to Etna[6] the news-papers appear to have published my intentions before I had made up my private mind on the matter. I think it is probable that I shall go. There are points about a total solar eclipse[7] which still demand solution, and on this account, if on no other, I should like to see it. The journey will not be without its inconvenience, but I think it ought to be accomplished.

I had a quiet dinner with Sir Henry Holland on Sunday. He is a great traveler and he recommends me to go by the Rhine,[8] by Munich, by Innsbruck, over the Brenner Pass[9] to Verona,[10] thence through to Florence and thence to Naples where I hope a Government ship will be in readiness to take us on to Messina.[11] I shall probably start about the 10th of December.

I am truly sorry to hear of your illness, this weather is very trying: I have been smitten by a cold for some time and it still clings to me. I earnestly trust that you will soon be well again. Under the circumstances it was particularly kind of you to take the trouble that you have taken regarding Caleb Tyndall's family[12]

Believe me | always yours faithfully | John Tyndall.

There is a youth here called William Jeffers,[13] once a letter carrier in Bagnalstown[14] He is seeking employment, and seems to be in low circumstances. I have given him a pound. But a note from Dr. Trayer[15] causes me to doubt him, though he has good characters from clergymen in the neighbourhood of Bagnalstown. You do not know him of course

RI MS JT/1/TYP/10/3391
LT Typescript Only

---

1. *[c. 24 November 1870]*: this letter was sent in response to Steuart's letter of 20 November 1870 (letter 3315) and a letter sent shortly following that one (letter missing). Steuart's response to this letter was sent 28 November 1870 (letter 3322).
2. *your last letters*: letter 3315. The second letter is missing.
3. *John Tyndall*: Tyndall's cousin.
4. *letter*: letter missing.
5. *Leighlin*: Leighlinbridge (see letter 3266, n. 6).
6. *Etna*: see letter 3313, n. 3.
7. *solar eclipse*: see letter 3230, n. 2.
8. *Rhine*: see letter 3022, n. 5.
9. *by Innsbruck, over the Brenner Pass*: a city in Tyrol, Austria, about eighteen miles from the Brenner Pass, which provides passage through the Alps from Austria to Italy.
10. *Verona*: a city located on the Adige River in Veneto, Italy.
11. *Messina*: a sizable city on the island of Sicily, Italy.
12. *the trouble that you have taken regarding Caleb Tyndall's family*: Tyndall and Steuart corresponded regarding his uncle Caleb and Caleb's children, Tyndall's cousins. See letters 3310 and 3322.
13. *William Jeffers*: not identified.
14. *Bagenalstown*: a small town on the River Barrow, in County Carlow, Ireland, where John Tyndall spent his youth.
15. *Dr. Trayer*: James John Trayer.

From Henry Bence Jones 24 November [1870][1] 3319

Dear friend

Thanks for your note[2] but the ventilation is your child[3] not mine I hope it will do well

I am as I was

Y[ou]rs very truly | H Bence Jones | Nov[ember] 24

RI MS JT/1/J/92
RI MS JT/1/TYP/2/723

1. *[1870]*: although this letter did not include a year, the RI building was undergoing renovation in the autumn of 1870 and the ventilation was included in this work. See letter 3278.
2. *your note*: letter missing.
3. *the ventilation is your child*: see note 1.

From Charles Lyell 24 November 1870 3320

73 Harley Street | 24. Nov[embe]r 1870—

Dear Tyndall

I am sorry I cannot respond to your summons to dine at the Philosophical Club[1] to-day & am much concerned to hear that there is any danger of the peace of some of the Members being disturbed[2]—I live so out of the world that I know nothing of the expected commotion to which you allude.

I have just come back from Murchison's[3] and found Bence Jones there who was trying to keep him from the excitement of seeing friends—his mind is quite clear & active & the paralysis would not be alarming if he were not in his 80th year[4]

believe me | ever truly y[ou]rs | Cha[rles] Lyell

RI MS JT/1/L/54
RIMS JT/1/TYP/3/855

1. *Philosophical Club*: the Philosophical Club of the RS (see letter 3317, n. 2).
2. *the peace . . . disturbed*: see letter 3317.
3. *Murchison*: Roderick Impey Murchison.
4. *alarming if he were not in his 80th year*: this is reference to the stroke that Murchison suffered in November 1870. He partially recovered but died of bronchitis on 22 October 1871 (*ODNB*).

To Editor of the *Daily News*[1] 25 November 1870 3321

PROFESOR TYNDALL ON PROFESSOR HUXLEY. | TO THE EDITOR OF THE DAILY NEWS.

SIR,—In your second leading article of this day you refer to certain words of mine uttered at a recent meeting of the ratepayers of Marylebone[2] apropos of Mr. Huxley's candidature.[3] For the sake of those who take an interest in this question, and they seem many, would you permit me to publish in the *Daily News* what I really did say on the occasion referred to? Here it is:—"Mr. Huxley has a large knowledge of men and things, and great capacity to apply that knowledge practically. He knows well that where opinions are so divided as they are in England, a large forbearance is often necessary in the working out of legislation. He will not disguise his sentiments, but he is no *doctrinaire,* and while avowing his own opinions with clearness, to secure a practical end, I know no man more likely to give due weight to the opinions of others. If elected to the *London School Board,*[4] he will be found no pedant, no scoffer, no strainer-out of gnats, no bigot; but, on the contrary, an able, honest, liberal, and courteous colleague. To recur to the most delicate point of all, I can give you the most emphatic assurance that he does not hate the Bible. Like myself, he rather loves and reverences the grand old book. I once heard him say, in regard to the education of his own children, that if nobody else could be found to do it, he would teach them the Bible himself. In a competitive examination upon the Scriptures, canonical and uncanonical, I would back Mr. Huxley against nine-tenths of the clergy. I say 'uncanonical' because he has also the Apocrypha[5] at his fingers' ends. I never heard a heartier Amen! than I did from him in response to some passage in the Baptismal Service at the christening of one of his boys. He had no notion that I should tell you this; and bear well in mind that I do not mention it to bamboozle or hoodwink you into the notion that either of us has any sympathy with the current theological notions regarding either the Bible or the Baptismal Service. I mention it to illustrate to you that large-heartedness and freedom from bigotry which, even now, enable Mr. Huxley to feel and respond to much that is noble and beautiful in the creed of his youth."[6] It certainly never occurred to me that an able Liberal newspaper would detect in these words the least flavor of a dread of the *odium theologicum*.[7]—I am, Sir, your obedient servant,

JOHN TYNDALL. | Athenaeum Club, November 25.

London Daily News, 25 November 1870

---

1. *Editor of the* Daily News: at this time the editor of the *Daily News* was Frank Harrison Hill (1830–1910), an English journalist (*ODNB*).

2. *Marylebone*: neighborhood in the West End of London.
3. *Mr. Huxley's candidature*: Huxley ran for the London School Board, with George Busk, John Lubbock, and Tyndall on his campaign committee (*X Club*, p. 300).
4. London School Board: the School Board for London (1870–1903) was created by the 1870 Elementary Education Act and aimed to provide elementary schooling for the poor of London. Huxley was elected to the Board and to chair a committee established to draw up a curriculum (R. Betts, *Dr Macnamara, 1861–1931* (Liverpool: Liverpool University Press, 1998), p. 58; P. White, *Thomas Huxley: Making the "Man of Science"* (Cambridge: Cambridge University Press, 2003), pp. 121–9; *X Club,* pp. 300, 322–3). The London School Board was open to women, and Huxley was elected in the same year as Elizabeth Garrett Anderson (1836–1917), the first woman to qualify in Britain as a surgeon and physician, who received the most votes of all candidates.
5. *Apocrypha*: a writing or statement of doubtful authorship or authenticity; *spec.* those books included in the Septuagint and Vulgate versions of the Old Testament, which were not originally written in Hebrew and not counted genuine by Jews, and which, at the Reformation, were excluded from the Sacred Canon by the Protestant party, as having no well-grounded claim to inspired authorship (*OED*).
6. *Mr. Huxley . . . creed of his youth*: this passage is partially reproduced in the *Pall Mall Gazette*, 26 November 1870, p. 6.
7. Odium theologicum: hatred of the kind which proverbially characterizes theological disputes (Latin; *OED*).

From Elizabeth Dawson Steuart 28 November 1870 3322

Nov[ember]. 28. 70

My dear John,

I have disposed of £5– as you desired,[1] which leaves me without any cash of yours. The person you mentioned,[2] Mr Jeffers,[3] was in employment as collector of taxes, and I believe letter carrier also, but he embezzled the public money and in order to avoid punishment absconded, and I fancy will not venture into this neighbourhood again. He certainly does not deserve encouragement from any one. It is really most difficult to find those who merit kindness, for instance, you may recollect arranging that your poor Mother's furniture was to be divided between Sarah (now Miller)[4] and her sister Dora.[5] The latter came here a few days ago to make a complaint, that Sarah will not give her anything, if this be true, I regret it much, as I always had such a favourable opinion of Sarah, and if not true, it is not likely that one would invent such a tale of the other. I intend making enquiries, as the statement has left a very unpleasant impression on my mind. I shall feel exceedingly anxious about you, dear John, during your long journey,[6] at this inclement season, and if possible will ask you for a line, to say you are safe and well. Wishing every success, and with kindest regards,

believe me, as ever, | most sincerely yours | E. D. Steuart

RI MS JT/1/TYP/10/3392
LT Typescript Only

---

1. *as you desired*: see letter 3318.
2. *you mentioned*: see letter 3318.
3. *Mr Jeffers*: William Jeffers. Not identified.
4. *Sarah (now Miller)*: Tyndall's cousin Sarah Miller (née Tyndall), daughter of Caleb Tyndall who married David Miller in 1869 (Irish Genealogy).
5. *Dora*: Tyndall's cousin Dorah (or Dorothea) Tyndall, daughter of Caleb Tyndall (Irish Genealogy).
6. *your long journey*: see letter 3230, n. 2

To Margaret Ginty 30 November 1870 3323

Royal Institution | 30th. Nov[ember]. 1870.

My dear Mrs Ginty,

In the midst of my rush of preparations for the eclipse expedition[1] your letter[2] comes to me and takes me away from my duties. Surely I must be a very tender-hearted man to be thus drawn aside from my work to please you.

I did not take the prescription, and though my cold has not entirely disappeared the cough is gone.

I fear in this life we must often limit our desires. Nothing would please me better than to see you in possession of means to keep a house and garden in the country; but nothing would please me worse than seeing you exceeding your means in the purchase of pleasure. I hope you will be able to carry out your wishes in a moderate way.

I am ready to be coaxed into any thing but an infraction of the clear duty of my position—There I think coaxing would fail.

I start on Monday and shall be away for a month. I am going to Gibraltar and thence to Algiers.

Yours ever faithfully | John Tyndall.

RI MS JT/1/TYP/11/3662
LT Typescript Only

---

1. *eclipse expedition*: see letter 3230, n. 2.
2. *your letter*: letter missing.

From Mary Adair 30 November [1870][1] 3324

Heatherton Park | Nov[embe]r 30th—

My dear Dr. Tyndall,

Again I have to thank you for a kind gift. I shall read with great pleasure the Norwich Lecture[2] which I have never come across in any form till now. I think the critiques in "Pros and Cons"[3] unfair and ungenerous—What right have they to say or imply of anyone that he is "Too proud to worship, & too wise to feel?"[4]

There's an old house that I have seen in Scotland with this inscription over the arch "They have sayd. What have they sayd? Let them say."———which is I suppose the highest philosophy after all!

Yours always very sincerely | H. M. Adair.

RI MS JT/1/A/15

1. *[1870]*: this letter has been dated through its reference to 'Pros and Cons Touching the First Edition', in J. Tyndall, *Essays on the Use and Limit of the Imagination in* Science, 2nd edn (London: Longmans, Green, and Co., 1870), pp. 1–12.
2. *Norwich lecture*: see letter 3279, n. 3.
3. *"Pros and Cons"*: see n. 1.
4. *"Too proud to worship, & too wise to feel?"*: from Reginald Heber's (1783–1826) poem 'Palestine (1803)' (*The Poetical Works of Reginald Heber, Late Lord Bishop of Calcutta* (Philadelphia: E. H. Butler & Co., 1870), pp. 43–58, on p. 54).

To Elizabeth Dawson Steuart [*c.* 30 November 1870][1] 3325

My dear Mrs Steuart,

The enclosed[2] only too faithfully confirms what you have said to me regarding the family of Caleb Tyndall.

It causes me to regret that I ever expended a shilling upon the writer who signs herself Dora Tyndall,[3] and of whom I have no recollection.

It would gratify me if you would notify to Mr Corke[4] that his action in this matter though doubtless well meant is mistaken.

Unless you have made very distinct arrangements to the contrary I wish the furniture to remain wholly in the hands of Sarah Tyndall,[5] who has never troubled me

Yours very faithfully | John Tyndall

It is somewhat exasperating, pressed as I am at present to have my time taken up with these barbarians.

A list of furniture returned to Sarah Tyndall by Mrs Steuart.[6]

One Sofa
4 parlour chairs
1 Arm chair
1 glass press
2 ½ (5) blankets
2 patch quilts & rug
1 feather bed
1 pillow
1 bolster
2 matrices
6 kitchen chairs
5 tables
5 sheets
1 table cloth
2 covers
2 pillow cases
1 bolster case
1 clock

1 looking glass
2 boxes
3 tumblers
2 glasses
2 bedsteads
1 carpet & house rug
5 cups and saucers
7 plates & 1 dish
2 jugs, one decanter
2 saltstands
1 Tea Tray
1 glass sugar bowl
1 cream ewer
2 candlesticks
1 tin can & two pots
2 tin saucepans
1 iron bedstead

One glass press
Six kitchen chairs
Four parlour chairs
Two feather beds
Two boxes
Five tables
One sofa
One set of fire-irons
One clock
One picture
One press and a dresser
Five cups and saucers
One bowl
Seven plates and one dish
Two jugs
One decanter
Three tumblers and two glasses
One pair of salt-stands
One tea-tray and a glass jug
Three pairs of blankets
Three quilts and a pair of pillows
One rug
Two bolsters
Two pair of sheets. One table cloth
Two towels
One table-cover
One oil-cloth
Two candlesticks

One tin can
Two pots
Two saucepans
One carpet and hearth-rug

The following is what Mrs Tyndall wants:—

Fifteen yards of calico for Chemise
One pair of sheets 10 yards
Two pillow covers 3 yards
One small shawl for Mrs Tyndall
One bellows & one basket
Miss Tyndall's dresses
4 dresses
2 flannel petticoats
2 other lo.
3 chemises
3 shawls
1 black cloak
1 cloth jacket
1 silk jacket and vest
2 pair of boots
2 pair of stockings

Mrs. Tyndall's dresses.

2 flannel dresses
2 calico dresses
2 day chimeses
2 flannel petticoats

Received from Miss Moore:[7]

3 dresses
2 flannel petticoats
1 quilted petticoat
1 stuff petticoat
2 flannel waistcoats
3 bonnets
1*[1–3 words excised]* and night wrapper
1 bed and bolster
1 pair of blankets
1 quilt and pillow
1 cloak
2 shawls
1 small handkerchief
1 bolster case
1 pair of sheets twilled
1 dress.

Mrs McAssey[8] got the following:

1 bed
1 bolster
1 pillow
1 pr. Blankets
1 Pr. Sheets
1 quilt.

RI MS JT/1/TYP/10/3395–7
LT Typescript Only

---

1. *[c. 30 November 1870]*: the date of this letter is unknown. It likely came between letter 3322 of 28 November, where Steuart informs Tyndall of a disagreement between Sara and Dora Tyndall regarding his mother's furniture, and 5 December, when Tyndall left for his eclipse journey.
2. *enclosed*: letter missing.
3. *Dora Tyndall*: see letter 3322, n. 5.
4. *Mr Corke*: not identified.
5. *Sarah Tyndall*: see letter 3322, n. 4. Note that while referred to as 'Sarah Tyndall' in the letter and 'Mrs. Tyndall' in the list of items, she was married as of 1869, and thus Sarah Miller.
6. *A list . . . Mrs Steuart*: this list does not seem to be part of the original letter but may have been typed by LT to accompany the letter.
7. *Miss Moore*: possibly one of the Moore sisters (see letter 3097, n. 1 and n. 15).
8. *Mrs McAssey*: in December 1867, Tyndall informed Steuart that he intended his mother's bedding to go to John McAssey (see letter 2707, *Tyndall Correspondence*, vol. 10). Tyndall's mother's maiden name was McAssey (sometimes spelled Macassey or Macasey) and one of her brothers was named John. This passage presumably describes items going to Tyndall's uncle and his wife.

To John Herschel 3 December [1870][1] 3326

Royal Institution | 3rd Dec[ember]

My dear Sir John,

Your last card to me has just turned up and it reminds me that I did not thank you for your last letter,[2] which was so full of suggestion. I am on the point of starting for Oran.[3] I go on Monday at 3 P.M. to Portsmouth.[4] Would that I had an opportunity of talking with you about the Corona beforehand.

Yours ever faithfully | John Tyndall

I have been examining water from various sources; but I find in our common tap water a more delicate blue than in most others. Now this blue, unlike the others, cannot be quenched by a Nicol.[5] Do you not think that in the autumn the scattering of leaves into rivers might be sufficient to produce perceptible fluorescence? I must examine the matter fully when I return.

I did not quite seize your meaning about the horse-chestnut bark. It surely does fluoresce splendidly.[6]

RI MS JT/1/TYP/2/575
LT Typescript Only

1. *[1870]*: the year of this letter has been determined through Tyndall's mention of his final preparation to view the solar eclipse of 22 December 1870 (see letter 3230, n. 2).
2. *your last letter*: see letter 3304.
3. *Oran*: see letter 3230, n. 2.
4. *Portsmouth*: a port city in Hampshire, England.
5. *Nicol*: Nicol prism.
6. *It surely does fluoresce splendidly*: see letter 3304.

To Thomas Archer Hirst 5 December 1870 3327

My dear Tom.

I leave in your hands the printed copies of my correspondence with M^r^. Gassiot,[1] and one or two of the last notes which passed between us in manuscript.

His contemplated publication[2] will I hope not render it necessary that a single copy of what I now send you should leave your hands.

This you must judge as soon as you have seen what Gassiot is about to publish.

If you think no rejoinder necessary pray write to Spottiswoode & Co Newstreet Square,[3] and ask them to pull down the type.

Of course if you allow a copy at all to leave your hands the first ought to be sent to Gassiot himself.

And now I think I have said my say on this subject. And having also left a well ordered will & testament, I commit myself to the gods, and bid you an affectionate good bye.[4]

John Tyndall

T. A. Hirst Esq[ui]^re^. | 5th Dec[ember]. 1870

RI MS JT/1/T/682
RI MS JT/1/HTYP/574

1. *my correspondence with M*r. *Gassiot*: Tyndall exchanged several letters with Gassiot on the subject of whether he supported Gassiot's nomination for the treasurership of the RS (see letter 3272, n. 7). A printed version has not been found.
2. *His contemplated publication*: J. Gassiot, *Remarks on the Resignation of Sir Edward Sabine, K.C.B., of the Presidency of the Royal Society* (London, 1870). This was a privately circulated pamphlet in very limited edition, in which Gassiot defended the elderly Sabine from the accusation that he had only resigned as President of the RS, because Gassiot had not been appointed Treasurer (M. B. Hall, *All Scientists Now: The Royal Society in the Nineteenth Century* (Cambridge: Cambridge University Press, 1984, p. 107). See also letter 3317.

3. *Spottiswoode & Co, Newstreet Square*: one of two Spottiswoode publishing houses in London, the other being Eyre and Spottiswoode. The two printers had been operated by William Spottiswoode's father Andrew but upon his death in 1866 ownership split between two branches of the family. William operated Eyre and Spottiswoode, the Queen's printer.
4. *good-bye*: Tyndall was leaving to join the expedition travelling to Algeria to view the solar eclipse of 22 December 1870 (see letter 3230, n. 2).

To the Editor of the *Times*[1] [13 December][2] 1870 3328

[Dr. Tyndall, who has gone out with the eclipse expedition,[3] has sent home from Cadiz[4] some account of the passage out. He writes:—] "After sighting Cape Finisterre[5] we went straight towards Lisbon, and here we were shaken with a vengeance. Day sank (on the 8th) amid squalls and clouds. The wind was high, became higher; hard work to keep oneself from pitching out of berth. The wind rose higher—first half a gale, then a gale, then a strong gale. Some mariners on board would call part of the gale a hurricane. Terrible crash of furniture and glass. The tiller ropes became loose. Hard work to tighten them. Meanwhile the ship was practically rudderless. She heeled round, lay in the trough of the sea, and lurched awfully. One wave smote the stern, and crumpled up a boat which was lashed there. The ship's carpenter went to the second lieutenant and reported that the sternpost was gone and the ship sinking. He was a greenhorn, and did not know the scientific use of the imagination. I dressed with difficulty and went on deck. It was indescribably grand, but very terrible. It was a gale and no mistake. The screw, tossed out of the water, would turn in the air, then plunge again, and shake the vessel horribly. I must say that, though very earnest, I thought, as far as I was concerned, that a worse end might befall me than going down in the performance of a duty. I clung for a time to fixed objects on deck, and watched the waves, and saw the storm take them by the top and carry them away in foam. At length the helm was righted, and the ship, answering to it, was brought round to face the waves. Our speed sank from ten to two knots, and for three days we had a head wind dead against us. About thirty hours had been thus lost, but happily a margin was allowed. We have now good hopes of reaching Oran[6] in time.[7]

Times, 22 December 1870, p. 12

---

1. *the Times*: this letter was copied and published in the *Glasgow Evening Citizen*, 23 December 1870, p. 2, and *Leeds Mercury*, 23 December 1870, p. 3.
2. *[13 December]*: this letter was sent from Cadiz. Tyndall describes this experience in a letter to Hirst of 14 December 1870, a Wednesday, mentioning that he had arrived in Cadiz on Tuesday. See letter 3329.

3. *the eclipse expedition*: see letter 3230, n. 2.
4. *Cadiz*: a port and city in south-western Spain.
5. *Cape Finisterre*: a peninsula on the west coast of Galicia, Spain.
6. *Oran*: see letter 3230, n. 2.
7. *in time*: the published letter did not include a closing quotation mark.

To Thomas Archer Hirst 14 December 1870 3329

Gibraltar[1] 14th. Dec[ember] 1870

My dear Tom

Before I quit "Gib" as the sailors call it I must send you a line to report progress thus far. We quitted Portsmouth[2] at 3 on Tuesday week having been delayed by a fog which prevented the ship's compass from being adjusted. We steamed down the channel,[3] and on the morn[in]g of the second day found ourselves opposite Ushant.[4] Here we committed ourselves to the Bay of Biscay,[5] and on Friday morning emerged from it in front of Cape Finisterre[6]—the "Landsend"[7] of Spain I suppose. All Friday the wind was strong and towards even[in]g it became very wild indeed. It was difficult to dine through the pitching of the plates from side to side of the table. When we went to bed on Friday night both sky and sea looked very wild. I slept for a time, but towards 1 o'C[lock] was roused by the violent lurching of the ship. The timbers creaked and in the cabin various objects were heard spinning to and fro. The screw laboured heavily; sometimes lifted quite out of the sea it twirled in the air, again it plunged and laboured in the waves, shaking the ship violently throughout. At length a blow was delivered which seemed to crumple us up. It did crumple up a strong boat which was slung astern. The cabin was reduced to chaos furniture crashed glasses rattled and the ripest sailor showed signs of unquietude if not of alarm. The tiller ropes got loose, and it was a hard job to tighten them. During this time the ship was practically rudderless. I dressed and went on deck. Holding on like grim death to an iron bolt I watched the sea and felt the storm. The waves might readily be exaggerated to mountains, and now and then those seized upon by the storm were blown far away as foam through the air. Across the full moon a wild scud was driven. The ship had turned her length parallel to the rollers, hence the lateral vibration was for a time of enormous amplitude. Sufficient to bring to mind the fate of the Captain, which was lost not far from the position which we then occupied. I watched the wildness without me, and the workings of my own mind with equal interest. But after 19 hours the wind lulled tho' it still kept ahead. We were again assailed in rounding Cape St Vincent,[8] but here the assault was of comparatively short duration: We got into Cadiz[9] on Tuesday morning—and last night we arrived here.

The weather has changed to that of sunny summer. All today I have been wandering over the rock;[10] on its highest point stands a lightning-riven tower—called O Hara's Folly.[11] I got to the highest point & climbed the tower, so higher I could not go.

We leave this place tomorrow morning & hope that a day & a half's steam[in]g will bring us to Oran.[12]

I have not been ill for a single moment.

Yours affectionately | John.

RI MS JT/1/T/683
RI MS JT/1/HTYP/576

1. *Gibraltar*: a British territory located at the southern tip of the Iberian Peninsula.
2. *Portsmouth*: see letter 3326, n. 4.
3. *the channel*: the English Channel.
4. *Ushant*: a French island at the south-western end of the English Channel.
5. *Bay of Biscay*: a gulf located along the western coastlines of France and Spain.
6. *Cape Finisterre*: see letter 3328, n. 5.
7. *Land's End*: a headland in Cornwall, England.
8. *Cape St Vincent*: a headland in the Algarve, southern Portugal.
9. *Cadiz*: see letter 3328, n. 4.
10. *the rock*: the Rock of Gibraltar, a limestone promontory located in the British territory of Gibraltar.
11. *O Hara's Folly*: originally O'Hara's Tower, this was a watchtower built on the Rock of Gibraltar in the late eighteenth century by the then governor of Gibraltar, Charles O'Hara, that proved unsuccessful as a means of spotting enemies coming from the Mediterranean Sea, since mountainous landscape actually blocked the view of the tower from Cadiz.
12. *Oran:* see letter 3230, n. 2.

From Louis Lortet[1] [late 1870][2] 3330

MUSÉUM D'HISTOIRE NATURELLE DE LYON |
Cabinet du Directeur

Lyon, le

Monsieur,

Mon frère m'écrit de /Rougemont/ au Nord Est de Besançon. Nous sommes toujours très occupés. Demain matin toute /l'ambulance/ part pour aller du côté de Belfort. Moi et quatre autres restons ici pour soigner de nombreux blessés. Nous avons aussi beaucoup de malades, varioles, typhus, fluxions de poitrine. Et comme nous manquons presque entièrement de literie, c'est un spectacle horrible de voir ces malheureux n'ayant plus figure humaine entassés les uns contre les autres sur de la paille infecte. Le pays est tellement ravagé par les */prussiens/* qu'on ne trouve plus même de la paille pour servir de */literie/* à tous ces malheureux. C'est un spectacle plus horrible que tous ceux que j'ai vu jusqu'à présent, etc.

Vous le voyez, la guerre dans notre malheureux pays devient de plus en plus épouvantable. C'est une destruction radicale. Bien que les Anglais aient pour */principe/* « Aide toi et Dieu t'aidera » C'est de ce côté que nous tournons nos regards en espérant la paix—Il faut que l'Angleterre nous donne un coup de main pour nous débarrasser de l'avalanche prussienne.

Quel triste spectacle pour l'homme qui réfléchit: Deux chiens */galeux/* se cherchent querelle pour un os à ronger et aussitôt deux grands peuples prennent fait et cause pour ces grands seigneurs et se font une guerre à mort. Ne serait-il pas temps de faire entre les hommes une véritable ligue de la paix? Les peuples ne doivent plus se laisser mener comme des troupeaux à la boucherie.

Il me tarde de reprendre ma palette et mes pinceaux.—Je ne puis voir le beau Mont Cervin */sans/* penser à votre merveilleuse ascension.

Adieu cher Monsieur Tyndall et croyez à l'expression de tous mes sentiments au milieu de cette tempête. | L. Lortet | peintre

MUSEUM OF NATURAL HISTORY OF LYON[3] |
Office of the Director

Lyon, the[4]

Sir,

My brother wrote to me from */Rougemont/* in the North East of Besançon.[5] We are always very busy. Tomorrow morning all the */ambulance/* will leave to Belfort.[6] Myself and four others will remain here to take care of the many wounded. We also have a lot of sick people, with smallpox, typhus, inflammation of the lungs. And since we lack bedding almost entirely, it is a horrible sight to see these poor people not having a human form as they are huddled against each other on infected straw. The country is so devastated by the Prussians that we cannot find even straw to be used as bedding for the unfortunate ones. It is a more horrible sight than I have ever seen before, etc.

You see that the war in our poor country is becoming more and more terrible. This is an extreme destruction. Although the English had for */motto/*: "Help yourself and God will help you." It is in that direction that we turn our gaze in hope for peace. England must give us a hand to free ourselves of the Prussian avalanche.

What a sad sight for a man who ponders: Two mangy dogs quarrel over a bone and immediately two great nations take up the cause of these great lords and enter a war till death. Is it not time to make a true allegiance of peace between men? People should not be led like sheep to the slaughter.

I cannot wait to take up again my palette and brushes. I cannot lay eyes on the beautiful Matterhorn[7] without thinking about your wonderful ascent.

Farewell dear Mr Tyndall and believe in the expression of my feelings in the midst of this storm. | L. Lortet | peintre[8]

RI MS JT/1/L/37

---

1. *Louis Lortet*: Louis Lortet (1836–1909), a French physician, botanist, and zoologist. Lortet was lecturer in zoology prior to being appointed director of the Museum of Natural History of Lyon in 1869. During the Franco-Prussian War he served as a surgeon.
2. *[late 1870]*: this letter is undated, but the events it describes (the siege of Belfort) indicate that it was written in late 1870 or early 1871.
3. *MUSEUM OF NATURAL HISTORY OF LYON*: Lortet was director of the Museum (see n. 1).
4. *MUSEUM . . . Lyon, the*: Museum headed notepaper.
5. *Besançon*: Besançon is a city in the Bourgogne-Franche-Comté région in eastern France.
6. *Belfort*: a city in the Bourgogne-Franche-Comté region in eastern France. The siege of Belfort took place from November 1870 to February 1871 during the Franco-Prussian War (see letter 3238, n. 16).
7. *Matterhorn*: see letter 2987, n. 17.
8. *Peintre*: a commune in the Jura department in the Bourgogne-Franche-Comté region in eastern France.

1871

To Louisa Baring 6 January 1871 3331

You surely know that I have been on the Coast of Africa.[1] I returned last night and found both your notes.[2] Assuredly I should have gone to you had I been here, and risked the danger attendant in the act.

Give my warm love to May:[3] I hope she will allow me to call her May for a little while.

Always yours | J. Tyndall

National Library of Scotland, Ashburton Papers Acc.11388 N0.101

1. *Coast of Africa*: Tyndall had just returned from his trip to Oran in Algeria to view the solar eclipse (see letter 3230, n. 2).
2. *your notes*: letters missing.
3. *May*: Mary Florence Baring.

To Alfred Russel Wallace 6 January 1871 3332

Royal Institution of Great Britain | 6th Jan[uary]. 71

My dear Wallace.[1]

Nothing gives me greater pleasure than to see my views[2] appreciated by a man like yourself. They are to create in the mass of the public such an interest in scientific things as will enable the leaders of science to obtain fair play for it in all our educational institutions.

To effect this we must appeal to the public in a language they can understand and in a manner with which they can sympathise. This I try to do.

In air and in ocean particles are universally diffused—They never *[1 word illeg]*.

With regard to the "criticism"[3] the cause I assign is a true & demonstrable cause, and renders the flight to a purely hypothetical cause unnecessary. The blue of heaven is by no means a constant quantity. Nor are the morning and the evening red. They vary just as one might suppose them to vary if caused as I suppose them to be caused.

Excuse this scrawl—I have just come from Africa[4] & have a host of letters to answer.

Yours faithfully | John Tyndall

BL Add MS 46439, f. 67

1. *My dear Wallace*: this letter appears to be the response to a letter that is missing.
2. *my views*: Tyndall had likely sent to Wallace his lecture from the 1870 BAAS meeting in Liverpool, 'On the Scientific Use of the Imagination' (see letter 3235, n. 11).
3. *"criticism"*: presumably in Wallace's missing letter, he voiced a disagreement with Tyndall concerning the colour of the sky. In his lecture at the BAAS meeting in Liverpool, Tyndall discussed his own use of imagination in investigating how the blue colour of the sky (and its red colour at dawn or dusk) is a result of the scattering of light through foreign particles in the air (J. Tyndall, *On the Scientific Use of the Imagination* (London: Longmans, Green, and Co., 1870), pp. 16–28).
4. *come from Africa*: see letter 3230, n. 2.

From Henrietta Huxley [6] January [1871][1] 3333

26 Abbey Place | St. John's Wood | January

Dear Brother John,

Can you, will you, dine with us on Friday the 20th. Just to put the New Year on the right line? You need our youthful society to cheer you after your eclipse[2]—If you think that you cannot sufficiently conform to the usages of Europeans—especially of that civilized portion inhabiting 26 Abbey Place (I believe you are still imbued with the dark superstition of Africa), then name a day later than the 20th. We, I especially, will make every allowance for your benighted condition.

Wishing you health, happiness, and improvement in this New Year

I am | Too sincerely Yours | Nettie Huxley.

RI MS JT/1/TYP/9/2943
LT Typescript Only

1. *[6] January [1871]*: Tyndall returned from Algeria on 5 January 1871. On 7 January 1871, Huxley wrote to Tyndall, responding to his suggestion of 26 January for dinner (letter 3335). This letter presumably preceded that one, and Tyndall's response is missing.

2. *your eclipse*: Tyndall travelled with the British expedition to Algeria to view the solar eclipse of 22 December 1870 (see letter 3230, n. 2).

From Eliza Spottiswoode 6 January 1871 3334

50, Grosvenor Place, S.W.

Dear M^r^. Tyndall,

We are so rejoiced to hear that you are back safe[1] in England from the perils of the stormy deep, stormy I fear with a vengeance Do do dine with us on Wed[nesda]^y^ next the 11th to meet Strasse Mme Viardot Garcia,[2] the singer, the sister of Malibran![3]

Hughie[4] intends invading you tomorrow to wish you a happy new year Either before or after the lecture at the R[oyal]. I[nstitution].[5] So if you are busy with lots of things send me word then that you can & will come at 5-30 on Wed[nesday] next.

Y[ou]^rs^ very sincerely | L. Spottiswoode. | Friday Jan[uary] 6. 1871

RI MS JT/1/S/199

---

1. *you are back safe*: Tyndall had recently returned from an expedition travelling to Algeria to view the solar eclipse of 22 December 1870 (see letter 3230, n. 2).
2. *Mme Viardot Garcia*: (Michelle Ferdinande) Pauline Viardot (née García, 1821–1910), Mezzo-soprano, the younger daughter of Manuel García and a highly successful singer noted for her roles in dramatic parts. In 1840 she married Louis Viardot (1800–83). From 1843 she played a leading role in the transmission of Russian music to the west. She also composed operettas and songs (S. Sadie (ed.), *The Norton/Grove Concise Encyclopedia of Music* (London & New York: Norton, 1994), pp. 856–7; M. Steen, *Enchantress of Nations: Pauline Viardot: Soprano, Muse and Lover* (Thriplow: Icon, 2007)).
3. *Malibran!*: Maria Felicia Malibran (1808–36), Mezzo-soprano, sister of Pauline Viardot (1821–1910). She had a significant influence on the Romantic generation of writers and is considered the most charismatic singer of her time. The highlight of her short career was her final season at La Scala (1835–1836). She died from a riding accident at the height of her career (S. Sadie (ed.), *The Norton/Grove Concise Encyclopedia of Music* (London & New York: Norton, 1994), p. 492).
4. *Hughie*: William Hugh Spottiswoode.
5. *the lecture at the R[oyal.] I[nstitution].*: presumably one of the Christmas lectures for juveniles, which William Odling delivered on 27, 29, and 31 December 1870, and 3, 5, and 7 Janaury 1871, on the topic of 'Burning and Unburning' (*Roy. Inst. Proc.*, 6 (1870–72), p. 187).

From Henrietta Huxley — 7 January [1871][1] — 3335

26 Abbey Place | St. John's Wood | 7th January

Dear Brother John

The 26th will suit us admirably—Therefore I have put you down for that day—Did you hear what the youngest Spottiswoode[2] inquired when he heard of the wreck of the Psyche?[3] "Is jolly old Tyndall lying at the bottom of the sea Mother?" And this in the most cheerful tone! Naturally in just such a cheerful spirit should I have received that lock of hair you alluded to. Hal[4] is glad to hear that our American cousins have printed so many copies of his essays.[5] He only hopes that they may <*1 word missing*> "please to remember the author".

Do you know a Mrs Berkley[6] in Hamilton Terrace—She says you dined there the other day—So near and yet so far! Is Mrs Berkley a desirable woman as a woman to know? I dont care to make fresh acquaintances unless they are of some good stuff. She called here yesterday and is anxious to be intimate therefore I ask you.

I have to reproach you for Mr Spencer's[7] evil and pernicious doings this Xmas—at which I have been much troubled. He brought a huge, rich unwholesome cake and boxes of sweeties for the children—and when I inveighed against this—said it was all owing to you. Of course, as you justly observed, had you a wife you would have known better. Allow me to point out how necessary it is to be careful that we sin not, for not only does it harm ourselves and those against whom we offend but that our example—may lead from the paths of rectitude even such a man as Herbert Spencer—A man who has stupendously evolved the best course of hygienic treatment, and indicated the loftiest aberrations of the soul, to offer plum cake at the shrine of helpless and unoffending infancy! How are the mighty fallen!

Yours more in sorrow than in anger | Nettie Huxley. | Write and confirm the 26th.

RI MS JT/1/TYP/9/2944
LT Typescript Only

---

1. *[1871]*: the year is given by relation to letter 3333.
2. *the youngest Spottiswoode*: Cyril Andrew Spottiswoode (1867–1915), son of William and Eliza Spottiswoode, would later serve as director of the printing firm Eyre and Spottiswoode Ltd.
3. *the wreck of the Psyche*: the H.M.S. *Psyche* was part of British efforts to observe the 1870 solar eclipse. It ran aground on a rock near Catania (Sicily) on 15 December 1870. Cyril Spottiswoode was mistaken, as this was not the ship that Tyndall traveled on to observe the eclipse (see letter 3230, n. 2).

4. *Hal*: nickname for Thomas Huxley.
5. *copies of his essays*: *Lay Sermons, Addresses, and Reviews* (see letter 3217encl.).
6. *Mrs Berkley*: possibly the wife of George Berkley of 32 Upper Hamilton Terrace, a British civil engineer who was a Manager of the RI at this time (*Minutes of Proceedings of the Institution of Civil Engineers*, 115 (1894), pp. 382–5).
7. *Mr Spencer*: Herbert Spencer.

To Elizabeth Dawson Steuart 9 January 1871 3336

Royal Institution | 9th Jan[uary]. 1871.

My dear Mrs Steuart,

I reached London safely[1] a couple of days ago. On our outward voyage we passed the Bay of Biscay[2] without violent tossing, but between Cape Finisterre[3] and Oporto[4] we encountered a storm[5] which to the landsmen at least was both grand and terrible. It endured more or less throughout two days; but on one night it reached its climax and really matters looked very threatening. I was not in the least sick during this storm so was able to go upon the poop or higher seat, cling to a fixed object and watch the storm. It was really wonderful and terrible at the same time. After we had rounded Cape St. Vincent[6] the wind batted. We reached Cadiz,[7] landed a party there and proceeded to Gibraltar.[8] "The Rock"[9] is full of interest: and I scampered almost over the whole of it. Seeing its galleries and its cannon, and its precipices. We had gone from England in cold winter weather: at Gibraltar we found it summer.

Thence we proceeded through the Mediterranean to Oran.[10] On the day we reached Oran the temperature was like that of a July day in England. The sun was shining in a blue heaven. But the wind changed, heavy clouds were rolled upon us and it was under these unfavourable auspices that we erected our instruments and practised the methods of observation. Two of the authorities comforted us by the assurance that the bad weather would not last, and that on the day of the eclipse we should probably have the air limpid and the sky clear. Well the day came, but the clouds came with it. I had my instrument mounted on the roof of a bastionet, on the wall erected to defend Oran from the Arabs. I had glorious glimpses of the advancing moon, and the shining solar crescent. But as the lights became dimmer the cloud became denser, and a few minutes before totality the sun and moon were totally hidden. We were in short defeated. But the darkening of the earth was still very fine. I immediately occupied my thoughts with other matters which interested me, and these drove disappointment and despondency from my heart.

The population of Oran is strikingly various and strikingly picturesque, French, Spaniards, Jews, Arabs and negroes. I was immensely interested in the place. We quitted it on the 24th as a storm was raging on the 23rd. On the 25th we were much knocked about, and the Captain ran in for shelter into

Almiera Bay[11] under the Sierra Nevada.[12] This mountain chain is far grander than I supposed it to be. On the 26th we steamed to Gibraltar where we spent two days in coaling, during which I examined the celebrated St. Michael's cave,[13] and many other things. I send you a bit of the stalagmite from this cave. The stalactitic pillars were truly grand, and the drapery of "congealed stone" as the people call it wonderful.

We halted at Cadiz for a day to see the city: finally got fairly off, and with some tossing in the Bay of Biscay reached Portsmouth[14] on Thursday morning; on Thursday night I was in London.

I came wonderfully well out of all the tossing.

Yours ever truly | John Tyndall

I have to thank you for replying to Mr Hirst[15] in my absence. If J.T.[16] applies to me[17] through you and if you sanction the application, I shall be ready to do for him what I promised, and wished to do before his untruth compelled me to decline all direct communication with him.

RI MS JT/1/TYP/10/3400–1
LT Typescript Only

1. *I reached London*: Tyndall had just returned from Algeria, where he traveled to observe the solar eclipse of 22 December 1870 (see letter 3230, n. 2).
2. *Bay of Biscay*: see letter 3329, n. 5.
3. *Cape Finisterre*: see letter 3328, n. 5.
4. *Oporto*: a city located along the Atlantic coast of Portugal.
5. *encountered a storm*: see letter 3329.
6. *Cape St. Vincent*: a headland in the municipality of Vila do Bispo, in the Algarve, southern Portugal.
7. *Cadiz*: see letter 3328, n. 4.
8. *Gibraltar*: see letter 3329, n. 1.
9. *The "Rock"*: the Rock of Gibraltar is a monolithic promontory that is the most significant physical feature of Gibraltar.
10. *Oran*: see letter 3230, n. 2.
11. *Almeria Bay*: located at the Sierra Nevada mountain range on the southern coast of Spain.
12. *Sierra Nevada*: a mountain range in the regions of Andalusia, Málaga, and Almería in Spain.
13. *St. Michael's cave*: located in Gibraltar, St. Michael's cave is a network of limestone caves.
14. *Portsmouth*: see letter 3326, n. 4.
15. *Mr. Hirst*: Thomas Hirst.
16. *J.T.*: John Tyndall, son of Caleb Tyndall and cousin of Tyndall.
17. *applies to me*: see letters 3315 and 3318.

To Margaret Ginty 11 January 1871 3337

Royal Institution | 11th. Jan[uary]. 1871.

My dear Mrs Ginty,

Next week will be a fearfully busy one with me; but if you are here on the Saturday I should be able to give you all the time you wish for.[1]

But any other day—say Wednesday, I can give you half an hour—Wednesday at 2 o'clock you can have half an hour, or indeed any other day at 2 o'clock except Friday.

Yours most faithfully | John Tyndall.

Every mother must expect her boy to become a man.[2] Her pleasure and her happiness ought to consist in seeing her son doing his duty bravely and successfully in the world—not in keeping him like a domestic pet at home.

Write down the name of the day you choose and send it to me. I shall then be sure to be here.

RI MS JT/1/TYP/11/3663
LT Typescript Only

---

1. *give you all the time you wish for*: Tyndall served as Ginty's trustee after the death of her husband and also advised her in family and financial matters. See letters 3062, 3113, 3191, and 3323.
2. *her boy to become a man*: Ginty's son William had studied at Queenswood. Ginty occasionally sought advise from Tyndall on his education and his career prospects. See letter 3191.

From Elizabeth Dawson Steuart 11 January 1871 3338

January 11. 1871.

My dear John,

I cannot express how truly glad I felt this day, when your letter[1] reached me, and hasten to thank you for it, and the very nice specimen you so kindly sent me, which I highly value, coming from one in whom I have taken such sincere interest from his childhood. It is curious, how often I have <u>wished</u> for a keepsake from you, one that I could <u>wear</u>, and say it was <u>your</u> gift! My wishings are seldom realized, and I am therefore <u>doubly</u> pleased in this instance, so I shall have the stone set as a brooch, and feel quite <u>proud</u> of it! I am truly rejoiced and thankful that you are safe in London.[2] Having read of your misfortune in the "Psyche",[3] I felt very anxious and uneasy about you, and fully intended writing to Mr Hirst[4] to enquire if he had heard from you. You had indeed, a most providential escape and I trust the change and relaxation have benefitted your health. Our weather here has been unusually severe, and numbers are feeling ill effects from it, but I hope after a little time, Spring will bring us sunshine and flowers. I do not think that J. Tyndall[5] will apply to me: he is aware of my dislike to untruthful, unsteady people and also of how unwilling I always am, that your money should be given to help those who are too idle and good for nothing to help themselves. He has a situation, and it will be his own fault, if he do not keep it. If you have a friend who is not too busy, and would kindly do a little matter for me, I should feel truly obliged: if the least inconvenience, do not mind it. You will see by the enclosed the nature of it.[6]

With best wishes for many happy new years
most sincerely yours | E. D. Steuart.

RI MS JT/1/TYP/10/3402
LT Typescript Only

1. *your letter*: see letter 3336.
2. *safe in London*: Tyndall had just returned from Algeria, where he traveled to observe the solar eclipse of 22 December 1870 (see letter 3230, n. 2).
3. *your misfortune in the "Psyche"*: the H.M.S. *Psyche* was part of British efforts to observe the 1870 solar eclipse. It ran aground on a rock near Catania (Sicily) on 15 December 1870. Steuart was mistaken, as this was not the ship that Tyndall traveled on to observe the eclipse (see letter 3230, n. 2).
4. *Mr. Hirst*: Thomas Hirst.
5. *J. Tyndall*: John Tyndall, son of Caleb Tyndall and cousin of Tyndall. See letters 3315, 3318, and 3336.
6. *the enclosed the nature of it*: enclosure missing.

From James Joseph Sylvester 11 January 1871 3339

Athenaeum | 11th Jan[uar]y 1871

My dear Tyndall,

I enclose herein Shadworth Hodgson's[1] letter[2] giving his full description to enable you to execute your kind intention of proposing him for the Athenaeum.[3]

Yours very truly | J. J. Sylvester

RI MS JT/1/S/294
RI MS JT/1/TYP/4/1511

1. *Shadworth Hodgson's*: Shadworth Hollway Hodgson (1832–1912), an English philosopher, was educated at Rugby School and Corpus Christi College, Oxford (1850–54). Hodgson was a member of the Metaphysical Society throughout its existence and was president of the Aristotelian Society from 1880 to 1894, its popularity increasing significantly during his stewardship. He was a founding member of the British Academy (elected 1901), an honorary fellow of Corpus Christi College (elected 1882), and was awarded an honorary LLD from the University of Edinburgh (*ODNB*).
2. *letter*: enclosure missing.
3. *proposing him for the Athenaeum:* Hodgson did not become a member of the Athenaeum Club until 1887. He was proposed by Tyndall and seconded by Sylvester (Jennie De Protani, Archivist, the Athenaeum Club, personal communication).

From Emilio Ruiz de Salazar[1] [17 January 1871][2] 3340

Monsieur John Tyndall

Professeur de l'Institute Royal de Londres

Monsieur et très honorable confrère: la bonne renommée a *[ses] [charges]* et a présent j'ira vous imposer une en vous priant de me répondre si tôt que vous soit possible.

Je suis depuis six années Professeur agrégé et Faculté des sciences à l'Université de Madrid et en qualité de Professeur et au nom de la confraternité scientifique j'espère que vous seriez *[très] [bien] [pour me faire]* plaisir de me satisfaire.

J'ai besoin de faire deux travaux dans peu de temps l'un Un Programme raisonné de Physique Mathématique et l'autre Une mémoire pour les sources du connaissance et méthodes d'enseignement dans cette science.

Connaissant vos travaux très excellents je vous prie instamment que vous me conseillez ce que je dois faire pour réunir dans mes propres—

Ainsi en vous priant que vous *[consenttez]* et que vous croyerez au surplus de *[convenance]* je désire obtenir vottre réponse dans les demandes suivants, bien entendu que vous me feriez un grand service.

1º Que est ce qu'on entend par Un Programme raisonné et par un mémoire sur la méthode de l'enseignement et sources de connaissance

2º La meilleure *[aupteur] [au] [raport] [de]* Physique Mathématique et la manière de la developpem[ent]

3º Qu'on entend par Physique Mathématique ses *[faces]*—ses *[deditions]*—ses fondements rationals—ses lois—ses particularités, ses *[1 word illeg]*, son état actuel et sa relation avec les autres sciences.

4º Laquelle sur les *[sources]* des connaissances et la méthode d'enseignement dans la Physique Mathématique—

5º Votre historique et bibliographique de la même

6º *[Degré]* de développement de cette science et nature des établissements ou ses *[1 word illeg]* son enseignement dans le monde scientifique: avec quel caracter et avec quel *[1 word illeg]* et noms des Professeurs

Je vous prie de me répondre et ce sera un très grand service que je n'oublierai jamais.

Je *[me]* *[1 word illeg]* à vottre *[disposition]* et vous m'avez à Madrid *[disposé]* à vous servir en tout.

Veuillez Monsieur l'assurance de ma profonde considération très distinguée | Emilio Ruiz de Salazar | Professeur à l'Université de Madrid

Mon adresse—calle del Horno de la Mata nº 12.2—ou bien Directeur du Journal de l'Instruction publique El Magisterio Español—calle del *[Olivo]*—11—10

Mister[3] John Tyndall

Professor of the Royal Institution of London

Sir and most honourable colleague: a good reputation has its burdens, and at present I am imposing one on you by asking you to respond as soon as is possible for you.

For the past six years I have been a teaching Professor and in the Faculty of Science at the University of Madrid, and as a Professor and in the name of scientific collegiality I hope that you will be well disposed *[to give me]* the pleasure of satisfying my request.

I must do two pieces of work in a short amount of time, the one A Rational Program of Mathematical Physics and the other An Essay on the Sources of Knowledge and the Methods of Teaching this Science.

Knowing your work to be excellent, I ask you to advise me quickly on what to do to bring together my own—

So, in asking you to *[agree]* and that you believe in the surplus of *[propriety]* I wish to obtain your response to the following questions, with the understanding that you will be doing me a great service.

1º What do we mean by a Rational Program and by an essay on the method of teaching and sources of knowledge

2º The best *[author]* where Mathematical Physics and the manner of its development is concerned

3º What we mean by Mathematical Physics, its *[faces]*—its *[1 word illeg]*—its rational foundations—its laws—its particularities, its *[1 word illeg]*, its current state and its relationship with the other sciences.

4º The one on the *[sources]* of knowledge and the method of instruction in Mathematical Physics

5º Your history and bibliography of the same

6º The degree of development of this science and nature of its establishments or its *[1 word illeg]* its instruction in the scientific world: with what character and with what *[1 word illeg]* and names of Professors

I ask you to respond to me and it would be a very great favour that I would never forget.

I *[1 word illeg]* myself at your disposal and you have me in Madrid ready to serve you in everything.

Be assured, Sir, of my most profound and distinguished consideration. | Emilio Ruiz de Salazar | Professor at the University of Madrid

My address—calle del Horno de la Mata nº 12.2—or instead the Director of the Journal of Public Instruction El Magisterio Español—calle del *[*Olivo*]*—11—10

RI MS JT/1/D121

1. *Emilio Ruiz de Salazar*: Emilio Ruiz de Salazar (1843–98), Professor of Mathematics at the University of Madrid. Strongly interested in the promotion of education, de Salazar founded the journal *El Magisterio Español* in 1867. He received his licentiate in 1869 and taught a range of abstract mathematical subjects, including mathematical physics, before moving to the chair of analytical mathematics in 1877. In 1878 he founded the Society for the Protection of Animals and Plants (dbe.rah.es/biografias/30285/emilio-ruiz-de-salazar-usategui [accessed 10 March 2020]).
2. *[17 January 1871]*: on 21 March 1871 Salazar wrote to Tyndall inquiring whether his letter of 17 January had been received and repeating much of the content found here (RI MS JT/1/D/12; this will be published in the forthcoming twelve volume of *The Correspondence of John Tyndall*).
3. *Mister*: Salazar was not a native French speaker. Inaccuracies in the original have been retained in the English translation.

From Nicholas de Khanikoff[1] 18 January 1871 3341

18 Janvier 1871

Cher monsieur,

Je m'empresse de vous remercier de Votre *[*obligeante*]* permission d'assister à Votre conférence du vendredi soir. Celle de *[*Dundee*]* m'a laissée une trop bonne impression pour que je ne sois pas très sensible à cette *[*faveur*]* si *[*recherchée*]* par les nombreux admirateurs de Votre parole aussi éloquente qu'instructive.

Agréez Monsieur l'assurance de ma haute consideration. | Votre tout dévoué | N. de Khanikoff

18 January 1871 |

Dear Sir,

I hasten to thank you of your /kind/ permission to attend your discourse on Friday night.[2] Dundee's presentation[3] left me with such a favourable impression that I could not help but be touched by this opportunity, much sought after by the many admirers of Your discourse which is as eloquent as it is instructive.

Please accept Sir the assurance of my highest consideration. | Your devoted | N. de Khanikoff

RI MS JT/1/D/87

1. *Nicholas de Khanikoff*: Nicholas de Khanikoff (1819–78), a Russian specialist on Afghanistan and the Middle East who published work on Persia (1866), Central Asia (1861) and Constantinople. He received the *grande médaille d'or,* of the French Société de Géographie in 1861 ('Nicholas de Khanikoff', *The Athenaeum*, 30 November 1878, p. 690).
2. *your discourse on Friday night*: Tyndall's delivered a Friday Evening Discourse at the RI on 20 January 1871, 'On the Colour of Water, and on the Scattering of Light in Water and in Air' (*Roy. Inst. Proc.,* 6 (1870–72), pp. 189–98).
3. *Dundee's presentation*: on 5 September 1867, Tyndall gave a lecture on 'Matter and Force' to working-class men at Kinnaird Hall, Dundee, where the annual meeting of the BAAS was held (*Brit. Assoc. Rep. 1867*, p. lxxiii). See 'Lecture to Working Men by Professor Tyndall', *Dundee Advertiser*, 6 September 1867, pp. 2–3; and 'Professor Tyndall's Lecture to Working Men. Matter and Force', *Dundee Courier and Argus*, 6 September 1867, p. 3. This appears as chapter four of *Fragments of Science*, pp. 69–94.

From Silicated Carbon Filter Company[1] 18 January 1871 3342

PRIZE MEDAL | 1867 | PARIS EXHIBITION

SILICATED CARBON FILTER COMPANY, | CHURCH ROAD, BATTERSEA. | *LONDON, S.W.*[2]

18. Jan[uar]y 1871
Prof. Tyndall | Royal Institution

Sir

At the request of[3] Dr H[enr]y Bence Jones and Dr R. Bowles[4] we forward for your inspection a bottle of the ordinary water supplied by the Southwark & Vauxhall Water Co.,[5] which has been passed once through a Silicated Carbon Filter—

We believe that this sample of filtered water will be found in a state of purity unattainable by any process except distillation, and as we are given to understand that this subject is just now engaging your attention, we take the liberty of submitting it to you—

Y[ou]r obed[ien]t serv[an]ts. | The Silicated Carbon Filter Co

RI MS JT/1/S/75

1. *Silicated Carbon Filter Company*: increasing awareness and understanding of water-borne diseases led to the establishment of companies offering water filters. This company was located in Battersea, London.
2. *Prize Medal… London, S. W.*: company headed notepaper.
3. *At the request of*: Robert Bowles was involved in a legal dispute with the Waterworks Company about the quality of Folkestone water and had solicited an analysis from Jones ('Office of Exchequer', *Journal of Gas Lighting, Water Supply & Sanitary Improvement*, 24 (14 July 1874), pp. 50–1).
4. *R. Bowles*: Robert Bowles.
5. *Southwark & Vauxhall Water Co.*: formed in 1845, the company supplied water to premises and homes in London.

From Edward Frankland 19 January 1871 3343

14 Lancaster Gate | Hyde Park W. Jan[uary]. 19/71

My dear Tyndall

Herewith I send you the promised pamphlet[1] which please return when done with. Dr. A. Wood[2] & Dr. Littlejohn[3] are high authorities on sanitary matters; the latter is Medical Officer of Health for the city of Edinburgh & a man of very different spirit from the Medical Officer of the city of London.[4] I send you also a copy of my letter[5] to Dr. Latham[6] who tells me by this mornings post that he has sent a reply to Letheby's[7] letter. Letheby is at the present moment a most serious obstacle in the way of sanitary reforms. You will always find him on the side of joint stock companies and against the public—companies pay well the public does not pay,—voila! Last session he condemned before a parliamentary committee[8] a water for the supply of the town of Brechin[9] because it had 110 of hardness and recommended another water as very suitable because it had only 3.8° of hardness. The following is the hardness of Cambridge water:—

As now supplied—25.3°

After Clark's process[10]—5.7°

Thus you see, for an adequate fee, he condemns a water as too hard which is less than half as hard as the Cambridge supply, and recommends in the place of it another which is softer than the Cambridge after Clark's process.

The following is an extract from my report to the Registrar General[11] on the Canterbury water Nov. 26/70. "I know of no better water than this in the United Kingdom. xxx The chalk, oolite and greensand around the Metropolis contain abundance of water, which is of unsurpassed purity after being submitted to this simple process."

As regards the "previous animal contamination" of these waters my meaning has been so grossly misrepresented by Letheby & others that I ask your attention to the accompanying memorandum[12] on the subject which will show you what I really have said.

Can you spare me a few more tickets[13]—6 if possible?

Yours sincerely | E. Frankland.

P.S. Winchester Quart[14] samples of all the London waters will be collected today & delivered for you at the Royal Institution this afternoon.

RI MS JT/1/F/50

---

1. *promised pamphlet*: enclosure missing. See letter 3344.
2. *Dr. A. Wood*: Andrew Wood (1810–81), a Scottish physician. Wood succeeded his father in several public medical appointments, including as surgeon at Heriot's Hospital. He was elected fellow of the Royal College of Physicians of Edinburgh (1840), and later served as its president (1858–61). He represented the Royal College of Surgeons of Edinburgh on the Medical Council and protected the interests of the College. Wood strongly opposed allowing women to study medicine at the Edinburgh Medical School and blocked them from obtaining medical degrees ('A. Wood. Obituary', *Edinburgh Medical Journal*, 26:2 (1881), pp. 854–61).
3. *Dr. Littlejohn*: Henry Duncan Littlejohn (1826–1914), a Scottish physician, expert in forensic medicine, and medical officer for public health in Edinburgh. Littlejohn was appointed police surgeon in 1854 and proved himself a very effective medical examiner. After several years investigating the sanitary conditions of Scottish municipalities, he was appointed medical advisor to the Board of Supervision for the relief of the poor in 1873. In 1862 he was appointed Edinburgh's first Medical Officer of public health. Littlejohn held down several positions in public health administration simultaneously, edited the *Edinburgh Medical Journal* for several years, was elected president of the Royal College of Surgeons of Edinburgh in 1875, taught forensic medicine at the surgeon's hall and became professor of forensic medicine and public health at the age of seventy-one. His *Report on the Sanitary Condition of Edinburgh* (1865) highlighted the deficiencies in the city's sanitary practices and their sequelae (*ODNB*).
4. *Medical Officer of the city of London*: Henry Letheby (see n. 7).
5. *a copy of my letter*: enclosure missing.

6. *Dr. Letham*: probably Baldwin Latham (1836–1917), an English civil engineer and meteorologist. He designed and published papers on the sewerage, irrigation, and water works for fifteen towns, and is best remembered for his contributions to public water supply. He was a member of the Royal Sanitary Institute (1876), the Geological Society (1876) and the Royal Meteorological Society (1876), where he served as president from 1890 to 1891 (*ODNB*).
7. *Letheby's*: Henry Letheby (1816–76), an English analytical chemist and public health official. He was professor of Chemistry in the London Hospital (1848) and Medical Officer of Health for the City of London (1855–74). He was one of the principal scientific consultants of the London water companies. Because his reports were underwritten by the companies, his views of the water supply were often positive, and in 1867 he was the only expert who did not blame the water for the cholera outbreak in east London (*ODNB*).
8. *he condemned before a parliamentary committee:* while no published report has been identified, there was a select committee 'Public Health (Scotland), Supplemental (so far as it relates to Brechin)' named in the 'Return of Number of Select Committees appointed, Session 1870', Parliamentary Paper 0.130 (1870).
9. *Brechin*: a town in Angus, Scotland.
10. *Clark's process*: a method of water softening invented by Scottish chemist Thomas Clark (1801–67) and patented in 1841. The method uses lime to merge free carbonic acid with bicarbonates of calcium and magnesium which form carbonates and hydroxides.
11. *Registrar General*: this later became the Registrar General of Births, Deaths, and Marriages. The Registrar General produced 'Reports on the Metropolitan Waters' as part of its public health and sanitation monitoring. Frankland worked as a water analyst for London from 1865 to 1876 and contributed to these reports. On Frankland and his role as a water analyst for London, see C. A. Russell, *Edward Frankland: Chemistry, Controversy and Conspiracy in Victorian England* (Cambridge: Cambridge University Press, 1996), pp. 362–410). On the history of water analysis in general, see C. Hamlin, 'Politics and germ theories in Victorian Britain: the Metropolitan Water Commissions of 1867–9 and 1892–3', in R. MacLeod (ed.), *Government and Expertise: Specialists, Administrators and Professionals, 1860–1919* (Cambridge: Cambridge University Press, 2003), pp. 110–27; and C. Hamlin, *A Science of Impurity: Water Analysis in Nineteenth Century Britain* (Berkeley: University of California Press, 1990).
12. *accompanying memorandum*: enclosure missing.
13. *spare me a few more tickets*: to Tyndall's Friday Evening Discourse at the RI on 20 January 1871 (see letter 3341, n. 2).
14. *Winchester Quart*: a type of glass bottle used for laboratory applications. It had a capacity that was half of an imperial gallon (2.273045 litres) and took its name from the old medieval Winchester measurements.

From Edward Frankland 19 January 1871 3344

R.C.C.[1] Jan[uary]. 19/71

My dear Tyndall

I do not know whether you intend to allude to the conclusions of the Royal Commission on Water Supply[2] but if you do, you ought to glance over the accompanying document[3] which contains my answers to their strictures on my water reports.[4] The Commissioners have let slip a glorious opportunity of recommending the chalk water[5] although I repeatedly pointed out its excellence to Prestwich[6] before their Report was presented. However they have done one piece of good service in recommending a constant supply and the purchase of the rights of the private companies.[7] Their recommendation of river water for London will not be acted upon.

Please return the enclosed[8] when you send back the Edinburgh pamphlet[9] which I forwarded to you this morning by post.

Yours sincerely | E. Frankland.

On Tuesday last I inspected the conduit through which tidal Thames water was admitted at Battersea[10] into the service reservoirs of the Southwark & Vauxhall Company[11] in August September & October 1868 and distributed to consumers. See page 259 of enclosed pamphlet[12] & compare Dr. Letheby's[13] reports of this water at that time on the same page. The Board of Trade[14] will now probably compel the abolition of this conduit, which is situated close to the mouth of one of the largest sewers—King's Scholars Pond Sewer.[15]

RI MS JT/1/F/51

1. *R.C.C.*: the Royal College of Chemistry was established in October 1845 through the efforts of James Clark, the Queen's physician, and Albert, the Prince Consort. It was based in Oxford Street, London, and in 1853 it was incorporated with the Government School of Mines and of Science Applied to the Arts into the Metropolitan School of Science Applied to Mining and the Arts. Frankland succeeded August Wilhelm Hofmann (1818–92) as lecturer there in 1865 (E. Frankland Armstrong, 'The Royal College of Chemistry', *Nature,* 156 (3 November 1945), pp. 524–27; G. K. Roberts, 'The establishment of the Royal College of Chemistry: An investigation of the social context of early-Victorian chemistry', *Historical Studies in the Physical Sciences*, 7 (1976), pp. 437–85).
2. *Royal Commission on Water Supply*: this commission was constituted in 1866, recommissioned in 1867, and reported in 1869. Its mandate was to inquire into the best means of preventing the pollution of rivers and the domestic water supply of Great Britain. It produced one lengthy report with a large body of supplementary evidence ('Royal Commission

on Water Supply. Report of the commissioners', C 4169, 4169-I, 4169-II, Parliamentary Papers (1868–9)).

3. *the enclosed*: enclosure missing. This may be the draft of Frankland's contribution to the water commissioners' report (see n. 2).
4. *my water reports*: from 1866 Frankland as served the 'government analyst' of the water supply of London. See letter 3343, n. 11. His report is in the Minutes of Evidence of the water commissioners' report (see n. 2), on pp. 338–54.
5. *chalk water*: see letter 3343.
6. *Prestwich*: Joseph Prestwich (1812–69), a British geologist. He worked in the wine trade and conducted geological research in his spare time. From 1840 onward he published in a number of academic journals and became a recognized expert on water supply. On the basis of this expertise, he was appointed to serve on the Royal Commission on Water Supply (*ODNB*).
7. *the private companies*: there were several private companies supplying water in London. This recommendation is in the water commissioners' report (see n. 2), p. cxxvii, para. 262.
8. *the enclosed*: enclosure missing.
9. *Edinburgh pamphlet*: see letter 3343.
10. *Battersea*: a district of South West London, England.
11. *Southwark & Vauxhall Company*: see letter 3342, n. 5.
12. *enclosed pamphlet*: enclosure missing.
13. *Letheby*: Henry Letheby (see letter 3343, n. 7).
14. *Board of Trade*: see letter 2937, n. 6.
15. *King's Scholars Pond Sewer*: a brick sewer constructed in the 1850s within the route of the culverted River Tyburn in London.

From Edward Sabine 25 January 1871 3345

13. Ashley Place | Jan[uar]y 25th. 71

Dear Tyndall:

I have been sent the enclosed[1] for my perusal, and in the hope that I might be interested and aid in the publication of eight lectures on the Aspects of Nature in East Greenland,[2] illustrating the peculiar claims of the country of which they form a part, on the physicist & the Naturalist, as well as on the lovers of nature, in all its varieties. It presents a very striking and I doubt not; a very faithful picture of a country which would repay the future traveller, and—as I read it, I could not forbear wishing that it should be read by you, who from your alpine experience, are so prominently qualified to recognize its distinctive characteristics.

Will you kindly return it to me when read.

Sincerely yours | Edward Sabine

RI MS JT/1/S/25
RI MS JT/1/TYP/4/1331

1. *the enclosed*: enclosure missing.
2. *Aspects of Nature in East Greenland*: officers of the two ships and 'scientific gentlemen who accompanied them' on the Second German Arctic Expedition of 1869–70 delivered a series of lectures after their return to Germany. One of them, Adolf Georg Pansch (1841–87), a German anatomist and naturalist (*NDB*), studied and published on the natural history and anthropology of East Greenland, including a summary of the lectures: A. Pansch, 'Notes on East Greenland', *Fraser's Magazine*, 85 (NS 5) (1872), pp. 148–55. For a list of publications resulting from this expedition, see 'Geographische Gesellschaft in Bremen', in *Arctic Bibliography: Prepared for and in Cooperation with the Department of Defense, Under the Direction of the Arctic Institute of North America*, 16 vols (Washington: Department of Defense, 1853), vol. 1, pp. 853–4. See also C. Koldewey, *The German Arctic Expedition of 1869–70, and Narrative of the Wreck of the "Hansa" in the Ice* (London: Sampson Low, Marston, Low, & Searle, 1874); and L. M'Clintock, 'Resumé of the Recent German Expedition, from the Reports of Captain Koldewey and Dr. Laube', *Proceedings of the Royal Geographical Society of London*, 15:2 (1870–1), pp. 102–14.

To Thomas Archer Hirst 26 January 1871 3346

University of London, | Burlington Gardens, W. | Jan[uary] 26th 1871

My dear Tom.

I came in expecting to see you.

For I want you to relieve me of a difficulty

I want you to go to the Phil[osophical]. Club[1] dinner today and to act my part towards my guests Prof. Young[2] and Prof. Langley[3] and to explain to them that an engagement to which I had previously solemnly pledged myself, and which I had forgotten prevents me from meeting them.

I hope to have that pleasure someday soon.

This day in fact the annual dinner comes off at Huxley's[4] & since his marriage I have never once omitted it.

Yours ever aff[ectionate]ly | J. Tyndall.

Could not you & the two Professors dine with me at Albemarle Street[5] tomorrow at 6½ or 7 & go to Odling's lecture[6] afterwards? If you can come ask them.

RI MS JT/1/T/685
RI MS JT/1/HTYP/577

1. *Phil[osophical.] Club*: the Philosophical Club of the RS (see letter 3317, n. 2).
2. *Prof. Young*: Charles Augustus Young (1834–1908), an American astronomer. Son of the astronomer and mathematician Ira Young, Charles Young studied theology at Andover Seminary before accepting an appointment to teach mathematics and natural science at Western Reserve College in Ohio. He served briefly during the American Civil War and afterwards assisted the local and federal government with astronomical data for time-keeping and surveying. In 1866 he became professor of astronomy and natural science at Dartmouth College, a position that had been previously held by his father and maternal grandfather, where he worked on solar spectroscopy, solar photography, and sunspots. In 1870, he travelled to Jerez, Spain, to observe the solar eclipse and apparently stopped in London on his return trip (*ANB*; *CDSB*).
3. *Prof. Langley*: Samuel Pierpont Langley (1834–1906), an American astrophysicist. Langley began studying astronomy in 1864 after several years of working as a draftsman and engineer. In 1866 he was appointed professor of mathematics and charged with re-establishing the U.S. Naval Observatory. In 1867 he moved to the Allegheny Observatory at the University of Western Pennsylvania (now University of Pittsburgh), where he established himself as a pioneer in astrophysics. In 1887 he became secretary of the Smithsonian Institution. Langley was primarily interested in solar physics, science education and aviation and is best remembered for his unsuccessful attempts to build an airplane. Like Young, he traveled to Jerez, Spain, to observe the 1870 eclipse, passing through London on his return (*ANB*; *CDSB*).
4. *Huxley's*: Thomas and Henrietta Huxley. See letters 3333 and 3335.
5. *Albemarle Street*: location of the RI.
6. *Odling's lecture*: William Odling delivered a Friday Evening Discourse at the RI on 27 January 1871, 'On Improvements in the Manufacture of Chlorine' (*Roy. Inst. Proc.*, 6 (1870–72), pp. 199–205).

To Elizabeth Dawson Steuart 26 January 1871 3347

26th Jan[uary]. 1871.

My dear Mrs Steuart,

Until this morning I had no pause in my work since I last wrote to you.[1]

I wished to go myself to Mr Halliday,[2] instead of sending to him. So this morning I walked to his place of business. He was absent, but I laid the whole matter[3] before his office keeper, who will communicate what I said to Mr Halliday. I left my name and address requesting to be informed as to the cause of the irregularity.

I must send you when I have a little breathing time a better present than that little stone.[4] I have no thought of these things, but had I known <u>your</u> thoughts I should have sent you some little curiosity long ago.

Yours ever faithfully | John Tyndall.

RI MS JT/1/TYP/10/3403
LT Typescript Only

1. *I last wrote to you*: see letter 3336.
2. *Mr Halliday*: not identified.
3. *the whole matter*: Steuart wrote to Tyndall on 11 January 1871 asking him if he had a friend who would 'do a little matter for me'. See letter 3338.
4. *that little stone*: Tyndall had sent Steuart a 'bit of stalagmite' from St Michael's cave, in Gibraltar which he had collected during his expedition to Algeria to view the solar eclipse of 22 December 1870 (see letter 3230, n. 2).

To the Editor of the *Times* 30 January 1871 3348

PROFESSOR TYNDALL ON THE LONDON WATER SUPPLY. | TO THE EDITOR OF THE TIMES.

Sir,—

Your courteous correspondent, "A Biologist," need fear no "violence" from me. I would lay violent hands on nothing that has been consecrated by use.

During the late Eclipse Expedition[1] part of my time was occupied with the writings of Dr. William Budd, of Bristol, on typhoid, or, as he calls it, "intestinal fever." These writings alone, which are models of clear statement and strong scientific logic, would lead me to an entire agreement with the proposition that the defilements in water which are most dangerous to man are those which emanate from man himself. But I would ask your correspondent to consider where such defilements are most likely to be found; in our rivers which receive the washings and drainage of the adjacent towns and soils, or in the body of our chalk strata, where the water is demonstrably free from organic matter?

And with regard to the question of quantity, suppose the chalk unable to furnish a supply for 3,000,000 people; suppose it only able to supply the wants of a million, or half a million, or even one-tenth of a million—why should not the precious gift be utilised to this extent? Granting the utmost restriction as regards quantity which your correspondent would venture to put upon the chalk water, let us have even that quantity. No Government aid is required, but simply liberty to construct the necessary works and to bring the water to London. Let London, then, decide between it and the Thames water. An attempt on a small scale of demonstrable utility will enable us to solve the larger problem, whether the chalk is or is not capable of supplying the whole of London. It may be that the needs of three Londons might be supplied from this source.

A great number of inquiries[2] have been addressed to me regarding the softening of water. Would you permit me to reply to them *en masse*, by stating that "Clark's process,"[3] as applied to domestic purposes, has been described by Dr. Frankland in the weekly returns of the Registrar-General[4] for the week ending October 29, 1870. The publishers are Eyre and Spottiswoode,[5] and the cost of the Return is 1 ½ d.?

I am, Sir, your obedient servant, | JOHN TYNDALL. | January 30.

Times, 31 January 1871, p. 4
RI MS JT/6/5/1/62–3

1. *Eclipse Expedition*: see letter 3230, n. 2.
2. *A great number of inquiries*: see letters 3342, 3343 and 3344.
3. *"Clark's process"*: see letter 3343, n. 10.
4. *Registrar General*: see letter 3343, n. 11.
5. *Eyre and Spottiswoode*: see letter 3327, n. 3.

To Mrs. Carpenter[1] 31 January 1871 3349

THE ATHENAEUM | 31st. Jan[uary]. 1871

My dear Mrs. Carpenter

I have just walked into London from St Albans.[2] To my regret I shall be engaged on Saturday evening.

Yours faithfully | John Tyndall

Oxford Bodleian Library, MS. Autogr. b.9. No. 86

1. *Mrs. Carpenter*: possibly Louisa Carpenter (née Powell, 1812–87), the daughter of a local merchant, Joseph Powell, and the granddaughter of the iron puddler, Henry Cort. She married William Benjamin Carpenter in 1840, and together they had five sons, including Joseph Estlin Carpenter (1844–1927) and Philip Herbert Carpenter (1852–91) (*ODNB*).
2. *St Albans*: a city and a region in Hertfordshire, England.

To Heinrich Rudolf Vieweg[1] 4 February 1871 3350

Royal Institution of Great Britain | 4th Feb[ruary]. 1871

Gentlemen.

D. K. B. Hofmann.[2] Privatdocent in Vienna has written to ask my permission to translate my lecture at Liverpool. I have referred him to you.

I hope within the coming year to publish in one volume a collection of the occasional essays and lectures which I have published in English magazines; in which collection the Liverpool lecture[3] w[oul][d]. be included.

If you have no intention to translate the lecture pray give D[r]. Hofmann the necessary persmission

Yours faithfully | J. Tyndall

TU Braunschweig/Universitätsbibliothek/UABS V 1 T : 63

1. *Heinrich Rudolf Vieweg*: while the letter is addressed to 'Gentlemen' the recipient is most likely Heinrich Rudolf Vieweg, who took over the family publishing firm, Friedrich Vieweg und Sohns, from his father Eduard Vieweg (1797–1869).
2. *Dr. K. B Hofmann*: Karl Berthold Hofmann (1842–1922), an Austrian chemist and physician. Hofmann was physician at the Vienna psychiatric asylum from 1866 to 1869 and later became Professor for Physiological and Pathological Chemistry (1873) and Applied Medicine (1879) at the University of Graz. Hofmann was also interested in the history of chemistry and published on this topic (*NDB*).
3. *Liverpool lecture*: Tyndall's lecture for the BAAS meeting in Livepool in September 1870 (see letter 3235, n. 11).

From Rudolf Clausius 5 February 1871 3351

Bonn, 5/2/71

Lieber Tyndall,

Ich muss Dir endlich einmal den letzten Bericht über die Verwendung Deiner freundlichen Gabe für die Verwundeten erstatten.

Es blieben bei meinem vorigen Berichte noch 90 Thaler zu verwenden. Hiervon habe ich gegeben: (1) an ein Lazareth, in welchem es während der kalten Zeit sehr an warmer Fussbekleidung fehlte, 10 Reichsthaler; (2) bei einer Sammlung für die zurückgelassenen Frauen und Kinder der im Felde befindlichen Landwehrmänner 20 Rt (3) als Beitrag für die in den Lazarethen eingerichteten Weihnachtsbescheerungen durch welche den Verwundeten das für sie so traurige Weihnachtsfest etwas verschönt werden sollte, 20 Rt; (4) Es waren durch die unerwartet lange Dauer des Krieges endlich allen hiesigen wohlthätigen Vereinen die Mittel ausgegangen, und es wurde daher eine neue allgemeine Sammlung veranstaltet, wobei nicht nur Geld, sondern auch beliebige schöne oder nützliche Gegenstände welche jeder etwa entbehren konnte, angenommen wurden, welche letzteren nächstens verloost werden sollen. Bei dieser Sammlung habe ich in Deinem Namen die letzten 40 Rt gezeichnet.

Auf diese Weise, glaube ich, ist das Geld gut angewandt und hat vielen

Nutzen und grosse Freude bereitet. Alle Empfänger lassen Dir herzlich danken.

Bei der oben erwähnten letzten Sammlung sind, ausser den Geldbeträgen, bis jetzt schon über 2000 zum Theil sehr werthvolle Geschenke eingegangen, indem viele Personen die kein Geld mehr geben konnten, Luxusgegenstände, die sie entbehren konnten, hergegeben haben. Diese Gegenstände werden nun öffentlich ausgestellt, und dann verloost, und es ist zu hoffen, dass dadurch wieder dem Geldmangel soweit abgeholfen werden wird, dass die Verpflegung der Verwundeten und Kranken ohne Unterbrechung bis zum Ende des Krieges, welches nun hoffentlich nahe bevorsteht, vor sich gehen kann.

Ich schicke Dir gleichzeitig mit diesem Briefe, eine Abhandlung in welcher ich versucht habe, den zweiten Hauptsatz der mechanischen Wärmetheorie der bisher nur auf einem hypothetisch angenommenen Grund basirt war, ebenso auf allgemeine mechanische Principien zurückzuführen, wie es mit dem ersten Hauptsatze schon längst geschehen ist. Beim zweiten Satze ist die Sache freilich schwieriger als beim ersten, indessen denke ich doch, dass es auf diese Weise geht.

Mit herzlichen Grüssen von meiner Frau und mir an Dich und Hirst | Dein | Clausius.

Bonn, 5/2/71.

Dear Tyndall,

I must finally send you the last account of how your kind donation[1] for the wounded has been spent.

In my last report to you,[2] there were 90 thalers remaining to spend. Of these, I have given: (1) to a military hospital where during the cold season there was a great lack of warm footwear, 10 thalers; (2) at a collection for the wives and children of home guardsmen still on the battlefield 20 thalers (3) as a contribution for Christmas gifts for the wounded at the military hospitals, to brighten up the otherwise so sad Christmas celebration, 20 thalers; (4) Due to the unexpectedly long duration of the war, all the local charitable associations exhausted their resources, and a new general collection campaign was therefore organised, in which not only money, but also any other beautiful or useful objects that people could spare were accepted, which will be raffled off before long. I have donated the remaining 40 thalers to this collection on your behalf.

In this way, I believe, the money is put to good use and has benefitted and brought great joy to many. All beneficiaries send you their heartfelt thanks.

At the above-mentioned collection, in addition to the monetary funds, over 2000 gifts have been received, some very valuable, as many people who could not contribute any more money donated luxury items that they could do without.

These items will now be publicly displayed and then raffled off, and it is to be hoped that the shortage of funds will thereby be remedied to the point where the provisioning for the wounded and sick can be assured without interruption until the end of the war, which hopefully is close at hand.

Along with this letter, I am sending you a paper[3] in which I have tried to trace the second law of thermodynamics, which so far has been based solely on a hypothetically assumed principle, back to general mechanical principles, as was done with the first law long ago. With the second law it was certainly more difficult than with the first one; however, I still think that it works this way.

With warmest wishes from my wife[4] and me to you and Hirst[5] | Your | Clausius.

RI MS JT/1/TYP/7/2320–1
LT Typescript Only

---

1. *your kind donation*: Tyndall sent Clausius £50 to aid those wounded during the Franco-Prussian War (see letter 3238, n. 16). See letters 3283, 3297, and 3305.
2. *my last report to you*: letter 3305.
3. *a paper*: probably R. Clausius, 'Ueber die Zurückführung der mechanischen Wärmetheorie auf allgemeine mechanische Prinzipien', *Annal. Phys. und Chem.*, 142 (1871), pp. 433–61.
4. *my wife*: Adelheid Clausius.
5. *Hirst*: Thomas Hirst.

From Agatha Russell[1] 6 February 1871 3352

Pembroke Lodge, | Richmond Park.[2] | Feb[ruary]. 6/ 71

Lady Russell[3] presents her compliments to Professor Tyndall—She & Lord Russell[4] are anxious to have the pleasure of seeing him— & hope he will either come to dinner or luncheon <u>here</u> any day before next Monday—If to dinner, they hope he will sleep here—If it is impossible or inconvenient to him to do this, they hope he will dine with them in London on the 15th of this month—

RI MS JT/1/R/44

---

1. *Agatha Russell*: Agatha Russell (1853–1933), daughter of John Russell, 1st Earl Russell, and Lady Frances (Fanny) Russell. Agatha Russell was co-editor of *Lady John Russell: A Memoir* (see n. 3).
2. *Pembroke Lodge, Richmond Park*: a lodge granted to Russell by the Queen, located in the London borough of Richmond upon Thames.

3. *Lady Russell*: Frances Anna Maria (Fanny) Russell (1815–98), the second wife of Lord John Russell and the daughter of Gilbert Elliot Murray Kynynmound, second earl of Minto (1782–1859) and his wife Mary Brydone (1786–1853). The Russells lived at Pembroke Lodge in Richmond Park, a home provided for them by the Queen, where they entertained the political, social, and scientific elite, including Tyndall, 'one of the most genial and delightful of her guests' (D. MacCarthy and A. Russell, *Lady John Russell: A Memoir* (New York: John Lane Company, 1911), p. 261; *ODNB*).
4. *Lord Russell*: John Russell, first Earl Russell, previously Lord John Russell (1792–1878), Prime minister and author. While abroad in 1813 he was returned to parliament as a Whig MP for the borough of Tavistock and in addition to his political duties he became a prolific author. He was at the forefront of the campaign for parliamentary reform hoping to enfranchise the large industrial towns at the expense of the small boroughs whose vote was given in exchange for government patronage. He served as Prime Minister from 1846 to 1852 and 1865 to 1866 (*ODNB*). In 1866, while Prime Minister, Russell attended Tyndall's opening Friday Evening Discourse at the RI, 'On Radiation and Absorption with reference to the Colour of Bodies and their State of Aggregation' (*Ascent of John Tyndall*, p. 195; *Roy. Inst. Proc.*, 4 (1862–66), pp. 487–92).

To the Assistant Secretary, Harbour Dept., Board of Trade[1] 7 February 1871 3353

Royal Institution, 7th February 1871.

Sir,

On the 8th of last October I had the honour of forwarding to you a report[2] in which were embodied the results of a visit to Ireland, made with the view of determining whether a gas flame is applicable to revolving lights.[3]

The question thus raised was answered in the affirmative. A gas burner of 28 jets, being almost identical in size and sensibly equal in luminous power to the Trinity 4-wick lamp, proved, as might be expected, quite as applicable as the lamp to a revolving apparatus.

It was shown in a former report‡ that in Ireland the cost of the 28-jet gas flame was less than that of the oil-lamp. This saving, which had reference to the fixed light of Howth Baily,[4] would, other things being equal, also accrue from the employment of the gas in revolving lights.

But in my report of October 8th I referred to a still further possible saving to be secured by the periodic extinction of the gas flame, the Rockabill light[5] being referred to in illustration. In reference to this point the Board of Trade took the opinion of the able and experienced engineers the Messrs. Stevenson,[6] which opinion is "that the occasional obscuration of the gas is not applicable to revolving lights."§

In accordance with the wish expressed in your letter of November 15th[7] (H. 4,713), I will now endeavor to explain to the Board the considerations on which my conclusion was founded. With this view I beg to forward a tin model (*see diagram on next page*) a reference to which will render my meaning clearer than it could be made by mere verbal description.

The model consists of a star of 8 rays moveable round a central pivot. The central octagon from which the rays issue is intended to represent the 8-paneled revolving apparatus. The points of the stars are to be regarded as the centres of the beams issuing from the respective panels. The blackness of the disc underneath the star is intended to denote the darkness of night, while the circle round the disc represents the horizon.

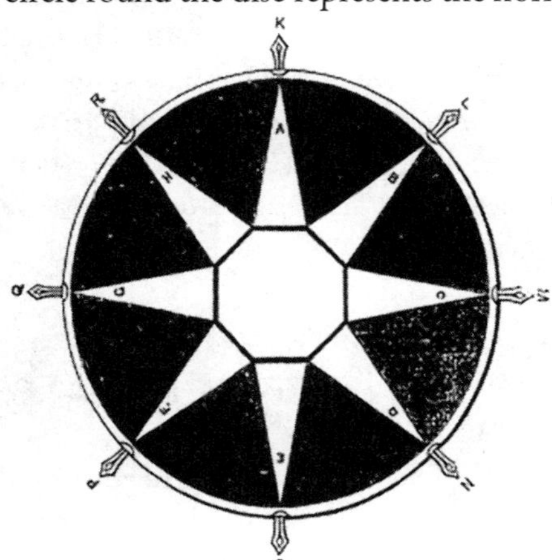

Every Part of the horizon receives a flash every second minute, the gas being lighted one minute during which the lens is moved one eighth round and each part receives a flash, then the gas is extinguished one minute, and so on in succession.

Thus half the consumption of gas is saved.

Let the star be turned so that the beam A shall point to and illuminate the station K on the horizon. The ray B would at the same moment illuminate L; C would illuminate M; D, N; E, C; F, P; G, Q; while the ray H from the 8th panel would illuminate R, the eighth station.

Let the star be turned, or in other words let the apparatus be supposed to revolve, until A reaches L. During this time B travels from L to M; C from M to N; D from N to O; E from O to P; F from P to Q; G from Q to H; and H from R to K. Thus, through the passage of A from K to L, in other words during the rotation of the apparatus through one eight of a revolution, *every point on the horizon is once illuminated.*

If the flame continue burning the same effect is produced during every succeeding eighth of a revolution. Every point of the horizon is once illuminated. At Rockabill the time of a complete revolution is 96 seconds; hence, the number of panels being eight, as in the model, each point of the horizon receives flashes which succeed each other in intervals of 12 seconds.

But suppose at the moment the beam A points towards L the supply of gas to be cut off, and the apparatus permitted to revolve in darkness until A reaches M. During this eighth of a revolution no gas is consumed, and no flash is received by any point on the horizon.

When A reaches M let the gas be relighted. During the succeeding eighth of a revolution every point of the horizon would be once illuminated as before. It is quite manifest that this process may be continued indefinitely; and the gas being lighted, and every point of the horizon once illuminated, during every *alternate* eighth of a revolution.

It is also plain that the intervals of darkness between the flashes instead of being, as they now are, 12 seconds, would be 24 seconds.

This reasoning, which as far as it goes is of the character of a mathematical demonstration, has, as stated in my report of the 8th of October, been verified by actual observations on the Rockabill light.

There is, as far as I can see, but one drawback to the perfection of this scheme; and this I will now try to point out with distinctness. Let the beam A point as at the outset towards K; at a certain moment we start from K towards L with the gas lighted. According to the new scheme K ought to be in absolute darkness for 24 seconds. But just as A reaches L, and before the gas is cut off, a glimmer is seen at K; this glimmer being only 12 seconds distant from the preceding flash. The same is true of the other seven points of the horizon faced by the panels the moment before the flame is extinguished. In all the remaining sweep of the horizon this secondary glimmer is absent.

Seeing that this glimmer was to be expected, and knowing that it had actually shown itself in my own observations at Rockabill, Captain Hawes[8] and Mr. Wigham,[9] at my request, observed it more strictly. I extract the following description of the observed effect from a letter addressed to me by Mr. Wigham:

"Our first care was to arrange, by means of a mark on shore, a certain point at which we might be sure of witnessing the extinction of the lamp exactly at one of the 'eight points.' The extinction was effected by the man who did it for you. For more than two hours we witnessed the effect, not only at that point, but at all others as we steamed round Rockabill. The result is that at these eight points, and not elsewhere, we found that a short flash was visible, not such a flash as is produced by the ordinary revolution of the lens, but such as might proceed from a mast-head light, and which Captain Hawes says no sailor could possibly mistake for a light exhibited from a light-house." Subsequent trials reduced this irregular light to a faint twinkling.

As already stated, I noticed the glimmer. I should not be inclined to ascribe to it any practical importance. On the isolated points where it is seen it is so masked by the superior brilliancy of the true flash, that no mariner could be deceived by it.

By the experiments made to mitigate if not abolish this small flash the flexibility, if I may use the term, with which gas lends itself to lighthouse purposes was still further demonstrated, and what appears to be an important modification of an intermittent light hit upon. In a letter addressed to me on the 27th of January Mr. Wigham writes thus:—

"While carefully looking into this part of the subject it occurred to me that we might avail ourselves of this peculiarity of gas-light, to enable us to receive, while the lens was passing the eye, three or four bright flashes, instead of only one. I tried it and found that we had hit upon one of the most striking and distinguishing lights. I asked Captain Hawes to meet me last night, when he told me that he had been thinking of something of the same kind. We then jointly looked at the light, and Captain Hawes is perfectly charmed with the effect. The impression that you receive when you first look at it is that there is like in it, and that it is actively exerting itself to warn the mariner of some danger."

The observations of this new form of intermitted light have been assiduously continued. The result is described in the following letter which has been addressed to me by Captain Hawes:

[. . .][10]

The substance of the foregoing report may be thus stated:—

1. A gas flame of the same size as an oil flame may, in a revolving apparatus, be employed instead of the oil flame. The flames are sensibly equal in luminous power, and the experiments hitherto made in Ireland show that the gas there is cheaper than the oil.

2. At the Rockabill lighthouse, by making the interval of darkness 24 seconds instead of 12 seconds, an intermittent light might be produced with half the expenditure of gas necessary for a fixed light. This is to be understood as a statement of fact, not as a recommendation.

3. Should the more recent observations, and the judgment pronounced regarding them, be correct, a very important form of intermittent light has been discovered.

Should the Board of Trade[11] think the subject worthy of further attention, I shall be happy, as soon as my duties here permit, to go again to Ireland, and to subject the observations of Mr. Wigham and Captain Hawes to strict personal examination.

I reserve for a future occasion such further remarks as might even now be made upon this subject. The Board may I think feel confident that before finally recommending any change in our system of lighthouse illumination, I shall assure myself by suitable experiments that the change would be an improvement optically or economically. That gas is capable of manifold application in lighthouses is even now evident. We have to deal in fact with two elements, the quantity of the light and its distribution. On the former score gas, in Ireland, competes successfully with oil; but it is to the ease and promptness with which gas lends itself to various modes of distribution that the chief manifestations of its superiority are to be looked for. I anticipate great ameliorations and advances from its future use.

I have the honour to be, Sir, | Your obedient Servant, | JOHN TYNDALL.

The Assistant Secretary | Harbour Department, Board of Trade.

*Ante, p. 19.

†Ante, p. 20.

‡See Papers on this subject presented to Parliament by Command in 1869, page 19.

§The report of Messrs. Stevenson here referred to is dated 9th Nov. 1870, and is as follows: "We have perused and considered the proposal of Professor Tyndall, contained in his report of 8th October last, to apply periodic flashes of gas in the illumination of revolving lights, and we have now to report that with such a combination of revolving lenses and periodic extinctions of the light, we do not see how it would be possible to maintain the characteristics of the light *uniform* in all azimuths, and therefore we are of opinion that the occasional obscuration of the gas is not applicable for revolving lights. A revolving light of uniform character in every azimuth can be produced by periodic extinctions of gas in a fixed light apparatus, but at an excessive sacrifice of power."

'Further Papers relative to a Proposal to substitute Gas for Oil as an Illuminating Power in Lighthouses', HC 282, Parliamentary Papers (1871), pp. 30–3
National Archives MT 10/131 H64B 2309–13 (unfinished, handwritten draft)

---

1. *Assistant Secretary, Harbour Dept., Board of Trade*: C. Cecil Trevor.
2. *a report*: letter missing. Report not identified.
3. *revolving lights*: Tyndall visited Ireland in late September 1870 to view lighthouses for the Board of Irish Lights. See letters 3262, 3264, and 3267. On Tyndall's work on lighthouses, see letter 2937, n. 6.
4. *Howth Baily*: see letter 3019, n. 3.
5. *Rockabill light*: see letter 3124, n. 7.

6. *Messrs. Stevenson*: engineering firm of David and Thomas Stevenson.
7. *your letter of November 15th*: letter missing.
8. *Captain Hawes*: Elliot H. Hawes.
9. *Mr. Wigham*: John Richardson Wigham.
10. *[. . .]*: letter 3124 copied here.
11. *Board of Trade*: see letter 2937, n. 6.

From Agatha Russell[1] 9 February 1871 3354

Pembroke Lodge, | Richmond Park, | S.W. | Feb[rua]ry 9 1871

Lady Russell[2] presents her compliments to Professor Tyndall & rejoiced to find by his note[3] that there is a prospect of seeing him here before next Wed[nesda]y—But in case any obstacle should unluckily occur she writes to tell him that Lord[4] & Lady Russell have decided not to go to town before Easter, & that they therefore hope he will do them the great pleasure of dining & sleeping here on Wed[nesda]y 15th—Dinner 1/h before 8—but should Professor Tyndall be able to come early in the afternoon they will be very glad. Sunday afternoons they are always to be found here.

RI MS JT/1/R/45

1. *Agatha Russell*: see letter 3352, n. 1.
2. *Lady Russell*: see letter 3352, n. 3.
3. *his note*: letter missing. See letter 3352.
4. *Lord*: John Russell (see letter 3352, n. 4).

From William Valentin 9 February 1871 3355

Royal College of Chemistry[1] | Febr[uar]y 9. 1871.

Professor Tyndall

Dear Sir

The following are the spec[ific]-Gr[avitie]s to Vapor Densities (as far as I can find) of the organic bodies a list of which you left with me

	Sp[ecific]. Gr[avity].	Vap[our]. Dens[ity].
Nitrate of Methyl at 22°C	1·182	2·64*
Iodide of Allyl	1·789	?
Iodide of Isopropyl at 15°C	1·70	?
Nitrite of Butyl	?	?

(I cannot find that this ether has been prepared even.)

Hydride of Caproyl at 15°	·678	2·98
Acetate of Methyl at 0°C	·9562	2·563

I imagine that the determinations of the Vapor density of such readily decomposable substances as Nitrates & Nitrites and of Allytic Iodide must be scanned very carefully, and should think that you would do better to take the calculated Vapor densities as the basis of any calculations you may wish to make

I am Dear Sir | Yours very faithfully | W[illia]m Valentin

* Other authorities say that no vapor density is known because the ether explodes at a high temperature

RI MS JT/8/3

1. *Royal College of Chemistry*: letterhead from the Royal College of Chemistry and embossed with the stamp, 'ROYAL SCHOOL OF MINES | JERMYN STREET S. W.' On the Royal College of Chemistry, see letter 3344, n. 1.

To Rudolf Clausius 10 February 1871 3356

10th. Feb[ruary]. | 1871.

My dear Clausius.

Hirst[1] was with me when your note[2] reached me, and he sends you and Mrs. Clausius[3] all kind greetings in return for those that you constantly send to him.

It gives me exceeding pleasure to find that you have been able to do so much good with the money I sent to you.[4] I wish I could send you more, but my purse is bone dry. Some time ago we heard very heart-rending accounts of the state of the peasantry around Paris. I tried to get a letter[5] into Regnault's[6] hands, but failed—at all events he has not replied to me. Then came the appeal of the Bishop of Versailles,[7] and I sent to Huxley, who is on the Committee,[8] the money I intended for Regnault.

I wish you could be induced to write some day an unmathematical history and commentary of your researches. It would be of immense advantage. And you could state the physical results in language clearer & better than anybody else.

I wish when you have a little time that you would cast your eye over a paper by Mr. J. W. Strutt[9] in the last number of the Philosophical Magazine. He has singled out a passage from a lecture of mine for somewhat Sharp Criticism. With regard to the word 'reflection' I have employed instead of it the word 'scattering' for nearly two years.

I hope the children are all well, and with love to them & kindest regards to M[rs]. Clausius.

Believe me ever | Yours /faith[full]y/ | John Tyndall

It is earnestly to be hoped for the sake of Germany, France & the world that this war[10] is now coming to an end. Had I been younger I trust I should have taken arms with the Prussians at the beginning. But now every body is sick of the war.

RI MS JT/1/T/227

---

1. *Hirst*: Thomas Hirst.
2. *your note*: letter 3351.
3. *Mrs. Clausius*: Adelheid Clausius.
4. *the money I sent to you*: Tyndall sent Clausius £50 on 12 October 1870 to aid those affected by the ongoing Franco-Prussian War (see letter 3238, n. 16). See letter 3283.
5. *a letter*: letter missing.
6. *Regnault*: Henri Victor Regnault (1810–78), a French chemist, physicist, and photographer. A pupil of Liebig's at the University of Giessen, he was professor of physics at the Collège de France, 1841 to 1854, and director of the famous porcelain factory at Sèvres, 1854 to 1870. During the Franco-Prussian War, in 1871, his laboratory was destroyed and his son Henri Regnault (1843–71), a painter, was killed in battle on 19 January (*CDSB*).
7. *appeal of the Bishop of Versailles*: due to the distress in Paris and the surrounding region caused by its siege during the Franco-Prussian War, the Bishop and the clergy in the Department of the Seine issued an appeal to the English public to raise funds for the relief of the non-combatant population ('The Distress around Paris', *Times*, 19 January 1871, p. 6).
8. *on the Committee*: 'The French Relief Fund in Aid of Paris and the Suffering Districts throughout France' was established at the Mansion House upon the motion of John Lubbock. Huxley was a member of the committee formed to administer the fund ('The Distress around Paris', *Times*, 19 January 1871, p. 6).
9. *paper by Mr. J. W. Strutt*: John William Strutt (1842–1919), a British experimental and mathematical physicist, was educated at Eton and Harrow schools. He graduated senior wrangler from Trinity College, Cambridge (first place in the mathematical tripos, 1865), received the first Smith's prize, and was elected a fellow (1866). He resigned his fellowship in 1871, married, and became interested in psychical research. He was elected FRS (1873) and served as secretary of the RS (1885–96). From 1879 to 1884 he was Cavendish professor of experimental physics at Cambridge and from 1887 to 1905 professor of natural philosophy at the RI (*ODNB*). The paper mentioned is J. W. Strutt, 'Light from the Sky, its Polarization and Colour', *Phil. Mag.*, 41:271 (1871), pp. 107–19.
10. *this war*: the Franco-Prussian War (see letter 3238, n. 16).

From Jane Barnard 11 February 1871 3357

Barnsbury Villa, | 320, Liverpool Road, | N.

My dear Dr Tyndall

I think Smith[1] has made a mistake.[2] I cannot remember that I ever said anything about discontinuing the porters books;[3] & I think it very unlikely from the anxiety I have always felt, to keep everything as nearly as possible in the order my uncle[4] established, both during the time I was acting for him & afterwards.

My impression is, that the three books came up every Saturday evening for inspection till the time we left R[oyal]. I[nstitution]. in Dec[embe]^r^ 1867; but at any rate, if they dropped out of sight before that time, it was in consequence of the unsettled state we were in.

My dear uncle, while he was able, was strict in his oversight & wrote his remarks at the end of the weekly reports. I, from the difficult position I held, could not keep this up, but I always endeavored as far as I was able, to carry out all his rules & arrangements; & I feel very glad you have given me this opportunity of clearing myself from the seeming presumption of altering any of them.

My Aunt[5] desires her kind regards; I need not tell you that whenever you like to come to see her, you are always very welcome.

I am | yours very sincerely | Jane Barnard

Dr Tyndall | &c &c &c | 11 Feb[ruary]. 1871

RI MS JT/1/B/47

---

1. *Smith*: not identified; possibly Basil Woodd Smith (1831–1901), an English lawyer and amateur astronomer who served as a Visitor of the RI ('Basil Woodd Smith', *Monthly Notices of the Royal Astronomical Society*, 62:4 (1902), p. 245).
2. *made a mistake*: this letter appears to be the response to a letter that is missing.
3. *porters books*: porters books from the RI. These appear to be records of the duties of the RI's porters.
4. *my uncle*: Michael Faraday.
5. *My Aunt*: Sarah Faraday.

From Alfred M. Mayer[1] 13 February 1871 3358

Lehigh University | South Bethlehem | Pennsylvania | U.S.A. | February 13. 1871

To | Prof. John Tyndall D.C.L. F.R.S.

My dear Sir;

I take the liberty of sending you (by book-post of this date) several papers[2] of researches on magnetism and electricity, which I have recently made—Before reading them I hope you will preface them with these words,—so true—found in "The Scientific Limit of the Imagination"[3] "There is no discovery so limited as not to illuminate something beyond itself"[4] Limited indeed they are—namely attempts at scientific investigation and in fact the paper on "electric- conductivities"[5] is an experimental task—self-imposed—to educate me in patience & precision.

Whatever merit then works may contain—I lay at the feet of one to whom—after my father in Science, Prof. Joseph Henry—I owe most to a judicious advice from the attentive study of his writings & example—

With the highest considerations of respect I remain | Yours truly | Alfred M. Mayer

P.S. I enclose photographs[6] of "lines of magnetic force" made by a method recently devised by me—As follows:—A plate of this glass is covered with a solution of shellac in alcohol as you flow a plate with collodion in photography. After drying finely—the plate—is placed over the magnet or magnets &c and fine iron filings, prepared by "draw-filing" very soft Norway-iron, is sifted uniformly over the film, then by vibrating the plate, by letting fall vertically a light piece of copper wire the "spectrum" is formed—The plate is now lifted vertically off the magnet &c and placed upon a past[e]board cylinder which serves as a support in bringing the glass plate close to the under side of a hot metal plate.—The radiant heat is absorbed by the filings which sink into the shellac and are "fixed". The plate can now be used (1) as a slide in the lantern (2) as a "negative" giving in the "printing-frame" light lines in a dark ground or (3) (and the most important) serves as a plate in which measures of precision can be made. I will as soon as I can send you a series of prints from negatives when prepared—and a paper describing in detail the process—which is now in press—together with one of the glass plates.

I hope you will excuse the scrawling writing of this letter—but my hand was disabled by a large electro-magnet falling upon it.

RI MS JT/1/M/72

1. *Alfred M. Mayer*: Alfred M. Mayer (see letter 3099, n. 1).
2. *several papers*: possibly A. Mayer, 'On a Simple Method of Measuring Electrical Conductivities by Means of Two Equal and Opposed Magneto-Electric Currents or Wavers', *American Journal of Science*, 50:2 (1870), pp. 307–18; and 'Researches on Electro-Magnetism', *American Journal of Science*, 50 (1870), pp. 195–212. It is not certain which papers Mayer sent to Tyndall and whether he is referring to published papers, but in 1871 Tyndall facilitated the publication of Mayer's work on 'magnetic spectra' in the *Phil. Mag.* (A. M. Mayer, 'On a Method of Fixing, Photographing, and Exhibiting the Magnetic Spectra', *Phil. Mag.*, 41 (June 1871), pp. 476–80). See letter from Mayer to Tyndall of 10 February 1872 in the forthcoming twelfth volume of *The Correspondence of John Tyndall*.
3. *"The Scientific Limit of the Imagination"*: J. Tyndall, 'Scientific Limit of the Imagination', in *Essays on the Use and Limit of the Imagination in Science* (London: Longmans, Green, and Co., 1870), pp 52–65. See also letter 3235, n. 11.
4. *"There is no discovery so limited as not to illuminate something beyond itself"*: 'Scientific Limit of the Imagination', p. 54. See n. 3.
5. *paper on "electric- conductivities"*: see n. 2.
6. *I enclose photographs*: enclosure missing.

From François-Napoléon-Marie Moigno 15 February [1871][1] 3359

LES MONDES. Revue Hebdomadaire | DES SCIENCES et de leurs applications aux arts et à l'industrie | par M. L'ABBÉ MOIGNO | RÉDACTION | 2, rue d'Erfurth, 2. | Paris, le 15 février 1870

Mon cher Mr Tyndall

Un de mes amis M^r^ le docteur Tamin qui va à Londres, veut bien se charger d'un petit mot pour vous, que je suis heureux de vous donner ce signe de vie; alors surtout que j'ai échappé comme par miracle à un *[*obus*]* qui a tout fracassé dans ma petite chambre à coucher.

Qu'il me tarde aussi de voir *[*finir*]* le blocus horrible qui nous sépare du monde entier. Je ne sais pas plus que si j'étais *[*mort*]* depuis six mois, ce qui s'est passé en Angleterre dans le domaine de la science. Avec quelle *[*joie*]* je recevais les premiers journaux anglais.

Je vais reprendre la publication de mes Mondes et de mes actualités scientifiques. J'ai traduit pendant le *[*siège*]* vos notes sur la lumière, ou les résumés de vos douze leçons.

Je vais faire des efforts pour augmenter en Angleterre le nombre des abonnés à mes Mondes, pour pouvoir sortir de mes Idées qui sont grandes comme celles de tous.

Daignez accueillir avec bonté M^r^ le docteur Tamin qui sans être savant prend un grand intérêt à la science.

A vous de coeur | F. Moigno

LES MONDES[2] | Paris, 15 February 1870

My dear Mr Tyndall,

One of my friends, Dr Tamin[3] who is going to London, has agreed to pass along a brief note to you, and how happy I am to give you this sign of life; especially that I was miraculously saved from a *[shell]* that smashed everything in my small bedroom.

I cannot wait to see the end to this horrible blockade[4] that separates us from the entire world. I know as much about what is happening in England in the field of science, as if I had been dead for six months. How happy I was when I received the first English newspapers.

I will resume the publication of my Mondes[5] and my scientific news.[6] During the siege[7] I translated your notes on light,[8] or the summaries of your twelve lessons.

I will make an effort to increase the number of subscribers to my Mondes in England, to be able to put out my ideas, which are as grand as anyone's.

Please welcome Dr Tamin with kindness; he is very interested in science without being a scientist himself.

Yours truly | F. Moigno

RI MS JT/1/M/118

---

1. *[1871]*: although Moigno wrote '1870', the events he describes took place later, during the Franco-Prussian War. The editors have therefore decided to date the letter according to content and assume inattention on the part of the writer.
2. *LES MONDES*: LES MONDES | Weekly Review | OF THE SCIENCES | and their applications | to the arts and the industry | by | FATHER MOIGNO | EDITORIAL OFFICES | 2 Erfurth street, 2. Headed notepaper.
3. *Dr Tamin*: possibly Onésime Tamin-Despalles (1838–83), a French physician practicing in Paris in the 1860s, but it is uncertain whether this is the Dr. Tamin Moigno is referring to.
4. *this horrible blockade*: the siege of Paris, part of the Franco-Prussian War (see letter 3238, n. 16), lasted from 19 September 1870 to 28 January 1871. The blockade led to intense food scarcity in the city, with inhabitants forced to slaughter all animals for food. Although the Franco-Prussian War did not officially end until 10 May, with the signing of the Treaty of Frankfurt, a settlement was signed on 16 February 1871.
5. *Mondes*: see letter 3226, n. 5.
6. *my scientific news*: presumably Moigno's review journal, *Cosmos. Revué encyclopédique hebdomadaire progrès des sciences et de leurs Applications*.
7. *the siege*: see n. 4.

8. *your notes on light*: *Notes of a Course of Nine Lectures on Light* (see letter 3097, n. 7). The translation appeared as J. Tyndall, *La lumière: notes d'un cours de neuf leçons sur le rôle scientifique de l'imagination par M. John Tyndall*, trans. F.-N.-M. Moigno (Paris: M. Gauthier-Villars, 1872).

From C. Cecil Trevor [16 February] 1871[1] 3360

Professor Tyndall F.R.S. | Royal Institution | Albemarle Street

Sir,

I am &c of the 7 Inst[an]t. in w[hic]h. you transmit a further report[2] upon the applicability of gas to revolving Lights & in thanking you for the same, I am to state that the B[oard] of T[rade]. will be indebted to you, for the result of the further investigations you propose to make in Ireland, & to request that you will undertake those investigations whenever your other duties will permit.

I am &c

The B[oar][d] propose to send you | A copy of the Report when printed | Signed by Mr Trevor | *[10]*/2/71 CCT | GAB 15/2/71 | Done | 16/2/71[3]

National Archives MT 10/131 H64B 2318

1. *[16 February]*: this working draft contains several dates, noting its place in the signature process. The 'done' date is 16 February 1871. The first date is 10 February.
2. *you transmit a further report*: see letter 3353.
3. The letter has several other sets of initials with dates which are not identified.

To [Philip Stanhope][1] 17 February [1871][2] 3361

Royal Institution of Great Britain

My dear Lord

With fine pleasure I am yours for the 1st.

And faithfully ever | John Tyndall | 17th. Feb[ruary]

Stanhope of Chevening Manuscripts; Letters from Sir Charles Lyell, Prof Tyndall, Sir Wm Stirling Maxwell, Lord Houghton, Froude, Iain Rogers 1847–1871; Letter from Prof Tyndall to the 5th Earl Stanhope, 17 Feb 1871
Courtesy of Kent Archive and Local History Service, U1590/C471/1
By kind permission of the Board of Trustees of the Chevening Estate

1. *[Philip Stanhope]*: although the recipient of this letter is not personally named, it is presumably Philip Stanhope, since the letter was found in the collection of the Stanhope estate. Philip Stanhope, Lord and fifth Earl Stanhope (1805–75), an English historian, married to Emily Harriet, née Kerrison (d. 1873). He was elected chairman of the National Portrait Gallery in 1857 and member of the Royal Commission on Historical Manuscripts from its foundation in 1869 until his death. His publications include, *Life of . . . William Pitt,* 4 vols (1861–2), and *The History of England, Comprising the Reign of Queen Anne until the Peace of Utrecht* (1870) (*ODNB*).
2. *[1871]*: the year on this letter was added by another hand.

To Thomas Carlyle 20 February [1871][1] 3362

Royal Institution | 20th Feb[ruary].

My dear Friend.

I have been thinking of you much of late, and indeed writing about you a little. I wish much to see you once more; but I want you to allow me to bring down to you two fine intelligent American gentlemen,[2] both of whose hearts have been warmed by your works. Could we come to see you on Tuesday or Wednesday evening?

Yours ever | John Tyndall

RI MS JT/1/T/159
RI MS JT/1/TYP/1/192

1. *[1871]*: LT's proposed date of 1871 is supported by Tyndall's mention of 'two fine intelligent American gentlemen' visiting in late January. See letter 3346.
2. *American gentlemen*: Tyndall's guests were Charles Young (see letter 3346, n. 2) and Samuel Pierpont Langley (see letter 3346, n. 3).

To Charles Darwin 23 February [1871][1] 3363

Royal Institution | 23rd. Feb[ruary].

My dear Darwin,

I have been recently working at Respirators for firemen,[2] and have devised one which permits a man to remain an hour in smoke, a single inhalation of which, without the respirator, would almost choke him.

I first tried Cotton-wool; but it let the pungent smoke of resinous pine through freely. A friend then suggested the moistening of the wool with glycerine. Glycerine had been previously used by Pouchet[3] to make particles stick to a plate of glass. I afterwards added fragments of charcoal and lime and in this way have obtained an excellent respirator.

I have had firemen here and they say that they could remain for any length of time in the dense smoke that I had prepared for them.

But what I write to you about is this. The glycerine was a great improvement: even without the coal it does much to quell the most intolerable smoke: it intercepts with extraordinary effect all the floating matter of the air.

Now the human nose with its hair and mucous is a respirator less perfect then mine, it is true, but still to a great extent effectual.

Supposing a savage tribe to be afflicted with epidemic disease which is undoubtedly propagated by floating particles, and that some of them had by a good fortune hairs within their noses: suppose them to breath through the nose, as many savages do. The chance of survival would certainly be with those possessing the hairs within the noses. Hairy nosed men would thus be left behind after men without this protection had been cut off. And hence we should have the hairs propagated as we now have them.

I believe this question of hairs within the nose has been a difficulty: but it certainly falls under your general principle.

Yours ever | John Tyndall.

Perhaps I am writing about what is already known to you.[4]

RI MS JT/1/TYP/9/2809
LT Typescript Only

1. *[1871]*: this letter was dated through reference to Darwin's response in letter 3365. The Darwin Correspondence Project has also dated the letter to 1871; the letter is published in F. Burkhardt, et al. (eds), *The Correspondence of Charles Darwin* (Cambridge: Cambridge University Press, 2012), vol. 19, pp. 97–9; this is letter 7508 in the online edition of the Darwin correspondence.
2. *working at Respirators for firemen*: for Tyndall's work on this, see J. Tyndall, 'On Dust and Smoke' (*Roy. Inst. Proc.*, 6 (1870–72), pp. 365–76, on pp. 372–6; *Fragments of Science*, pp. 334–5; and J. Tyndall, 'On Some Recent Experiments with a Fireman's Respirator', *Roy. Soc. Proc.*, 22 (1874), pp. 359–61.
3. *Pouchet*: Felix Pouchet. The Darwin Correspondence Project notes that Pouchet 'was interested in airborne particles as possible contaminants in experiments that he carried out in order to prove the existence of spontaneous generation'.
4. At the end of the typescript, written by hand (probably by LT), is: 'How is the negro's nose?'.

From Jules Jamin 26 February 1871 3364

Pornic 26 février 1871

Mon cher ami

Pendant tout le temps du siège de Paris j'ai été absolument sans nouvelles de ma famille; aussi me suis-je empressé de la venir retrouver à Pornic aussitôt que cela a été possible. J'ai trouvé tout mon monde en bonne santé, et ce n'est pas sans attendrissement que j'ai lu deux lettres de vous dans lesquelles vous mettez à la disposition de ma femme votre bourse et votre cordiale hospitalité. Vous avez dû recevoir en leur temps les remercîments qu'elle vous a adressés. Elle n'avait heureusement aucun besoin d'argent, et elle est demeurée parfaitement tranquille dans ce coin de terre où la guerre n'est point parvenue; mais je ne puis vous dire à quel point elle a été touchée ainsi que moi de *[la]* *[bonne]* *[amitié]* et de l'intérêt que vous nous avez témoignés dans cette circonstance. Si l'ennemi avait pénétré jusqu'à Pornic ma famille n'aurait point hésité à aller se mettre en Angleterre sous votre protection, c'est vous dire avec quelle confiance elle avait reçu vos offres de service. Pour moi, mon cher ami je ne l'oublierai jamais.

Et maintenant que va devenir la France? Elle paiera bien cher l'agression *[étourdie]* à laquelle on l'a conduite; elle n'aura plus aucun rôle dans le monde, et comme il est permis de dire qu'elle a largement contribué à la <u>civilisation</u> Européenne, je crois qu'on doit regretter l'impuissance matérielle et morale où elle se trouve aujourd'hui réduite. Depuis 1815 à part quelques nuages la France et l'Angleterre avaient marché d'accord et l'influence de deux nations serait accrue par leur alliance: je me persuade que l'Angleterre va perdre de sa force n'ayant plus la France à côté d'elle.

Je ne suis pas de ceux qui croient que la moralité et la vertu ont disparu de notre malheureux pays. Il y a sans doute à Paris des vices très apparents comme *[1–2 words illeg]* dans toutes les grandes villes. Mais le gros de la nation est honnête et moral; vous le savez bien vous qui nous connaissez; nous ne méritons pas d'être traités comme des habitants d'une Babylone corrompue, d'être rayés de la carte d'Europe à l'applaudissement de tous et nous pensons que les divers états nous viendront en aide pour mettre un terme aux exigences allemandes et aux rapines *[éhontées]* dont notre pauvre pays est aujourd'hui la victime, rapines qui laissent chez nous un éternel souvenir de haine et qui déshonorent les vainqueurs.

Mais nous causerons longuement de ces choses quand vous viendrez nous voir. Tâchez que ce soit bientôt et laissez-moi vous imposer cette fois mon hospitalité. Vous avez offert la vôtre c'est me donner des droits sur vous. Il y a une chambre qui vous est réservée à la maison et une place à la table de famille.

Cela ne gênera pas votre liberté et ce sera pour nous une grande satisfaction d'amitié.

Ma femme et mes enfants se rappellent à votre souvenir. Croyez à ma bien sincère amitié | J. Jamin

Pornic 26 February 1871

My dear friend,

For the entirety of the siege of Paris,[1] I was completely without news from my family; I hurried to meet them in Pornic as soon as it was possible. I found everyone to be in good health, and it was not without emotion that I read your two letters[2] in which you put at the disposal of my wife[3] your purse and your warm hospitality. You should have received her expressions of gratitude[4] at that time. Fortunately she did not need money, and she resided peacefully on this piece of land that the war had not reached, but I cannot tell you how touched she and I were by the kind *[*friendship*]* and the concern that you have shown us in this circumstance. If the enemy had invaded as far as Pornic, my family would not have hesitated to go to England under your protection. This shows how confidently she received your offers of help. As for me, my dear friend, I will never forget it.

And what will now become of France? It will pay dearly for the *[*whirlwind*]* aggression to which it was brought. It will not have a role in the world anymore, and as one can say that France greatly contributed to the European Civilization, I think that we should lament the material and moral helplessness that had befallen it today. Since 1815,[5] despite minor incidents, France and England have worked together and their alliance led to the growth of both nations' influence. I am convinced that England will lose some of its strength without having France by its side.

I am not one of those who believe that morality and virtue have disappeared from our unfortunate Country. In Paris, there are undoubtedly many vices committed *[1–2 words illeg]* in all large cities. But the majority of this nation is honest and morally upstanding; you know this well, since you know us. We do not deserve to be treated like the residents of a corrupt Babylon,[6] to be wiped off the map of Europe to general applause, and we think that the other states will help us to put an end to the German demands and shameless pillage, pillage that will leave an eternal legacy of hate and bring dishonour to the victors, which plague our poor country today.

But we will speak at length on this subject when you will come to see us. Try to do so soon and allow me to impose my own hospitality this time. You offered yours, which gives me rights over you. There is a room that is reserved for you at our house and a seat at the family table. This will not interfere with your freedom and it will give us a great satisfaction of friendship.

My wife and my children send their regards. With my sincere friendship | J Jamin

RI MS JT/1/J/15

1. *Siege of Paris*: see letter 3359, n. 4.
2. *your two letters*: letters missing.
3. *my wife*: Eudoxie Jamin (see letter 3309, n. 1).
4. *her expressions of gratitude*: letter 3309.
5. *Since 1815*: 1815 marked the end of the Napoleonic wars and the start of a new era in European relations organised at the Congress of Vienna.
6. *corrupt Babylon*: in the Book of Isaiah, Babylon is identified as signifying both false religion and corrupt civilization.

From Charles Darwin [27 February 1871][1] 3365

6. Queen Anne Street | Monday.

My dear Tyndall,

I called yesterday for the improbable good luck of finding you at home on a Sunday at the R[oyal] Institution, to thank you for your note[2]—What a very interesting & practical discovery you have made in saving the poor firemen from being choked by noxious fumes[3]—By an odd chance I was wondering some months ago about the hairs within our nostrils, & it would be fine if you have explained it. At present, however, I can hardly think, except in a few anomalous individuals, that the hairs are thick enough together. It would be very advisable to ascertain how far the character is present in the hairless races of man. also under a different point of view with animals which have great powers of scent.

I hope to see tomorrow Dr. Ogle[4] who has been attending to powers of smelling, & I will tell him your idea, & he may have opportunity, as a Doctor, of observing different races of man in Hospitals.

Many thanks for your note. | Yours very sincerely | Ch[arles]. Darwin

RI MS JT/1/TYP/9/2810
LT Typescript Only

1. *[27 February 1871]*: this letter was dated by the Darwin Correspondence Project through its relationship to letter 3363 from Tyndall to Darwin and Darwin's sojourn in London from 23 February to 2 March 1871. This letter is published in F. Burkhardt, et al. (eds), *The Correspondence of Charles Darwin* (Cambridge: Cambridge University Press, 2012), vol. 19, p. 110; this is letter 7518 in the online edition of the Darwin correspondence).

2. *your note:* letter 3363.
3. *saving the poor firemen from being choked by noxious fumes*: see letter 3363, n. 2.
4. *Dr. Ogle*: John William Ogle (1824–1905), an English physician. Ogle worked at St. George's Hospital on morbid anatomy and was curator of the anatomical museum. In 1861 he became lecturer on pathology and in 1866 full physician, a position he resigned in 1876 due to depression. After his recovery he held numerous positions in professional organisations and worked as a consulting physician (*ODNB*).

From William Budd [February 1871][1] 3366

Another point[2] of great practical importance is, as far as possible, to appraise the respective shares which air and water take in the great act of distribution. That both cholera and typhoid fever are sometimes disseminated by drinking water has been amply proved.[3] I have myself related many instances of the fact, and have in my notes the record of many others still more striking. But that water is the sole or even the chief vehicle of cholera and typhoid is a notion which, if I may trust my own experience, facts do not warrant.

Limiting myself, for the moment, to the case of typhoid, I am in a position to state that all the worst and most widespread outbreaks of that fever which I have ever witnessed, have occurred among communities supplied by drinking water which was absolutely blameless. Two illustrations will suffice.

I live in a town in which the divorce between sewage and drinking water has long been consummated. Bristol[4] is supplied with drinking water which from its source in the Mendips[5] to the tap from which it is delivered under high pressure to the consumer, flows through conduits out of all reach of sewage contamination.

And yet typhoid fever has not only not ceased to exist in Bristol, but about eight or ten years ago (before the appointment of a health officer), there occurred in the Parish of St. James' one of the worst outbreaks of this fever which I have ever seen in the city. In the course of a circuit which I took one morning with the late Dr. Pring[6] (at that time Poor Law Medical Officer)[7] I saw within a comparatively small area more than eighty cases of the disease.

Now, with the exception of a single household, all the patients were drinking the Mendips water; the very same water which, as far as fever is concerned, more than 150,000 of their fellow-citizens outside the infected area were drinking with absolute impunity.

Some four or five years ago I was sent for to advise what measures should be taken to stay an outbreak of typhoid fever which had occurred in a large convent about two miles from Bristol.[8] The inmates were divided into three distinct divisions; the largest being a reformatory for girls, who occupied the central block of the building. Into this reformatory the fever was brought by a girl already suffering from it, and who had contracted it in a seaside place more than twenty miles off. From this girl the disease spread until, at the date of my visit, more than fifty girls we lying ill of it. From first to last the fever was confined to the reformatory girls, and to persons in immediate attendance upon them.

Now, the facts as to the drinking water were these.

1. The water was proved by examination of the well, and by chemical analysis, to be entirely free from sewage contamination.

2. The inmates of another large division of the convent, who remained entirely free from fever, drank the same water as the girls among whom fever was raging like a plague.

3. From the very time when disinfection was brought to bear on the excreta the disease ceased to spread, although the inmates of the infected division continued to drink the same water as before.

Lastly, nothing has since been done to the well—the water remains what it was—but no fever has occurred in the convent since.

The evidence in both these cases is, as you see, of that crucial, decisive order that admits of no reply.

They show, at least, that typhoid fever may do its worst where drinking water takes absolutely no part in the distribution of the poison.

But if water be excluded, the air is the only other possible vehicle by which a poison generated in the living body can find its way back to other living bodies on a scale sufficiently large to cause the resulting disease to assume an epidemic form.

I may remark further that the infection of the air in these two cases was obviously not the work of chance, but only represented the effect of agencies which are always in operation where this fever prevails.

The phenomenon is, in fact, merely the expression of a general law.

The germs cast off in the liquid excreta of contagious diseases rise into the air by no power of their own, but in virtue of the very same physical conditions which cause the germs of the great tribe of Infusoria, which, as their name bespeaks, breed in liquids, to rise in swarms into the same medium.

If there were time or need, I could show, by evidence quite as decisive, that all these statements apply equally to cholera also.

I do not know that there is anything in these data to suggest additional matter for your essay, but I have though it worth while to bring them under your notice, harmonising as they do with your own investigations which show by such striking phenomena that air and water are equally objects of distrust.

As to the germ theory itself, that is a matter on which I have long since made up my mind. From the day when I first began to think of these subjects,

I have never had a doubt that the specific cause of contagious fevers must be living organisms.

It is impossible, in fact, to make any statement bearing upon the essence or distinctive characters of these fevers, without using terms which are of all others *the most distinctive* of life. Take up the writings of the most violent opponent of the germ theory, and, ten to one, you will find them full of such terms as "propagation," "self-propagation," "reproduction," "self-multiplication," and so on. Try as he may—if he has anything to say of those diseases which is characteristic of them—he cannot evade the use of these terms, or the exact equivalents of them. While perfectly applicable to living things, these terms express qualities which are not only inapplicable to common chemical agents, but, as far as I can see, actually inconceivable of them.[9]

J. Tyndall, 'Addition to Dust and Disease', in *Fragments of Science*, pp. 343–6

---

1. *[February 1870]*: the date of this letter can only be roughly estimated. Tyndall includes this letter as an addendum to his paper 'Dust and Disease' in *Fragments of Science*. He did not mention it in his Friday Evening Discourse on 'Dust and Disease' at the RI on 21 January 1870 (*Roy. Inst. Proc.*, 6 (1870–72), pp. 1–14), nor in a letter to the editor of *Times* of 31 January 1871 (letter 3348), despite mentioning Budd in both, but he did read Budd's work while on his eclipse journey in December (letter 3348). The preface to *Fragments of Science* indicates that this volume was with the publisher by March 1871. It is likely Tyndall received this letter between the *Times* letter and finalizing *Fragments of Science*, probably in early February.
2. *Another point*: this letter is a fragment of a letter from Budd. We do not have the full letter. See also n. 9 below.
3. *That both . . . amply proved*: John Snow (1813–58), an English physician, had figured out that cholera was spread by contaminated water, and not foul air, by 1855 (see J. Snow, *On the Mode of Communication of Cholera*, 2nd edn (London: Wilson and Ogilvy, 1855)). Budd began to focus on clean drinking water in Bristol to stem cholera as well, and, by 1866, deaths from the disease had dropped from 2000 in 1849 to 26 in 1866. On Snow and cholera, see S. Johnson, *The Ghost Map: The Story of London's Most Terrifying Epidemic—and How It Changed Science, Cities, and the Modern World* (New York: Penguin, 2006); and P. Vinten-Johansen, et al., *Cholera, Chloroform, and the Science of Medicine: A Life of John Snow* (New York: Oxford University Press, 2003).
4. *Bristol*: a town in the southwest of England, close to the coast, and on the River Avon. During the nineteenth century, the city experienced a major population boom.
5. *Mendips*: the Mendips Hills are limestone hills located to the south of Bristol in Somerset, England. The Bristol Water Works Company, founded in 1846 by William Budd and others, supplied water from sources surrounding the city including the Mendips.
6. *Dr. Pring*: John Pring (*c.* 1820–58) was the Poor Law Medical Officer in Bristol and

member of the Royal College of Surgeons. He died of typhoid fever contracted while carrying out his duties during an outbreak in Bristol (W. Budd, *Typhoid Fever: its Nature, Mode of Spreading, and Prevention* (London: Longmans, Green, and Co., 1873), p. 114; 'Deaths', *Medical Times & Gazette*, 20 March 1858, p. 308).

7. *Poor Law Medical Officer*: with the passing of the Poor Act in 1834 in England, there was a need for medical care as well. A few years after this, the Poor Law Commissioners provided for doctors to care for the poor and be paid for this, setting up Poor Law Medical Officers. See R. G. Hodgkinson, 'Poor Law Medical Officers of England 1834–1871', *Journal of the History of Medicine and the Allied Sciences*, 11:3 (July 1956), pp. 299–348.
8. *Some four or five years ago . . . Bristol*: in February of 1864 there was a typhoid outbreak at the Convent of the Good Shepherd at Arno's Court (W. Budd, *Typhoid Fever* (see n. 6), pp. 128–35).
9. *As to the germ theory . . . inconceivable of them*: Tyndall found this passage impressive enough to repeat verbatim in J. Tyndall, 'On Dust and Smoke', *Roy. Inst. Proc.,* 6 (1870–72), pp. 365–76, on pp. 369–70; and J. Tyndall, 'On Dust and Smoke', *Nature* (15 June 1871), pp. 124–8, on p. 125.

BIOGRAPHICAL REGISTER

This register contains the names and biographical details of people who are mentioned three or more times in the letters included in this volume. Biographical information about people mentioned only once or twice is contained in the notes appended to the appropriate letter(s).

Conventions

The sources for the biographical entries and the letter annotations are standard biographical dictionaries, such as the *Complete Dictionary of Scientific Biography*, the *Oxford Dictionary of National Biography*, Poggendorff's *Biographisch-literarisches Handwörterbuch der exacten Wissenschaften*, *Historisches Lexikon der Schweiz*, *Neue Deutsche Biographie*, *American National Biography*, *The Dictionary of Nineteenth-Century British Scientists*, G.E.C.'s *Complete Peerage of England, Scotland, Ireland, Great Britain*, and the biographical database of the Bibliothèque nationale de France. These have not been specified in the entries unless necessary. Additionally, the editors have found material from the Tyndall correspondence and from the journals of Tyndall and T. A. Hirst very useful. Where additional sources add important material, they are given in parentheses at the end of the entry.

Adair, Henrietta Mary (1843–1921), was the youngest daughter of Alexander Adair (1791–1863) and Harriet Eliza Atkinson (1811–78). She attended Tyndall's lectures at the RI and they conversed about scientific matters, as shown by letters in this volume. Tyndall fell in love with Mary and proposed to her in late 1869. She declined but they remained friends (*Ascent of John Tyndall*, 257).

Airy, George Biddell (1801–92), was an English astronomer. Airy was educated at Cambridge University, graduating senior wrangler in 1823. In 1826 he was appointed Lucasian Professor of Mathematics and in 1828 to the Plumian Chair of Astronomy and directorship of the Cambridge University observatory. In 1835 he was appointed astronomer royal, a position he held until 1881, and set about improving the instrumentation and research activities of the Royal Observatory, Greenwich. He pursued research on a wide range of topics, including optics, sound, electricity, calculating the Earth's mean density, and analyzing the inequalities in the motions of the Earth and Venus. Airy served as a sounding board for Tyndall's theories and observations.

Allen, Robin (1820–98), was a British poet who was appointed as a clerk in the secretary's office at Trinity House in 1837. From 1867 to 1881 he was the secretary of Trinity House ("The Trinity House," *Times*, 25 April 1881, 9). Tyndall reported to Allen the results of his observations on British and Irish lighthouses.

Baring, Louisa Caroline (née Stewart-Mackenzie, 1827–1903), Lady Ashburton, was a philanthropist and art collector. In 1858 she married the extremely wealthy William Bingham Baring, 2nd Baron Ashburton (1799–1864). After her husband's death, Baring amassed a significant collection of art, which she displayed in her numerous residences. She was also known for hosting gatherings attended by Britain's social and cultural elite, including Tyndall and Thomas Carlyle.

Baring, Mary Florence "May" (or "Mazy") (1860–1902), was the daughter of Louisa Baring, Lady Ashburton, and William Bingham Baring, 2nd Baron Ashburton. She married William Compton, 5th Marquess of Northampton in 1884, and held the title Marchioness of Northampton from 1897.

Barnard, Jane (1832–1911), was a niece of Michael and Sarah Faraday. She was the daughter of Sarah's brother, John Barnard, and Michael's younger sister Margaret. Like her aunt and uncle, she was a member of the Sandemanian Church. After Faraday's death she cared for her aunt.

Bastian, Henry Charlton (1837–1915), was a British physician and neurologist. He studied at University College, London, completing his MD in 1866. Following a post as assistant physician and lecturer in pathology at Saint Mary's Hospital, Paddington, Bastian returned to University College as professor of pathological anatomy in 1867. He was appointed physician there in 1878, and chair of medicine from 1887 to 1898. An early interest in flora and fauna saw him elected FRS in 1868. Bastian supported the theory of evolution and was a prominent promoter of spontaneous generation. This led to confrontational disputes with numerous scientists and clinicians, including Tyndall, Thomas Huxley, Joseph Lister, and Louis Pasteur, who supported the increasingly popular germ theory, as illustrated by several of Tyndall's letters in this volume.

Bennett, George Wheatley (dates unknown), was a member of the London Dialectical Society, an association founded in 1867 and dedicated to investigating psychical and spiritualist phenomena. Bennett was honorable secretary for its committee to investigate the "phenomena alleged to be spiritual manifestations." In 1869 Bennett asked Tyndall, at the request of the committee, to attend their séances so he could advise on further testing to be done regarding the reported spiritual phenomena (*Report on Spiritualism, of The Committee of the London Dialectical Society* [London: Longmans, Green, Reader and Dyer, 1871]).

Berkley, George (1821–93), was a British civil engineer who was a manager of the RI at the time covered in this volume (*Minutes of Proceedings of the Institution of Civil Engineers*, 115 [1894], pp. 382–85).

Bonaparte, Louis Napoléon (1808–73), also Napoléon III, emperor of France (1852–70), and president of the Second French Republic (1848–52). He was the nephew of Napoleon I and son of Louis Bonaparte (king of Holland, 1806–10). As a Bonaparte he spent his early life in exile. He returned to France during the 1848 revolution, and was elected president of the Second French Republic. Constitutionally ineligible for a second term as president, he seized power on 2 December 1851 and was subsequently proclaimed emperor by plebiscite in November 1852. Ambitious to bolster France's international reputation, he hoped to reorganize the European system installed at the Congress of Vienna, 1815. The Prussian defeat of Austria in 1866 further threatened France's position and a dispute over the candidacy of Leopold of Hohenzollern-Sigmaringen for the Spanish throne led to the outbreak of the Franco-Prussian War (1870–71). Following France's defeat, Louis was deposed and lived the rest of his life in England.

Bonaparte/Buonaparte, Napoléon (1769–1821), French general, first consul (1799–1804) and emperor of France (1804–14/15), rose to prominence during the French Revolution (1789–99) and several campaigns during the French Revolutionary Wars (1792–1802). A controversial figure whose achievements include legal reform and the abolition of serfdom, he was also infamous for rampant use of propaganda and the institutionalized plunder of conquered territories. After he was forced to abdicate in 1814, he attempted to regain power the following year, which culminated in the savage battle at Waterloo against British troops led by the Duke of Wellington and reinforced by the Prussians. Napoleon was ultimately defeated and exiled to the remote island Saint Helena in the South Atlantic, where he wrote a spurious memoir, *Mémorial de Sainte-Hélène*, and died in 1821.

Bowles, Robert Leamon (1834–1913), was a member of the Royal Colleges of Surgeons of Great Britain and Ireland (MRCS). He lived in Folkestone and was an acquaintance of Henry Bence Jones.

Brown, William (dates unknown), was a porter at the RI.

Brücke, Ernst Wilhelm von (1819–92), was a German physiologist. Brücke trained at the University of Berlin, graduating as a doctor of medicine and surgery in 1842. He became professor of physiology at the University of Königsberg (Kalinengrad) in 1848 and the University of Vienna the following year. Brücke's research interests ranged from the study of cells and muscles to plant physiology, optics, and theories of color. He contributed to the physiology of the eye and digestion, as well as studying language and speech. In 1845 he founded, with Emil du Bois-Reymond and Hermann Helmholtz, the Physical Society of Berlin. He was given the noble title von Brücke in 1873 by Emperor Franz Joseph I.

Budd, William (1811–80), was an English physician and epidemiologist. Born in Devon, Budd studied medicine in Paris, London, and Edinburgh from 1828 to 1837. His particular interest was in the prevention of epidemic disease, and he was singular in his interest in epidemic diseases in animals, which he saw as analogous to those experienced by humans. He also advocated for disinfection as a preventative measure. His most significant work, *Typhoid Fever: Its Nature, Mode of Spreading, and Prevention* (1873), challenged the widespread view that typhoid was caused by a putrefactive process. Tyndall campaigned for Budd to be made a fellow of the RS, and he was elected on 8 June 1871.

Bulwer Lytton, Rosina Anne Doyle (née Wheeler, 1802–82), was an Anglo-Irish novelist. She married writer and politician Edward George Earle Bulwer-Lytton (1803–73), 1st Baron Lytton, in 1827, with whom she had two children. Financial strain, infidelity, and domestic abuse affected their marriage. They were separated in 1836, citing an "incompatibility of temper." The children stayed with their father and Rosina was not permitted access. She published several novels between 1839 and 1860 to supplement her meager allowance, but also published public condemnations of her husband and the plight of separated wives. In retaliation Edward had her confined to a private insane asylum in Brentford, but her friends facilitated her release three weeks later. Her longstanding efforts to draw attention to the injustices experienced by women contributed to later legislation to protect their interests. She died in poverty in 1882 and was buried in an unmarked grave.

Bunsen, Robert Wilhelm (1811–99), was a German chemist. Bunsen held a number of positions, including professor at the University of Marburg (1839–51), the University of Breslau, and, from 1852, the University of Heidelberg. He made many contributions to chemistry, including the discovery of cesium and rubidium, and developed the battery and gas burner that bear his name. A renowned researcher, he was also an extremely popular teacher, mentoring both Tyndall and Edward Frankland at Marburg. He received the RS's Copley Medal in 1860, and its Davy Medal, with Gustav Robert Kirchhoff, in 1877.

Carlyle, Thomas (1795–1881), was a Scottish essayist, novelist, historian, and one of the most notable intellectual figures in the Victorian era. Tyndall was heavily influenced by Carlyle and avidly read his publications. After reading *Past and Present* (1843), Tyndall commented that Carlyle "must be a true hero. My feelings towards him are those of worship 'transcendental wonder' as he defines it" (Journal, 18 July 1847, RI MS JT/2/13a/231). During the 1860s, Tyndall became a good friend of Carlyle's, supporting him in the Eyre controversy and accompanying him to Louisa Baring's home in Mentone, France, in December 1866. By 1869–70 they were close, with Tyndall referring to Carlyle as "the General" and doting on the aging gentleman.

Carpenter, William Benjamin (1813–85), was an English biologist, comparative anatomist, and administrator. Carpenter was elected FRS and appointed to the Fullerian professorship in physiology at the RI in 1844. In 1856 he was appointed registrar of the University of London, a position he held for twenty-three years. His publications *The Principles of General and Comparative Physiology* (1839) and *The Principles of Human Physiology* (1842) went through several editions, and he received the RS's Royal Medal in 1871 for his physiological research. The same year, Carpenter and Tyndall had a very public disagreement over who was the first to recognize the work of physicist Robert Mayer (1814–78). Carpenter died tragically in 1885, after suffering severe burns from an accident with a vapor bath (*Memorials of William Benjamin Carpenter* [London: H. K. Lewis, 1885]).

Cassell, Messrs and Co., was a British book publishing house, founded in 1848 by its namesake John Cassell (1817–65). It is considered one of the most innovative publishing houses of the period, with a sense of diverse publics and occupying a politically radical position. After Cassell's death, the business was run by his partners, Thomas Dixon Galpin and George William Petter. Messrs Cassell and Co. got on Tyndall's cantankerous side in 1869, when they repackaged and published some of his old lectures without consultation (*The Story of the House of Cassell* [London: Cassell, 1922]).

Chasles, Michel Floréal (1793–1880), was a French mathematician noted for his work on the history of geometry. His main experimental work was in kinematics and dynamics. In 1841 Chasles became professor at the École polytechnique, where he remained until 1846, when a chair was created for him at the Sorbonne. In 1865 he was awarded the RS's Copley Medal, "for his historical and original research in pure geometry." Chasles was interested in the history of science, and letters in this volume discuss his loss of reputation after he was duped by the legendary forger Denis Vrain-Lucas in 1867. Chasles invested considerable personal funds in counterfeit letters from Isaac Newton, Blaise Pascal, and other scientific luminaries, which he then presented as genuine to the French scientific community.

Clausius, Adelheid (née Rimpau, 1833–75), was the wife of Rudolf Clausius. They married in Zurich in 1859 and had six children. Both she and her husband corresponded with Tyndall. They named their son Rudolf John, after Tyndall, and he became his godfather.

Clausius, Alfred (1868–1950), was the son of Adelheid and Rudolf Clausius.

Clausius, Rudolf John "Johnny" (1864–1915?), was the son of Adelheid and Rudolf Clausius. He was named after Tyndall, who was also his godfather. He later completed a doctorate of law in 1890 at Friedrich Wilhelms University Bonn (now University of Bonn), and became a lawyer in Cologne.

Clausius, Rudolf Julius Emanuel (1822–88), was a German mathematician and physicist. His paper on the theory of heat, "Ueber die bewegende Kraft der Warme," published in the *Annalen der Physik* in 1850, is considered one of the foundational papers of modern thermodynamics. Clauisus held a professorship in mathematical physics at the Polytechnikum in Zurich from 1855, moved to the University of Würzburg in 1867, and finally to the University of Bonn in 1869. Tyndall translated many of his major works into English for publication in the *Phil. Mag.* and *Scientific Memoirs* in the early 1850s. The friendship between the two was cemented during the Mayer-Joule priority dispute in the early 1860s, and they became close friends. In 1864 Clausius asked Tyndall to be godfather to his newborn son and christened the baby Rudolf John. Tyndall was tremendously flattered by the request, and it cemented the friendship between the two, as evidenced by the beginning of their use of the informal German form of address "Du" in their letters and the otherwise very formal German showing his gossipy side. See letter 2147 in Michael D. Barton, Piers J. Hale, Nathan Kapoor, and Elizabeth Neswald, eds., *The Correspondence of John Tyndall*, vol. 8, *The Correspondence, June 1862–January 1865* (Pittsburgh: University of Pittsburgh Press, 2021).

Clifford, William Kingdon (1845–79), was an English mathematician and philosopher. Friends with Frederick Pollock, he attained second wrangler at the University of Cambridge in 1867. In 1868 he was elected fellow at Trinity College. In the same year he gave a public discourse at the RI, applying Herbert Spencer's philosophy of evolution to the development of mind. In 1870 Clifford traveled with the solar expedition to Sicily. His ship wrecked on the return voyage; no one was killed.

Colville/Colvile, (Frances) Elinor (née Grant, 1838–1919), was the eldest daughter of John Peter Grant, of Rothiemurchus, who served as lieutenant governor of Bengal and governor of Jamaica. Her sister was Jane Strachey. In 1857 she married James William Colville (1810–80, knighted in 1848) in Calcutta Cathedral and they returned to Britain in 1859. At their Hyde Park residence they entertained a select group of London's cultural elite.

Cottrell, John (dates unknown), was Tyndall's senior laboratory assistant at the RI from 1866 to 1885. He assisted Tyndall with observations on the Aletsch glacier in 1875.

Crookes, William (1832–1919), was a British chemist and editor of science journals. He worked as a personal assistant to August Wilhelm Hofmann at the Royal College of Chemistry, London, from 1849 to 1854. At the RS, he was elected fellow (1863), recipient of the Davy and Copley Medals (1888, 1904), and made foreign secretary (1908–13) and president (1913–15). Crookes found financial success through the establishment of several journals, including the *Quarterly Journal of Science*, which he edited from 1864 to 1879. His research on the electrical discharge in vacuum tubes and gas-filled tubes led to the discovery of cathode rays. Crookes was also interested in spiritualism and psychical phenomena. He was knighted in 1897.

Darwin, Charles (1809–82), was an English naturalist and geologist, author of *On the Origin of Species* (1859) and the more famous contributor to the theory of biological evolution by natural selection, the other being Alfred Russel Wallace. Earlier in Tyndall's career, Darwin encouraged his research on glaciers and supported him in his disagreement with J. D. Forbes. Darwin was awarded the RS's Copley Medal in 1864.

Davy, Humphry (1778–1829), was a chemist and inventor born in Cornwall, England. At the RI, he was director of the laboratory (1801–25), professor of chemistry (1802–12), and honorary professor (1813–23). He achieved fame as a lecturer and for his chemical research that isolated such elements as sodium, potassium, and chlorine. The invention of the miner's safety lamp is often attributed to Davy. He served as president of the RS from 1820 to 1827. Michael Faraday began his RI career as Davy's assistant.

Debus, Heinrich (1824–1915), was a German chemist. Debus studied at Marburg (1845–48), where he met Tyndall, Edward Frankland, and Thomas Hirst. He moved to England in 1851 and taught chemistry at Queenwood College until 1867, when he began teaching at Clifton College, Bristol. Later Debus lectured at Guy's Hospital Medical School (1870–1888) and became professor of chemistry at the Royal Naval College, Greenwich (1873–1888). Tyndall assisted and advised Debus in his negotiations for the position at Guy's College in 1870 and facilitated his appointment.

de la Rive, Auguste Arthur (1801–73), was a Swiss physicist known for his foundational work on the electrochemical theory of batteries. He served as professor of natural philosophy at the Academy of Geneva and worked principally on electricity, including publishing the three-volume *Traité d'électricité théoretique et appliquée* (Paris: J.-B. Baillière et H. Baillière, 1854–58). De la Rive also investigated the specific heats of gases and calculated the temperature of the Earth's crust. He lived in Geneva and was Tyndall's primary contact after Tyndall injured his leg while vacationing in the Alps.

Douglass, James Nicholas (1826–98), was an English civil engineer and lighthouse designer. Employed by Trinity House, he designed the experimental lighthouse at Blackwall. He was later knighted for designing the fourth Eddystone lighthouse (1882). Douglass and Tyndall exchanged ideas about lighthouse lighting and foghorns from the 1860s into the 1880s. He was elected FRS in 1887 (T. Williams, *The Life and Times of Sir James Nicholas Douglass* [London: Longmans, Green, 1900]).

du Bois-Reymond, Emil Heinrich (1818–96), was a German physician and physiologist who studied with Johannes Müller at the University of Berlin. Under Müller's guidance, he developed a lifelong interest in animal electricity. Upon Müller's death in 1858, du Bois-Reymond was offered the position of professor of physiology at Berlin. He wrote extensively on animal electricity, publishing the first volume of his *Investigations of Animal Electricity* in 1848 and the second in 1884. In 1877, after his intensive lobbying, the Prussian government finally funded the provision of a modern physiological laboratory. Du Bois-Reymond and his colleagues Carl Ludwig, Hermann Helmholtz, and Ernst Brücke were instrumental in shaping the methods and direction of German physiology in the second half of the nineteenth century, integrating approaches from physics and engineering into the analysis of organic systems.

Dumas, Jean-Baptiste (1800–84), was a French organic chemist and professor at the Sorbonne. In 1824 he founded the journal *Annales des sciences naturelle*. Politically active after the February Revolution of 1848, he had a political career as minister of agriculture and minister of commerce in 1850–51. Dumas was subsequently appointed senator in Napoleon III's Second Empire, where he oversaw the infrastructural modernization of Paris. He focused on applied chemistry but did some research into physiological chemistry and famously feuded with Justus Liebig. Elected foreign member of the RS in 1840, Dumas received its Copley Medal in 1844. He is most recognized for his work in organic chemistry and his *Leçons sur la philosophie chimique*, which traces the development of chemical hypotheses.

Egerton, Mary Alice "May" (1848–1924), was the eldest daughter of Edward and Mary Egerton. She married Beauchamp Tower in 1902 ("Births," *Times*, 19 February 1848, 8; "Wills and Bequests," *Times*, 11 June 1924, 15).

Egerton, Mary Frances (née Pierrepont, 1819–1905), was the daughter of Charles Pierrepont (1778–1860), 2nd Earl Manvers. In 1845 she married Edward Christopher Egerton (1816–69), with whom she had five children: Mary Alice, Georgina, Charlotte, Hugh Edward, and Charles Augustus. Egerton attended Tyndall's lectures and was a frequent correspondent of Tyndall's, especially after the sudden death of her husband in 1869.

Emerson, Ralph Waldo (1803–82), was an American author, poet, and lecturer. He was a leading figure in the transcendentalism movement, which he heavily influenced through such essays as "Self-Reliance" (1841), "Nature" (1836), and "The American Scholar" (1837). He was highly productive during the 1860s, editing posthumous works of Henry David Thoreau (1817–62), publishing the collection of poems *May-Day and Other Pieces* (1867), and delivering hundreds of lectures across the United States. Tyndall greatly admired his works and cited him often.

Faraday, Michael (1791–1867), was a natural philosopher, Fullerian Professor of Chemistry at the RI, and a member of the Sandemanian Church. His extensive research focused on electricity and encompassed electrochemistry, electromagnetism, diamagnetism, and magneto-optics (the Faraday effect). Faraday also held the positions of director of the laboratory (1825–67) and superintendent of the house (1852–67) at the RI. He was Tyndall's patron and helped him obtain the position of professor of natural philosophy at the RI in 1853. Despite significant differences in their views on religion and on the nature of magnetism, they held the highest mutual regard for each other.

In 1867 Tyndall succeeded Faraday as superintendent of the house and director of the laboratory at the RI. Tyndall's *Faraday as a Discoverer* (1868) was his tribute to his mentor.

Faraday, Sarah (née Barnard, 1800–79), was the daughter of Mary Boosey and Edward Barnard, a successful London silversmith and Sandemanian. She was the wife of Michael Faraday, whom she likely met at the Sandemanian meeting house in Paul's Alley. After they married in 1821, she moved into his rooms at the RI, and in later years she would take on many of his administrative responsibilities at the RI.

Francis, William (1817–1904), was a chemist, translator, and editor. He was the illegitimate son of the printer and publisher Richard Taylor. Francis learned the printing trade in Germany in the 1830s before training as a chemist with Justus Liebig and Gustav Rose in Giessen, where he obtained his doctorate in 1842. He made many translations for his father's *Scientific Memoirs* and the *Phil. Mag.*, and became the editor of the latter in 1851. In 1852 he began a partnership with his father, establishing the Taylor & Francis publishing firm. He first met Tyndall in 1850 and became one of Tyndall's close confidants.

Frankland, Edward (1825–99), was an English chemist and one of Tyndall's closest friends. He and Tyndall met at Queenwood College before they studied together at the University of Marburg in 1848. After holding a professorship at Owens College, Manchester (1851–57), Frankland moved to London and held a number of positions, including at Saint Bartholomew's Hospital and the East India Company's military college. He was professor of chemistry at the RI (1863–68) and afterward at the Royal College of Chemistry (1868–85). In 1864 Frankland joined Tyndall, Thomas Hirst, Joseph Hooker, Thomas Huxley, and four others as members of the X Club, a monthly dining club in which they promoted science in British society.

Froude, James Anthony (1818–94), was an English historian and editor, educated at Oxford University. His *The Nemesis of Faith* (1849) was highly controversial at Oxford, and a copy was set on fire by subrector William Sewell. A pivotal meeting with Thomas Carlyle in 1849 developed into a lifelong close friendship between the two men, with Froude not only editing Carlyle's posthumous *Reminiscences of Jane Welsh Carlyle* (1881) but also writing a multivolume biography. In 1856 he began work on his twelve-volume *History of England from the Fall of Wolsey to the Death of Elizabeth* (1856–70), and in 1860, after the death of his first wife, he moved to London and took on the editorship at *Fraser's Magazine.* Froude was a member of the Metaphysical

Society, established in 1869, where he would have encountered fellow member John Tyndall. In the 1870s he presented controversial views on the Irish Question and imperial policy in South Africa.

Gassiot, John Peter (1797–1877), was an electrician and a member of the wine merchant firm Martinez, Gassiot & Co. He maintained a private laboratory at his house on Clapham Common. In 1837 he was a founding member of the London Electrical Society, and in 1841 was elected FRS. He was active in the RS's financial affairs in the second half of the nineteenth century. Gassiot published papers on the voltaic pile (battery) and showed that a spark could be produced before the circuit of a battery was closed. In 1859 he helped found a relief fund to assist scientific men in financial difficulties and established a fund in 1872 to support the Kew observatory following the decision of the BAAS to stop supporting it.

Gautier, Dr. (dates unknown), was the Geneva physician who treated Tyndall in the autumn of 1869, after an accident in the Alps left him with a severely infected leg wound.

Ginty, Margaret (née Roberts, b. ca. 1826), was the widow of Tyndall's friend William Ginty. She married William in Ireland in 1846, after which they relocated to Manchester. In 1854 the family moved to Rio de Janeiro, where William Ginty died in 1866. After William's death, Tyndall served as Margaret's trustee and supervised the education and employment of her children. The letters in this volume show Tyndall continuing to assist her in legal matters concerning the estate (*Ascent of John Tyndall*, 198).

Goethe, Johann Wolfgang von (1749–1832), was the outstanding German literary figure of the late eighteenth and early nineteenth centuries. His extensive writings included his theory of colors and a study of the structure and function of plants. During his time in Marburg, Tyndall became familiar with some of Goethe's works. He was, along with Alfred Tennyson, Ralph Waldo Emerson, and Thomas Carlyle, one of the major literary and philosophical-poetic influences on Tyndall. Tyndall paid homage to the spirits of Goethe and Friedrich Schiller during his visit to Weimar in 1869.

Hawes, Elliot H. (dates unknown), Captain, R.N., was the assistant inspector of lights for the Commissioners of the Irish Lights. He succeeded Captain E. F. Roberts, R.N., as inspector upon his death (J. Tyndall, "The Story of our Lighthouses," *Nineteenth Century* 24 [July 1888], p. 73). In some publications, including Parliamentary Papers, Hawes's middle initial is given as *W*.

Helmholtz, Anna (née Mohl, 1834–99), was a Berlin salonnière and translator of scientific works. She married Herman Helmholtz in 1861. With Clara Wiedemann, she translated Tyndall's *Heat* and *Sound* into German.

Helmholtz, Hermann Ludwig Ferdinand (1821–94), was a German physicist and physiologist. Educated in military medicine at the Friedrich-Wilhelms-Universität in Berlin, he taught anatomy at the Berlin Academy of Arts in 1848. His academic career took him from professor of physiology at the University of Königsberg (1849) to professor for anatomy and physiology at the University of Bonn (1855), where he turned his attention to physics; professor of physiology at the University of Heidelberg (1858); and, finally, professor of physics at the University of Berlin (1871). Helmholtz made significant contributions to numerous fields in physics, including thermodynamics, electrodynamics, and the energy conservation law, and in physiology, the physiology of sound, vision, and nerve conduction. He was instrumental in promoting the foundation of the German Imperial Physical Technical Institute, a non-university research institution with ties to industry, and served as its first president.

Henry, Joseph (1797–1878), was an American physicist. Following his studies at Albany Academy he held various roles there, eventually becoming professor of mathematics and natural philosophy in 1826. His early research focused on electromagnetism. Henry was appointed to the chair of natural philosophy at the College of New Jersey in 1832 and elected to the American Philosophical Society in 1835. A trip to Europe in 1837 brought about an opportunity for him to meet many leading scientists, including Michael Faraday. In 1846 he was appointed the first secretary of the new Smithsonian Institution.

Herschel, John Frederick William (1792–1871), was an English mathematician and astronomer. Son of the famous astronomer William Herschel (1738–1822), he himself became one of the most eminent scientists in Victorian Britain. He graduated from Cambridge University in 1813 as senior wrangler, received two RS Copley Medals (1821 and 1847), and was awarded a knighthood (1831), followed by a baronetcy (1838). His *Preliminary Discourse on the Study of Natural Philosophy* (1830) was widely read and greatly admired. In 1849 he published *Outlines of Astronomy*, a popular presentation of astronomy that went through eleven editions by 1871. He made many contributions to the science of photography and also edited the Admiralty's *Manual of Scientific Enquiry* (1849), a science field guide for the use of the Royal Navy. From 1850 to 1855 he served as master of the mint, and on his death was buried next to Isaac Newton in Westminster Abbey.

Hilgard, Julius Erasmus (1825–91), was a German American engineer and geodesist. Born in Zweibrücken, Rhine-Palatinate, Hilgard immigrated to the United States with his family in 1835. From 1844 to 1860 he worked for the US Coast Survey in the computing department and in 1867 was appointed the agency's assistant superintendent, a position in which he supervised the work of the Office of Weights and Measures and drafted legislation for the implementation of the metric system in the United States. Hilgard was elected president of the American Association for the Advancement of Science in 1877, the same year he won a gold medal for his design of an instrument to measure the density of seawater. He died in 1891 due to complications from Bright's disease ("Julius Erasmus Hilgard," *Proceedings of the American Academy of Arts and Sciences* 26 [May 1890–May 1891], pp. 370–73).

Hirst, Emily Anna "Lilly" (1853–83), was the niece of Thomas Archer Hirst, with whom she was very close. Her parents were William and Rachel Hirst (William H. Brock and Roy M. MacLeod, eds., *Natural Knowledge in Social Context: The Journals of Thomas Archer Hirst FRS* [London: Mansell microfiche, 1980], index, p. 50).

Hirst, Thomas Archer (1830–92), was an English mathematician, educator, and Tyndall's closest friend and correspondent. His friendship with Tyndall began in Halifax, Yorkshire, in 1845, when they were both working for the surveyor Richard Carter. Encouraged by Tyndall—who acted as a sort of mentor to him—Hirst studied at the universities of Marburg, Göttingen, and Berlin between 1849 and 1852. In 1853 he replaced Tyndall as mathematics teacher at Queenwood College, resigning in 1856 to look after his sick wife, Anna, who died in Paris in 1857. During the 1860s Hirst established himself as an active and respected member of the London scientific community. He taught mathematics at University College London (1860–64) and subsequently held professorships of physics and mathematics (1865, 1867–70). He joined the RS council (1864); was a founder of the X Club, a monthly dining club that promoted science in British society; and served as general secretary of the BAAS (1866–70).

Holland, Henry (1788–1873), was an English physician and travel writer educated at the University of Edinburgh. He accompanied Sir George MacKenzie of Coul and Richard Bright on an expedition to Iceland in 1810, and his work on Icelandic diseases formed the basis of his doctoral thesis. He became a licentiate of the Royal College of Physicians in 1816 and a fellow in 1828, opening a practice in London in 1816 that served members of London's wealthy and fashionable society. He became physician-extraordinary to Queen Victoria (1837) and physician-in-ordinary to Prince Albert (1840). In

1853 Holland was made a baronet, after previously declining the offer in 1841. In his role as president of the RI, he promoted lectures as a way of popularizing science and supported Faraday and Tyndall in their efforts to improve the RI. He established the Holland Fund for the promotion of scientific research.

Hooker, Joseph Dalton (1817–1911), was a British botanist who specialized in taxonomy and the geographical distribution of plants. He was one of the most famous British botanists of the Victorian era. The younger son of William Jackson Hooker (1785–1865), director of the Royal Botanic Gardens at Kew, he studied medicine at the University of Glasgow. He became a naval surgeon and spent the years 1839 to 1843 on James Clark Ross's geomagnetic survey of the southern oceans, where he collected plants in addition to serving as surgeon. Subsequently, he explored and collected plants in India (1847–51), achieving considerable mountaineering success, before joining his father as assistant director of Kew Gardens in 1855. In 1864 Hooker joined Huxley, Tyndall, and others in founding the X Club, a monthly dining club that promoted science in British society. He succeeded his father as director of Kew (1865), served as president of the BAAS (1868), was knighted several times, and received many scientific awards, including the RS's Royal Medal (1854) and Copley Medal (1887).

Huxley, Henrietta Anne "Nettie" (née Heathorn, 1825–1915), was born in the West Indies to Henry Heathorn and Sarah Henrietta Richardson. In 1843 she, her mother, and her half-sister moved to Australia to join her father. In 1847 she met Thomas Huxley, who was then a surgeon and zoologist on HMS *Rattlesnake.* After an eight-year engagement they married in London. She assisted her husband in his work by translating German and drawing diagrams for his lectures. She was also a poet, and at the age of eighty-six published a volume of her collected poems, *Poems of Henrietta A. Huxley with Three of Thomas Henry Huxley* (1913). As with many of the wives of this small group of scientific friends, Henrietta was very much a part of their community. She corresponded with the Darwins as well as with Tyndall ("Death of Mrs. Huxley," *Times*, 6 April 1915, p. 12).

Huxley, Thomas Henry (1825–95), was a British zoologist, educator, science popularizer, and one of Tyndall's closest friends. Following medical training at Charing Cross Hospital in London, he was a ship's surgeon aboard the HMS *Rattlesnake* expedition to Australia (1846–50) under Captain Owen Stanley. While in Australia he met Henrietta Heathorn, whom he married in July 1855. He became professor of natural history at the Royal School of Mines, London (1854–71) and moved with the school to South Kensington, where he taught in a new laboratory until 1885. He joined Tyndall twice at the RI as

Fullerian Professor of Natural History (1855–58, 1866–69). In 1864 Huxley joined Tyndall, Edward Frankland, Thomas Hirst, Joseph Hooker, and four others as members of the X Club, a monthly dining club in which the members promoted science in British society.

James, Bartlett (dates unknown), was manager, or "gerente," of the Imperial Gas Works Company in Rio de Janeiro in Brazil, where he worked with Tyndall's friend William Ginty, who was the engineer, or "engenheiro" (*Almanak Administrativo, Mercantil e Industrial da Corte e Provincia do Rio de Janeiro para o anno de 1865* [Rio de Janeiro: Eduardo & Henrique Laemmert, 1865], pp. 414–15). Ginty died in 1866, and according to a 5 December 1882 letter from Ginty's daughter Fanny Unwin to Tyndall, James had taken on the role of executor to her late father's will (RI MS JT/1/TYP/11/3682).

Jamin, Jules Célestin (1818–86), was a French physicist, educated at the École normale supérieure. He was professor of physics at the École polytechnique (1858–81), and from 1863 professor of physics at the faculté des sciences, Paris. He worked in many branches of physics, including optics, acoustics, electricity, magnetism, and radiation, and in 1858 he was awarded the RS's Rumford Medal.

Jamin, Thérèse Joséphine Eudoxie Lebrun (1832–80), was the wife of Jules Jamin.

Jeffers, William Nicholson (1824–83), was an American naval officer and author. Jeffers was a key naval figure during the American Civil War (1861–65) due to his technical and tactical expertise. In the years following the war, he wrote *Nautical Surveying* (1871) and *Care and Preservation of Ammunition* (1874) in addition to two revisions of *Ordnance Instructions of the U.S. Navy*. The application of his work on the modernization of US naval ordnance was particularly useful during the Spanish-American War (1898).

Jones, Henry Bence (1813–73), was a wealthy English physician and medical chemist. He was educated at Harrow School and Trinity College, Cambridge, before studying medicine at Saint George's Hospital, London. Attracted by the potential applications of chemistry to clinical medicine, he studied organic chemistry with Justus Liebig at Giessen (1841–43). He then went on to a career as a successful private physician and physician at Saint George's Hospital. In May 1842 he married his cousin, Lady Millicent Acheson, daughter of the 2nd Earl of Gosford, with whom he had seven children. He published important research on protein chemistry. In 1853 he became a manager (trustee) of the RI and was elected secretary in 1860, a position he held until

1873. He frequently entertained Tyndall at both his London home and at Folkestone, where he kept a summer house. In 1869 Tyndall convalesced at Bence Jones's Folkestone home after injuring his leg in the Alps.

Joule, James Prescott (1818–89), was an English brewer, physicist, and mathematician. The son of a brewer, he studied for a brief time under John Dalton before focusing his own work on electromagnetism. In 1838 Joule published his first paper on electromagnetic engines. This research led to his formulation of the mechanical equivalent of heat in 1842, which he presented to the BAAS in 1843, and for which the RS awarded him a Royal Medal in 1852. In the 1860s Tyndall supported the German physician Robert Mayer, who had also formulated the mechanical equivalent in 1842, in his priority dispute with Joule. While Tyndall aimed to give appropriate credit to both, the dispute disrupted the scientific community, but by the period in this volume tensions between Tyndall and Joule had largely eased.

Kingsley, Charles (1819–75), was a British clergyman, historian, and novelist. Kingsley was sympathetic to the Chartists and active in the formation of a Christian Socialist movement. He wrote several historical novels that promoted masculine Christian social views, most famously *Alton Locke* (1850), *Westward Ho!* (1855), and *The Water-Babies* (1863). In 1860 he was named regius professor of modern history at Cambridge University, but resigned in 1869 due to the strain. Kingsley corresponded with both Charles Darwin and Thomas Huxley (J. Conlin and J. M. I. Klaver, eds., *Charles Kingsley: Faith, Flesh, and Fantasy* [New York: Routledge, 2021]; *ODNB*).

Kirchhoff, Gustav Robert (1824–87), was a German physicist. In Königsberg Kirchhoff studied physics under Franz Neumann and mathematics under Friedrich Julius Richelot, submitting his doctoral dissertation, which included his formulation of the laws of electrical circuits in 1847. During an associate professorship at the University of Breslau in 1850, he met Robert Bunsen, who made sure that Kirchhoff followed his move to Heidelberg in 1854, where they worked on thermal radiation and spectrum analysis. Regarding his law of thermal radiation, Kirchhoff would fight a long battle with the Scottish physicist Balfour Stewart (1828–87) for recognition of priority. Kirchhoff and Tyndall were close friends and colleagues. Tyndall recommended him for the RS's Rumford Medal, which he received in 1862. Kirchhoff was unable to attend the ceremony and Tyndall accepted it in his place.

Knoblauch, Karl Hermann (1820–95), was a German physicist best known for his studies of radiant heat. He was one of the six founding members of the Deutsche Physikalische Gesellschaft at Berlin on 14 January 1845. As a researcher and teacher at the University of Marburg (1849–53), he produced experimental demonstrations on the nature of diamagnetism. Tyndall was Knoblauch's student and collaborated on the diamagnetism work during his time at Marburg. They kept up a correspondence for over twenty-five years.

Lister, Joseph (1827–1912), was a British surgeon and founder of a system of antiseptic surgery. He married Agnes Syme (1834–93) in 1856, and she assisted him in his research and experiments throughout their marriage. He wrote numerous papers, but the work he published on antiseptics in the late 1860s/70s received the greatest recognition and generated the most controversy. In 1860 Lister was appointed chair of surgery at Glasgow University and made FRS. From 1869 to 1877 he was professor of surgery at Edinburgh. After Lister's move to London in 1877, Tyndall and Huxley nominated him for membership at the Athenaeum Club. He received a baronetcy in 1883 and was raised to the peerage as Baron Lister of Lyme Regis in 1897.

Lubbock, Harriet Charlotte (née Wood, d. 1878), was the first wife of Nevile Lubbock (1839–1914), brother of John Lubbock. They were married at North Cray Place, Kent, and together had six sons and three daughters. In February 1872 she sang "Waiting and Watching" at the Soirée musicale d'invitation before a room "overflowing by an aristocratic audience." She died on board RMS *Arno* in 1878 off the coast of Santo Domingo in the Caribbean (Robert Birkbeck, *Notes on the History and Genealogy of the Family of Lubbock* [London: Mitchell and Hughes, 1891], pp. 38–39; "Concerts Various," *Musical World* 50, no. 8 [24 February 1872], p. 125).

Lubbock, John (1834–1913), 1st Baron Avebury, was an English banker, anthropologist, scientist, and politician. In his youth he befriended Charles Darwin and John Tyndall. In 1865 he published the highly influential *Prehistoric Times: As Illustrated by Ancient Remains,* in which he coined the terms *paleolithic* and *neolithic.* He was also an amateur biologist and in the 1870s published works on insects and plants. In 1856 he married Ellen Frances "Nelly" Horden (1835–79), and the couple were among Tyndall's closest friends. He was the youngest member of the X Club and joined Tyndall on his climbing adventures in the Alps on a number of occasions.

Lyell, Charles (1797–1875), was a highly influential Scottish geologist. Trained as a lawyer, Lyell became professor of geology at King's College. He was most noted for his controversial three-volume work *Principles of Geology* (1830–33), in which he argued that the only legitimate type of causal explanation for the past had to be of the kind seen in operation in the present; natural laws were nothing if not consistent. In consequence of this philosophical commitment, Lyell was forced to argue that the earth was in fact many millions of years older than even the most generous of his contemporaries had hitherto been willing to admit. The "catastrophism" of Georges Cuvier was then the dominant philosophy among British geologists. Although controversial, Lyell was highly regarded, both as a philosopher of science and as a geologist. He became a mentor to Charles Darwin upon his return to England from his voyage on HMS *Beagle* in 1836, and in 1858, along with Joseph Hooker, had pressed Darwin to publish an abstract of his big species book, which became *On the Origin of Species* (1859). By the 1860s Lyell was one of the elder statesmen of science.

Lyell, Rosamond Francis Anne (1856–1904), was the daughter of Colonel Henry Lyell and Katherine Murray Lyell. She was the niece of Charles Lyell, and sister of Leonard Lyell. Tyndall was quite fond of her.

Magnus, Heinrich Gustav (1802–70), was a German chemist and physicist, and professor of technology and physics at the University of Berlin from 1833 until his death. He established a physics laboratory at his home in Berlin to which he invited young physicists to work, including Tyndall in 1851. Although his early research was primarily in chemistry, by the 1850s he had begun to investigate a range of physical problems, including the absorption of heat by gases, as his letters to Tyndall demonstrate. Tyndall and Magnus had several professional disagreements, which they managed cordially.

Marcet, François (1803–83), was a Swiss physicist in Geneva best known for his work on superheating. Marcet was the son of Alexandre John Gaspard Marcet (1770–1822), a Genevan chemist and physician who lived in London, and Jane Marcet (née Haldimand, 1769–1858), an enormously successful author of popular science works.

Maxwell, James Clerk (1831–79), was a Scottish physicist and mathematician. He graduated in the Cambridge mathematics tripos as second wrangler in 1854, after which he held professorships at Marischal College, Aberdeen (1856–60), King's College, London (1860–65), and from 1871 the Cavendish Professorship at the University of Cambridge. Maxwell is best known for his contributions to the study of electromagnetic radiation. He developed Faraday's field theory and demonstrated that electricity, magnetism, and light were instances of the same phenomenon. Maxwell's work culminated in his *A Treatise on Electricity and Magnetism* (1873).

Mayer, Alfred M. (1836–97), was an American physicist. He studied at the University of Paris and became professor of physics and chemistry at Pennsylvania College of Gettysburg. Gettysburg awarded him a doctorate in 1866 and the following year he assumed the role of professor of physics and astronomy at Lehigh University. In 1869 he led a mission to study the solar eclipse. He visited Tyndall in 1873 during a tour of England and was an organizer of Tyndall's farewell dinner held in New York in February 1873 at the end of Tyndall's four-month American lecture tour (Alfred G. Mayer and Robert S. Woodward, *Biographical Memoir of Alfred Marshall Mayer, 1836–1897* [Washington, DC: National Academy of Sciences, 1916], p. 260).

Mayer, Julius Robert (1814–78), was a German physician and physicist. His work on the conservation of energy contributed to the foundation of thermodynamics. Tyndall supported Mayer's priority claim for being the first to articulate the mechanical equivalent of heat, praising both Mayer and James Joule for their contributions and championing the significance of Mayer's work in the discovery of the conservation of energy.

Miller, William Allen (1817–70), was an English chemist and chair of chemistry at King's College, London from 1845. He was the author of the popular textbook series *Elements of Chemistry, Theoretical and Practical* (London: J. W. Parker and Son, 1855), which went into several editions, and conducted many experiments in spectrum analysis including the effects of atmospheric water vapor on the solar spectrum (1845). Elected FRS in 1845, he served as treasurer of the RS from 1861 until his death in 1870. After Miller unexpectedly passed away, Tyndall dove deeply into the institutional politics surrounding the new occupation of his position as treasurer.

Moigno, François-Napoléon-Marie (1804–84), was a French Catholic priest, physicist, writer, and editor. Educated at a Jesuit seminary, he studied science alongside theology and was given the mathematical chair at Jesuit House in Paris in 1836. After leaving the order in 1843 he became the science editor of *L'époque*, and later editor of the scientific journal *Cosmos* (1852–62). He founded the popular journal of science and industry *Les mondes* in 1863 and undertook the translation and publication of many of Tyndall's works into French, including *Radiation* (1867), *Le son* (1869), *La chaleur* (1864), and *Faraday inventeur* (1868) ("L'Abbé Moigno," *Nature* 30 [1884], pp. 291–92).

Morren, Jean François Auguste (1804–70), was a French physician, member of the Academy of Marseilles, and dean of the faculty of sciences at the University of Rennes. He published several papers on the phenomena of light.

Murchison, Roderick Impey (1792–1871), was a Scottish geologist and geographer who published over 180 scientific papers as well as several books. He was elected fellow of the Geological Society in 1825 and was twice its president. He was one of the founders of the Royal Geographical Society of London (1830) and from 1843 served as its president for sixteen years. He used his role in the RGS in particular to promote imperial exploration. He was also one of the founders of the BAAS (1831) and was president of the geographical and geological sections ten times in total.

Odling, William (1829–1921), was an English chemist. He studied medicine at Guy's Hospital in 1846, before switching to chemistry. He was appointed director of the chemical laboratory at Guy's, then lecturer in chemistry, and finally professor of practical chemistry (1856). Elected a fellow of the Chemical Society in 1848, he served in several capacities, including president from 1873 to 1875. In 1859 he was elected FRS and became a fellow of the Royal College of Physicians. He was appointed joint professor of chemistry at Saint Bartholomew's Hospital, assuming the role of sole lecturer in 1864. In 1868 he succeeded Michael Faraday as Fullerian Professor of Chemistry at the RI. He was appointed Waynflete Professor of Chemistry at the University of Oxford in 1872.

Pasteur, Louis (1822–95), was a French chemist and microbiologist and professor of chemistry at the University of Strasbourg. He was a member of the Parisian Académie des Sciences and was awarded membership in France's Légion d'honneur. Pasteur is best known for the process of pasteurization and his work was central to debates around germ theory and spontaneous generation. In order to support his germ theory, Pasteur actively sought to experimentally disprove theories of spontaneous generation. Tyndall supported Pasteur's position and promoted it in England through his own research on airborne particles and through public lectures and writings in the 1870s.

Peel, Emily (née Hay, 1836–1924), was the daughter of George Hay (1787–1876), 8th Marquess of Tweeddale, and Lady Susan Montagu (1797–1870). In 1856 she married Robert Peel (1822–95), third baronet. They had three daughters and one son before she left him to live in Geneva and later Florence. Peel attended Tyndall's lectures regularly.

Phillips, John (1800–74), was an English geologist. Phillips was raised by his uncle, William Smith (1769–1839), who created the first geological map of Britain. He learned surveying and geology from his uncle, was appointed keeper of fossils at the Yorkshire Museum in 1825, and had established himself as a geologist by 1829. He helped organize the first meeting of the BAAS in 1831 and a year later became its secretary, a position he held for thirty years. Phillips's career took him to the University of London, the Geological Survey, Trinity College Dublin, and Oxford University. The author of numerous books and popular essays on geology and paleontology, Phillips opposed Darwin's theory of evolution by natural selection and instead studied the history of life on earth through the lens of natural theology (Jack Morrell, *John Phillips and the Business of Victorian Science* [Burlington, VT: Ashgate, 2005], ch. 13).

Plateau, Joseph (1801–83), was a Belgian physicist. Born in Brussels, Plateau studied at the University of Liège, where he graduated in 1829 with a doctoral degree in mathematics and physics. In 1832 Plateau invented the phenakistoscope, a device that could produce the illusion of a moving image. In 1835 he was appointed professor of experimental physics at the University of Ghent. Plateau is best known for his experimental work in which he investigated what is now known as the "Plateau problem," a mathematical problem concerning the calculation of surfaces. He derived his result through experiments he conducted from 1840 onward to demonstrate that a minimal surface could be formed from a soap film made of a solution of soapy water and glycerin that was stretched on a wire frame.

Pollock, Frederick (1845–1937), was a third baronet, jurist, and the eldest son of Juliet and William Frederick Pollock. He attended Eton College (1858–63) before entering Trinity College, Cambridge, where he obtained his BA (1867) and MA (1870). He was admitted to Lincoln's Inn in 1868, was called to the bar in 1871, and married Georgina Harriet Deffell in 1873. He was a distinguished jurist and the author of many legal treatises, and became professor of jurisprudence at Oxford University (1883–1903).

Pollock, Jonathan Frederick (1783–1870), was a British judge, politician, and the father of William Frederick Pollock (1815–88). He was educated at Cambridge University, graduating in 1806 as senior wrangler. He was member of parliament for Huntingdon (1831–44) and served twice as attorney general (1843–45, 1841–44). He was lord chief baron of the exchequer from 1844 to 1866 and received a baronetcy on his retirement.

Pollock, Juliet (née Creed, 1819–99), was the daughter of the Reverend Henry Creed, vicar of Corse, Gloucestershire. She married lawyer and author William Frederick Pollock in 1844, and they had three sons. She and her husband were staunch supporters of the RI and frequently entertained Tyndall at their homes. She was the author of several books, including *Julian and His Playfellows* (London: Grant and Griffith, 1852), *New Friends* (London: John W. Parker and Son, 1858), and *Macready as I Knew Him* (London: Remington, 1885). She and Tyndall were intimate friends and regular correspondents.

Pollock, William Frederick (1815–88), second baronet, was a British lawyer and author. The eldest son of Jonathan Frederick Pollock (1783–1870), and Frances Rivers (d. 1827), he married Juliet Creed in 1844. He was admitted to Trinity College, Cambridge, in 1832, where he obtained his BA (1836) and MA (1840). Admitted to the Inner Temple in 1833 and called to the bar in 1838, he then became a master of the court of exchequer in 1846 and was appointed queen's remembrancer in 1874. Pollock's two-volume *Personal Remembrances* (London: Macmillan, 1887) illustrates how the family circulated among London's elite and entertained at their Montagu Square home. Both William and Juliet were loyal supporters of the RI and Tyndall was frequently among their guests.

Pouchet, Felix (1800–72), was a French naturalist and a leading proponent of spontaneous generation. He was director of the Rouen Museum of Natural History and in 1838 was appointed a professor at the School of Medicine at Rouen. His most significant work, *Hétérogénie* (1859), details the chemical process that could give life to inorganic matter. He invented an apparatus called the "aeroscope" to analyze dust and study airborne microbes as part of his attempt to resolve the question of spontaneous generation.

Rees, George Owen (1813–89), was a Welsh-Italian physician. He was born in Smyrna, Turkey, and was the second child of Josiah Rees, a Levantine merchant and British consul at Smyrna of Welsh descent. He studied medicine at Guy's Hospital in 1829 and earned his MD at the University of Glasgow in 1837. Rees was one of the first to make a serious study of the chemistry of urine and the nature and shape of blood cells, for which he was elected FRS. He was appointed assistant physician at Guy's Hospital in 1843 and was promoted to physician and lecturer in medicine from 1856 until 1873.

Rothschild, Charlotte von (1819–84), was a German philanthropist and the eldest of the five children of Carl Mayer von Rothschild (1788–1855), originally of Frankfurt, and his wife, Adelheid Herz (1800–53). She married her first cousin Lionel Nathan (1808–79), 2nd Baron Rothschild, in 1836 and together they had five children. Their primary residences in London were Hill Street, Mayfair, and Piccadilly, with country estates in Hounslow and the Vale of Aylesbury. Lionel was the first Jewish member of parliament, a position he held from 1858 to 1874. In addition to her charitable work and political campaigning for her husband, Charlotte was an adored hostess and prominent art patron.

Ruskin, John (1819–1900), was an influential English art critic and social critic. Ruskin was one of the leading intellectuals in Victorian Britain. At Oxford University, he was appointed Slade Professor (1869) and founded the Ruskin School of Drawing and Fine Art (1871). He was a prolific writer, with *Modern Painters* (1843–60), *The Seven Lamps of Architecture* (1849), and *The Stones of Venice* (1851–53) among his most popular works. Ruskin's views on architecture and, especially, his praise of Byzantine and Gothic periods to the detriment of the Renaissance, had a tremendous effect on Victorian architecture. His idealization of a harmonious, ordered pastoral past fueled his criticism of industrialization and social disharmony. By the late 1860s Ruskin's focus had shifted to the political economy of art, education, and social criticism, his physical and mental health were deteriorating, his unconsummated marriage had been annulled, and he was in the midst of the unsuccessful courtship of a woman several decades his junior.

Sabine, Edward (1788–1883), was an Anglo-Irish astronomer, geophysicist, and military officer who led the British effort to establish magnetic observatories around the globe. A graduate of the Royal Military Academy at Greenwich, he reached the rank of general in the 1850s. He accompanied J. C. Ross's Arctic expedition (1818) and W. E. Parry's expedition (1819). Sabine received the RS's Copley Medal in 1821 for his Arctic magnetic observations. He was closely connected with the administration of the RS and served as its treasurer (1850–61) and president (1861–71).

Sauerwald, Ferdinand (dates unknown), was an expert German scientific instrument maker whose workshop was located in Berlin. He constructed apparatuses for Gustav Magnus, John Tyndall, Emil du Bois-Reymond, and Hermann Helmholtz, among others (David Cahan, *Helmholtz: A Life in Science* [Chicago: University of Chicago Press, 2018], pp. 186, 264).

Savage, Sarah (dates unknown), was housekeeper at the RI.

Schultze, Max (1825–74), was a German zoologist and cytologist. He lectured in anatomy at the University of Halle before accepting a chair at the University of Bonn in 1859. He was subsequently appointed director of that university's Anatomical Institute in 1872. He was a renowned histologist and produced several texts on the nerve endings of sense organs. In 1865 he founded the journal *Archiv für mikroskopische Anatomie*, and served as its editor until his death.

Seeley, John Robert (1834–95), was an English historian and essayist. He published on a myriad of topics ranging from Jesus Christ to British imperialism to Johann Wolfgang von Goethe. He was a classics professor at University College London in the 1860s. He married Mary Agnes Phillott in 1869, the same year he was appointed regius professor of modern history at Cambridge. His inaugural lecture was the following year. In 1894 he was knighted, less than a year before his death ("John Robert Seeley," *English Historical Review* 10 [July 1895], pp. 507–14; Ian Hesketh, "John Robert Seeley, Natural Religion, and the Victorian Conflict between Science and Religion," *Journal of the History of Ideas* 79 [2018], pp. 309–29).

Soret, Jacques-Louis (1827–90), was a Swiss chemist and professor of chemistry and medical physics at the University of Geneva. He was an adept spectroscopist. His achievements included determining the chemical composition of ozone (1865), the discovery of the chemical element holmium (1878), and "the Soret band," a band in the absorption spectrum of hemoglobin (1883). He married Clémentine Odier (1831–89), and their son Charles Soret (1854–1903) later became a physics professor at the University of Geneva, where he worked collaboratively with his father (J. K. Platten and P. Costesèque, "Charles Soret. A Short Biography," *European Physical Journal* 15 [2004], pp. 235–39; *Nature* 70 [14 July 1904], pp. 251–52).

Spencer, Herbert (1820–1903), was an English biologist, philosopher, journalist, and sociologist. Through family connections, at the age of twenty-eight, he secured a post as subeditor at the *Economist*, which put him at the heart of literary London. He published his first book, *Social Statics*, in 1850 and his second, *Principles of Psychology*, in 1855. Spencer was a respected philosopher, but one of his most lasting contributions was his *Principles of Biology* (1864) in which he coined the phrase "survival of the fittest." Along with Tyndall and seven others, Spencer was a member of the X Club, a monthly dining club whose members promoted science in British society (Spencer was the only member who was not an FRS). In 1865 he joined the Jamaica Committee, which advocated the prosecution of Governor Edward Eyre for his violent repression of the Morant Bay rebellion. Tyndall, like Thomas Carlyle and Charles Kingsley, joined others in Eyre's defense.

Spottiswoode, Eliza "Lise" Taylor (née Arbuthnot, 1837–94), was the daughter of William Urquhart Arbuthnot, member of the Council for India. She married mathematician William Spottiswoode in 1861. The Spottiswoodes were known for their gatherings at their Grosvenor Place home, which were attended by the scientific, political, and social elite and frequently included scientific demonstrations. Their country residence was in Combe Bank, Sundridge, near Sevenoaks.

Spottiswoode, William (1825–83), was an English mathematician and physicist and partner of the Eyre & Spottiswoode printing firm, the queen's printer. In 1861 he married Eliza Taylor Arbuthnot. Parties at the Spottiswoode home were considered the height of the London season, attracting men of science, politics, and society. In 1864 he, along with Tyndall, Huxley, Spencer and five others, formed the X Club, a monthly dining club in which they promoted science in British society. He was treasurer of the BAAS (1861–74), the RI (1865–73), and the RS (1871–78), and served as president of the BAAS (1878) and the RS (1878–83).

Spottiswoode, William Hugh "Hughie" (1864–1916), was the son of Eliza and William Spottiswoode and an editor. He founded *Printers' Pie* and *Winter's Pie*, publications that supported the Printers' Pension Corporation. After his father's death he became a partner in the firm Eyre & Spottiswoode. In 1899 he donated his father's collection of physical apparatuses to the RI where he also worked as a manager ("Literary Gossip," *Athenaeum*, 28 August 1915, p. 146).

Stanley, Arthur Penrhyn (1815–81), was an Anglican clergyman. Educated at Rugby School and Oxford University (1837), where he obtained a fellowship, he accepted a canonry at Canterbury (1851) before becoming professor of ecclesiastical history at Oxford (1856). In 1863 he married Lady Augusta, one of Queen Victoria's women of the bedchamber. They were subsequently frequent guests at Windsor Castle. From 1864 to 1881 he was dean of Westminster, during which time he continued to publish lectures and sermons.

Stegmann, Friedrich Ludwig (1813–91), was a German mathematician. Stegmann was professor of mathematics in Marburg and was advisor to both Tyndall and Thomas Archer Hirst. Tyndall wrote his dissertation at Marburg under Stegmann's supervision.

Steuart/Stewart, Elizabeth Dawson (née Duckett, 1802–93), was born into a very wealthy family with an estate of twelve thousand acres and a Gothic mansion (Duckett's Grove, now a ruin) situated several miles northeast of Carlow. In 1820 she married William Richard Steuart and they lived at

Steuart's Lodge, Leighlinbridge. She took a keen interest in Tyndall's career and was probably his most consistent and long-lived patron. Letters in this volume show that Steuart helped Tyndall arrange financial support for his relatives and acquaintances in Ireland and manage family affairs.

Stevenson, Thomas (1818–87), was a Scottish civil engineer, meteorologist, and commissioner of northern lighthouses. He was the youngest surviving son of Robert Stevenson (1772–1850), and his wife, Jane (c. 1779–1846). In 1838 his brothers David (1815–86) and Alan (1807–65) partnered with their father to form the engineering firm Robert Stevenson and Sons. In 1846 Thomas Stevenson joined the firm as a junior partner, and it was soon known as Messrs Stevenson and eventually as D. and T. Stevenson. The company carried out extensive work in the construction of lighthouses. Through his research and development at the firm, Stevenson became a leading authority on lighthouse illumination and harbor engineering. The *Third Report of Messrs Stevenson* (1869) suggested that the gas burners that Tyndall defended were much more expensive than the oil used in their standard four-wick Trinity House–approved burner.

Stokes, George Gabriel (1819–1903), was an Anglo-Irish mathematician and physicist. Born in Dublin, he graduated from Cambridge University as senior wrangler (1841) and served as Lucasian Professor of Mathematics at Cambridge University (1849–1903). He served as secretary of the RS (1854–85), during which time he was responsible for refereeing papers for the *Phil. Trans.*, a task for which he frequently sought Tyndall's assistance. In the 1850s Stokes turned his attention to optics, and in 1852 he published on fluorescence, a term he coined, showing that the phenomenon was the result of changes in the wavelength of light. For this work he was awarded the RS's Rumford Medal (1852). He was later awarded the RS's Copley Medal (1893) for his contributions to science.

Strachey, Richard (1817–1908), was a scientist, military engineer, and British colonial administrator in India. Educated at the East India College, he joined the Bombay Engineers (1836) and Bengal Engineers (1839), where he worked on the Jumna Canal irrigation project before serving in the Anglo-Sikh War (1845–46). Traveling to the Himalayas to recover his health, he began to study botany and geography. He explored Tibet in 1848, collecting botanical specimens and later publishing much-lauded descriptions of the geography, ethnology, and botany he observed. Strachey held numerous appointments in India, including secretary for public works (1862–65) and inspector-general of irrigation (1866), before returning to England in 1871.

Sylvester, James Joseph (1814–97), was an English mathematician, educated at Cambridge University, where he was second wrangler in the mathematical tripos in 1837. However, because he was Jewish, he was not permitted to take the degree or hold fellowships. He subsequently held professorships at University College, London (1838–1841) and the University of Virginia (1841), and earned a BA and MA from Trinity College, Dublin (1841). From 1844 to 1854 he worked as an actuary at a life insurance company in London while at the same time pursuing his mathematical research on invariants. He was professor of mathematics at the Royal Military Academy, Woolwich (1855–70) and later at Johns Hopkins University (1876–80).

Tait, Peter Guthrie (1831–1901), was a Scottish mathematician and physicist, who was senior wrangler in the mathematics tripos at Cambridge University in 1852. He was professor of mathematics at Queen's College, Belfast (1854–60), and held the chair of natural philosophy at Edinburgh University from 1860 until shortly before his death. During the 1860s Tait forged a friendship with William Thomson and began writing their *Treatise on Natural Philosophy* (1867). During the 1860s Tait and Thomson had been in a dispute with Tyndall over his claims about Robert Mayer's priority over Joule in developing the mechanical equivalent of heat. This dispute spilled over into popular journals before they resolved their differences at the Dundee meeting of the BAAS in 1867, but they never warmed up to one another.

Tennyson, Alfred (1809–92), was poet laureate from 1850, in succession to William Wordsworth. Born in Lincolnshire, he was educated at Trinity College, Cambridge. In 1850 he published *In Memoriam* (London: Edward Moxon). In 1853 he moved to Farringford on the Isle of Wight, where he published *Maud* (London: Edward Moxon, 1855). Tyndall greatly admired Tennyson's poetry and met him in 1858. He subsequently developed a close friendship with the Tennysons and stayed with them at their home in Farringford. Tennyson declined honors or titles from Queen Victoria four times between 1865 and 1880 but accepted a barony in 1883.

Thomson, William (1824–1907), was a distinguished Irish-Scottish mathematician, physicist, and inventor. He was educated at the University of Glasgow and Peterhouse College, Cambridge University, where he was second wrangler and Smith's Prizeman in 1841. He was professor of natural philosophy in Glasgow from 1846, paying frequent visits to London. Thomson contributed foundational research to the field of thermodynamics and was instrumental in the development of the transatlantic telegraph. He first met Tyndall at the Edinburgh meeting of the BAAS in 1850, and from the start the two physicists had a strained relationship. For much of his career Tyndall regarded Thomson as a scientific rival; however, the two physicists resolved their differences by the Dundee meeting of the BAAS in 1867. In 1892 Thomson assumed the title Lord Kelvin and became 1st Baron of Kelvin.

Trayer, James John (c. 1816–77), was a doctor in Bagenalstown, County Carlow, Ireland. He was educated at Trinity College, Dublin where he earned a BA (1837) and MB (1839). In 1855 he married Jane Carey.

Trevor, Charles Cecil (1830–1921), was an English civil servant who was assistant secretary in the Harbour Department of the Board of Trade from 1867 to 1895 ("Sir Cecil Trevor," *Times*, 20 September 1921, p. 20).

Tyndale, George Hector (1821–80), was an American businessman, Union general, and cousin of John Tyndall. He was the son of Tyndall's uncle, Robinson Tyndall (1775–1842), who had emigrated to Philadelphia from Leighlinbridge around 1804 and founded a glass and ceramics import business. Hector inherited the business and first met his physicist cousin in 1854 on one of his frequent business trips to Europe. They immediately struck up an intellectual rapport and began an occasional correspondence. During the American Civil War (1861–65), Tyndale rose to the rank of brigadier general, and after his resignation in 1864 due to ill health, he was promoted to brevet major general in acknowledgment of his service (1865).

Tyndale/Tyndall, William (1494–1536), was an English religious reformer. Tyndale made the first published English translation of the New Testament, thus making it accessible in the vernacular. A Lutheran and an extremely influential figure in the English Reformation, he was taken captive by British forces in Antwerp, tried for heresy, and martyred. John Tyndall liked, rightly or wrongly, to trace his roots back to this illustrious historical figure.

Tyndall, Emily "Emma" (1817–1904), was John Tyndall's sister. She does not appear to have married and after the death of their mother suffered from a mental illness that manifested as an extreme form of religiosity. In 1868 Emma was sent to live with a relative, John Tyndall of Gorey, while she recovered from her breakdown. Emma's health is a frequent subject of Tyndall's correspondence with his Irish correspondents.

Tyndall, John (c. 1815–75), was a relative of Tyndall's in Gorey, County Wexford, Ireland. A doctor and member of the Royal Historical and Archaeological Society of Ireland, he died in 1875 of acute bronchitis. Tyndall's sister, Emily, stayed with him in through the years in this volume, while she recovered from a breakdown following the death of their mother ("Obituaries," *Lancet*, 105 [1 May 1875], p. 633).

Tyndall, John (dates unknown), was the son of Caleb Tyndall and first cousin of John Tyndall. This branch of the family and their endless need for money caused Tyndall considerable headaches in 1870 and was the subject of several letters with Elizabeth Steuart, who managed his mother's estate after her death and defended it from their demands.

Tyndall, William (c. 1826–1902), was the son of Caleb Tyndall and Dorothea Shirley and first cousin of John Tyndall. He was a civil bill officer and married Frances Stone in 1860.

Valentin, William G. (1829–79), was an English chemist. Born in Neuenbürg, Germany, he moved to England in 1855 and studied at the Royal College of Chemistry. He became senior laboratory assistant to August Hofmann and later Edward Frankland at the Science Schools, South Kensington. He also held the post of gas examiner to the Great Western Gas Company and was chemical advisor to Trinity House until his death.

Vieweg, Heinrich Rudolf (1826–90), was a German publisher. He was the son of Eduard Vieweg (1797–1869) and grandson of Friedrich Vieweg (1761–1835), who had founded the family publishing business in Berlin (1786) and later Brunswick (1799). Vieweg studied natural sciences in Heidelberg before he joined his father as a partner in the family firm in 1853. He took over the firm in 1866, upon which he continued his father's emphasis on scientific works and published German translations of a number of Tyndall's books (Paul Forman, "The Archive of Friedr. Vieweg & Sohn, Braunschweig: Correspondence of Nineteenth-Century Physical Scientists, Especially German Chemists," *Isis* 60 [1969], pp. 384–86).

Wallace, Alfred Russel (1823–1913), was a British naturalist known for independently formulating a theory of evolution by natural selection around the same time as Charles Darwin. Wallace traveled in the Amazon (1848–52) and in the Malay Archipelago (1854–62) collecting specimens and studying the flora and fauna. It was during the second expedition that Wallace arrived at his theory of natural selection, which he forwarded to Darwin. Darwin was thus prompted to publish his own work on evolution, *On the Origin of Species* (1859), which he had begun formulating over twenty years earlier. Wallace was awarded the RS's Royal Medal in 1868 for his contributions to zoology. He was also interested in spiritualist phenomena.

Wheatstone, Charles (1802–75), was an English experimental physicist, inventor, and businessman. He had been a musical instrument maker before becoming professor of experimental philosophy at King's College, London (1834–75), where he was a reluctant lecturer. There he developed the electric telegraph with William Fothergill Cooke. He also developed a number of optical instruments, including the stereoscope that led him into controversy with David Brewster. He and Cooke established the Electric Telegraph Company, which introduced electric telegraphy in Britain. Wheatstone married Emma West in 1847 and together they had five children. The family settled in Hammersmith, where Wheatstone frequently entertained Tyndall, whom he first met in 1852.

Whitworth, Joseph (1803–87), was an English mechanical engineer and machine tool manufacturer. He developed a uniform system of screw threads that was in general use by 1860. His tools and inventions were noted at the Great Exhibition in London (1851) and in 1853 he served on a royal commission to the New York Industrial Exhibition. He was elected president of the Institution of Mechanical Engineers (1856), made FRS (1857), and created baronet (1869), among other honors. Following the death of his first wife, Fanny Ankers (1800–70), he married Mary Louisa (1829–96) in 1871. Tyndall stayed at Whitworth's Manchester residence in 1870 and attended the laying of the first stone at Owens College. After Whitworth died in 1887 leaving no heirs, the bulk of his fortune was bequeathed to charitable and educational purposes, notably, Whitworth Park and Institute, Owens College, and the Manchester Technical School.

Wigham, John Richardson (1829–1906), was a British lighthouse engineer. Born in Edinburgh, in 1843 he went to Dublin, where he served an apprenticeship in a hardware and manufacturing business under his brother-in-law, Joshua Edmundson. After Edmundson's death, the Edmundson's Electricity Corporation grew rapidly under Wigham's stewardship and focused on the provision of gas lighting to private and public buildings. Wigham was engineer to the Commercial Gas Company of Ireland and served as director of the Alliance and Dublin Consumers' Gas Company from 1866 and director and vice-chairman of the Dublin United Tramways Company from 1881. He was a member of the council of the Dublin chamber of commerce and served as its secretary (1881–93) and president (1894–96). He was best known as an inventor of powerful gas burners that were adopted by the commissioners for Irish lights. A system he invented, with an on-site gas works, was installed in Baily lighthouse in Howth, County Dublin.

Wynne, Frank George (dates unknown), was the third son of Anne (née Osborne, 1808–64) and George Wynne. He became a civil engineer.

Wynne, George (1804–90), was an Irish surveyor, engineer, and military officer. Born in Dublin, he enlisted in the Royal Engineers (c. 1825) and was commissioned as lieutenant in 1826. Wynne met Tyndall while employed in the Irish Ordnance Survey (1835–40). Wynne later served with distinction in Greece, China, and the West Indies, rising to general by the time he retired in 1877 ("The Late General Wynne," *Times*, 12 July 1890, p. 11).

Wynne, Henry "Harry" Le Poer (dates unknown), was the son of Anne (née Osborne, 1808–64) and George Wynne.

Wynne, Lucy Harriette (1840–99), was the daughter of Anne (née Osborne, 1808–64) and George Wynne. She married Aubrey Stephen O'Brien (1837–98) on 10 August 1871.

Youmans, Edward Livingston (1821–87), was an American scientific writer, editor, and popularizer of science. Youmans was a conduit for British popular science in the United States through the publisher Appleton and was crucial for arranging for the American publication of the works by Herbert Spencer, Tyndall, and others, for a general audience. He also lectured and published on popular science himself and founded *Popular Science Monthly* in 1872.

INDEX

The index refers to the letters using their serial numbers. The letter 'n' denotes a footnote reference and appears if an item is not otherwise mentioned in the letter to which the footnote pertains. A number in **bold** signifies a letter to or from a correspondent.